The great apes (chimpanzees, bonobos, gorillas and orangutans) are our closest living relatives, sharing a common ancestor only a few million years ago. We also share key features such as high intelligence, omnivorous diets, prolonged child-rearing and rich social lives. The great apes show a surprising diversity of adaptations, particularly in social life, ranging from the solitary life of orangutans, through patriarchy in gorillas, to complex but different social organizations in bonobos and chimpanzees. As great apes are so close to humans, comparisons yield essential knowledge for modeling human evolutionary origins. *Great Ape Societies* provides comprehensive up-to-date syntheses of work on all four species, drawing on decades of international field work, zoo and laboratory studies. It will be essential reading for students and researchers in primatology, anthropology, psychology and human evolution.

Great Ape Societies

To Evered, Gigi and Ntologi,
who taught us so much

Great Ape Societies

EDITED BY William C. McGrew

AND Linda F. Marchant
Miami University, Oxford, Ohio, USA

AND Toshisada Nishida
Kyoto University, Kyoto, Japan

CAMBRIDGE
UNIVERSITY PRESS

Published by the Press Syndicate of the University of Cambridge
The Pitt Building, Trumpington Street, Cambridge CB2 1RP
40 West 20th Street, New York, NY 10011-4211, USA
10 Stamford Road, Oakleigh, Melbourne 3166, Australia

First published 1996

Printed in Great Britain at the University Press, Cambridge

A catalogue record for this book is available from the British Library

Library of Congress cataloguing in publication data available

ISBN 0 521 55494 2 hardback
ISBN 0 521 55536 1 paperback

WV

Contents

Contributors

BOESCH, CHRISTOPHE
Institut für Zoologie, Universität Basel, Rheinspring 9, 4051 Basel, Switzerland.

CHAPMAN, COLIN A.
Department of Zoology, University of Florida, Gainesville, FL 32611, USA.

CLARK-ARCADI, ADAM P.
Department of Anthropology, Cornell University, Ithaca, NY 14853, USA.

DORAN, DIANE M.
Department of Anthropology, State University of New York at Stony Brook, Stony Brook, NY 11789, USA.

FRUTH, BARBARA
Max-Planck-Institut für Verhaltensphysiologie, Seewiesen, D-82319, Starnberg, Germany.

FURUICHI, TAKESHI
Laboratory of Biology, Meiji-Gakuin University, 1518 Kamikurata, Totsuka, Yokohama, Japan, 244.

GOODALL, JANE
Jane Goodall Institute, PO Box 599, Ridgefield, CT 06877, USA.

HEMELRIJK, CHARLOTTE K.
Institut für Informatik der Universität Zürich, Winterthurerstr. 190, CH-8057 Zürich, Switzerland.

HOHMANN, GOTTFRIED
Max-Planch-Institut für Verhaltensphysiologie, Seewiesen, D-82319, Starnberg, Germany.

VAN HOOFF, JAN A. R. A. M.
Ethologie en Socio-oecologie, Utrecht Universiteit, P.O. Box 80086, 3508 TB Utrecht, The Netherlands.

HOSAKA, KAZUHIKO
Department of Zoology, Faculty of Science, Kyoto University, Sakyo-ku, Kyoto, Japan 606-01.

IDANI, GEN'ICHI
The Center for African Area Studies, Kyoto University, Yoshida-Shimoadachi-cho, Sakyo-ku, Kyoto, Japan 606-01.

IHOBE, HIROSHI
Department of Zoology, Faculty of Science, Kyoto University, Sakyo-ku, Kyoto, Japan 606-01.

ISABIRYE-BASUTA, GILBERT
Department of Zoology, Makerere University, P.O. Box 7062, Kampala, Uganda.

ITANI, JUNICHIRO
20 Nishiumenoki-cho, Shimogamo, Sakyo-ku, Kyoto, Japan 606-01.

KANO, TAKAYOSHI
Primate Research Institute, Kyoto University, Kanrin, Inuyama, Aichi, Japan 484.

KURODA, SUEHISA
School of Human Cultures, The University of Siga Prefecture, 2500 Hassakacho, Hikone, Siga, Japan 522.

MARCHANT, LINDA F.
Department of Sociology, Gerontology and Anthropology, Miami University, Oxford, OH 45056, USA.

MARUHASHI, TAMAKI
Department of Human and Cultural Sciences, Musashi University, 1-26-1, Toyotamakami, Nerima, Tokyo, Japan 176.

MATSUZAWA, TETSURO
Primate Research Institute, Kyoto University, Kanrin, Inuyama, Aichi, Japan 484.

MCGREW, WILLIAM C.
Department of Sociology, Gerontology and Anthropology, and Department of Zoology, Miami University, Oxford, OH 45056, USA.

MITANI, JOHN C.
Department of Anthropology, University of Michigan, Ann Arbor, MI 48109, USA.

MOORE, JAMES
Department of Anthropology, University of California, San Diego, La Jolla, CA 92093-0101, USA.

MWANZA, NDUNDA
Centre de Recherches en Sciences Naturelles, Lwiro, D.S. Bukavu, Zaïre.

NISHIDA, TOSHISADA
Department of Zoology, Graduate School of Science, Kyoto University, Sakyo-ku, Kyoto, Japan 606-01.

NISHIHARA, TOMOAKI
Department of Zoology, Faculty of Science, Kyoto University, Sakyo-ku, Kyoto, Japan 606-01.

OKO, RUFIN A.
Ministrie de Forestrie Economique, B.P. 13794, Brazzaville, Congo.

SAVAGE-RUMBAUGH, E. SUE
Department of Biology, Georgia State University, Atlanta, GA 30303, USA.

VAN SCHAIK, CAREL P.
Department of Biological Anthropology and Anatomy, Duke University, Durham, NC 27708-0383, USA.

SUZUKI, SIGERU
Department of Zoology, Faculty of Science, Kyoto University, Sakyo-ku, Kyoto, Japan 606-01.

TAKAHATA, YUKIO
Naruto University of Education, Takashima, Naruto-cho, Naruto, Japan 772.

TUTIN, CAROLINE E. G.
Station d'Etudes des Gorilles et Chimpanzés, B.P. 7847, Libreville, Gabon.

DE WAAL, FRANS B. M.
Department of Psychology, Emory University, Atlanta, GA 30322, USA.

WATTS, DAVID P.
Department of Anthropology, University of Michigan, Ann Arbor, MI 48109, USA.

WHITE, FRANCES J.
Department of Biological Anthropology and Anatomy, Duke University, Durham, NC 27708-0383, USA.

WILLIAMS, SHELLY L.
Department of Biology, Georgia State University, Atlanta, GA 30303, USA.

WRANGHAM, RICHARD W.
Peabody Museum, Department of Anthropology, Harvard University, Cambridge, MA 02138, USA.

YAMAGIWA, JUICHI
Primate Research Institute, Kyoto University, Kanrin, Inuyama, Aichi, Japan 484.

YUMOTO, TAKAKAZU
Center for Ecological Research, Kyoto University, Shimosakamoto, Otsu, Shiger, Japan 520-01.

ZIHLMAN, ADRIENNE
Department of Anthropology, University of California, Santa Cruz, CA 95064, USA.

Preface

WILLIAM C. MCGREW, LINDA F. MARCHANT AND TOSHISADA NISHIDA

This book emerged from a conference, 'The Great Apes Revisited,' sponsored and hosted by the Wenner-Gren Foundation for Anthropological Research, and held in Cabo San Lucas, Baja California Sur, Mexico, 12–19 November, 1994. The organizers were William McGrew and Toshisada Nishida, and the monitor was Linda Marchant. But the result is much more than conference proceedings, and in the 20 months from meeting to publication most chapters went through notable transformations. For example, although only 23 primatologists took part in the conference, 40 contributed to the book.

The origins of the conference lie two decades earlier, in another Wenner-Gren conference, 'The Great Apes,' organized by Jane Goodall and David Hamburg, and held in Burg Wartenstein, Austria, in July, 1974. Five years later, the results were published, along with additional invited chapters, as *The Great Apes* (Hamburg & McCown, 1979). Of the participants in the 1974 conference, only a handful are still active in research on great apes (including the two organizers), while others have died, retired, or shifted to other interests.

The two conferences can be contrasted in several ways. In 1974, there were no papers discussed on bonobos (*Pan paniscus*) nor lowland gorillas (*Gorilla gorilla gorilla*); too little was known. All but one of the field studies of chimpanzees were of the eastern subspecies (*Pan troglodytes schweinfurthii*), given the lack of knowledge then of the other geographical races in central western (*Pan troglodytes troglodytes*) and far western (*Pan troglodytes verus*) Africa. Instead, five of the participants had studied wild orangutans (*Pongo pygmaeus*), more than those who focused on all other apes except *P. t. schweinfurthii*. In 1994, chimpanzee studies still predomi-

nated, but all subspecies were represented, as was the case for gorillas, and even the number of bonobo contributions dwarfed those on orangutans. (These proportions mirror changes in scientific interest in apes over that period; for a quantitative analysis, see Hebert & Courtois, 1994.)

On a more theoretical front, in 1974 the hottest topic in primatology was 'ape language,' whether with American Sign Language, plastic disks, or Yerkish lexigrams. Sociobiology had yet to emerge (E.O. Wilson's book was published a year later), though Irven DeVore proselytized vigorously at Burg Wartenstein in advance of its appearance. Behavioral diversity in apes was beginning to be appreciated, but it was labeled as preculture or protoculture rather than culture. Imitation by apes was taken for granted, but social cognition had yet to be invented. Laboratory workers still did deprivation studies on apes. Fieldwork was still largely natural history, and problem-oriented and theory-driven behavioral ecology or socio-ecology had yet to be applied to primates. Perusal of the table of contents of this book will show how things have changed since 1974. (These developments in behavioral primatology are also revealed in Hebert & Courtois's 1994 study: the four areas of research with the greatest growth have been sex and reproduction, feeding ecology, social relations, and migration and other aspects of ranging.)

Even the make-up of the participants changed over the 20 years: In 1974, most participants had been trained in the USA, while in 1994, two-thirds had been trained in Europe or Japan. In 1974, the only three women present were the Leakey-chosen trio of Dian Fossey, Biruté Galdikas and Jane Goodall; in 1994, there were nine women. In 1974, the most prominent disciplinal

Preface Illustration I. Wenner-Gren Conference on 'The Great Apes Revisited,' Cabo San Lucas, November 1994. *Left to right* (seated): Takayoshi Kano, Suehisa Kuroda, Juichi Yamagiwa, John Mitani, Linda Marchant, Charlotte Hemelrijk, Sue Savage-Rumbaugh; *left to right* (standing): Laurie Obbink, Tetsuro Matsuzawa, Bill McGrew, Jane Goodall, Toshisada Nishida, Frans de Waal, Adrienne Zihlman, Richard Wrangham, Sydel Silverman, Jim Moore, Barbara Fruth, Carel van Schaik, Diane Doran, David Watts, Christophe Boesch, Caroline Tutin, Mark Mahoney, Frances White.

affiliations of participants were biology and psychology; in 1994, these were clearly superseded by anthropology. The most telling contrast, however, was in the breadth of knowledge: In 1974, only three participants (Richard Davenport, Junichiro Itani and Toshisada Nishida) had studied more than one species of ape. By 1994, most participants had studied at least two, and four participants (Diane Doran, Suehisa Kuroda, John Mitani and Adrienne Zihlman) had studied three or more!

The overall aim of the book is to synthesize important, current knowledge, and to exemplify state-of-the-art developments in the ethology and ecology of the great apes. In effect, therefore, the goals are to update, extend, review, evaluate, clarify, explore, criticize and celebrate the hard-won findings of primatologists over the last 20 years. The main emphasis is on spontaneous behavior as it can be observed in natural and naturalistic settings; this requires input from both the natural and social sciences, so the framework is comparative and cross-disciplinary. An important, but concluding aim of the book is to show how such knowledge may help elucidate the origins of another hominoid line, the genus *Homo*.

How to justify such a tight focus on only four species of great ape? Of all terrestrial forms besides ourselves, their lives are the most complex and varied. Whether grounded in the challenges of subsistence or the demands of sociality, their intelligence is strategic. Even if they were *not* our evolutionary cousins, we could not resist trying to figure out their coalition formation, tactical deception, sophisticated communication, elementary technology, etc. Therefore, that they *are* our nearest living relations is but icing on the cake. The simple point that chimpanzees are likely to be genetically closer to

Preface Illustration II. Wenner-Gren Conference on 'The Behavior of Great Apes,' Burg Wartenstein, July 1974. *Left to right*: Mark Konishi, Roger Fouts, Peter Rodman, David Horr, Bill Mason, Dick Davenport, David Hamburg, Jane Goodall, Robert Hinde, Elizabeth McCown, David Bygott, Biruté Galdikas-Brindamour, Pat McGinnis, Toshisada Nishida *back row*: Irven DeVore, Emil Menzel, Bill McGrew, Dian Fossey, John MacKinnon *in front*: Jo van Orshoven, Junichiro Itani, Lita Osmundsen

humans than to gorillas remains staggering. The question of which of the living African apes most closely resembles the common hominoid ancestor guarantees long-running debate. By the same token, the orangutan, as a relative outlier, becomes even more enigmatic, as an Asian analogue.

Given these two points – complexity and closeness – we are bound to make more use of the great apes in the reconstructive modeling of human origins and evolution. The still-prevalent tendency among paleoanthropologists to minimize or ignore apes in evolutionary scenarios has worn thin, e.g. lithic technology is customarily invoked in interpreting the extractive foraging of extinct hominoids, yet reference is rarely made to stone tool use in living but non-human hominoids. Continuing to deny advanced mental and communicatory abilities, such as symbol use, to non-humans, as is the case with many cognitive scientists, is not only outdated, but scientifically embarrassing. Yet those who study great apes must guard against complacency or even arrogance, and against assuming a null hypothesis of no-difference

across hominoids. It is worth keeping in mind Goodall's early dictum: 'We will only be able to appreciate real differences between humans and apes when we know their similarities.'

The bulk of the book is in six parts: Part I synthesizes current knowledge of three species, emphasizing cross-subspecific comparisons. Not all gorillas live in the mist, and orangutan solitariness is not as simple as it seems. Part II keys in on the dynamic interface between making a living and coping with others, and not just within a species, but across them. (It is convenient for such analyses that some populations of chimpanzee and gorilla are sympatric, but others are not.) Part III luxuriates in the social lives of apes, from demography to politics. Face-to-face relations in apes appear to be more transactional than interactional, that is responses to social overtures are individualistically contingent on context and history. Yesterday's rival may be today's ally and tomorrow's scapegoat. Copulation is therefore negotiation. Part IV tackles intelligence, and the chapters reflect the broad pattern of methods currently in use. Hard-nosed empiricism is tempered by intuitional inference; the traditional dichotomy of field versus laboratory is seen for what it is: a confounded constraint. Simplistic parsimony may die hard, but scientific advances sometimes require paradigmatic risk-taking, even to the point of working only in the virtual reality of the computer. Part V emphasizes what apes have in common – the stem-traits or universalities of apehood, upon which the frills of diversity are firmly based. These chapters examine variations on the foundational themes of locomotion, shelter, communication and laterality. Finally, Part VI touches on issues of self-absorption: can we humans understand from whence we came by looking at the apes of today? Can the data from molecules to fossils be reconciled to allow the piecing together of the whole picture of human origins?

Potential readers should also be told what this book is *not* about: it is strongly biased toward studies of apes in nature (or at least what passes for nature on this crowded planet infested by one overwhelmingly dominant species), so studies in captivity are neglected. The main reason for this is the limits on space needed to keep down the purchase price of the book, but the editors admit to ethical uneasiness about the justification of confining apes for any reason. (For a wide-ranging discussion of these issues, see Cavilieri & Singer, 1994.) The book is also restricted to basic as opposed to applied primatology, that is it is about science and not conservation. The latter deserves a book in its own right, one that covers the spectrum from individual welfare to species survival, from issues of protection and rescue to habitat preservation to rehabilitation and restoration. Readers interested in pursuing the conservational points raised in the foreword are referred to Peterson & Goodall (1993).

ACKNOWLEDGMENTS

The editors are grateful to the Wenner-Gren Foundation, and especially to Sydel Silverman (President), Laurie Obbink and Mark Mahoney, not only for the conference, but also for the publication grant that enabled the publication of this book; at Miami University, we thank Joy Donovan, Cary Berryman, Amanda Ensminger, Carol Kist, Sarah Karpanty, Melanie Peterson, Kathy Robinson and Jennifer Weghorst for essential help on all practical aspects of manuscript preparation; Melanie Peterson also drew the wonderful cover illustrations; we are thankful for the solid and unwavering support from Cambridge University Press, especially Tracey Sanderson, our editor, and from Myra Givans, our copy editor. Finally, some people made special and unique contributions: John Mitani spurred us to undertake the project, both in Dar-es-Salaam and in Ann Arbor, where he hosted a crucial planning meeting in March 1993; Carel van Schaik and Evelyn Vineberg added vital information and advice to the foreword; and Tetsuro Matsuzawa saved our eleventh hour sanity by solving a trans-Pacific problem in information technology.

REFERENCES

Cavilieri, P. & Singer, P. (eds.) 1994. *The Great Ape Project*. London: Fourth Estate.

Hamburg, D. A. & McCown, E. R. (eds.) 1979. *The Great Apes*. Menlo Park: Benjamin/Cummings.

Hebert, P. L. & Courtois, M. 1994. Twenty-five years of behavioral research on great apes: Trends between 1967 and 1991. *Journal of Comparative Psychology*, **108**, 373–80.

Peterson, D. & Goodall, J. 1993. *Visions of Caliban: On Chimpanzees and People*. New York: Houghton Mifflin.

Foreword: conserving great apes

JANE GOODALL

The Great Apes, in Africa and Southeast Asia, are endangered. My expertise relates to chimpanzees, and so I will concentrate on them and on forest conservation in Africa. In most cases, the same issues also apply in the conservation of bonobos and gorillas. At the turn of the century, there were hundreds of thousands of chimpanzees living in what are now 25 African nations throughout a more or less unbroken belt of forest from the far west coast of Africa to the eastern shores of Lake Tanganyika. There are still many chimpanzees in the area that was the heartland of their historic range – in the contiguous forests of Cameroon, Central African Republic, Congo, Equatorial Guinea, Gabon and Zaïre. There are still vast areas of forest in some of these countries. But for how much longer? Forests and woodlands worldwide are disappearing as a result of commercial logging and the ever-increasing growth of human populations that tend to compete with and to destroy, rather than to live in harmony with, the natural world. When an area of forest is despoiled, the chimpanzees who live there disappear too: the territorial nature of wild chimpanzees ensures that communities squeezed out by habitat destruction cannot easily integrate with others, away from the devastated areas. Chimpanzees have already become extinct in four of the nations where they once lived and their numbers have shrunk alarmingly in many others.

In western Africa, chimpanzees and their forest habitats are particularly endangered. Recent DNA analyses suggest that the far western subspecies, *Pan troglodytes verus*, separated from the *Pan troglodytes troglodytes* of central Africa about 1.6 million years ago. If later research corroborates this finding of two species of chimpanzee instead of one, then this will be a further argument for intensifying conservation efforts in many west African countries. The eastern or long-haired chimpanzees, *Pan troglodytes schweinfurthii*, of eastern Africa, are almost extinct in some countries; in Burundi, no more than 300–400 remain. The population in Tanzania has shrunk to no more than 2000, if that. And everywhere, except in the great forests of central western Africa, the remaining chimpanzee populations are increasingly fragmented, forced to exist in smaller and smaller areas, and with ever-decreasing opportunity for gene exchange between neighboring groups.

All across Africa, there are pathetic remnants of chimpanzee communities hanging on in strips or pockets of forest in areas so steep and rugged that even desperate humans can see no way of clearing them for their own purposes. But, even if these forest fragments cannot be cultivated, villagers will utilize them as they have for hundreds of years: for firewood, building poles, charcoal, and hunting. Often this involves the setting of snares, and although chimpanzees may not be the intended victims many will nevertheless be caught.

The single most crucial issue with regard to the continued existence of chimpanzees, bonobos and gorillas in their wild state in Africa is the explosive growth of human populations. Yet only comparatively recently have conservationists dared to mention this issue in public, mainly because of rumors widely believed in many African countries that the developed world hoped to retain its political and economic domination by suppressing population growth in the developing nations. The terminology used at one time – *population control* – did not help. *Family planning* is less paternalistic, but even that has been contaminated in the heated controversy. Now, however, increasing numbers of politicians

Foreword Figure I. Adult male chimpanzee (right) was alpha male at Gombe in 1974; pirouetting infant is alpha male in 1995; adult female (rear) is sister of the first and mother of the second. (Photo by J. Moore)

in more and more African countries have begun to admit the need for preserving areas of wilderness, as they face up to the devastating effect of mushrooming human populations on natural resources. It makes little sense to search for ways of achieving sustainable development unless the issue of voluntary population stabilization is also part of the agenda.

The fact is that deforestation, for whatever reasons, is likely to lead to soil and water erosion, changing climate and desertification that in time create conditions that are equally inhospitable to human and non-human primates alike. Even with full understanding of the results of poor land utilization, it is often difficult to come up with conservation measures that will work in rural areas. The subsistence farmer in Africa is usually well aware that loss of trees will lead to loss of soil fertility in years to come, but he cannot concern himself with the future when it is a struggle to feed himself and his family today. And if the numbers of people living in an area are increasing, where can they grow the extra crops they need unless they cut down the forest? How can they build houses without building poles, cook without firewood, feed their cattle and goats without pasture? For those who have not been to Africa, who have not seen conditions for themselves, the crippling poverty in which so many people live is hard, almost impossible, to comprehend. Millions of peasants lead a hand-to-mouth existence depending on their own efforts for survival – they cannot buy food from the local supermarket, for there is none, and there is no such thing as welfare or social security payments.

If one is concerned about the conservation of the great apes, with protecting their forest environment, then it is necessary to understand that any plan must include the people who are living in and around the area to be preserved. Until relatively recently, conservation was linked only to the protection of wildlife; the welfare of the local people living in the vicinity of a national park or reserve was seldom on the agenda. This has changed. Increasingly, local people, thanks to their traditional understanding of the land and to new education programs, realize that the final destruction of these forests will also blight their own ultimate survival. Provided they are consulted, provided they themselves stand to gain if they stop felling the forests, and provided alternative ways of making a living are available, peasants will cooperate and participate in conservation efforts. In many countries today, there are imaginative rural development programs such as agro-forestry (growing trees along with other crops for building poles, firewood, charcoal, silage, and so on), game farming and the crafting of products that can be sold to *buy* food rather than grow it all. Increasingly popular in donor nations are projects that give women more education and more control over their lives. The introduction of eco-tourism brings in foreign exchange to the central government and bolsters the local economy, particularly when local villagers are involved, as in Uganda and Gabon. Most important is to shed the paternalistic attitude to Africa that has characterized so many development plans. The developed world must also acknowledge its past, and sometimes continuing, exploitation of the African continent, its people and their wealth.

Many African countries, concerned by rapid deterioration of their natural resources, have introduced stringent conservation measures, which is a crucial first step, even if there are not always mechanisms in place to enforce regulations. (This, of course, is also a common failing in the developed world.) In Burundi and Uganda, once flourishing centers for illegal traffic in chimpanzees and other wildlife, governments have clamped down on the trade, and seeing orphan chimpanzees for sale is almost a thing of the past. Burundi is fighting to save the last of its unique montane rain forest in the north of the country. Only about 350 chimpanzees live there, but the forest is contiguous with a larger area of similar habitat in Rwanda, so the total number of chimpanzees may be around 500. Uganda has designated its nine main forest blocks as reserves and is actively encouraging chimpanzee-focused tourism in place of commercial logging. In Congo, the magnificently pristine northern forests are being preserved in the newly designated Nouabalé-Ndoki National Park, and the government has expressed interest in creating another large park in the coastal rainforests in the south. The government in Ivory Coast has sent the army to combat poaching and mining in the beleaguered Taï Forest. These are just a few examples of conservation measures recently introduced in some African countries.

For the subsistence farmer, living in harmony with chimpanzees is not always possible. Around Gombe, where small fragmented groups of chimpanzees have been isolated and confined to rugged ravines of gallery forest totally surrounded by cultivation and human dwellings, there are clashes between farmers and the chimpanzees. With rapidly decreasing supplies of wild foods, the apes increasingly abandon their traditional dietary conservatism and make forays into the fields. There are reports of children being killed by enraged (and probably terrified and desperate) adult chimpanzees. A similar situation in southern Burundi was tackled by Rob Clauson, who employed a small group of well-trained staff to protect the villagers from the chimpanzees, and the chimpanzees from their own ill-advised behavior. These men monitored the whereabouts of the chimpanzee groups, and when it seemed that they were getting too close to a village or its crops, they drove them away.

Both Wrangham and Boesch have described how groups of chimpanzees set out to raid commercial crops. In Uganda, it is banana plantations that are raided; in the Taï Forest, the chimpanzees go on long excursions to feast on cocoa beans. In Guinea, Kortlandt described the raiding of a grapefruit plantation, although in that case, it was permitted and approved of by the owner of the plantation.

However, at Bossou, Guinea, there seems to be some kind of accord between villagers and the chimpanzees, as studied first by Kortlandt and then by Sugiyama and Matsuzawa. But, this is an unusual situation: there are only 18 chimpanzees remaining on a sacred hill. And even there, the attitude of the villagers to the apes is changing as a result of the influx of refugees from Liberia and the increased pressure on diminishing resources. Similarly, the bonobos of Wamba have long enjoyed a peaceful coexistence with humans thanks to a local taboo against killing and eating bonobos. However, old traditions now compete with alternative views, as villagers come into contact with people who lack the taboo. Researchers are challenged to devise ways to encourage maintenance of traditional values without disparaging local peoples' desires for more modern ways.

Unfortunately, in most places where humans and chimpanzees are close neighbors, the chimpanzees are doomed. They are either hunted for food or for the live animal trade, shot as vermin, or caught in snares usually set to capture other mammals for food. Boesch, Wrangham and others report a disturbing percentage of the surviving population in and around their study areas who have snares around wrist and ankle, or who have lost parts of their limbs from gangrene and other infections caused by snares. In Zaïre, both at Wamba and Lomako, researchers encounter bonobos with ensnared limb parts. While bonobos are not the target of capture, wire snares may cripple or kill these unintended victims.

In our conservation efforts, should we completely ignore the plight of remnant groups of chimpanzees in order to concentrate only on areas that still have significant numbers? Although many would answer this question in the affirmative, I believe that our attitude towards the conservation of the great apes should be influenced by the fact that they are so close to humans, not only in their genes, anatomy and physiology, but also in their intellect and emotional expression and needs. Although areas having larger numbers may have to be given priority on practical grounds, I do not believe that the protection of these amazing beings should be debated simply in terms of numbers. From a humanistic and ethical point of view, the plight of the endangered great apes should perhaps be viewed as similar to the plight of beleaguered human populations. Just because there is a larger population of chimpanzees in Congo than in Mali, does this mean that the tiny remnant in Mali is not worth trying to save? Is it more important to protect the many chimpanzees in the Ndoki region of Congo and Central African Republic, than the much smaller and embattled groups further south?

There are many who believe that conservation should be strictly concerned with the survival of species and that any consideration of individuals and their 'rights' is sentimental, frivolous or even unethical in view of the undisputed need to conserve species. For these critics, there is a valid scientific argument for trying to protect remnant populations: each such group has a unique gene pool, along with unique ways of adapting to ecological and social pressures. Each group has its behavioral traditions. At the very least, these apes should be surveyed and studied, and this alone is often a step towards protection.

Great ape field researchers, if they care about the beings they study and particularly if they are interested in longitudinal observations in the natural habitat, can be a major force in conservation efforts. Simply being present in the field can make a significant contribution. The research will bring in some foreign exchange, it will provide employment for local people and it will focus attention on the area. If the area is already designated as protected, the research project and its staff can be a source of practical support and knowledge to the custodians of the area. If the area is not protected, the results of the study or survey may persuade local authorities that it should be.

At Gombe, we employ local people not only to cut trails, but also to collect ecological and behavioral data. These men are proud of their jobs, and they care about the chimpanzees as individuals and talk about them to their families and friends in the villages. There is virtually no poaching at Gombe. In other words, our staff act as ambassadors for chimpanzees to the villages. Some of the Tanzanian field assistants recently visited local schools to talk about wildlife and conservation as it relates to village life. The four most senior of them recently travelled to the capital city, Dar es Salaam, and spoke to school children there. It was morale-boosting for the assistants and fascinating and highly informative for the students. It is particularly important to develop conservation programs for children, and we now bring small groups of older pupils to Gombe in order to show them something of the wonder of their natural heritage.

Field researchers can also become involved in conservation efforts by helping to develop carefully controlled eco-tourism. Even small numbers of visitors can make a difference to a local economy, especially if villagers are employed as guides or in other ways. Some national parks receive a good deal of revenue, these days, from tourists. The best-known example of the way in which foreign visitors can help to protect a highly endangered form of ape is mountain gorilla tourism in Rwanda. Before the tragic civil war, such tourism was the biggest foreign exchange earner in the country. In Eastern Zaïre, a group of chimpanzees at Tongo was habituated by Annette Lanjouw for the express purpose of starting tourism for chimpanzees in an attempt to preserve them. The initiative has been very successful. Dean Anderson, working for the Jane Goodall Institute (JGI), initiated a similar project in the Kibira Forest of northern Burundi; the project is being carried on by Ally Wood. G. Isabirye-Basuta and Richard Wrangham have been cooperating with the Ugandan authorities to develop chimpanzee tourism in the Kibale Forest, and a number of Peace Corps volunteers are habituating chimpanzees in other Ugandan forest blocks where the tourist industry is being established to replace logging. Useful and interesting information about the distribution and habits of the chimpanzees is accumulating during these habituation exercises.

Of course, tourism is not without its dangers: too many visitors, if not properly controlled, may increase levels of stress among the creatures they have come so far to see. The great apes are susceptible to human infectious diseases and, when they are habituated to humans, tourism can pose a health hazard. In some cases, there are other dangers: chimpanzees, gorillas and orangutans are very strong, and visitors have been injured by both gorillas and chimpanzees in the forest. Sometimes tourism or the presence of researchers has undesirable side effects, making a once low-visibility area more salient. For example, people from neighboring villages and towns are sometimes attracted when researchers are present, because of rumors of opportunities to obtain prized commodities such as cloth and salt. The more people there are, the more hunting, snaring and cutting of the forest.

Individual great apes may act as ambassadors for others of their kind. I refer here to orphans, those whose mothers have been killed by hunters. In many countries, the relevant governmental departments have now agreed to confiscate infant apes who are being held or sold illegally, provided there is a responsible organization prepared to take charge of the orphans. There are several such sanctuaries already established for chimpanzees and for gorillas in Africa, and for orangutans in Indonesia and Malaysia. In Congo, there are two sanctuaries for confiscated chimpanzee infants, one near Pointe Noire, operated by JGI; the other is near the border with Gabon, operated by Aliette Jamart and M. Pique with

Foreword Figure II. Infant western lowland gorilla, orphaned by hunters and held in captivity. (Photo by J. Moore)

the help of the French organization, H.E.L.P. Between 1989–92, more than 40 infants were received by JGI, and Jamart and Pique are caring for more than 30. It is estimated that for each infant that reaches its destination alive, at least 10 chimpanzees die: the mothers, the infants who die when their mothers are shot or during the journey to the market place, and the other group members who are shot as they try to defend victims. That means that the 70–80 confiscated infants in Congo represent the deaths of 700–800 other chimpanzees. The infants offered for sale are only a fraction of the total number of infants that are taken from the forest. Some, whose mothers were killed for the bush meat trade, may be kept for a while, then eaten. Very small ones, who live only a few days or weeks without specialized care,

are given to village children as playthings, or, if they are older, are kept chained up as mascots or are fattened for the pot. Others are sold to dealers for the international trade in live animals. Often these contraband apes must be smuggled out of the country, but this is easy, as the great Zaïre River, with its unguarded banks, runs between Congo and Zaïre along wild, uninhabited borders.

Sometimes poachers do not even bother to cross the river at unpatrolled borders, but instead disembark at Brazzaville, Congo, with an orphaned bonobo or two destined for the market. Because bonobos are indigenous to Zaïre, wild-caught individuals cannot be removed legally from the country without valid CITES permits. Yet it happens. Some bonobos have been confiscated in Congo and are cared for at the Projêt de Protection des Gorilles, primarily a gorilla orphanage, run by Mark Attwater and Helen Hudson.

Elsewhere in Africa are other chimpanzee sanctuaries. The oldest is Abuko-Baboon Islands, in The Gambia, started by Eddie and Stella Brewer, later joined by Janice Carter; another, Chimfunshi, was built and is operated by David and Sheila Siddle in Zambia; and another is being established by Peter Jenkins and Liza Gadsby in Cameroon. The JGI, in addition to its Congo sanctuary, is also helping orphaned chimpanzees in Burundi, Kenya, Tanzania and Uganda. These sanctuaries hope to attract paying visitors, so that they will become self-supporting as well as provide economic benefits to the surrounding villages. Most importantly, sanctuaries today incorporate conservation education programs. The chimpanzees then serve as a focal point, in explaining the nature and importance of overall wildlife conservation. Urban Africans seldom get a chance to see wild animals in their natural habitat, but visits to sanctuaries provide stimulating opportunities, especially for classes of school children.

In Zaïre's capital city, a local Zaïrean, Kizito, was trained in art by Delfi Messenger. Messenger maintains a shelter for bonobo orphans at the Institut National de Recherche Biologique (INRB) and has embarked on several innovative conservation education projects. One resulted in an educational package about elephants and gorillas through the popular medium of comic books written in Swahili and distributed in eastern Zaïre. Kizito's drawings of indigenous wildlife such as okapi and bonobos have also been printed on the covers of children's notebooks distributed in Kinshasa and major

towns in Zaïre. These small-scale conservation projects can have far-reaching effects at the local level.

Are we ethically justified in spending money to create good environments for captive chimpanzees, when efforts to save wild populations and their forest habitats are so desperately in need of funding? Some people feel that confiscated infants should be euthanized. In my view, however, individual great apes matter, just as individual humans matter, and so they deserve our help.

Almost all field primatologists today find themselves involved, internationally or otherwise, in conservation efforts, especially when studying our closest relatives, the great apes, with their long lives and complex behavior. We still have so much to learn, particularly in the sphere of traditional and cultural variation in behavior in the different parts of each species' range. But time is running out – for them and for us. Only a handful of the remaining apes in the wild are being studied, and every year there will be fewer and fewer populations to choose from as the unobserved, unknown, yet utterly fascinating apes disappear. They will be gone, exterminated, taking their secrets with them. We can never do more than guess at the total range of ape adaptations – both ecological and social – across their historic ranges. Even between my writing these words, in January 1995, and the publication of this book, thousands of square kilometers of forest will have fallen to the pit saw and the logging companies.

But we should not give up hope. As recounted above, there are signs of changed attitudes and more importantly, of concrete, positive results. And we can take comfort in the knowledge that every research project helps to protect the great apes and their forest homes, provided that local people are involved, and provided that researchers cooperate with governments to develop some controlled tourism, when appropriate, and help in conservation education efforts.

ACKNOWLEDGMENTS

This paper was originally prepared for the Wenner-Gren conference 'The Great Apes Revisited,' held November 12–19, 1994 in Cabo San Lucas, Baja, Mexico. I am grateful for assistance in the preparation of this chapter to Carel van Schaik and the editors, but especially to Evelyn Ono Vineberg.

- PART I

Apes overviewed

1 • Toward an understanding of the orangutan's social system

CAREL P. VAN SCHAIK AND JAN A. R. A. M. VAN HOOFF

INTRODUCTION

The aim of primate socioecology is to clarify how social structure and organization are influenced by the interaction between ecological factors and biological constraints. A coherent theoretical framework is gradually being developed to account for the interspecific and intraspecific variation found among primates (Wrangham, 1980, 1987; Dunbar, 1988; Watts, 1989; van Schaik, 1989; van Schaik & Dunbar, 1990), and specific models are increasingly being subjected to empirical tests. These tests will lead to further refinements in these models, which therefore promise to make important contributions to our understanding of the origins of human social behavior and social systems.

Orangutans (*Pongo pygmaeus*) stand out among diurnal primates in several ways. [Adults are largely solitary, have low interaction and association rates, show extreme sexual dimorphism in body size and appearance, show pronounced bimaturism among sexually mature males, engage in forced copulations (especially non-resident males), display female mating preferences in favor of some males and in opposition to others, and range in seasonal or irregular movements] which seem to make it impossible to distinguish spatially discrete communities (social units). All of these unique or unusual features have yet to be reconciled with existing socioecological and sociobiological ideas. It is likely that they are interrelated in as yet unknown ways.

In order to develop good theoretical models and to perform valid tests, we must have good descriptive information on all relevant features of a population's social system. Three models are currently available to describe the orangutan social system. MacKinnon (1974, 1979)

argued that some form of [spatially dispersed, age-graded male group characterizes their social organization, with adult males acting as range guardians for their reproductively active subadult male relatives.] Rodman & Mitani (1987), in contrast, characterized it as a [dispersed social organization with a promiscuous mating system with strong male–male competition] Schürmann & van Hooff (1986) and van Hooff (in press) proposed a variant of this model, which emphasizes the [pivotal role of female choice.] Here we scrutinize these models, point to gaps in our understanding, and suggest directions for future research.

In this review of the orangutan social system, we follow the now classic approach which postulates that female distribution and relationships are above all a response to the distribution of risks and resources in the environment, whereas male distribution and relationships reflect above all the spatiotemporal distribution of mating opportunities (Ims, 1988; Davies, 1992). Thus, mammalian mating systems are widely thought to reflect the female's response to environmental factors, unless the need for male parental care is unusually great (Clutton-Brock, 1989).

Rarely considered by students of non-human primates, however, is the influence male behavior can exert on female behavior. Wrangham (1986) postulated that vulnerability to hostility by conspecifics is the principal ultimate cause for male–female bonds in apes (cf. van Schaik & Dunbar, 1990). Intersexual interactions and relationships can thus not be ignored.

Accordingly, we examine the orangutan social system in three steps. First, we consider female spatial distribution and relationships, especially as these are thought to relate to environmental conditions. Second, we examine male distribution and relationships, and ask whether

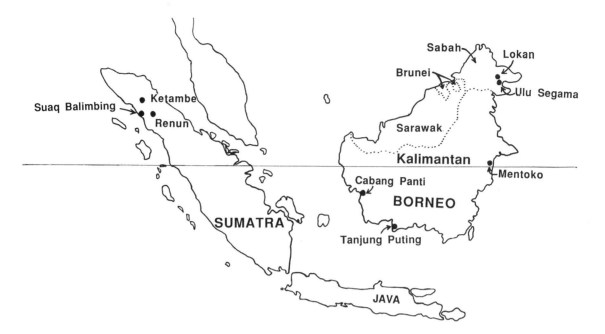

Fig. 1.1. Places where orangutan field studies have been done. Information on sites not mentioned explicitly in the text is in the following sources: Renun: MacKinnon, 1974; Cabang Panti: Leighton, 1993; Lokan: Horr, 1975; Ulu Segama: MacKinnon, 1974.

or not they reflect the distribution of mating opportunities. Third, we look for indications of female mating preferences, and examine the possible benefits to females from these preferences and the possible implications for female distribution and sexuality.

Figure 1.1 shows the study sites mentioned in this review. Almost all information is published, except for the preliminary results of a study recently started by the first author in a Sumatran peat swamp forest at Suaq Balimbing.

FEMALE DISTRIBUTION AND RELATIONSHIPS

Why females are solitary

Diurnal primates are usually gregarious. Predation risk and feeding competition may have provided the selective pressures toward their gregariousness (Wrangham, 1980; van Schaik, 1983). Orangutans face significant predation risk only on the ground, as suggested by the rarity of terrestrial travel in Sumatra, where tigers occur. The apes can reduce this risk by being arboreal. Predation risk may also be significant at night, but by building nests away from obviously risky spots, such as popular fruit trees, females with vulnerable infants minimize the risk of being attacked by arboreal predators such as the clouded leopard (Sugardjito, 1983). Thus, the lack of serious predation risk obviates the need for gregariousness as a means to reduce it.

Feeding competition in alliances over high-quality food patches might also favor group-living, especially in frugivores (Wrangham, 1980). However, for feeding competition to favor gregariousness as a means of increasing food intake through coalitions over access to food patches, mean group size must be at least two. Under most conditions, orangutans could not afford to live even in such small groups. The high energetic costs of association are expressed in changes in the activity budgets of individuals that form travel parties: they travel more and eat less (Galdikas, 1988), or travel more and rest less (Mitani, 1989). In Mitani's study, subadult males did not change their activity patterns when in association, indicating that it is mainly fully adult males

Fig. 1.2. Adult female orangutan moves arboreally from one liana to another. (Photo by C. van Schaik)

and adult, reproducing females that suffer the costs of such association.

The costs of association for adults emanate from both feeding competition and from differences in male and female feeding styles. Orangutans are selective feeders that need enormous amounts of food. Their diets are dominated by fruits (see Rodman, 1988), and dietary correlations indicate that orangutans prefer fruits over leaves and bark (Sugardjito, 1986; Galdikas, 1988). Galdikas (1988) noted that orangutans have a tendency to exhaust the patches in which they feed, even though Leighton (1993) showed that orangutans select fruit patches largely on the basis of crop size. Rodman (1979) was the first to note that males and females differ in the patterning of their patch use. Sugardjito's (1986) data indicated that adult males have much longer feeding bouts on average than females. Obviously, were adult males and females to associate this would impose a severe cost on the sex that is forced to compromise most (probably the male).

As expected, aggregations of orangutans are seen mainly in large fruiting trees with superabundant fruit crops, such as the strangling figs (e.g. Rijksen, 1978), whereas travel bands form only during periods of high fruit abundance, i.e. a high density of fruit patches (Sugardjito et al., 1987). Hence, the solitary nature of orangutan females is consistent with current theory on the ecological benefits of gregariousness in diurnal primates.

Female ranges

The consequences of solitary female ranging for male behavior should depend at least in part on the females' spatial distribution. Horr (1975) and Rodman (1973) presented data suggesting that resident females live in small ranges of 65 ha and 40–60 ha, respectively. However, subsequent studies of longer duration at the site of Rodman's initial study, Mentoko, by Mitani (1985b) and Suzuki (1992) showed that females lived in much larger ranges of at least 150 ha. Rijksen's (1978) mean home range of some 150 ha at Ketambe is consistent with these observations (see also te Boekhorst et al., 1990). Galdikas (1988) reported even larger female home ranges, between 350 and 600 ha, at Tanjung Puting. Preliminary observations at Suaq Balimbing also indicate large female ranges of well over 100 ha, for at least two different females. These contrasts in range size are independent of population density and may depend on the spatial scale of habitat patchiness. Unlike the other sites, both Tanjung Puting and Suaq Balimbing are mosaics of peat swamp and dryland forests.

These large ranges, relative to the females' movements, appear to make female territoriality impossible. Mitani and Rodman (1979) developed an index of range defensibility, $D = d/ \sqrt{(4A/\pi)}$, where d = day journey length and A = area of range. The patterns they found across primate species suggested that D should be at least 1 to make territories possible. D equals 0.27 for the females of Tanjung Puting (Galdikas, 1988), and 0.42 at Mentoko (Mitani & Rodman, 1979; probably an overestimate – see above). Thus, range defense among females is not expected. Nonetheless, Horr (1975) and Rodman (1973) noted that female ranges barely overlap, but as noted above such small ranges with little overlap are probably relatively uncommon and transient, possibly when females have young offspring. In contrast, Galdikas (1988) reported, and Suzuki's (1992) maps indicated, considerable overlap among female ranges, although there is much variation in the degree of overlap. Instances of exceptionally high overlap, where female ranges are almost entirely encompassed in those of others, are likely to reflect close genetic ties (Galdikas, 1988). Early results from Suaq Balimbing also suggest high overlaps: ten adult females and at least six nulliparous females have been seen in the central 200 ha portion of the study area.

With this spatial distribution, there should be frequent opportunities for female encounters. When females meet there is evidence for decided dominance relationships (e.g. Rijksen, 1978), but females rarely associate (e.g. Rijksen, 1978; Galdikas, 1984).

Natal dispersal

Little is known about dispersal. Rodman (1973) hypothesized that maturing females tended to remain near the natal area, while males moved away. The number of developing juveniles who stay in the study area in long-term studies provide the best estimate of a sex bias in philopatry, but such direct observations are few. Rijksen (1978) described how the adolescent female Jet stayed in the range of her mother (Joosje) after she died in 1972 (and in the mid-1990s Jet is still using the same study area), whereas another adolescent, the male Mos, moved away as he grew older. Galdikas (1988, p. 27) also noted that adolescent females 'traveled mostly within their mothers' ranges.' At Ketambe, two rehabilitant females, released as juveniles at Ketambe in the late 1970s, still use the same study area, as do two adult daughters of one of them. In contrast, several young males disappeared forever after increasingly long intervals between visits to the feeding ground. These females stayed on after the rehabilitation program closed and all artificial feeding ceased in 1979.

There is also circumstantial evidence. First, in long-term studies, new individuals are more likely to be sub-

adult males than females. Moreover, unidentified individuals, most likely those wandering over large areas, are often subadult males (Galdikas, 1984; Mitani, 1985a). Second, there are clear sex differences in the attendance records of rehabilitants observed at Ketambe during the 1970s. As individuals grew older their attendance declined, but with an important sex difference. Males often disappeared for weeks or even months, sometimes never to reappear, while females remained more faithful visitors (van Schaik & van Noordwijk, unpublished). If males simply ranged more widely than females, they would occasionally still show up in the study area. Since they do not, one must assume that at least some of them emigrated. Thus, both sets of findings are consistent with a predominance of female philopatry.

A stronger philopatric tendency of females than males is in agreement with the trend seen in most other solitary mammals (Waser & Jones, 1983). However, in solitary species a greater degree of opportunism in philopatry and dispersal is expected, as compared with gregarious ones, because opportunities to settle at or near the natal range must be more variable for lone females (e.g. depending on whether the mother is still alive), whereas females in groups can almost invariably stay in their natal groups.

Conclusion

In sum, although the data are sparse, they evoke the image of thoroughly solitary females, with a tendency toward philopatry in ranges that are not actively defended against other females and thus tend to overlap widely. There is no indication of active female bonding through affiliation, grooming, agonistic support, or even some dispersed form of spatial association. Thus, there is no indication of supportive social relationships among females other than the mother–offspring dyad.

MALE DISTRIBUTION AND RELATIONSHIPS

Male options

Males are expected to map themselves onto the female distribution, and male–male relationships should reflect competitive or cooperative strategies over mating access to females. Theoretically, many different outcomes remain possible when, as in orangutans, females are solitary and dispersed. Clutton-Brock (1989) argued that, in the absence of female gregariousness and male care for offspring, the observed social organization is determined by the defensibility of the female's range, or barring that, of the female herself. This leaves the following possibilities:

1. *Monogamy.* A male can defend access to the range of a single female only, and facultative monogamy ensues, in which the two sexes share one range but do not coordinate travel in close spatial proximity. This requires low range overlap among females.

2. *Resource defense polygyny.* A single adult male can defend access to several females' ranges, where he is able to exclude other males from this range, at least to the extent that he controls matings in this area. Subordinate adult males and smaller (subadult) males live in peripheral areas not inhabited by females.

3. *Male-bonded promiscuity.* In a situation with highly overlapping female ranges, a group of related males can sometimes achieve what a single male cannot: defending the range of several females (Rodman, 1984). Such a system may well have evolved from a roving promiscuity system where males converge on estrous females (cf. te Boekhorst, 1991).

4. *Lekking.* Female ranges are not defensible, but during breeding time females tend to congregate at predictable places that can be defended by males.

5. *Roving male promiscuity.* Males cannot defend access to female ranges and females do not congregate at predictable areas. Thus, all males have large and widely overlapping ranges within which they search for receptive females.

Orangutan male ranging

The observed mating system is thought to depend on whether or not the male can defend the range of one or more females. Defensibility usually is inferred from the observed outcome, since it is difficult to test for directly (Davies, 1992, p. 265). However, again using the defensibility index of Mitani & Rodman (1979), we can get an impression. Thus, if we take the observed day journey lengths of males, this should point to the maximal area a male is able to defend in his usual mode of ranging (when D becomes 1; cf. van Schaik & Dunbar, 1990). Given the mean adult male day journey length of 688 m (Rodman 1979) and 850 m (Galdikas 1988), this suggests maximally defensible areas of 37–57 ha, much smaller than most of the observed female range sizes (see above).

This result eliminates facultative monogamy and resource defense polygyny as possible mating systems for orangutans. It is admittedly possible that males could dramatically increase their mobility. However, even in that case, short detection ranges in dense forest may preclude complete defense of an area against males that quietly trespass. Rodman (1973, 1984), who found the smallest female ranges, speculated that adult males were able to exclude other males from their ranges. However, all subsequent studies indicate that adult male ranges overlap and that mating is promiscuous (see below; cf. Rodman & Mitani, 1987).

Suzuki (1989) has claimed that the orangutan system is male-bonded, because male ranges seem to share common boundaries. However, a male-bonded system also requires that adult males associate in flexible parties and show mutual affiliation. Adult male orangutans do neither, although subadult males do some of both (e.g. Galdikas, 1985b). The absence of male-bonding in orangutans presents a paradox, because chimpanzees, with similar diets and similar female body size are male-bonded (e.g. Goodall, 1986), even in rain forest habitats (cf. Rodman, 1984). Perhaps, the orangutans' regular movements between habitats (te Boekhorst *et al.*, 1990; M. Leighton, personal communication; van Schaik, unpublished) preclude the communal defense by males of a more or less exclusive range in orangutans.

Lekking or small male territories in areas of high female traffic are not observed in primates, unlike the situation in ungulates (Clutton-Brock, 1989). It is possible that predictable concentrations of fertile females (the basis for leks) are precluded in primates, due to their lower densities, lower mobility and long interbirth intervals.

By a process of elimination, we are thus left with the roving male promiscuity system (cf. Rodman & Mitani,

Table 1.1. *Numbers of fully adult males seen in a single study area, during limited periods*

N of males	Period	Site	Park	Source
3	3 years	Ketambe	GL	Rijksen (1978)
5	1 year	Ketambe	GL	van't Land (1990)
5	1 year	Suaq Balimbing	GL	van Schaik (unpublished)
12	4 years	Camp Leakey	TP	Galdikas (1985a)

Note: GL, Gunung Leuser National Park, northern Sumatra; TP, Tanjung Puting National Park, Central Kalimantan.

1987). If so, we should see males using very large and widely overlapping ranges. Little is known on male range size, except that it must be larger than many study areas. Suzuki's (1992) sketch maps suggest a male home-range size of 500–700 ha, three or four times the size of female ranges; van't Land's (1990) maps suggest somewhat smaller ranges, but still several times larger than those of females.

Male ranges must overlap considerably, because all studies except Rodman's (1973) report several adult males in the same area (Table 1.1). However, most researchers (e.g. Rodman, 1984; Galdikas, 1985a; Mitani, 1985a; te Boekhorst *et al.*, 1990) distinguish between resident and wandering adult males, with usually only a single male being the resident. Nonetheless, multiple males routinely use the same area. Galdikas (1985a) considers two individual males resident. It is also not clear whether the distinction between residents and wanderers represents a true dichotomy in ranging or is an artifact of relatively small study areas with large ranges and wide overlaps.

Resident males are not always dominant (Fig. 1.3). Rijksen (1978) reports that the resident male OJ was subordinate to both Gorm and Mozes, who were occasional visitors. Mitrasetia & Utami (1994) report that dominance relationships among the several males observed in the Ketambe area between 1987 and 1992, while decided, were not transitive.

While the resident adult male is relatively sedentary, sometimes he disappears from the area for a long time, while other resident individuals stay behind. Mitrasetia & Utami (1994) report how OJ left the area for a few years after he had been deposed as the local dominant, and Galdikas (1985a) reports how TP left the study area for two years, for no apparent reason.

In sum, adult male ranges are neither exclusive nor

Fig. 1.3. A solitary, resident adult male orangutan. (Photo by C. van Schaik)

stable and the most resident male need not be the dominant male of the neighborhood. All of this is consistent with the notion of roving male polygyny.

Intermale competition

Despite high range overlaps, contest competition among adult males is implied by their extreme mutual intolerance, by spatial avoidance and by their large size. When they are in each other's vicinity, there is mutual or one-sided avoidance (e.g. Rijksen, 1978; personal observations), and when they come close, there is invariably some agonistic interaction. Not surprisingly, adult males have the highest incidence of disfigurements (MacKinnon, 1974; Galdikas, 1985a), presumably from injurious fights.

Adult males give 'long calls,' which carry for up to a kilometer over level terrain (e.g. Mitani, 1985a). Only fully adult males with cheek flanges and big throat sacs produce these long calls. Some long calls are given in direct response to social stimuli: when the male approaches another individual, often a female or a subadult male, or after he has defeated another adult male (Rijksen, 1978; Galdikas & Insley, 1988).

Although the impact of long calls on females is still debated (see below), there is abundant evidence that males either approach or avoid them (van't Land, 1990). Mitani (1985a) has confirmed this function through playback experiments. Rijksen (1978) and van't Land (1990) have suggested that at Ketambe, males are more likely to call near the boundary of their range, but it is difficult to exclude confounding effects: males may avoid calling near noisy rivers or prefer to call from ridges (away from rivers).

The flanges may act as bullhorns that concentrate the call's energy in the direction in which the caller faces (cf. Rodman & Mitani, 1987). Preliminary observations at Suaq Balimbing indicate that males always call in the direction in which they will travel, and our impression is that they are more likely to call before they travel over long distances. Thus, the call may act to announce their arrival in an area. Accordingly, van't Land (1990) noted that days with more long calls were days in which the males were more likely to travel in a straight line. This timing and directionality in long calls make them efficient spacing mechanisms.

Adult male–subadult male relationships

Adult males dominate subadult males. They will tolerate subadult males, for instance, when they are consorting with an adult female, provided that the subadults maintain a respectful distance. Subadult males among themselves are fairly tolerant, although dominance relationships can probably be distinguished among them (Schürmann 1982; Galdikas 1985b; van Schaik, preliminary observations).

The sociosexual strategies of adult and subadult males clearly contrast. Because subadult males suffer little from association (Mitani, 1989), they can spend more time in association than can other reproductively adult individuals (Rijksen, 1978; Galdikas, 1985b). Accordingly, these mobile animals maximize the time they stay with females. Adult males, in contrast, suffer energetically from association and are extremely solitary most of the time (Rijksen, 1978; Galdikas, 1988). Hence, they can engage in consortships only briefly, but they maximize their fertilization success by doing so when a female is most likely to conceive and by excluding rival males.

Subadult males can remain arrested in the subadult state for many years (te Boekhorst et al., 1990; Uchida, 1994). There is long-standing speculation about whether the presence of dominant adult males in the same area is responsible for this (Kingsley, 1982).

Mating behavior

Most matings take place in voluntary or involuntary consortships, when a male and a female associate, with the male essentially guarding mating access to the female. When male consorts change, adult males always displace subadults and within age classes dominants displace subordinates (Schürmann, 1982; Galdikas, 1985a). When adult males meet in the presence of receptive females, escalated contests are likely (Galdikas, 1985a). Thus, severe mating contest among orangutan males is evident.

Subordinate, subadult males sometimes trail the consort pair at a safe distance, attempting to mate when the consorting male is distracted (e.g. Schürmann, 1982; Galdikas, 1985b). Females may also mate with more than one adult male in one reproductive cycle, which lasts several weeks. Moreover, females usually take several menstrual cycles to conceive (e.g. Schürmann, 1982). Hence, mating is promiscuous.

Females strongly resist mating with some males, although copulations usually ensue. The males forcing these copulations are usually subadults, but adults also do so (MacKinnon, 1974; Schürmann & van Hooff, 1986; Djojosudharmo, personal communication). Parous females are especially likely to resist mating attempts. This resistance is unique among non-human primates. It may indicate that the conditions favoring sexual coercion (Smuts & Smuts, 1993) hold for orangutans: lack of allies for females and superior male strength. However, this is unlikely. Females of other solitary and promiscuous species do not seem to be harassed by males (aye-ayes, *Daubentonia madagascariensis:* Sterling, 1993; grey kangaroos, *Macropus giganteus:* Jarman & Southwell, 1986). Orangutan females could likewise minimize sexual coercion if they experienced short, clearly advertised receptive periods.

On the other hand, females also show evidence of positive preference. Females normally avoid close contact with adult males, which they can do because males regularly emit long calls (e.g. Galdikas, 1988). Yet, receptive females willingly engage in consortships with adult males. This suggests that when they are receptive females derive some benefits from approaching adult males. Primiparous, adolescent females actively seek out consortships with adult males (Schürmann, 1982; Galdikas, 1985a; Schürmann & van Hooff, 1986) and show extremely proceptive behavior, up to the point that the females actively effect intromission and perform pelvic thrusting (Schürmann, 1982; cf. Nadler, 1988). Such persistent proceptive behavior is uncommon among primates (cf. Janson, 1984).

Discussion

The information on mating behavior does not support MacKinnon's (1979) hypothesis that fully adult males act as 'range guardians' in which their subadult relatives (sons?) can mate without being displaced by strange adult males (cf. Schürmann & van Hooff, 1986; van Hooff, in press). When observations were made around the probable date of conception, adult males were the most likely father (Schürmann, 1982; Galdikas, 1988; Mitani, 1985b). Moreover, in view of male natal dispersal, it is unlikely that fathers and sons are in each others' vicinity.

Among the mating system models presented earlier, roving male promiscuity provides the best fit with observations. There is no spatial exclusion of males, so an estrous female attracts multiple males who compete vehemently for sexual access to her and she often ends up mating promiscuously.

Nonetheless, unparalleled in primates, female orangutans express strong mating preferences for and against certain (classes of?) males (*pace* Rodman & Mitani, 1987).

These preferences are not necessarily inconsistent with roving male promiscuity (cf. van Hooff, in press). If the genetic quality of the males was the criterion for mate choice, females are expected to select males on the basis of revealing handicaps (cf. Andersson, 1994): physical traits such as the size and symmetry of flanges or the length of the hair on forearms and back, or behavioral traits such as the rate, pitch and duration of long calls. The male bimaturism suggests that full adult male size itself may also be a revealing handicap, if time needed to attain full adulthood is determined by genetic quality. Thus, convergence on a small set of adult males with the preferred phenotypical traits is expected.

However, if different females prefer different adult males, male services or male-mediated access to resources might be involved. Some observations suggest that individualized relationships might exist. First, females occasionally avoid matings with adult males who then force copulations. These adult males are not the resident adults that spend most of their time in the area (MacKinnon, 1974; Galdikas, 1979; Mitani, 1985b; Djojosudharmo, personal communication), but nonetheless possess fully adult features, and may even be dominant to the resident male(s). Second, females may show different responses to subadult males, resisting copulations with some much more strongly than with others (personal observations). If, as these observations suggest, female mating preferences are individual-specific, this should lead to the rejection of the original roving male promiscuity model.

MALE–FEMALE INTERACTIONS AND RELATIONSHIPS

If all interactions between males and females concern only mating, the spatiotemporal distribution of females, largely a product of the species' ecology, will determine the mating system, and the social system will be determined overwhelmingly by the relationships within the two sexes. However, if there are other significant interactions between the sexes, the social system (and possibly the mating system) will reflect this. Non-mating interactions between the sexes hardly seem to characterize the thoroughly solitary orangutans. However, MacKinnon (1974, 1979) suggested that orangutans form social units in which one or two adult males and several subadult males are associated with a number of adult females and their young. Thus, socially distinct, if spatially overlapping, communities could exist.

Evidence for the community model

Several observations are consistent with the community model. Many researchers suggest that their area harbors both residents and transients or wanderers (see above). The existence of male wanderers makes sense under the roving male promiscuity model, but the only formal analysis so far, by te Boekhorst *et al.* (1990), suggested that a relatively fixed set of males and females moved into the floodplain forest at Ketambe during the peak fruiting season only to move out again at its end, consistent with the notion of communities.

The data presented by te Boekhorst *et al.* (1990) indicate that the disappearance of two former dominant males, Mozes and Gorm, was accompanied by the disappearance of several adult females and their offspring. At least some of these females had a friendly relationship with a male that disappeared. For instance, Mif associated with Mozes, who was significantly weakened and approaching death, just before he died. She was rarely seen afterward. Finally, Suzuki (1992) noted that male range boundaries tend to coincide. While his interpretation of a male-bonded system is not plausible, the basic observation is consistent with MacKinnon's model. All these observations suggest the existence of some form of community.

If individualized social relationships exist, this explains the puzzling observation that most adult males give long calls. All adult males in Suaq Balimbing emitted long calls in the central study area, although not all at the same time (van Schaik, unpublished). Likewise, van't Land (1990) found that three of four adult males gave long calls at Ketambe, and that the calling males gave long calls at about the same rate. This is remarkable, since emitting long calls increases a male's risk of being detected and chased by a dominant male in the area (observations at Suaq Balimbing). But it makes sense if male long calls also function in male–female communication. Differential tracking and responding to long calls would enable females to express their partner preferences among adult males. Although long calls do not elicit explicit responses from females in most cases (Mitani, 1985a), most studies have noted that females

occasionally respond to the call by immediately approaching it (Rodman, 1973; MacKinnon, 1974; Horr, 1975; Galdikas, 1985a). In one case, where a female was involved in an involuntary consortship with a subadult male in which two forced copulations had already taken place, she immediately took off in the direction of a long call, but was intercepted and forcibly restrained by the consorting male (van Schaik, unpublished).

The community model also explains why most of the resident adult males stay in an area for years without wandering away when reproductive opportunities seem reduced. Such persistent residence is inconsistent with roving male promiscuity.

There are also problems with the community model, however. First, adult males twice left the area after the females they had mated with became pregnant (Nur at Ketambe; Utami, personal communication) or gave birth (Galdikas, 1985a). If these males had fathered offspring, and were absent during the periods of high infant vulnerability, these observations should lead us to reject the protective function of adult males. A second problem is that we might expect that females actively seek promiscuity and blur the ovulation signalling, since this tends to reduce the risk of infanticide (Hrdy & Whitten, 1987). The strong resistance to mating with some males is not easily reconciled with this idea, although it is possible that females only show resistance to males from different communities.

THE FUNCTION OF FEMALE MATING PREFERENCES

There seems to be enough evidence to take seriously the possibility that females have individual social relationships with males. These bonds may enable interactions or associations that benefit the females (and secondarily the males). The services provided by males could involve care for offspring, acquisition of food, protection against predators, or protection against hostile conspecifics. Male care for offspring is minimal in orangutans, as the almost exclusively arboreal females do not need males as sentinels against predators (*pace* MacKinnon, 1974), and males do not make inaccessible food available to females, so relationships based on these services can be excluded from consideration. However, mating conflict is common, because males generally benefit from mating with females, even if females prefer not to cooperate. Where females are unable to refuse such matings, or can do so only at great cost, this mating conflict can take the form of harassment and forced matings of receptive females (Wrangham, 1979; Smuts & Smuts, 1993). Where infants are vulnerable to male aggression, mating conflict can take the form of male infanticide (Smuts & Smuts, 1993).

At least among primates, the female's best defense against this harassment is male allies and, ironically, it is the best harassers that are also the best protectors (Smuts & Smuts, 1993). Female primates are especially vulnerable to infanticide by males. Primate infants can not run, hide or fight back, can easily be located by males, and remain vulnerable for a long time due to their slow development. Infanticide avoidance has been implicated as the selective pressure that made it advantageous for a solitary female to associate closely with a male, thus giving rise to bonded monogamy in primates (van Schaik & Dunbar, 1990). In fact, this idea can also account for the ubiquity of permanent male–female associations in gregarious primate species (van Schaik & Kappeler, unpublished). As predicted, this association has evolved mainly among those species in which females carry their infants with them and thus benefit from associating with one or more males in order to reduce the risk of infanticide.

However, the association seems to have been lost again in some species, in particular the largest-bodied arboreal frugivores (apes and spider monkeys, *Ateles* spp.). Perhaps, constant association between a solitary female and an adult male is impossible due to the high energetic costs of spatial association above a certain body size (roughly that of gibbons). Are there alternative solutions to the problem of male harassment? It is tempting to argue that in male-bonded species, such as chimpanzees, the male groups who defend a communal territory effectively protect the resident females by communally patrolling the borders and imposing prohibitively high costs on male intruders (death!). Formal tests of this hypothesis need to be developed (cf. Rodman, 1984).

With their extremely slow life histories, orangutans should in theory be among the primate species most vulnerable to infanticide, just like the other great apes (chimpanzees: Hiraiwa-Hasegawa, 1987; gorillas: Watts, 1989). Yet, surprisingly, no attempts at infanticide have

ever been seen in wild orangutans, despite the orangutan females' apparent vulnerability. However, the possibility of female–male relationships points to a way out of the enigma of the orangutan as the only truly solitary higher primate. It is conceivable that, despite the usual spatial separation, certain males and females maintain some spatial coordination which allows social contact if needed. Since the need rarely arises, such contact is rarely seen.

COMPETING MODELS FOR THE ORANGUTAN'S SOCIAL SYSTEM

This review leaves us with two plausible models of the orangutan social system, both partly consistent with the published evidence: a roving male promiscuity system, in which there is no higher-level social unit; or a spatially dispersed but socially distinct community organized around one or more large adult males.

Before critical tests can be done, extensive additional observational work is needed to assess whether or not these models correctly describe the situation in orangutans. Both models require that paternity is highly concentrated in the adult males and paternity tests are needed to establish this. Both models also need to clarify the role of subadult males. More detailed descriptions of female–subadult male relationships are needed before we can assess how the subadult males fit into the two models, if at all.

The plausibility of a community model would be bolstered if observations were to demonstrate high levels of spatial association between certain males and certain females. Unfortunately, the resident-transient dichotomy used to support this model is also consistent with roving-male promiscuity, if it reflects simultaneous attraction to locally abundant food (some kind of ideal free distribution: Milinsky & Parker, 1991). Hence, simultaneous follows of multiple individuals are needed to demonstrate true association or coordination at a finer scale. Also, since it is possible that multiple adult males, along with several subadult males, are members of a single community, we need estimates of the genetic relatedness among the males frequenting a given area.

Further refinement of the roving male promiscuity model requires observations on male movements in relation to estrous females. This model also predicts that female mating preferences rely on male traits (see above), which can be tested experimentally. Perhaps, genetic heterozygosity or other, more direct measures of genetic quality can be estimated non-invasively to assess the validity of this idea.

The critical functional difference between the two models is that the roving male promiscuity model assumes that the female mating preferences observed among orangutans are an expression of choice for the best genes in the population, whereas the community model assumes that the preferred males somehow provide protection to the females against harassment, particularly infanticide. Detailed descriptive data on matings are needed, as is evidence for differential association between a female and a male depending on the female's reproductive state.

An indirect, but critical, test of this functional difference is that one expects very different female responses to playbacks of the long calls of strange adult males. The roving male model predicts the usual response to male long calls: females in estrus might approach the call, while females not in estrus should ignore it. The community model, however, predicts that females with infants should show strong evasion and should increase their tendency to stay close to their protector male. In general, any potentially threatening presence of other adult or subadult males should lead to stronger coordination between the female and her protector.

In conclusion, the key challenge for future work is to determine whether female mating preferences are based on genetic quality or on direct phenotypic benefits, and whether individualized male–female relationships exist or not. We suspect that the orangutan's social life is a lot more complex than is suggested by its solitary habits.

ACKNOWLEDGMENTS

This paper was originally prepared for the Wenner-Gren conference 'The Great Apes Revisited,' held November 12–19, 1994 in Cabo San Lucas, Baja, Mexico. Van Schaik's fieldwork on orangutans at Suaq Balimbing is generously supported by the New York Zoological Society – The Wildlife Conservation Society. The fieldwork at Ketambe has been supported by WOTRO (the Netherlands Foundation for the Advancement of Tropical Research) and by the Dobberke Foundation. We thank all major participants in these projects for help and discussion: S. Djojosudharmo, E. A. Fox, T. Mitrasetia, H. Rijksen, C. Schürmann, J. Sugardjito and S. S. Utami. Conference participants, Beth Fox, Charlie Nunn and Maria van Noordwijk gave useful comments on the chapter.

We are very grateful to the Indonesian Institute of Sciences, LIPI, for granting permission to do research in Indonesia, and Universitas Nasional and Universitas Indonesia for sponsoring the projects. The directorate general of Nature Conservation and Forest Protection, PHPA, gave permission to do the work in Gunung Leuser National Park. CvS thanks S. Poniran, J. Supriatna and Mike Griffiths for support.

REFERENCES

Andersson, M. 1994. *Sexual Selection.* Princeton, NJ: Princeton University Press.

te Boekhorst, I. J. A. 1991. Social structure of three great ape species: an approach based on field data and individual oriented models. Ph.D. thesis, Utrecht University.

te Boekhorst, I. J. A., Schürmann, V. L. & Sugardjito, J. 1990. Residential status and seasonal movements of wild orang-utans in the Gunung Leuser Reserve (Sumatra, Indonesia). *Animal Behaviour,* **39,** 1098–109.

Clutton-Brock, T. H. 1989. Mammalian mating systems. *Philosophical Transactions of the Royal Society of London B,* **236,** 339–72.

Davies, N. B. 1992. Mating systems. In: *Behavioural Ecology: An Evolutionary Approach,* ed. J. R. Krebs & N. B. Davies, pp. 263–94. Oxford: Blackwell Scientific Publications.

Dunbar, R. I. M. 1988. *Primate Social Systems.* Ithaca, NY: Cornell University Press.

Galdikas, B. M. F. 1979. Orangutan adaptation at Tanjung Puting Reserve: mating and ecology. In: *The Great Apes,* ed. D. A. Hamburg & E. R. McCown, pp. 195–233. Menlo Park, CA: Benjamin/Cummings.

Galdikas, B. M. F. 1984. Adult female sociality among wild orangutans at Tanjung Puting Reserve. In: *Female Primates: Studies by Women Primatologists,* ed. M. Small, pp. 217–35. New York: Alan R. Liss.

Galdikas, B. M. F. 1985a. Adult male sociality and reproductive tactics among orangutans at Tanjung Puting. *Folia Primatologica,* **45,** 9–24.

Galdikas, B. M. F. 1985b. Subadult male orangutan sociality and reproductive behavior at Tanjung

Puting. *International Journal of Primatology*, **8**, 87–99.

Galdikas, B. M. F. 1988. Orangutan diet, range, and activity at Tanjung Puting, Central Borneo. *International Journal of Primatology*, **9**, 1–35.

Galdikas, B. M. F. & Insley, S. J. 1988. The fast call of the adult male orangutan. *Journal of Mammalogy*, **69**, 371–5.

Goodall, J. 1986. *The Chimpanzees of Gombe.* Cambridge, MA: Harvard University Press.

Hiraiwa-Hasegawa, M. 1987. Infanticide in primates and a possible case of male-biased infanticide in chimpanzees. In: *Animal Societies: Theories and Facts*, ed. Y. Itô, J. L. Brown & J. Kikkawa, pp. 125–39. Tokyo: Japan Science Society Press.

van Hooff, J. A. R. A. M. In press. The orangutan: outsider and socio-ecological test case. In: *Orangutans*, ed. R. Nadler, B. Galdikas, L. Sheehan & N. Rosen. New York: Plenum.

Horr, D. A. 1975. The Borneo orang-utan: population structure and dynamics in relationship to ecology and reproductive strategy. In: *Primate Behavior: Developments in Field and Laboratory Research*, ed. L. A. Rosenblum, pp. 307–23. New York: Academic Press.

Hrdy, S. B. & Whitten, P. L. 1987. Patterning of sexual activity. In *Primate Societies*, ed. B. B. Smuts, D. L. Cheney, R. M. Seyfarth, R. W. Wrangham & T. T. Struhsaker, pp. 370–84. Chicago: University of Chicago Press.

Ims, R. A. 1988. Spatial clumping of sexually receptive females induces space sharing among male voles. *Nature*, **335**, 541–3.

Janson, C. H. 1984. Female choice and mating system of the brown capuchin monkey *Cebus apella* (Primates: Cebidae). *Zeitschrift für Tierpsychologie*, **65**, 177–200.

Jarman, P. & Southwell, C. J. 1986. Grouping, associations, and reproductive strategies in eastern grey kangaroos. In: *Ecological Aspects of Social Evolution*, eds. D. I. Rubenstein & R. W. Wrangham, pp. 399–428. Princeton, NJ: Princeton University Press.

Kingsley, S. 1982. Causes of non-breeding and the development of the secondary sexual characteristics in the male orang utan: a hormonal study. In: *The Orang utan. Its Biology and Conservation*, ed. L. E. M. de Boer, pp. 215–29. The Hague: Dr. W. Junk.

van't Land, J. 1990. Who's calling there? A study on the social context of the adult male orang-utan long call

and its relation with short-term ranging behaviour. M.Sc. thesis, Utrecht University.

Leighton, M. 1993. Modeling dietary selectivity by Bornean orangutans: evidence for integration of multiple criteria in fruit selection. *International Journal of Primatology*, **14**, 257–313.

MacKinnon, J. 1974. The behaviour and ecology of wild orang-utans (*Pongo pygmaeus*). *Animal Behaviour*, **22**, 3–74.

MacKinnon, J. 1979. Reproductive behavior in wild orangutan populations. In: *The Great Apes*, eds. D. A. Hamburg & E. R. McCown, pp. 257–73. Menlo Park, CA: Benjamin/Cummings.

Milinsky, M. & Parker, G. A. 1991. Competition for resources. In: *Behavioural Ecology: An Evolutionary Approach*, ed. J. R. Krebs & N. B. Davies, pp. 137–68. Oxford: Blackwell Scientific Publications.

Mitani, J. C. 1985a. Sexual selection and adult male orangutan long calls. *Animal Behaviour*, **33**, 272–83.

Mitani, J. C. 1985b. Mating behaviour of male orangutans in the Kutai Game Reserve, Indonesia. *Animal Behaviour*, **33**, 392–402.

Mitani, J. C. 1989. Orangutan activity budgets: monthly variations and the effects of body size, parturition, and sociality. *American Journal of Primatology*, **18**, 87–100.

Mitani, J. C. & Rodman, P. S. 1979. Territoriality: the relation of ranging pattern and home range size to defensibility, with an analysis of territoriality among primate species. *Behavioral Ecology and Sociobiology*, **5**, 241–51.

Mitrasetia, T. & Utami, S. S. 1994. Male dominance in orangutan (*Pongo pygmaeus*) society. Abstract, XVth Congress of the International Primatological Society, Bali, Indonesia, August 1994.

Nadler, R. D. 1988. Sexual and reproductive behavior. In: *Orang-utan Biology*, ed. J. H. Schwartz, pp. 105–16. New York: Oxford University Press.

Rijksen, H. D. 1978. *A Fieldstudy on Sumatran Orang utans* (Pongo pygmaeus abelii *Lesson 1827*). Wageningen: H. Veenman and B. V. Zonen.

Rodman, P. S. 1973. Population composition and adaptive organisation among orang-utans of the Kutai Reserve. In: *Comparative Ecology and Behaviour of Primates*, ed. R. P. Michael & J. H. Crook, pp. 171–209. London: Academic Press.

Rodman, P. S. 1979. Individual activity patterns and the solitary nature of orangutans. In: *The Great Apes*, ed. D. A. Hamburg & E. R. McCown, pp. 235–55. Menlo Park, CA: Benjamin/Cummings.

Rodman, P. S. 1984. Foraging and social systems of orangutans and chimpanzees. In: *Adaptations for Foraging in Nonhuman Primates*, ed. P. S. Rodman & J. G. H. Cant, pp. 134–60. New York: Columbia University Press.

Rodman, P. S. 1988. Diversity and consistency in ecology and behavior. In: *Orang-utan Biology*, ed. J. H. Schwartz, pp. 31–51. New York: Oxford University Press.

Rodman, P. S. & Mitani, J. C. 1987. Orangutans: sexual dimorphism in a solitary species. In: *Primate Societies*, ed. B. B. Smuts, D. L. Cheney, R. M. Seyfarth, R. W. Wrangham, and T. T. Struhsaker, pp. 146–54. Chicago: University of Chicago Press.

van Schaik, C. P. 1983. Why are diurnal primates living in groups? *Behaviour*, **87**, 120–44.

van Schaik, C. P. 1989. The ecology of social relationships amongst female primates. In: *Comparative Socioecology. The Behavioural Ecology of Humans and Other Mammals*, ed. V. Standen & R. A. Foley, pp. 195–218. Oxford: Blackwell Scientific Publications.

van Schaik, C. P. & Dunbar, R. I. M. 1990. The evolution of monogamy in large primates: a new hypothesis and some crucial tests. *Behaviour*, **115**, 30–62.

Schürmann, C. L. 1982. Mating behaviour of wild orang utans. In: *The Orang utan. Its Biology and Conservation*, ed. L. E. M. de Boer, pp. 269–84. The Hague: Dr. W. Junk.

Schürmann, C. L. & van Hooff, J. A. R. A. M. 1986. Reproductive strategies of the orang-utan: new data and a reconsideration of existing sociosexual models. *International Journal of Primatology*, **7**, 265–87.

Smuts, B. B. & Smuts, R. W. 1993. Male aggression and sexual coercion of females in nonhuman primates and other mammals: evidence and theoretical implications. *Advances in the Study of Behavior*, **22**, 1–63.

Sterling, E. 1993. Behavioral ecology of the aye-aye (*Daubentonia madagascariensis*) on Nosy Mangabe. Ph.D. dissertation, Yale University.

Sugardjito, J. 1983. Selecting nest sites by Sumatran orang-utans (*Pongo pygmaeus abelii*) in the Gunung Leuser National Park, Indonesia. *Primates*, **24**, 467–74.

Sugardjito, J. 1986. Ecological constraints on the behaviour of Sumatran orang-utans (*Pongo pygmaeus abelii*) in the Gunung Leuser National Park, Indonesia. Ph.D. thesis, Utrecht University.

Sugardjito, J., te Boekhorst, I. J. A. & van Hooff, J. A. R. A. M. 1987. Ecological constraints on the grouping of wild orang-utans (*Pongo pygmaeus*) in the Gunung Leuser National Park, Sumatra, Indonesia. *International Journal of Primatology*, **8**, 17–41.

Suzuki, A. 1989. Socio-ecological studies of orangutans and primates in Kutai National Park, East Kalimantan in 1988–1989. *Kyoto University Overseas Research Reports of Studies on Asian Non-human Primates*, **7**, 1–42.

Suzuki, A. 1992. The population of orangutans and other non-human primates and the forest conditions after the 1982–1983's fires and droughts in Kutai National Park, East Kalimantan. In: *Forest Biology and Conservation in Borneo*, ed. G. Ismail, M. Mohamed & S. Omar, pp. 190–205. Kota Kinabalu: Yayasan Sabah.

Uchida, A. 1994. Reevaluation of 'adulthood' definitions of male *Pongo pygmaeus* in behavioral and skeletal biology. Abstract, XVth Congress of the International Primatological Society, Bali, Indonesia, August, 1994.

Waser, P. M. & Jones, W. T. 1983. Natal philopatry among solitary mammals. *Quarterly Review of Biology*, **58**, 355–90.

Watts, D.P. 1989. Infanticide in mountain gorillas: new cases and a reconsideration of the evidence. *Ethology*, **81**, 1–18.

Wrangham, R. W. 1979. On the evolution of ape social systems. *Social Science Information*, **18**, 335–68.

Wrangham, R. W. 1980. An ecological model of female-bonded primate groups. *Behaviour*, **75**, 262–300.

Wrangham, R. W. 1986. Ecology and social relationships in two species of chimpanzee. In: *Ecological Aspects of Social Evolution: Birds and Mammals*, eds. D. I. Rubenstein & R. W. Wrangham, pp. 352–78. Princeton: Princeton University Press.

Wrangham, R. W. 1987. Evolution of social structure. In: *Primate Societies*, ed. B. B. Smuts, D. L. Cheney, R. M. Seyfarth, R. W. Wrangham & T. T. Struhsaker, pp. 282–97. Chicago: University of Chicago Press.

2 • Comparative socio-ecology of gorillas

DAVID P. WATTS

INTRODUCTION

Great apes have remarkably diverse social systems. For example, only in gorillas do females live in stable, cohesive groups and associate permanently with males and do males show conditional dispersal. However, all apes show female dispersal from natal groups or home ranges and are non-female bonded (*sensu* Wrangham, 1980). Female African great apes transfer directly between groups or communities, behavior unusual in gregarious mammals (Greenwood, 1980; Pusey & Packer, 1987). These and other contrasts and similarities raise questions about variation in the costs and benefits of group living, of permanent male–female association, and of dispersal and philopatry (Wrangham 1979, 1980, 1982, 1987; van Schaik, 1989).

Here, I review aspects of gorilla ecology, life history strategies and mating tactics, and social relationships, with a focus on the ecology of female social relationships and on female transfer. I mostly cover mountain gorillas (*Gorilla gorilla beringei*, 'MGs' below), represented by the 10–20% of the Virungas population studied at the Karisoke Research Centre, Rwanda, since 1967. Few comparable data exist on social behavior for western (*Gorilla gorilla gorilla*; 'WLGs' below) and eastern lowland gorillas (*Gorilla gorilla graueri*; 'ELGs' below), but subspecific differences in dispersal costs are probably small despite ecological variation.

Sex biases in mammalian dispersal

Explanations for why most males disperse, and why most females are philopatric, in most mammals, invoke multiple factors (Greenwood, 1980; Dobson, 1982; Moore & Ali, 1984; Dobson & Jones, 1986; Johnson, 1986; Pusey & Packer, 1987). Dispersers may face higher mortality or breeding delays due to predation, competition for territories or resources, or poor foraging in unfamiliar and perhaps marginal habitats. Ecological costs are usually higher for females than for males because female reproductive success depends more on foraging efficiency; this favors female philopatry. Mating competition is usually more intense among males who are more likely to disperse to find mates or who are forced to disperse by competitors. When one sex is typically philopatric or has breeding tenures longer than a generation, inbreeding avoidance may help explain dispersal by the other (e.g., chimpanzee female natal transfer; Pusey, 1980).

However, these accounts omit several factors. Ecological costs are lower if females can transfer directly between groups: emigrants need not enter marginal habitat or lose the anti-predator benefits of grouping, and they can gain information on resource distribution from their new groups. Where groups are non-territorial and home ranges overlap greatly, females have some familiarity with the foraging areas of neighboring groups. Conflicts of reproductive interest between the sexes, and cooperation between the sexes to counter negative effects like infanticide ('inter-sexual mutualism;' Wrangham, 1982), can strongly influence dispersal costs, residence patterns and social relationships (Wrangham, 1979, 1982, 1987; Rubenstein, 1986; Wrangham & Rubenstein, 1986; van Schaik, 1989; Smuts & Smuts, 1993).

Within-sex cooperation in competition also affects dispersal costs in many primates and is probably a major reason for the predominance of female philopatry. Females often cooperate with female relatives (and sometimes non-relatives) in within-group contest competition for food. When this significantly affects their fitness, emigrants would leave behind their allies, and resident females in other groups could cooperatively prevent immigration. Therefore, female transfer should be common only when contest competition is unimportant (van Schaik, 1989).

GORILLA ECOLOGY

Diet composition and diversity

Gorillas are primarily herbivores. They have relatively large caeca (although subspecific variation has not been assayed; Chivers & Hladik, 1980) and long gut retention times (Milton, 1984), and thus have much hindgut fermentation capacity. This, plus large body size, helps them to meet nutritional needs by processing large amounts of plant foods high in structural carbohydrates.

MGs are the most folivorous subspecies (Fig 2.1). Virunga MGs mostly eat non-reproductive parts (leaves, stems, shoots and pith) of the terrestrial herbaceous vegetation (THV) that blankets most of their habitat (Schaller, 1963; Fossey & Harcourt, 1977; Watts, 1984). Their diet, the narrowest of all well-studied gorilla populations, comes mostly from a few plant species (see Table 2.1).

WLGs are the most frugivorous subspecies (Sabater Pi, 1977; Williamson et al., 1990; Rogers et al., 1990; Tutin et al., 1992; Tutin & Fernandez, 1993; Kuroda et al., Chapter 6). Those at Lopé have the broadest docu-

Table 2.1. *Numbers of plant foods, species and familes in the diets of gorilla subspecies at three study sites, and species from which the gorillas eat fruit, leaves, and pith, shoots or stems*

	Lopé	Kahuzi-Biega		Virungas
		A	B	
Foods	213	> 160	129	106
Species	156	104	79	62
Families	37	42		28
Fruit	97		20	5
Leaves	56		51	36
Stems/pith/shoots	16		23	27

Note: Lopé, western lowland gorillas; data from Tutin, Chapter 5. Kahuzi-Biega, eastern lowland gorillas; data from A. Goodall, 1977 (montane forest) and B. Yamagiwa *et al.*, Chapter 7 (lowland forest). Virungas, mountain gorillas; data from Watts 1984 and personal observation.

mented diet (see Table 2.1). They eat much THV throughout the year, but also eat large amounts of fruit, obtained from more plant species than in the entire diet of Karisoke MGs (see Table 2.1). Most fruits eaten are succulent and come from primary forest trees that produce seasonal crops. ELGs in montane habitat at Kahuzi-Biega eat mostly THV and bamboo, but eat more fruit than MGs and have a dietary breadth intermediate between MGs and Lopé WLGs. Those in lower altitude habitat eat more fruit (Casimir, 1975; Goodall, 1977; Yamagiwa & Mwanza, 1994; Yamagiwa *et al.*, Chapter 7 Table 2.1).

Nutritional ecology

Gorillas feed selectively, as expected for large generalist herbivores, and their apparent rules for food choice resemble those of highly folivorous monkeys (e.g. *Presbytis melalophos* and *Presbytis rubicunda*; Davies *et al.*, 1988) and of the baboons *Papio ursinus* and *Papio anubis* (Whiten *et al.*, 1992). For example, Virunga MG food preferences were positively correlated with the ratio of protein content to that of acid detergent fiber and condensed tannin (Watts, 1990).

Leaves and some shoots (e.g. bamboo shoots) are most important to gorillas as protein sources (Casimir, 1975; Goodall, 1977; Waterman *et al.*, 1983; Calvert, 1985; Rogers *et al.*, 1990; Tutin *et al.*, 1992). At Lopé,

Fig. 2.1. Adult male mountain gorilla eats terrestrial herbaceous vegetation (THV) in the Virunga Volcanoes, Rwanda. (Photo by D. Watts)

fruit is most important for energy, mostly from non-structural carbohydrates (Rogers et al., 1990). Gorillas probably extract much energy from stems, which are high in fiber, especially cellulose (Rogers et al., 1990; Watts, Waterman, & Conklin, unpublished data), via hindgut fermentation. Stems may be the major MG energy source. Bark can provide protein (Tutin et al., 1992), but may be equally or more important for its effects on passage rates and digestive efficiency (Goodall, 1977; Watts et al., unpublished data).

Leaf food quality is higher for MGs than for various colobines, which are foregut fermenters that must select highly digestible, high protein foliage because of their small size (e.g. WLGs vs. *Colobus satanas* at Lopé: Rogers et al., 1990; Virunga MGs vs. several African and Asian colobines: Waterman et al., 1983). However, the non-leaf component of gorilla diets includes high fiber, less digestible staples that colobines could not process in adequate bulk and that, while not highly preferred, are highly abundant. Importantly, even 'high quality,' preferred leaf foods are abundant, if less so than stems, shoots and pith.

Food distribution, diet seasonality, and foraging effort

Except for primary forest fruit, gorilla foods are abundant and evenly distributed compared with those of most primates. Species composition, stem density and biomass of THV vary spatially, but THV foods are more evenly and densely distributed than are fruit and are perennially available (Vedder, 1984; Watts, 1984; Rogers & Williamson, 1987). MGs mostly eat THV, so their diet varies little over time. Except for heavy seasonal use of bamboo shoots by groups that have access to bamboo forest, their dietary variation reflects variation in habitat use, not in food availability. Habitat use also varies little seasonally, except for use of bamboo forest (Schaller, 1963; Fossey & Harcourt, 1977; Vedder, 1984; Watts, 1984, 1991a).

MGs' use of habitat tends to maximize their foraging efficiency. For example, they mostly use vegetation zones where high quality food is abundant, diet quality is high, day journeys are short and daily feeding time is relatively low despite relatively high food intake (Watts, 1988, 1991a).

The dense and even distribution of food means that MGs can usually meet daily intake needs in a small area

and that group size effects on foraging effort (scramble effects) are slight. The mean day's journey is only about 0.5 km (Fossey & Harcourt, 1977; Watts, 1991a). Group size has a significant positive effect on length of day journey, but the magnitude of the effect is small (Watts, 1991a), and group size does not influence feeding time (Watts, 1988). Per capita ecological costs of adding group members should thus be low.

Fruit intake varies greatly across seasons for WLGs at Lopé (Williamson et al., 1990; Rogers et al., 1990; Tutin & Fernandez, 1993) and elsewhere (Sabater Pi, 1977; Kuroda et al., Chapter 6). During troughs in succulent fruit production, WLGs switch to more fibrous fruit and to fruit from aseasonal understory shrubs, and eat more leaves, piths, stems and bark (Tutin & Fernandez, 1993). THV species are thus their fallback foods (Rogers & Williamson, 1987; Tutin & Fernandez, 1993).

Diets of ELGs vary considerably across habitats. Like MGs, ELGs in montane habitat at Kahuzi-Biega use THV heavily throughout the year, and use bamboo forest heavily and eat many bamboo shoots seasonally. Unlike Virunga MGs, they use seasonal fruit crops and those in lowland forest show seasonal fluctuations in frugivory and THV use like those of WLGs (Casimir & Butenandt, 1973; Casimir, 1975; Goodall, 1977; Yamagiwa & Mwanza, 1994; Yamagiwa et al., Chapter 7).

Foraging effort varies more over time for ELGs and WLGs than for MGs. Day journeys are longer for ELGs (Goodall, 1977; Yamagiwa & Mwanza, 1994) and WLGs (Tutin, Chapter 5) when they use seasonal fruit crops than when they eat mostly THV or bamboo. Seasonal frugivory probably increases foraging costs because of the energetic expense of climbing, but the returns from the harvested fruit may offset such costs (Yamagiwa & Mwanza, 1994).

Lone males travel little to meet daily nutritional needs, but sometimes go far to look for or to follow groups. Lone males have home ranges much larger than is nutritionally necessary because they must search for mates (Yamagiwa, 1986; Watts, 1994a).

Home range overlap and site fidelity

Gorilla home ranges have few discrete, limited but rich patches of high quality habitat and are not economically defendable. Low variation in habitat quality and, in

WLGs, the need to track seasonal fruit crops lead to extensive home-range overlap between groups and to gradual or abrupt home-range expansions and shifts (MGs: Fossey & Harcourt, 1977; Watts, 1994a; ELGs: Casimir & Butenandt, 1973; Yamagiwa *et al.*, Chapter 7; WLGs: Tutin, Chapter 5). Home range-shifts may also result from male attempts to search for mates or to avoid mating competition (Watts, 1994a). Site fidelity seems to have limited importance for females: they can transfer, but still use much of their former foraging areas, or can shift foraging areas without changing groups. The ecological cost of transfer should thus be low, although these may be higher in more frugivorous populations because of greater variation in the distribution of fruit than of THV.

DISPERSAL IN MOUNTAIN GORILLAS

Female natal and secondary transfer are common, but not universal (Harcourt, 1978; Watts, 1990). Twenty-one of 29 females (72%) of known origin made natal transfers. Nine have reproduced in natal groups (one after re-immigration) that contained sexually mature males other than their fathers, but seven left natal groups that had such potential mates. Four of seven females who gave birth in natal groups and who have had at least two infants transferred after their first births, but one female has had so far five infants in her natal group.

Female tenure in a group varies from a few weeks to more than 27 years. Twenty-two females (the majority in the study population) have transferred secondarily from 1–4 times each (median = 1, $N = 35$ transfers). Histories for many females are incomplete, so secondary transfer is probably more common than this indicates. The number of births per female per group varies from none to at least seven. Many females reproduce in more than one group: 15 of 27 females with at least two infants have reproduced in two or more groups.

Because some females reproduce in natal groups and others transfer to groups that contain relatives or transfer with relatives, more than 70% spend at least some of their reproductive careers with female relatives.

Most males either leave natal groups and become solitary, or become subordinate followers in natal groups where they may breed and may eventually become dominant. Natal dispersal seems less common for males than

females (Robbins, 1995), although the male sample size is small. Only two males, both immature, have transferred into breeding groups. Some immature males join all-male groups; many of these later become solitary. In all contexts, mature males try to attract females from groups (Harcourt 1978; Stewart & Harcourt, 1987; Yamagiwa, 1987; Sicotte, 1993; Robbins, 1995; cf. Tutin, Chapter 5, for WLGs).

MALE–FEMALE AND FEMALE–FEMALE SOCIAL RELATIONSHIPS

Male–female relations

Long-term social bonds between adult males and females are central in MG social groups (Fig. 2.2). Most adult females spend more time close to males than to each other. All, especially those with no female relatives in their groups, have relatively frequent affinitive interactions with males (Harcourt, 1979a; Stewart & Harcourt, 1987; Watts, 1992, 1994b).

Male aggression to females is also common, but intense aggression and wounding are rare in within-group interactions (Harcourt, 1979a; Watts, 1992). Females nearly always respond submissively, often with formal submissive signals they rarely give to other females, and they reconcile with males after receiving male aggression (Watts, 1992, 1995). Males often intervene aggressively to end conflicts between females (Harcourt, 1979a; Watts, 1991b). They often display at females and sometimes disrupt matings by subordinate males; this can be considered 'courtship' aggression (Watts, 1992; cf. Wrangham, 1979).

Male–female relationships in groups with two or more males vary with male age and rank, the number of females per group and female residence histories (Harcourt, 1979a; Stewart & Harcourt, 1987; Watts, 1992; Watts & Pusey, 1993; Sicotte, 1994). Some males invest differentially in relationships with females. For example, an old male may preferentially affiliate with mature daughters and granddaughters and give them agonistic support. Also, adolescent and subordinate adult males are usually more responsible than females for maintaining proximity; conversely, females are usually more responsible for staying close to dominant males, and spend more time near and interact affinitively more

Fig. 2.2. Typical resting group of mountain gorillas: silverback adult male, adult females and their offspring. (Photo by D. Watts)

with them than with subordinates. Recent immigrants may be particularly attracted to dominant males. Some females in large groups may have strong social ties to subordinate males, but perhaps more as by-products of female–female competition for access to dominant males than to attraction to subordinates *per se*. Subordinate males and those with uncertain dominance relationships seem most likely to groom females, which may help win their mating allegiance (Watts, 1992).

Females with infants spend more time than those without infants close to males, where they are presumably safer from outside threats like extra-group males. Female proximity tends to decrease with the age of their offspring, although this is clearer in small than in large groups (Harcourt, 1979a; Watts, 1992).

Relationships between females

When related females reside together, female–female relationships differentiate clearly along lines of relatedness. Maternal relatives are close associates and often interact affinitively. Serious aggression between them is rare. Most co-residents are unrelated; they associate little and have few affinitive interactions and relatively many aggressive ones, although some become friends (Harcourt, 1979b; Stewart & Harcourt, 1987; Watts, 1994b). Paternal relatives have intermediate interaction rates and proximity (Watts, 1994b).

Female relatives support each other in contests about as often as females in some female-bonded species (Harcourt & Stewart, 1987, 1989; Watts, 1991b, in preparation). However, female agonistic relationships distinguish MGs from female-bonded species. Contest competition for food and cooperation between relatives in contests have minor nutritional importance for MGs (Watts, 1985, 1994b), and females have egalitarian agonistic relationships (*sensu* Vehrencamp, 1983). In many or most dyads, most aggressive interactions are undecided (targets ignore or return aggression) and agonistic relationships are unresolved. Females usually do not form linear dominance hierarchies. Aggression, including initiation of damaging fights, is commonly bidirectional, often significantly so (the more aggression from female A to female B, the more from B to A; Watts, 1994a). As predicted, if female–female relationships have limited importance for female fitness, females do not reconcile after conflicts (Watts, 1995).

In some groups, females also show bidirectionality in interventions 'contra:' the more A intervenes against B, the more B does against A (Watts, in preparation), unlike cercopithecines with strict, nepotistic female hierarchies (de Waal & Luttrell, 1988). This works against establishment or reinforcement of rank differences and against nepotism. The absence of consistent 'winner support' by mothers for daughters, the effects of male agonistic interventions (Harcourt & Stewart, 1987, 1989), and the rarity of support between non-relatives also forestall nepotism.

GROUP SIZE, MATE CHOICE AND FEMALE REPRODUCTIVE SUCCESS

Group size should be directly correlated with the level of within-group feeding competition and may influence female competition for male services (e.g. male protection may be diluted in large groups). In the 'feeding competition' model (Moore, 1984; Dunbar, 1988), females live where competition for food is minimized, and both birth rate and infant survival vary inversely with group size. This model does not explain why females live in groups, but predicts that they transfer from larger to smaller groups and that no group is too small. In the 'predation protection' model (Dunbar, 1988), females form groups to enhance predator detection and defense. Group size affects birth rate negatively because of feeding competition, but infant survival positively; female reproductive success is maximized in medium-sized groups. Dunbar (1988) argues that the alternative 'male protection against male harassment [infanticide]' model predicts an inverted U-shaped relationship between birth rate and group size: females in small groups are with poor quality males and do poorly; good males attract many females, but eventually feeding competition reduces per capita birth rates.

The data on MGs given do not support the 'feeding competition' model as given by Dunbar (1988). Group size effects on foraging effort are small (as above). Large groups tend to lose females, small ones to gain them and medium-sized ones to remain stable. But immigration and emigration rates are independent of group size and some females transfer from small to very large groups (Watts, 1990). Also, in 24 cases where females transferred between groups known to differ in size, they went equally often to larger ($N = 10$) and smaller ($N = 14$) groups (Sign test, ns).

Still, group size may influence female reproduction. The number of females per group is negatively, but non-significantly related to birth rate ($r = -0.39$, df = 7, ns) and to infant survival to one year ($r = -0.43$, df = 7, ns; most mortality is in the first year). Together these lead to a nearly-significant negative relationship between female group size and the rate at which females produce young who survive to at least one year ($r = -0.55$, df = 7, $0.10 > p > 0.05$). This is not consistent with the 'predation-feeding competition' model, but may reflect a cost of scramble competition. Females in large groups sometimes harass new immigrants, as expected if they can gain by limiting group size. Females without dependent infants (potential emigrants) are responsible for most harassment (Watts, 1991b).

The 'male protection' model predicts two consequences of the death of a breeding male in single male groups, and both are supported by data: infants are then highly vulnerable to infanticide, and the groups fragment in ways consistent with females seeking to avoid infanticide (Watts, 1989). Females concerned only with detecting and defending against predators could stay together, join or accept solitary males, or transfer together to other groups. Simultaneous multiple transfer occurs, but only by females without infants or those who lose infants in the process. Other predictions of this model that are supported by data include female transfer to males who have killed their infants (Watts, 1989), temporary female transfer to solitary males (which may indicate sampling of those males; Watts, 1990) and lower infanticide risk in groups with two or more males than in single male groups (Sterck *et al.*, in preparation). The other two models predict that females should leave large groups with more than one male to avoid feeding competition, but female group size is positively related to the number of males per group (Robbins, 1995). Finally, male protection against infanticide and against predation are not mutually exclusive: males good at one should also be good at the other.

Much aggression among females occurs near males and is not clearly ascribable to resource competition, as would be expected if females also compete for proximity and social access to males. However, whether competition leads to consistent variation in spatial position and safety is unknown (Watts, 1994b, c). Frequent male interventions that end contests between females mitigate competition and contribute to egalitarian female agonistic relationships. Males good at controlling female conflicts may retain immigrants and intrinsically poor competitors who would otherwise go to smaller groups. But reproductive state, not intrinsic competitive ability, determines female responses to male deaths in single male groups (Watts, 1989); this suggests that male protection against infanticide is more important than protection against competitors. Also, infanticide protection, but not protection against competition, can benefit all females equally. Finally, male control abilities decline as groups gain more females; this in turn may lead some females to leave large groups (Watts, 1991b, 1994c).

MALE–MALE RELATIONSHIPS AND MATING COMPETITION

Encounters between groups, or groups and lone males, are contests in which males try to attract or retain females, to protect their own infants and, sometimes, to kill infants of other males (Harcourt, 1978; Yamagiwa, 1986, 1987; Watts, 1989, 1994a; Sicotte, 1993). Data summarized by Sicotte (1993) support two predictions based on consideration of the potential gains from escalated aggression (Harcourt, 1978; cf. Yamagiwa, 1986; Watts, 1994a). Male aggression is more intense, and the risk of fighting higher, when more potential female emigrants are present, and encounters between two long-established groups are usually shorter than those that either have solitary males or that are smaller, recently formed groups.

Aggression is more common than affiliation between co-resident adult males, between adults and adolescents, and between adolescent peers. Maturing natal males spend progressively less time near older males and receive progressively more aggression from them, but older males may be more tolerant of sons and full brothers than of more distant relatives and of much younger males than of those closer in age (Harcourt, 1979c; Harcourt & Stewart, 1981; Watts & Pusey, 1993; Sicotte, 1994; Robbins, 1995). Males have clear dominance relationships and subordinates sometimes reverse ranks with older males. They compete aggressively over estrous females and those showing post-conception, estrus-like behavior. Dominant males often interrupt mating attempts by subordinates and may guard females, although subordinates may steal some fertilizations, even while still immature and not yet maximally effective as allies in competition with outside males. Older males tolerate mating between daughters and young males, however (Harcourt et al., 1980; Watts, 1992; Robbins, 1995).

Still, a mature male may gain from tolerating a follower who is his ally against extra-group males so long as he loses few matings or gets inclusive fitness payoffs because his ally is a relative (Watts, 1990; Sicotte, 1993; Robbins, 1995). The follower can help him to protect infants and to herd females during encounters with outside males and, if physically mature, can prevent infanticidal attacks that otherwise would occur after the older male's death (Watts, 1989; Sicotte, 1993; cf. Berger, 1986 for herding by wild horse males, *Equus caballus*). Groups with more than one male tend to attract and retain more females than single male groups (Robbins, 1995), so tolerance of a follower may increase, or at least not decrease, a male's total number of reproductive opportunities. Natal philopatry may often be advantageous to followers, given that not all dispersers reproduce and that philopatric males probably start reproducing as soon or sooner than dispersers who do breed (Robbins, 1995).

A past-prime male who has completed most of his reproductive career would probably lose fewer fertilizations than a prime male and gain more from an ally's protective efforts, and thus is more likely to tolerate a follower than is a prime male. Infanticide risk makes acceptance of unrelated allies unlikely. All else equal, inclusive fitness gains and the influence of age make sons the best allies, and most two-male combinations are probably father–son pairs (Yamagiwa, 1986; Stewart & Harcourt, 1987). But many males die before any of their sons mature. Brothers, closer in age and with more similar power trajectories, probably compete more intensively and for the same set of mates. Philopatric males may try to evict younger male relatives who may also emigrate to avoid eviction or because they lack mating opportunities (Moore & Ali, 1984). Coalitions of brothers occur, but are probably uncommon. One unrelated pair with a sibling-like relationship (co-residence since immaturity) is also known (Watts, 1990; Sicotte, 1993). These strategic and demographic considerations explain why most groups have only one male.

DISCUSSION

Ecology of gorilla female social relationships and transfer

Group living and transfer have low ecological and social costs for female gorillas who also can gain little from residing with female relatives (Wrangham, 1979; Watts, 1985, 1994c; Stewart & Harcourt, 1987; Harcourt & Stewart, 1989). Data on age at first reproduction by female MGs also strongly support the low-cost argument: average age at first birth is the same for females who undergo natal transfer (10.1 years, $N = 8$) and those who do not (9.9 years, $N = 7$; t-test, $t = 0.29$, df = 13, ns).

But 'low cost' alone cannot explain group living or female transfer (Wrangham, 1987). Female need for male protection against infanticide probably led to male–female association in its present form (Wrangham, 1979, 1982; Stewart & Harcourt, 1987; Watts, 1989, 1990). Females can gain by residing and mating with males who minimize costs imposed on them by inter-male reproductive competition and, secondarily, by competition among females. All else equal, they should prefer smaller groups, where competition for food and for male services is lower, to larger ones, but group size is not the primary influence on transfer decisions. Females can sometimes gain from residence in large groups with more than one male, in which the males may compete to provide services and to develop bonds with them. Inbreeding avoidance influences natal transfer, but, unlike intersexual mutualism, explains neither secondary transfer nor why some females mate with relatives other than fathers in natal groups (Harcourt, 1978; Stewart & Harcourt, 1987; Watts, 1990).

Some other mammals, notably many colobine monkeys and gregarious equids, show permanent male–female association and female transfer combined with use of low quality but evenly distributed foods (leaves, monocotyledons) as staples or fallbacks. Their social dynamics resemble those of gorillas in many respects, although the resemblance is not exact (Watts, 1994c). For example, female capped langurs (Presbytis pileatus; Stanford, 1991) and Thomas' langurs (Presbytis thomasi; Sterck, 1992, personal communication), and, in some populations, wild horses (Berger, 1986) have infrequent feeding contests and weakly differentiated female agonistic relationships. Contest competition probably has little effect on their reproductive success (but see Lloyd & Rasa, 1989, for possible contest effects in Cape Mountain zebra, Equus zebra). Transfer is voluntary and secondary transfer is common in capped and Thomas' langurs (Stanford, 1991; Sterck, 1992) and in wild horses and ponies (Berger, 1986; Rubenstein, 1986; Rutberg, 1990; Stevens, 1990; Duncan, 1992). Permanent association with males helps protect females against predators and against harassment by conspecific males. Harassment interferes with female foraging efficiency in equids, and wild horse stallions on Shackleford Island and on the Outer Banks who effectively protect females against such harassment attract many females and enjoy high reproductive success (Rubenstein, 1986; Stevens, 1990). Stallions are a potential threat to fetuses and foals (Tyler, 1972; Berger, 1986), and infanticide is prevalent among colobines (Struhsaker & Leland, 1987; Stanford, 1991; Sterck, 1992).

Whether transfer costs are as low for female chimpanzees and bonobos as for gorillas is not clear. Their food supplies vary much more in space and time, but they join or leave parties in relation to food availability to reduce feeding competition (White, Chapter 3; Wrangham et al., Chapter 4). However, female chimpanzees, at least, compete to establish core areas, and residents resist attempts by immigrants to do so (Goodall, 1986; Nishida, 1989).

Low-cost female transfer in gorillas contrasts sharply with a second pattern of female dispersal in primates that characterizes high density populations of red (Alouatta seniculus) and mantled (Alouatta palliata) howler monkeys. Female dispersal in these species is 'non-adaptive' (Stenseth, 1983): poor competitors, usually juveniles, are forced into marginal habitat and face breeding delays, and they enter or form breeding groups with difficulty (Crockett & Pope, 1993; Glander, 1993).

How representative, or disturbed, are mountain gorillas?

Habitat loss and human-caused mortality may have influenced the frequency of certain events in the Virungas' MG population, but probably without substantive changes in social dynamics. For example, proportionately more groups have habitat edge as a home range boundary as habitat area decreases. Edge groups may meet fewer groups and solitary males than groups with home ranges entirely in the habitat interior; this may alter the payoff balances for dispersal and make female natal philopatry more likely. But many natal females disperse from edge groups, and some stay to reproduce in interior groups. Transfer is a conditional strategy, and variation in local demographic and socioecological circumstances should lead to within- and among-population variation in the proportion of females who transfer (van Schaik, 1989). Humans have killed some males in the population; this may influence the payoffs for male dispersal and the frequency of groups with more than one male. However, this is also just one of many sources of demographic variation. Deaths from human hunting can lead to group fragmentation and infanticide, but these also happen after natural male deaths in groups without a second mature male (Watts, 1989).

Basic similarities across and within subspecies and populations in the use of THV for fallback or staple foods underlie variation in frugivory and explain why gorillas can form cohesive, stable groups where sympatric chimpanzees cannot (Tutin & Fernandez, 1993). Even WLGs using discrete, limited fruit patches reduce competition by subgrouping (Tutin, Chapter 5; Kuroda et al., Chapter 6; Yamagiwa et al., Chapter 7). Average group size varies across habitats, probably due to variation in food distribution (Tutin, Chapter 5), but not consistently across sub-species. Female transfer occurs in WLGs (Tutin, Chapter 5) and ELGs (Yamagiwa, 1986), and modal group composition is similar across sub-species.

But recent WLG research shows apparent contrasts with MGs, so we should be wary of over-generalizing from MGs without better data on WLG social behavior. For example, between-group contests at fruit trees may affect WLG foraging efficiency at Lopé (Tutin, Chapter 5), which raises the possibility that contesting access to food is a male service for females. Such competition is absent in MGs. Also, temporary subgrouping, especially during foraging for fruit, is more common in WLGs. For a group whose members travel in coordinated fashion between nests sites to adjust party size and composition in response to food dispersion and to competition is not surprising. However, females need male protection and use THV heavily year-round, so they should forage in permanent association with males, as seems to happen at Ndoki (Kuroda et al., Chapter 6), rather than show chimpanzee-like solitary foraging.

Gorillas and the other great apes

Wrangham (1979) argued that because female gorillas use THV foods as staples and as primary fallbacks, gregariousness is less costly for them than for females of the other, more frugivorous great apes. Subsequent research has largely substantiated this argument, although we still face questions about the social effects of WLG frugivory and the extent and causes of within- and between-species variation in social organization (see White, Chapter 3; Wrangham et al., Chapter 4; Boesch, Chapter 8; Kano, Chapter 10).

Also, the sexes are more similar in behavioral ecology in gorillas (Watts, 1984, 1988, 1990) than in chimpanzees (Wrangham & Smuts, 1980) and orangutans (reviewed in Rodman & Mitani, 1987). Male gorillas are thus better able than males of these species to associate with and monopolize access to fertile females (Wrangham, 1979; Dunbar, 1988). Fertile females are sparsest and least clumped spatially (and maybe temporally) in orangutans, so prolonged or permanent association with a given female would be a poor reproductive tactic for a male even if it were ecologically feasible (Wrangham, 1979; Dunbar, 1988; van Schaik & van Hooff, Chapter 1). Long birth spacing and small foraging group size make permanent association disadvantageous for chimpanzees. Individual males could benefit by maintaining exclusive access to several, dispersed females, but this may be ecologically unfeasible (Dunbar, 1988). Instead, chimpanzee males cooperate in intense contest competition with extra-community males, while engaging in relatively low contest competition, but high sperm competition, within communities. Fertile females are most clumped in gorillas. Prime single males can monopolize access to female groups, although male philopatry and cooperative male defense of female groups can be advantageous in some demographic circumstances (cf. van Hooff & van Schaik, 1994). Like chimpanzees, male gorillas engage in intense between-group contest competition, but they lack the restraint on within-group contest competition and the emphasis on sperm competition. Dunbar (1988) argues that permanent association with females gives gorilla males more mating opportunities than would a strategy of wandering mate search. However, his argument is circular if we accept that females are clumped (live in cohesive groups) precisely because they need to associate with males: they accept 'permanent consortships' in exchange for protection (Wrangham, 1979).

Slow maturation and infrequent reproduction by female great apes means that most males have few mating opportunities. Infanticide and less costly forms of sexual coercion (Smuts & Smuts, 1993) can increase male reproductive success. In theory, females could benefit from male protection against conspecific males (Wrangham, 1979, 1982, 1986; Smuts & Smuts, 1993), especially because they usually lack available or effective female allies. Coercion, infanticide and male protection could have been fundamentally important in great ape social evolution, even if they do not have universally visible effects in current environments.

Male chimpanzees and gorillas are infanticidal. Infanticide has not been seen in bonobos, and its apparent absence in orangutans is anomalous (see van Schaik & van Hooff, Chapter 1). Milder sexual coercion is common in orangutans and chimpanzees (Smuts & Smuts, 1993) and can occur in multi-male gorilla groups (personal observations). Sexual dimorphism in body size limits the ability of female gorillas and orangutans to defend themselves against male aggression (although it gives female orangutans more arboreal agility than males). Size dimorphism is lower in chimpanzees and bonobos, but in chimpanzees, at least, male coalitions can overcome female defenses. Male chimpanzees, like male gorillas, protect females and infants in between-group aggression. Females sometimes transfer from communities with few males to neighboring ones with more, where the level of protection is higher. Male chimpanzees sometimes aggressively expand their ranges and add females to their communities at the expense of males in neighboring communities (Nishida *et al.*, 1985; Goodall, 1986). Males protect immigrant females against harassment by resident females (Goodall, 1986; Nishida, 1989). Future research on several important questions

could increase our understanding of the role of inter-sexual mutualism in great ape social evolution. These include how much female chimpanzees gain from male protection against other within-community aggression; how similar bonobos are to chimpanzees in these respects; and how important subtle forms of male protection are in orangutans (van Schaik & van Hooff, Chapter 1).

ACKNOWLEDGMENTS

This paper was originally prepared for the Wenner-Gren conference 'The Great Apes Revisited,' held November 12–19, 1994 in Cabo San Lucas, Baja, Mexico. I thank Bill McGrew and Toshisada Nishida for their extraordinary organizational efforts, and all the participants for their constructive criticism, intellectual stimulation and sense of fun.

I thank the Rwandan Office of Tourism and National Parks and Zaïre National Park authorities for permission to work in the Parc National des Volcans and Parc des Virungas, and thank the late Dian Fossey for permission to work at Karisoke. My fieldwork in Rwanda has been supported by NIMH grant 5T32 MH 15181–03, The Dian Fossey Gorilla Fund, The Eppley Foundation for Research, Inc., The Chicago Zoological Society, The Wildlife Preservation Trust International, The World Wildlife Fund USA and, especially, The L. S. B. Leakey Foundation. I thank all who have contributed to long-term demographic and life history data at Karisoke. I am particularly indebted to A. Nemeye, E. Rwelekana, E. Rukera, A. Banyangandora, K. Munyanganga, L. Munyanshoza, S. Kwiha, F. Barabgiriza, F. Nshogoza, J. D. Ntantuye, C. Nkeramugaba and M. Mpiranya, without whose expert field assistance I could not have done my research. I am also indebted to them and to their Rwandan colleagues at Karisoke for heroically keeping the work there, and conservation efforts in the Virungas, alive during an immensely difficult, dangerous and tragic time, and this chapter is dedicated to them and to their families.

REFERENCES

Berger, J. 1986. *Wild Horses of the Great Basin*. Chicago: University of Chicago Press.

Calvert, J. J. 1985. Food selection by western gorillas in relation to food chemistry. *Oecologia*, **65**, 236–46.

Casimir, M. J. 1975. Feeding ecology and nutrition of an eastern gorilla group in the Mt. Kahuzi region (Republic of Zaïre). *Folia Primatologica*, **24**, 81–136.

Casimir, M. J. & Butenandt, E. 1973. Migration and core area shifting in relation to some ecological factors in a mountain gorilla group (*Gorilla gorilla beringei*) in the Mt. Kahuzi region (Republic of Zaïre). *Zeitschrift für Tierpsychologie*, **33**, 514–22.

Chivers, D. & Hladik, C. M. 1980. Morphology of the gastrointestinal tract in primates: comparisons with other mammals in relation to diet. *Journal of Morphology*, **166**, 337–86.

Crockett, C. M. & Pope, T. R. 1993. Consequences of sex differences in dispersal for juvenile red howler monkeys. In: *Juvenile Primates*, ed. M. E. Pereira & L. A. Fairbanks, pp. 104–18. Oxford: Oxford University Press.

Davies, A. G., Bennett, E. L. & Waterman, P. G. 1988. Food selection by two southeast Asian colobine monkeys (*Presbytis rubicunda* and *Presbytis melalophos*) in relation to plant chemistry. *Biological Journal of the Linnean Society*, **34**, 33–56.

Dobson, F. S. 1982. Competition for mates and predominant juvenile male dispersal in mammals. *Animal Behaviour*, **30**, 1183–92.

Dobson, F. S. & Jones, W. T. 1986. Multiple causes of dispersal. *American Naturalist*, **126**, 855–8.

Dunbar, R. I. M. 1988. *Primate Social Systems*. Ithaca, NY: Cornell University Press.

Duncan, P. 1992. *Horses and Grasses*. New York: Springer.

Fossey, D. & Harcourt, A. H. 1977. Feeding ecology of free-ranging mountain gorillas (*Gorilla gorilla beringei*). In: *Primate Ecology*, ed. T. H. Clutton-Brock, pp. 415–77. London: Academic Press.

Glander, K. G. 1993. Dispersal patterns in Costa Rican mantled howler monkeys. *International Journal of Primatology*, **13**, 415–36.

Goodall, A. G. 1977. Feeding and ranging behavior of a mountain gorilla group (*Gorilla gorilla beringei*) in the Tshibinda-Kahuzi region, Zaïre. In: *Primate Ecology*, ed. T. H. Clutton-Brock, pp. 449–79. London: Academic Press.

Goodall, J. 1986. *The Chimpanzees of Gombe*. Cambridge, MA: Harvard University Press.

Greenwood, P. J. 1980. Mating systems, philopatry, and dispersal in birds and mammals. *Animal Behaviour*, **28**, 1140–62.

Harcourt, A. H. 1978. Strategies of emigration and transfer by primates, with particular reference to gorillas. *Zeitschrift für Tierpsychologie*, **48**, 401–20.

Harcourt, A. H. 1979a. Social relationships among adult female mountain gorillas. *Animal Behaviour*, **27**, 251–64.

Harcourt, A. H. 1979b. Social relationships between adult male and adult female mountain gorillas. *Animal Behaviour*, **28**, 325–42.

Harcourt, A. H. 1979c. Contrasts between male relationships in wild gorilla groups. *Behavioral Ecology and Sociobiology*, **5**, 39–49.

Harcourt, A. H. & Stewart, K. J. 1981. Gorilla male relationships: can differences during immaturity lead to contrasting reproductive tactics in adulthood? *Animal Behaviour*, **29**, 206–10.

Harcourt, A. H. & Stewart, K. J. 1987. The influence of help in contests on dominance rank in primates: hints from gorillas. *Animal Behaviour*, **35**, 182–90.

Harcourt, A. H. & Stewart, K. J. 1989. Functions of alliances in contests in wild gorilla groups. *Behaviour*, **109**, 176–90.

Harcourt, A. H., Fossey, D., Stewart, K. J. & Watts, D. P. 1980. Reproduction in wild gorillas and some comparisons with chimpanzees. *Journal of Reproduction and Fertility, Supplement*, **28**, 59–70.

van Hooff, J. A. R. A. M. & van Schaik, C. P. 1994. Male bonds: affiliative relationships among nonhuman primate males. *Behaviour*, **130**, 309–37.

Johnson, C. N. 1986. Sex biased philopatry and dispersal in mammals. *Oecologia*, **69**, 626–7.

Lloyd, P. H. & Rasa, O. A. E. 1989. Status, reproductive success, and fitness in Cape Mountain zebras (*Equus zebra zebra*). *Behavioral Ecology and Sociobiology*, **25**, 411–20.

Milton, K. 1984. The role of food processing factors in primate food choice. In: *Adaptations for Foraging in Nonhuman Primates*, ed. P. S. Rodman & J. G. Cant, pp. 249–79. New York: Columbia University Press.

Moore, J. 1984. Female transfer in primates. *International Journal of Primatology*, **5**, 537–89.

Moore, J. & Ali, R. 1984. Are dispersal and inbreeding avoidance related? *Animal Behaviour*, **32**, 94–112.

Nishida, T. 1989. Social interactions between resident and immigrant female chimpanzees. In: *Understanding Chimpanzees*, ed. P. G. Heltne & L. A. Marquardt, pp. 68–89. Cambridge, MA: Harvard University Press.

Nishida, T., Hiraiwa-Hasegawa, M., Hasegawa, T. & Takahata, Y. 1985. Group extinction and female transfer in wild chimpanzees in the Mahale National

Park, Tanzania. *Zeitschrift für Tierpsychologie*, **67**, 284–301.

Pusey, A. E. 1980. Inbreeding avoidance in chimpanzees. *Animal Behaviour*, **28**, 543–82.

Pusey, A. E. & Packer, C. 1987. Dispersal and philopatry. In: *Primate Societies*, ed. B. B. Smuts, D. L. Cheney, R. M. Seyfarth, R. W. Wrangham & T. T. Struhsaker, pp. 250–66. Chicago: University of Chicago Press.

Robbins, M. M. 1995. A demographic analysis of male life history and social structure of mountain gorillas. *Behaviour*, **132**, 21–48.

Rodman, P. S. & Mitani, J. C. 1987. Orangutans: sexual dimorphism in a solitary species. In: *Primate Societies*, ed. B. B. Smuts, D. L. Cheney, R. M. Seyfarth, R. W. Wrangham & T. T. Struhsaker, pp. 146–54. Chicago: University of Chicago Press.

Rogers, M. E. & Williamson, E. A. 1987. Density of herbaceous plants eaten by gorillas in Gabon: some preliminary data. *Biotropica*, **19**, 278–81.

Rogers, M. E., Maisels, F., Williamson, E. A., Fernandez, M. & Tutin, C. E. G. 1990. Gorilla diet in the Lopé Reserve, Gabon: A nutritional analysis. *Oecologia*, **84**, 326–39.

Rubenstein, D. I. 1986. Ecology and sociality in horses and zebras. In: *Ecological Aspects of Social Evolution*, ed. D. I. Rubenstein & R. W. Wrangham, pp. 469–87. Princeton, NJ: Princeton University Press.

Rutberg, A. T. 1990. Intergroup transfer in Assateague pony mares. *Animal Behaviour*, **40**, 945–52.

Sabater Pi, J. 1977. Contribution to the study of alimentation of lowland gorillas in the natural state, in Rio Muni, Republic of Equatorial Guinea (West Africa). *Primates*, **18**, 183–204.

van Schaik, C. P. 1989. The ecology of social relationships amongst female primates. In: *Comparative Socioecology*, ed. V. Standen & R. Foley, pp. 195–218. London: Blackwell Scientific Publications.

Schaller, G. 1963. *The Mountain Gorilla*. Chicago: University of Chicago Press.

Sicotte, P. 1993. Inter-group interactions and female transfer in mountain gorillas: influence of group composition on male behavior. *American Journal of Primatology*, **30**, 21–36.

Sicotte, P. 1994. Effect of male competition on male–female relationships in bi-male groups of mountain gorillas. *Ethology*, **97**, 47–64.

Smuts, B. B. & Smuts, R. W. 1993. Male aggression and sexual coercion of females in nonhuman primates and

other mammals: evidence and theoretical implications. *Advances in the Study of Behavior*, **22**, 1–63.

Stanford, C. 1991. *The Capped Langur in Bangladesh: Behavioral Ecology and Reproductive Tactics*. Basel: S. Karger.

Stenseth, N. C. 1983. Causes and consequences of dispersal in small mammals. In: *The Evolution of Animal Movement*, ed. I. R. Swinglund & P. J. Greenwood, pp. 63–101. Oxford: Oxford University Press.

Sterck, E. H. M. 1992. Timing of female migrations in the Thomas's langur (*Presbytis thomasi*). Paper to XIVth Congress, International Primatological Society, Strasbourg.

Stevens, E. F. 1990. Instability of harems of feral horses in relation to season and presence of subordinate stallions. *Behaviour*, **112**, 149–61.

Stewart, K. J. & Harcourt, A. H. 1987. Gorillas: variation in female relationships. In: *Primate Societies*, ed. B. B. Smuts, D. L. Cheney, R. M. Seyfarth, R. W. Wrangham & T. T. Struhsaker, pp. 155–64. Chicago: University of Chicago Press.

Struhsaker, T. T. & Leland, L. 1987. Colobines: Infanticide by adult males. In: *Primate Societies*, ed. B. B. Smuts, D. L. Cheney, R. M. Seyfarth, R. W. Wrangham & T. T. Struhsaker, pp. 155–64. Chicago: University of Chicago Press.

Tutin, C. E. G. & Fernandez, M. 1993. Composition of the diet of chimpanzees and comparisons with that of sympatric lowland gorillas in the Lopé Reserve, Gabon. *American Journal of Primatology*, **30**, 195–211.

Tutin, C. E. G., Fernandez, M., Rogers M. E., Williamson, E. A. & McGrew, W. C. 1992. Foraging profiles of sympatric lowland gorillas and chimpanzees in the Lopé Reserve, Gabon. In: *Foraging Strategies and Natural Diets of Monkeys, Apes, and Humans*, ed. A. Whiten & E. M. Widdowson, pp. 19–26. Oxford: Oxford University Press.

Tyler, S. J. 1972. The behavior and social organization of the New Forest ponies. *Animal Behaviour Monographs*, **5**, 85–196.

Vedder, A. L. 1984. Movement patterns of a group of free-ranging mountain gorillas (*Gorilla gorilla beringei*) and their relationship to food availability. *American Journal of Primatology*, **7**, 73–88.

Vehrencamp, S. 1983. A model for the evolution of despotic vs. egalitarian societies. *Animal Behaviour*, **31**, 667–82.

de Waal, F. B. M. & Luttrell, L. 1988. Mechanisms of social reciprocity in three primate species: symmetrical relationship characteristics or cognition? *Ethology and Sociobiology*, **9**, 101–18.

Waterman, P. J., Choo, G., Vedder, A. L. & Watts, D. P. 1983. Digestibility, digestion inhibitors, and nutrients from herbaceous foliage from an African montane flora and its comparison with other tropical flora. *Oecologia*, **60**, 244–9.

Watts, D. P. 1984. Composition and variability of mountain gorilla diets in the central Virungas. *American Journal of Primatology*, **7**, 323–56.

Watts, D. P. 1985. Relations between group size and composition and feeding competition in mountain gorilla groups. *Animal Behaviour*, **33**, 72–85.

Watts, D. P. 1988. Environmental influences on mountain gorilla time budgets. *American Journal of Primatology*, **15**, 295–312.

Watts, D. P. 1989. Infanticide in mountain gorillas: new cases and a reconsideration of the evidence. *Ethology*, **81**, 1–18.

Watts, D. P. 1990. Ecology of gorillas and its relationship to female transfer in mountain gorillas. *International Journal of Primatology*, **11**, 21–45.

Watts, D. P. 1991a. Habitat use strategies of mountain gorillas. *Folia Primatologica*, **56**, 1–16.

Watts, D. P. 1991b. Harassment of immigrant female mountain gorillas by resident females. *Ethology*, **89**, 135–53.

Watts, D. P. 1992. Social relationships of resident and immigrant female mountain gorillas, I. Male female relationships. *American Journal of Primatology*, **28**, 159–81.

Watts, D. P. 1994a. The influence of male mating tactics on habitat use in mountain gorillas (*Gorilla gorilla beringei*). *Primates*, **35**, 35–47.

Watts, D. P. 1994b. Social relationships of resident and immigrant female mountain gorillas, II: relatedness, residence, and relationships between females. *American Journal of Primatology*, **32**, 13–30.

Watts, D. P. 1994c. Agonistic relationships of female mountain gorillas. *Behavioral Ecology and Sociobiology*, **34**, 347–58.

Watts, D. P. 1995. Post-conflict social events in wild mountain gorillas, 1. Social interactions between opponents. *Ethology*, **100**, 158–74.

Watts, D. P. & Pusey, A. E. 1993. Behavior of juvenile and adolescent great apes. In: *Juvenile Primates*, ed. M. E. Pereira & L. A. Fairbanks, pp. 148–72. Oxford: Oxford University Press.

Whiten, A., Byrne, R. W., Barton, R. A., Waterman, P. G. & Henzi, S. P. 1992. Dietary and foraging strategies of baboons. In: *Foraging Strategies and Natural Diets of Monkeys, Apes, and Humans*, ed. A. Whiten & E. M. Widdowson, pp. 27–37. Oxford: Oxford University Press.

Williamson, E. A., Tutin, C. E. G., Rogers, M. E., & Fernandez, M. 1990. Composition of the diet of lowland gorillas at Lopé in Gabon. *American Journal of Primatology*, **21**, 265–77.

Wrangham, R. W. 1979. On the evolution of ape social systems. *Social Science Information*, **18**, 334–68.

Wrangham, R. W. 1980. An evolutionary model of female bonded primate groups. *Behaviour*, **75**, 262–300.

Wrangham, R. W. 1982. Kinship, mutualism and social evolution. In: *Current Problems in Sociobiology*, ed. R. W. Wrangham, D. I. Rubenstein, R. I. M. Dunbar & B. C. R. Bertram, pp. 269–90. Cambridge: Cambridge University Press.

Wrangham, R. W. 1986. Ecology and social relationships in two species of chimpanzees. In: *Ecological Aspects of Social Evolution*, ed. D. I. Rubenstein & R. W. Wrangham, pp. 325–78. Princeton, NJ: Princeton University Press.

Wrangham, R. W. 1987. Evolution of social structure. In: *Primate Societies*, ed. B. B. Smuts, D. L. Cheney, R. M. Seyfarth, R. W. Wrangham & T. T. Struhsaker, pp. 282–97. Chicago: University of Chicago Press.

Wrangham, R. W. & Rubenstein, D. I. 1986. Social evolution in birds and mammals. In: *Ecological Aspects of Social Evolution*, ed. D. I. Rubenstein & R. W. Wrangham, pp. 452–70. Princeton, NJ: Princeton University Press.

Wrangham, R. W. & Smuts, B. B. 1980. Sex differences in the behavioral ecology of chimpanzees in the Gombe National Park, Tanzania. *Journal of Reproduction and Fertility, Supplement*, **29**, 13–31.

Yamagiwa, J. 1986. Activity rhythm and the ranging of a solitary male mountain gorilla (*Gorilla gorilla beringei*). *Primates*, **27**, 273–82.

Yamagiwa, J. 1987. Intra- and inter-group interactions of an all-male group of Virunga mountain gorillas (*Gorilla gorilla beringei*). *Primates*, **28**, 1–30.

Yamagiwa, J. & Mwanza, N. 1994. Day journey length and daily diet of solitary male gorillas in lowland and highland habitats. *International Journal of Primatology*, **15**, 207–24.

3 • Comparative socio–ecology of *Pan paniscus*

FRANCES J. WHITE

INTRODUCTION

Pan paniscus, the pygmy chimpanzee or bonobo, has been studied at a number of sites in Zaïre including Lomako, Wamba, Ikela, and briefly at Yalosidi and Lake Tumba. Lake Tumba (Horn, 1980) is a swampy habitat that is visited by pygmy chimpanzees only seasonally. Studies in the mosaic of undisturbed, disturbed and swamp forest of Yalosidi (Kano, 1983; Uehara, 1988, 1990) and at Ikela in the Lilungu region (Sabater Pi *et al.*, 1993) have yet to yield detailed results on socio–ecology. In contrast, investigations at Lomako Forest (Badrian & Badrian, 1984; Badrian & Malenky, 1984; Thompson-Handler *et al.*, 1984; White, 1986, 1988, 1989a, 1989b, 1992a, 1992b; Malenky, 1990; Thompson-Handler, 1990; White and Burgman, 1990; White & Lanjouw, 1992; Hohmann & Fruth, 1993, 1994; Fruth & Hohmann, 1994, Chapter 17) and at Wamba (Kuroda, 1979, 1980, 1984; Kano, 1980, 1982, Chapter 10; Kitamura, 1983; Kano & Mulavwa, 1984; Furuichi, 1987, 1989, 1992; Idani, 1990, 1991; Ihobe, 1992a, 1992b) have both spanned several years and have provided much of the information currently available on the behavior of the pygmy chimpanzee, so that most of the comparisons presented here concentrate on the differences and similarities between the Lomako and Wamba study populations. There are differences in methods used and in the type of information available from each site. Provisioning at Wamba has facilitated habituation, so that more detailed behavioral data collection is possible. Studies at Lomako have emphasized long-term observation without provisioning.

As for many social species, the social organization of bonobos can be correlated with several ecological variables. Such relationships can be useful in explaining both the variation within the species as well as the divergence of the bonobo from other apes, especially the closely related chimpanzee, *Pan troglodytes*. It is also possible to examine specific behavioral patterns from a similar socio-ecological perspective. Some such behaviors highlight the relative differences between bonobos and chimpanzees, such as food sharing and inter-party coordination and communication, whereas others can be considered as truly unique adaptive strategies of the bonobo: one of the most obvious of these is genito-genital rubbing among females.

Like chimpanzees, bonobos share both plant and animal foods (Kuroda, 1984; Ihobe, 1992b; Hohmann & Fruth, 1993; White, 1994). Unlike chimpanzees, there is no evidence to date that bonobos actively hunt mammalian prey, as observations indicate that prey species such as squirrels (*Uromastyx* sp.) and duikers (*Cephalophus nigrifrons*, *Cephalophus dorsalis*) are taken only opportunistically as encountered. Once obtained, meat appears to be a highly-prized food item, given the relatively high frequency of begging and the low rate of sharing. Unlike chimpanzees, however, female bonobos can retain ownership of meat even from adult males, and male–male sharing is rare. Meat sharing is not consistent with the sharing-under-pressure hypothesis (Wrangham, 1975) that may explain meat sharing in chimpanzees, possibly because bonobos have weaker male–male bonds and bonobo females can cooperatively dominate males (Kano, 1992; Parish, 1994). As in chimpanzees, plant foods are shared between bonobo mothers and infants, but unlike chimpanzees, there is frequent sharing between adult bonobos, especially among females and between females and males, with females sharing more frequently than males. The occurrence of plant food sharing may be related to the relatively large fruits available in the bonobo habitat, including such fruits as *Treculia africana*, a single fruit of which can weigh 10–20 kg. Food sharing patterns appear to be consistent with ideas of reciprocal sharing (Trivers, 1971; de Waal, 1989), where the return currency may be equivalent sharing or, in the case of male–female sharing, future sexual access or food sharing to gain status (Moore, 1984).

Like chimpanzees, bonobos have a fission–fusion social organization in which individuals within a community associate in groups of variable size and composition. There seem to be real differences between the species, however, in intercommunity relationships (see below). Bonobo parties also change less frequently than chimpanzee parties at Taï (Boesch, Chapter 8) and at Gombe (Halperin, 1979), but not those at Bossou (Sugiyama, 1984). Parties of bonobos are mostly mixed in composition, containing both male and female adults and lower frequencies of groups of mothers (Boesch, Chapter 8), although all-female parties of females with or without offspring can be highly stable (White, 1988). Lone individuals are relatively rare, and unlike most chimpanzee populations, are usually males (White, 1988). There seem to be only quantitative differences between chimpanzees and bonobos in the way parties split and coalesce, especially when feeding (Chapman *et al.*, 1994). During periods of fruit abundance when several trees are in fruit in a given area, bonobos seem to be more cohesive, remaining together to visit the same large trees, whereas chimpanzees disperse among many neighboring trees. A similar difference has been seen within a single tree, where chimpanzees disperse while feeding so that nearest-neighbor distances increase with the size of the food tree (Ghiglieri, 1984), while bonobos remain more bunched, so that the size of the canopy explains relatively little variation in nearest-neighbor distances during feeding (White, 1986). Differences also exist in the cohesiveness of the sexes during feeding, with male–male spatial cohesion in chimpanzees versus male–male dispersion in bonobos (White & Chapman, 1994). Bonobo parties also seem to congregate at night to form larger sleeping parties (Furuichi, 1989; Fruth & Hohmann, 1994). Unlike chimpanzees, parties of bonobos seem not to comprise females meeting within their separate but overlapping core areas, since non-

estrous female bonobo core areas can be as large or larger than those of males (White & Lanjouw, 1992).

The vocalizations of Lomako's bonobos are very different from chimpanzee vocalizations (Hohmann & Fruth, 1994) and appear to be important in inter-party communication. Bonobo long calls do not appear to be suitable for long distance communication (de Waal, 1988; Hohmann & Fruth, 1994) and are only detectable in the Lomako Forest within about 500 m. The structure of these calls, however, may allow more precise detection of location (Hohmann & Fruth, 1994), thus allowing for coordination of separate parties. Vocalization rates vary during the day, with late afternoon vocalizations being associated with travel and occupation of nests, whereas morning vocalizations are associated with movements between and arrival at feeding sites (Hohmann & Fruth, 1994). These peaks of vocalizing in the morning correspond to peaks in feeding activity (White, 1986).

Genito-genital or GG rubbing is an affiliative behavioral pattern of female bonobos. Usually two, but sometimes more, females clasp each other ventro-ventrally and rapidly rub the anteriors of the external genitalia together with a repeated lateral motion. Associations between GG rubbing and feeding have been seen in the wild (Kano, 1980; Kuroda, 1980; Thompson-Handler *et al.*, 1984; White, 1986; White & Lanjouw, 1992) and in captivity (de Waal, 1987). GG rubbing seems to ease tension during group excitement (de Waal, 1987) and is associated with begging for food and heterosexual matings (Kuroda, 1984; Thompson-Handler *et al.*, 1984). GG rubbing occurs in situations of reduced feeding competition (White & Lanjouw, 1992) and may reflect cooperative alliances among females for food-patch defense (White, 1986; White & Lanjouw, 1992).

Although a number of features of the bonobo social system are consistent between Lomako and Wamba, several important differences demonstrate the variability of *Pan paniscus*. Most studies at Wamba have concentrated on social behavior (e.g. Kano, 1992) with most ecological work being done at Lomako (e.g. Malenky, 1990; White, 1992a). It is, therefore, not always possible to relate intersite differences to ecological differences. However, variation in some aspects of social organization can be related to variability within one site, and can be expanded to offer possible explanations for differences between sites. This chapter examines some issues that have not been documented at more than one site or that show intersite variation, and it attempts to relate these differences to ecological differences whenever possible.

There are both major differences and similarities in social organization between chimpanzees and bonobos. Female bonobos are more cohesive and affiliative with each other than are female chimpanzees, and the degree of male bonding observed in chimpanzees is not seen in bonobos. The greater sociality of female bonobos can be related to reduced feeding competition in this species. This reduced competition both allows bonobo females to be more social, and provides further benefits to females in the cooperative defense of food patches against other community members as well as other communities. It has been suggested that defense of food patches is mediated through interparty vocalizations (White, 1986), which may relate to the correspondence between peaks in vocalizations and morning feeding. Lower feeding competition in bonobos may be due to more than one factor, including the abundance, size and distribution of fruit trees, and the use of terrestrial herbaceous vegetation (THV). The differences in food patch utilization between species is complex and is discussed in more detail below, while the importance of THV is discussed by Wrangham *et al.* (Chapter 4).

METHODS

The data came from the Lomako Forest study site in the Province of Equateur in central Zaïre. Situated at 0° 51' N 21° 5' E, the site consisted of about 40 km² of a mosaic of forest types, principally climax, evergreen, polyspecific rain-forest, with some areas of secondary growth, slope and swamp forest (White, 1992a). The study subjects used all types of forest, but spent most of their time in the undisturbed climax forest (93% of focal animal sampling: White, 1992a). Parties were found by vocalizations, walking trails, vigils at fruit-trees and prior knowledge of location. Nest-to-nest follows were conducted whenever possible. Focal sampling started when individuals left their overnight nests in the morning and stopped when they settled in their overnight nests in the evening. Nests were then watched for at least 30 minutes. Parties were defined as all individuals in a behaviorally cohesive unit that maintained visual contact and coordinated movements and activities. Individuals

within the same party were often separated by as much as 100 meters in the trees. Changes in party size tended to be obvious, due to the extensive vocal and social exchanges between members of merging parties. Party fission was often harder to determine, since large parties would often rest dispersed throughout several trees. A change in party membership was considered to have occurred if a departure was observed, or if all previous members of the party could not be found by the observer and did not appear when the remaining party members reunited to eat or to travel. Party size was recorded at each 2- or 5-minute time-point (see below) and mean hourly party sizes were extracted from the data using only time-points that fell on the hour.

Individual recognition was based on distinctive characteristics and facial features. All individuals were classed by age, sex, and state of perineal swelling as described elsewhere (White, 1986). Data are presented from 2-minute focal subject sampling from October 1984 to July 1985 on the Rangers (Eyengo) and Hedon communities and on the Blob splinter group, and from 5-minute focal subject sampling from June to August 1991 on the Rangers and Hedon communities. The Blobs ceased to exist as a distinct unit by 1991. Interactions were recorded opportunistically. Dominance rank among individuals was determined using decisive agonistic interactions such as a dominant chasing a retreating subordinate. Observation time totaled 406.4 hours.

The nearest neighbor of the focal subject was recorded at each time-point and used to construct an association index based on relative time spent together (White & Burgman, 1990). These data were used to look for significant association patterns using Mantel (1967) tests as described below (see section on Community residence and the influence of mother–son bonds). As the data used did not conform to the assumptions of parametric tests, non-parametric statistical tests (Sokal & Rohlf, 1995) were used throughout.

RESULTS

Female–female versus male–female affiliation

There are differences in social organization between study sites (Kuroda, 1979; Kano, 1982; Kitamura, 1983; Badrian & Badrian, 1984; White, 1988, 1992b). At

Wamba, there is frequent affiliation between males and females in many aspects of social organization, including interactions and relationships (Kano, 1980; Kuroda, 1980, 1984; Furuichi, 1987, 1989) and party composition (Kuroda, 1979; Kano, 1982; Kitamura, 1983; Furuichi, 1987, 1989). There is also affiliation among adult males (Furuichi, 1989). Affiliation between adult males and their mothers has been documented (Kano, personal communication in Nishida & Hiraiwa-Hasegawa, 1987; Furuichi, 1989) and may be an important influence on the social status of males (Furuichi, 1989; Ihobe, 1992a). Studies at Wamba have found association, if not affiliation, among females, but this is generally thought to be secondary in importance to affiliation between the sexes (Kano, 1982; Kitamura, 1983; Furuichi, 1989).

Studies of the Lomako Forest population have also found a high degree of affiliation between males and females and among females, but not among males (Badrian & Badrian, 1984; White, 1988, 1989b, 1992b; White & Burgman, 1990). Studies of party membership, however, have shown associations among females as being more significant than associations between males and females (White, 1986; White & Burgman, 1990). The two study sites, therefore, differ in the relative importance of affiliation between males and females compared with affiliation among females.

Party size and affiliation rates

This difference in patterns of affiliation may be related to differences in party size between the sites (White, 1992b). At Lomako, patterns of interactions varied with party size and composition. Female–female affiliation predominated in small parties (2–6 individuals), while male–female affiliation was more common in intermediate (7–10) and large (11 or more) parties (White, 1992b; see Fig. 3.1). The monthly rate of affiliation among and between the sexes was calculated from the number of observed interactions divided by the number of focal subject time-points for that month. As some months had only female–female or male–female affiliation, bimonthly means were used to pool data and to avoid zeros in ratio calculations. The bimonthly rate of male–female affiliation increased with party size (Spearman correlation coefficient, $r_s = 0.97$, $N = 5$, $p < 0.005$), whereas the rate

of female–female affiliation did not ($r_s = 0.20$, $N = 5$, ns). As Lomako parties increase in size, there is more male–female affiliation relative to female–female affiliation. This pattern is similar to nearest neighbor and party membership association patterns found at Lomako (White & Burgman, 1990), where there is significant female–female affiliation in small parties, but at larger party sizes males join to maintain proximity with females.

Party sizes at Wamba are considerably larger than those at Lomako (see Table 3.1). Counts of party sizes are available for both sites and show that party size has varied somewhat more at Wamba than at Lomako. Means for party size based on focal subject sampling at Lomako yield larger party sizes than their corresponding means from counts (Wilcoxon two-sample test $U_S = 467989$, $t_s = -5.09$, $p < 0.001$), since larger parties remain stable longer than smaller parties (Kendall's coefficient of rank correlation, party duration vs. party size, $\tau = 0.33$, $N = 164$, $p < 0.001$). If the same variation in affiliation patterns observed at Lomako also occurs at Wamba, the differences in male–female to female–female affiliation may simply reflect larger party sizes at Wamba.

Variation in party size

At Lomako, party size varies with the dimensions of the food patch, measured as the radius of a tree or vine canopy (White, 1986; White & Wrangham, 1988). Small patches (of less than 10 m radius) contain limited amounts of food and feeding competition is high. Larger patches (of greater than 10 m radius) contain superabundant food and there is little or no apparent feeding competition (White & Wrangham, 1988). Larger parties, therefore, occur in large superabundant food patches. Some measures of affiliation can be directly related to measures of feeding competition, such as the rate of GG rubbing between females, which, although not correlated with the size of parties, is inversely correlated with the rate of feeding competition (White, 1986; White & Lanjouw, 1992).

The size and abundance of food patches varies throughout the year. At Lomako, party size varied monthly (Kruskal–Wallis test, differences in mean party size over ten months, measured at hourly time points, $H = 77.86$, df = 9, $p < 0.0001$). Party size is lowest

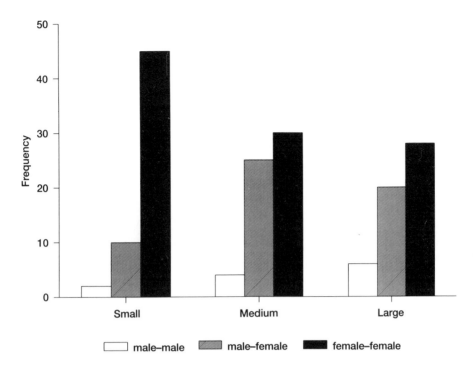

Fig. 3.1. At Lomako, differences in frequencies of male–male, male–female, and female–female affiliative interactions with increasing party size for small parties (2 to 6 individuals, excluding infants), medium parties (7 to 10 individuals, excluding infants), and large parties (11 or more individuals, excluding infants) (from White, 1992b).

Table 3.1. *Mean party sizes at Lomako and Wamba*

Location	Year(s)	Mean party size	N (method)	Source
Lomako	1984–85	6.2	114 (counts)	White, 1988
Lomako	1984–85	7.3	6825 (2-min time-points)	this study
Lomako	1991	5.6	23 (counts)	this study
Lomako	1991	7.2	715 (5-min time-points)	this study
Wamba	1974–75	16.9	147 (counts)	Kuroda, 1979
Wamba	1984–85	22.7	62 (counts)	Idani, 1991

during the months of October and November. This is about the period of lowest fruit production at Lomako, which occurs November through December (Malenky, 1990). This party size variation is significantly correlated with the average size (as measured by patch radius) of food patches used each month ($r_s = 0.64$, $N = 10$, $p < 0.05$, see Fig. 3.2).

Differences in party size between Wamba and Lomako may reflect differences in patch size, possibly

because one effect of provisioning is to provide a consistent, large food patch. Differences in party size are not consistent with differences due to habituation, since party sizes at Lomako have decreased with habituation therefore producing greater differences between the sites. Differences in non-human predation pressure also do not reflect intersite differences, as party size is expected to increase with predation threat. Parties at Lomako, where predators are common, are smaller than

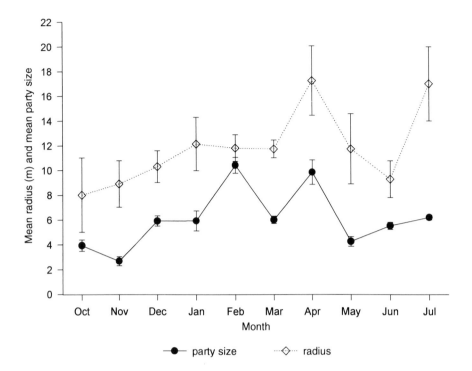

Fig. 3.2. At Lomako, mean monthly party size and mean monthly food patch radius (1984–85 data only). Party size means are calculated from party size recorded at hourly time-points and averaged by calendar month. Monthly food patch radii are from counts of individual patches visited by month and include repeated visits to the same food tree. Means and one standard error are plotted for each month.

at Wamba, where such predators are locally extinct. Inter-site differences may be due to pressures from human populations, as Wamba is more densely populated and has extensive habitat disturbance, whereas Lomako is not (White, 1992b; Appendix).

Fission–fusion

The fission–fusion social system of pygmy chimpanzees is characterized by temporary associations or parties. At Lomako, parties fission and fuse so that members of the same community may be separated by kilometers for days or weeks. All members of one community at Lomako have yet to be seen in one place at the same time. In contrast, at Wamba, there appears to be less fission and fusion, although it is hard to make comparisons due to differences in the methods of defining parties. At Wamba, individuals that traveled separately but maintained vocal contact were considered to be members of the same party (Kano, personal communication). Under the definition for the Lomako data used here, these loners would have been considered as separate parties. However, despite these difficulties, some differences remain. All members of the E1 community at Wamba usually formed one mixed party consisting of most of its members (Furuichi, 1989). Although this community fragmented during the day, they coalesced back into a single party to sleep at night. At Lomako, occasional gatherings of larger parties were observed to come together to nest, and hourly party size in the late afternoon to evening (15.00–18.00 hours) were significantly larger (8.5 individuals excluding infants) than those earlier (05.00–14.00 hours, mean party size = 6.8 individuals excluding infants) in the day (Fig. 3.3: Kruskal–Wallis $H = 6.59$, df = 1, $p < 0.02$).

There are no data available on food patches at Wamba. Since party size varies with the size of food patch at Lomako, the reduced rates of fission and fusion

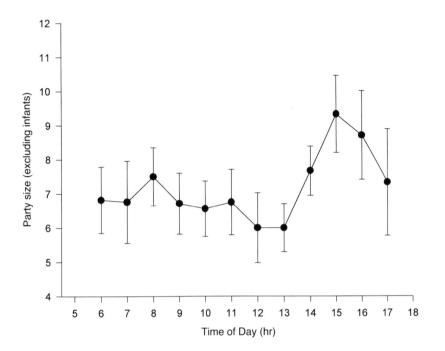

Fig. 3.3. At Lomako, mean hourly party size and time of day. Party size was recorded hourly during focal animal sampling and averaged for each hour across all observations. Means and one standard error are plotted for each hour. Focal subject sampling was conducted only on individuals out of nests (see methods). No data fell on the 05.00 hours time-point and there was only one time-point of an individual not in a nest by 18.00 hours.

at Wamba may simply reflect the larger party size selected for by provisioning, so that the pygmy chimpanzees no longer need to range widely in small parties to utilize efficiently small patches, but rather they have switched to a strategy of closer association for feeding in the artificial feeding site or when food is provided by observers in the forest. However, larger parties recorded by Kuroda (1979) were seen before provisioning was successful. Alternatively, these larger parties may be an adaptation to the greater predation pressure from local human populations at Wamba or from other impacts of the local humans, such as removal of forest resources or introduction of food crop species into areas used by the bonobos.

Intercommunity association and female transfer

At both Lomako and Wamba, peaceful associations between members of different communities have some- times been seen (Idani, 1990, 1991; Thompson-Handler, 1990; Doran, personal communication; personal observations). However, due to the lack of fission-fusion at Wamba, such aggregations appear to be the whole of two communities coming together most often at the artificial feeding site (Idani, 1990), whereas at Lomako these associations usually consist of only a few individuals of each community. During these associations at Wamba, young nulliparous females were seen to move between communities (Idani, 1990, 1991). At Lomako female transfer has also been observed. A splinter group known as the 'Blobs' that included two nulliparous females was seen as distinct from the other two communities from 1984 to 1985 (White, 1988); but by 1991 these two females had become parous females within the Hedon community and at least one of the Blob males was also observed in the Hedon community. Similarly, in the Ranger community, two highly peripheral nulliparous females were seen in 1984–85 occasionally associating with community members, and by 1991 one of them was a central parous female within the Ranger community.

Community residence and the influence of mother–son bonds

Since females transfer between communities, adult females within a community are unlikely to be related. Males, in contrast, appear to remain in their natal community (Kano, 1982; Furuichi, 1989). At Lomako, at least one juvenile male has risen to mating status within his community. At Wamba, the dominance rankings of adult males are strongly influenced by the presence of their mothers in the community (Furuichi, 1989; Ihobe, 1992a). At both Wamba and Lomako, old females occupy the highest ranks among the females (Furuichi, 1989; personal observations), but strong ties between old females and adult sons have yet to be demonstrated at Lomako.

It is possible to test for the statistical significance of such affiliations by using an association index that incorporates the time two individuals spend as nearest neighbors in parties (White & Burgman, 1990). The observed association indices for two communities and the splinter group for 1984–85 data were tested against two hypothetical matrices using a Mantel test (Mantel, 1967; Schnell et al., 1985). The hypothetical matrices tested were for important affiliations: (1) between old females and other adult females; and (2) between old adult females and adult males. This yields a Mantel test statistic, Z, that can be compared to its permutational distribution (see White & Burgman, 1990). For this test, 5000 random permutations were used.

While there were no significant associations between old females and other females in either community or in the splinter group, there was a significant association between old females and adult males in one community (Hedons; $Z = 0.29$, $p = 0.003$) but not in the other community (Rangers; $Z = 0.05$, $p = 0.236$). In the Hedon community, the males in association with old females did not include, however, the highest-ranking (alpha) male (Fig. 3.4). Thus, the evidence at Lomako of bonds between old females and adult males may represent bonds between mothers and their sons. This hypothesis may be tested in the future using DNA fingerprinting to determine interindividual relationships.

Relatedness and dominance hierarchies within communities

Paradoxically, communities of pygmy chimpanzees seem to be composed of unrelated females who are highly affiliative with each other (Fig. 3.5) and of related males who are not highly affiliative with each other. There is a dominance hierarchy among males at Wamba with a single dominant male and with adults outranking adolescents (Furuichi, 1991; Ihobe 1992a). At Lomako the top-ranking male in each community and in the splinter group was clear, based on a small sample of decisive aggressive interactions, but relative rankings among the other males were not obvious. In the Hedon and Ranger communities and in the Blob splinter group, the top-ranking male evicted or excluded other males from feeding trees at the start of feeding bouts when most matings occurred, and one of the dominant males interrupted mating attempts by other males. Adult male pygmy chimpanzees outrank adult females in that males will occasionally aggress against females, but females often ignore such male pressure. For example, unlike Pan troglodytes, female Pan paniscus at Lomako can retain a duiker carcass and not be forced to share it with begging males (White, 1994). Therefore, pygmy chimpanzees present a unique complex among apes of what is essentially female-bonding without female kinship, male kinship without male-bonding and relatively little ability of males to dominate females.

COMPARISON OF SOCIO-ECOLOGY IN PAN PANISCUS AND PAN TROGLODYTES

Earlier examinations of the differences in social organization of chimpanzees and pygmy chimpanzees supported the hypothesis that pygmy chimpanzees have a reduced level of feeding competition that permits the larger parties observed in pygmy chimpanzees (White, 1986; White & Wrangham, 1988). The level of feeding competition in trees shown by Lomako's pygmy chimpanzees was found to be lower than that for Gombe chimpanzees (White & Wrangham, 1988), supporting the hypothesis that bigger food trees for pygmy chimpanzees may permit greater female sociality. However, in a recent test of these ecological parameters for rain forest chimpanzees at Kibale, Chapman et al. (1994) found that

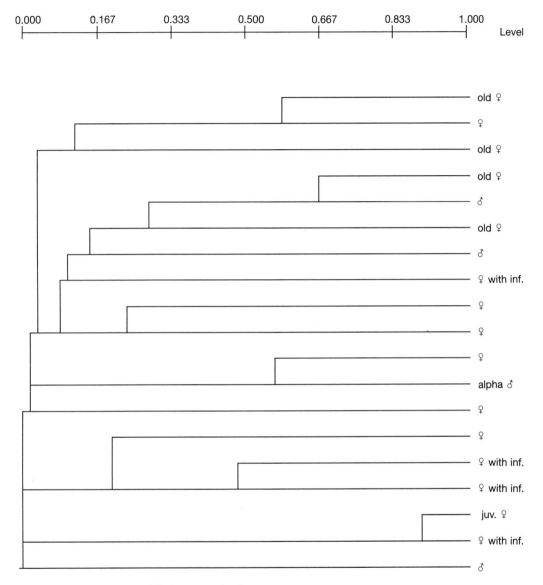

Fig. 3.4. At Lomako, dendrogram of Hedon community showing old females, other females, alpha male and other males (for methods see White & Burgman, 1990).

Kibale chimpanzees used patch sizes that were comparable to those at Lomako, although female–female affiliation at Kibale was similar to other populations of *Pan troglodytes* (Wrangham *et al.*, 1992). However, the relationship between patch size and feeding competition differed between the two species, in that the amount of food removed from large patches at Kibale was not dependent on the number of individuals present. The Kibale study recorded DBH (diameter at breast height) while the Lomako study recorded the canopy radius as the measure of tree size, so that these differences may be methodological. Chapman *et al.* (1994) concluded that variation in the levels of competition throughout the year may be the most important factor

Fig. 3.5. An adult female pygmy chimpanzee grooms another, as an infant plays overhead, at Lomako. (Photo by F. White)

there is more variation in food and distribution for *P. troglodytes* than is experienced by *P. paniscus*, suggesting that it is the level of variation in food availability rather than the absolute amount that is the critical difference selecting for differences in social organization between the two species. This interpretation of the socio-ecological differences between the species is supported by observations for another population of rain forest chimpanzees studied at Taï Forest, where party sizes of chimpanzees are also highly variable (Boesch, personal communication) and where during periods of food short-ages small parties are often observed (Doran, 1989).

CONCLUSIONS

The social organization of *Pan paniscus* with its female bonding among non-relatives is unique among non-human primates. Comparisons of the degree of female bonding in many aspects of social organization, including affiliative interactions, food sharing, GG rubbing, nearest neighbors and party composition, have shown variation that can be correlated with variation in feeding compe-tition in food patches. Less feeding competition is corre-lated with more female–female affiliation. This close tie between female bonding and ecological parameters in *P. paniscus*, together with the ecological differences between *P. paniscus* and *Pan troglodytes*, suggest that the social differences between the two species can be related to greater variability in feeding competition experienced by *P. troglodytes* compared with the relatively more stable levels seen by *P. paniscus*. Unlike *P. troglodytes*, the sociality of *P. paniscus* females and the lack of core areas in their ranging (White & Lanjouw, 1992) mean that *P. paniscus* males cannot cooperate to defend the range of more than one female. Single *P. paniscus* males, in

in determining interspecific differences in social bonding.

This interpretation is supported by comparisons of the variation in monthly mean party size. Party size for focal subjects was measured at each time-point (every 15 minutes for Wrangham *et al.*, 1992, every 2 minutes for Lomako 1984–85 and every 5 minutes for Lomako 1991), as shown in Table 3.2. Mean party size was smaller for more months at Kibale than at Lomako ($G = 3.98$, $p < 0.05$). The party size measured by this method also showed larger mean party sizes overall for *Pan paniscus* than for *P. troglodytes*. These party size differences are not due to differences in community size, since Kibale communities are bigger than Lomako communities (Table 3.2).

As at Lomako, there was at Kibale a significant corre-lation between mean monthly party size and food abun-dance (Wrangham *et al.*, 1992). These results imply that

Table 3.2. *Small and large parties of* Pan paniscus *and* Pan troglodytes

	Mean monthly party size less than 5	Mean monthly party size greater than 5	Number of months sampled	Average party size variation	Community size (including infants)
Kibale[a]	66.6%	33.3%	18	5.6 to 6.1	37
Lomako[b]	30.8%	69.2%	13	7.21 to 7.26	28[c], 35[d], 9[e]

Notes: [a]Data from Wrangham *et al.*, 1990.
[b]Data from 1984–1985 and 1991 combined ([c]Hedons, [d]Rangers, [e]Blobs).

Fig. 3.6. Resting party of pygmy chimpanzees at Lomako; adult females groom while an adult male reclines. (Photo by F. White)

contrast, are able to monopolize small cohesive groups of females and to exclude other males, so that in small parties there is typically only one male (Fig. 3.6). As party size increases, the proportion of males increases, as it is no longer possible for a single male to exclude others, so that in larger parties individual male–female relationships become more important.

ACKNOWLEDGMENTS

Field work at Lomako would not have been possible without the support and help of Richard Malenky, Nancy Thompson-Handler and Randall Susman. Permission to work at this site was given by the Institut de Recherche Scientifique (I.R.S.) of Zaïre. Field work was supported by NSF grant BNS8311251, The Boise Fund, The L. S. B. Leakey Foundation, Duke University Research Council and Conservation International. This paper was originally prepared for the Wenner-Gren conference 'The Great Apes Revisited,' held November 12–19, 1994 in Cabo San Lucas, Bajà, Mexico. I thank all the Wenner-Gren conference participants for stimulating and helpful comments and discussions. Special thanks go to Bill McGrew, Toshisada Nishida and Linda Marchant for all their work in organizing the conference and this volume, and to James Gerhart for his cooperation during the conference.

REFERENCES

Badrian, A. J. & Badrian, N. L. 1984. Group composition and social structure of *Pan paniscus* in the Lomako Forest. In: *The Pygmy Chimpanzee: Evolutionary Biology and Behavior*, ed. R. L. Susman, pp. 325–46. New York: Plenum Press.

Badrian, N. L. & Malenky, R. K. 1984. Feeding ecology of *Pan paniscus* in the Lomako Forest, Zaïre. In: *The Pygmy Chimpanzee: Evolutionary Biology and Behavior*, ed. R. L. Susman, pp. 275–99. New York: Plenum Press.

Chapman, C. A., White, F. J. & Wrangham, R. W. 1994. Party size in chimpanzees and bonobos: a reevaluation of theory based on two similarly forested sites. In: *Chimpanzee Cultures*, ed. R. W. Wrangham, W. C. McGrew, F. B. M. de Waal & P. G. Heltne, pp. 41–57. Cambridge, MA: Harvard University Press.

Doran, D. M. 1989. Chimpanzee and pygmy chimpanzee positional behavior: the influence of environment, body size, morphology, and ontogeny on locomotion and posture. Ph.D. dissertation, State University of New York, Stony Brook.

Fruth, B. & Hohmann, G. 1994. Comparative analyses of nest-building behavior in bonobos and chimpanzees. In: *Chimpanzee Cultures*, ed. R. W. Wrangham, W. C. McGrew, F. B. M. de Waal & P. G. Heltne, pp. 109–28. Cambridge, MA: Harvard University Press.

Furuichi, T. 1987. Sexual swelling, receptivity, and grouping of wild pygmy chimpanzee females at Wamba, Zaïre. *Primates*, **28**, 309–18.

Furuichi, T. 1989. Social interactions and the life history of female *Pan paniscus* in Wamba, Zaïre. *International Journal of Primatology*, **10**, 173–97.

Furuichi, T. 1992. The prolonged estrus of females and factors influencing mating in a wild group of bonobos (*Pan paniscus*) in Wamba, Zaïre. In: *Topics in Primatology, Vol. 2, Behavior, Ecology, and Conservation*, ed. N. Itoigawa, Y. Sugiyama, G. P. Sackett & R. K. R. Thompson, pp. 179–90. Tokyo: University of Tokyo Press.

Ghiglieri, M. P. 1984. *The Chimpanzees of the Kibale Forest*. New York: Columbia University Press.

Halperin, S. D. 1979. Temporary association patterns in free ranging chimpanzees: an assessment of individual grouping preferences. In: *The Great Apes*, ed. D. A. Hamburg & E. R. McCown, pp. 491–9. Menlo Park, CA: Benjamin/Cummings.

Hohmann, G. & Fruth, B. 1993. Field observations on meat-sharing among bonobos (*Pan paniscus*). *Folia Primatologica*, **60**, 225–9.

Hohmann, G. & Fruth, B. 1994. Structure and use of distance calls in wild bonobos (*Pan paniscus*). *International Journal of Primatology*, **15**, 767–82.

Horn, A. D. 1980. Some observations on the ecology of the bonobo chimpanzee (*Pan paniscus*, Schwarz 1929) near Lake Tumba, Zaïre. *Folia Primatologica*, **34**, 145–69.

Idani, G. 1990. Relations between unit-groups of bonobos at Wamba, Zaïre: encounters and temporary fusions. *African Study Monographs*, **11**, 153–86.

Idani, G. 1991. Social relationships between immigrant and resident bonobo (*Pan paniscus*) females at Wamba. *Folia Primatologica*, **57**, 83–95.

Ihobe, H. 1992a. Male–male relationships among wild bonobos (*Pan paniscus*) at Wamba, Republic of Zaïre. *Primates*, **33**, 163–79.

Ihobe, H. 1992b. Observations on the meat-eating behavior of wild bonobos (*Pan paniscus*) at Wamba, Republic of Zaïre. *Primates*, **33**, 247–50.

Kano, T. 1980. Social behavior of wild pygmy chimpanzees (*Pan paniscus*) of Wamba: a preliminary report. *Journal of Human Evolution*, 9, 243–60.

Kano, T. 1982. The social group of pygmy chimpanzees *Pan paniscus* of Wamba. *Primates*, **23**, 171–88.

Kano, T. 1983. An ecological study of the pygmy chimpanzees (*Pan paniscus*) of Yalosidi, Republic of Zaïre. *International Journal of Primatology*, **4**, 1–31.

Kano, T. 1992. *The Last Ape: Pygmy Chimpanzee Behavior and Ecology*. Stanford, CA: Stanford University Press.

Kano, T. & Mulavwa, M. 1984. Feeding ecology of the pygmy chimpanzees (*Pan paniscus*) of Wamba. In: *The Pygmy Chimpanzee: Evolutionary Biology and Behavior*, ed. R. L. Susman, pp. 233–74. New York: Plenum Press.

Kitamura, K. 1983. Pygmy chimpanzee association patterns in ranging. *Primates*, **24**, 1–12.

Kuroda, S. 1979. Grouping of the pygmy chimpanzees. *Primates*, **20**, 161–83.

Kuroda, S. 1980. Social behavior of the pygmy chimpanzees. *Primates*, **21**, 181–97.

Kuroda, S. 1984. Interactions over food among pygmy chimpanzees. In: *The Pygmy Chimpanzee: Evolutionary Biology and Behavior*, ed. R. L. Susman, pp. 301–24. New York: Plenum Press.

Malenky, R. K. 1990. Food choice in *Pan paniscus*. Ph.D. dissertation, State University of New York, Stony Brook.

Mantel, N. 1967. The detection of disease clustering and a generalized regression approach. *Cancer Research*, **27**, 209–20.

Moore, J. 1984. The evolution of reciprocal sharing. *Ethology and Sociobiology*, **5**, 5–14.

Nishida, T. & Hiraiwa-Hasegawa, M. 1987. Chimpanzees and bonobos: cooperative relationships among males. In: *Primate Societies*. ed. B. Smuts, D. L. Cheney, R. M. Seyfarth, R. W. Wrangham & T. T. Struhsaker, pp. 165–77. Chicago: University of Chicago Press.

Parish, A. R. 1994. Sex and food control in the 'uncommon chimpanzee': how bonobo females overcome a phylogenetic legacy of male dominance. *Ethology and Sociobiology*, **15**, 157–79.

Sabater Pi, J., Bermejo, M., Illera, G. & Vea, J. J. 1993. Behavior of bonobos (*Pan paniscus*) following their capture of monkeys in Zaïre. *International Journal of Primatology*, **14**, 797–804.

Schnell, G. D., Watt, D. J. & Douglas, M. E. 1985. Statistical comparison of proximity matrices: applications in animal behaviour. *Animal Behaviour*, **33**, 239–53.

Sokal, R. R. & Rohlf, F. J. 1995. *Biometry*, 3rd edn. New York: W. H. Freeman.

Sugiyama, Y. 1984. Population dynamics of wild chimpanzees at Bossou, Guinea, between 1976 and 1983. *Primates*, **25**, 381–400.

Thompson-Handler, N. 1990. The pygmy chimpanzee: socio-sexual behavior, reproductive biology and life history patterns. Ph.D. dissertation, Yale University.

Thompson-Handler, N., Malenky, R. K. & Badrian, N. 1984. Sexual behavior of *Pan paniscus* under natural conditions in the Lomako Forest, Equateur, Zaïre. In: *The Pygmy Chimpanzee: Evolutionary Biology and Behavior*, ed. R. L. Susman, pp. 347–68. New York: Plenum Press.

Trivers, R. L. 1971. The evolution of reciprocal altruism. *Quarterly Review of Biology*, **46**, 35–57.

Uehara, S. 1988. Grouping patterns of wild pygmy chimpanzees (*Pan paniscus*) observed at a marsh grassland amidst the tropical rain forest of Yalosidi, Republic of Zaïre. *Primates*, **29**, 41–52.

Uehara, S. 1990. Utilization of a marsh grassland within the tropical rain forest by the bonobos (*Pan paniscus*) of Yalosidi, Republic of Zaïre. *Primates*, **31**, 311–22.

de Waal, F. B. M. 1987. Tension regulation and nonreproductive functions of sex in captive bonobos (*Pan paniscus*). *National Geographic Research*, **3**, 318–35.

de Waal, F. B. M. 1988. The communication repertoire of captive bonobos (*Pan paniscus*), compared to that of chimpanzees. *Behaviour*, **106**, 183–251.

de Waal, F. B. M. 1989. Food sharing with reciprocal obligations among chimpanzees. *Journal of Human Evolution*, **18**, 433–59.

White, F. J. 1986. Behavioral ecology of the pygmy chimpanzee. Ph.D. dissertation, State University of New York, Stony Brook.

White, F. J. 1988. Party composition and dynamics in *Pan paniscus*. *International Journal of Primatology*, **9**, 179–93.

White, F. J. 1989a. Ecological correlates of pygmy chimpanzee social structure. In: *Comparative Socioecology: The Behavioural Ecology of Humans and Other Mammals*, ed. V. Standen & R. A. Foley, pp. 151–64. Oxford: Blackwell Scientific Publications.

White, F. J. 1989b. Social organization of pygmy chimpanzees. In: *Understanding Chimpanzees*, ed. P. G. Heltne & L. A. Marquardt, pp. 194–207. Cambridge, MA: Harvard University Press.

White, F. J. 1992a. Activity budgets, feeding behavior and habitat use of pygmy chimpanzees at Lomako, Zaïre. *American Journal of Primatology*, **26**, 215–23.

White, F. J. 1992b. Pygmy chimpanzee social organization: variation with party size and between study sites. *American Journal of Primatology*, **26**, 203–14.

White, F. J. 1994. Food sharing in wild pygmy chimpanzees, *Pan paniscus*. In: *Current Primatology, Vol. II, Social Development, Learning and Behaviour*, ed. J. J. Roeder, B. Thierry, J. R. Anderson & N. Herrenschmidt, pp. 1–10. Strasbourg: Université Louis Pasteur.

White, F. J. & Burgman, M. A. 1990. Social organization of *Pan paniscus*: multivariate analysis of intercommunity association. *American Journal of Physical Anthropology*, **83**, 193–201.

White, F. J. & Chapman, C. A. 1994. Contrasting chimpanzees and pygmy chimpanzees: nearest neighbor distances and choices. *Folia Primatologica*, **63**, 181–91.

White, F. J. & Lanjouw, A. 1992. Feeding competition in Lomako pygmy chimpanzees: variation in social organization with party size. In: *Topics in Primatology, Vol. 1, Human Origins*, ed. T. Nishida, W. C. McGrew, P. Marler, M. Pickford & F. B. M. de Waal, pp. 67–79. Tokyo: University of Tokyo Press.

White, F. J. & Wrangham, R. W. 1988. Feeding competition and patch size in the chimpanzee species *Pan paniscus* and *Pan troglodytes*. *Behaviour*, **105**, 148–64.

Wrangham, R. W. 1975. The Behavioural Ecology of Chimpanzees in Gombe National Park, Tanzania. Ph.D. thesis, University of Cambridge.

Wrangham, R. W., Clark, A. P. & Isabirye-Basuta, G. 1992. Female social relationships and social organization of Kibale forest chimpanzees, In: *Topics in Primatology, Vol. 1, Human Origins*, ed. T. Nishida, W. C. McGrew, P. Marler, M. Pickford, and F. B. M. de Waal, pp. 81–98. Tokyo: University of Tokyo Press.

- PART II
Social ecology

4 · Social ecology of Kanyawara chimpanzees: implications for understanding the costs of great ape groups

RICHARD W. WRANGHAM, COLIN A. CHAPMAN,
ADAM P. CLARK-ARCADI AND GILBERT ISABIRYE-BASUTA

Chimpanzees, *Pan troglodytes,* and bonobos, *Pan paniscus,* differ greatly in their social relationships and psychology, as many chapters in this book show (e.g. Takahata *et al.,* Chapter 11; de Waal, Chapter 12). Why they do so is not understood. Yet since these are the two closest relatives of humans, and since each species has a different set of similarities with humans, the question is especially important by virtue of its relevance to human behavior. Why, for instance, do humans and chimpanzees have similarly violent intergroup aggression? The answer will likely depend on understanding why bonobos do not.

In general, bonobos have more relaxed relationships than chimpanzees, with a more pervasive web of alliances or friendships linking community members, especially adult females. This set of differences is thought to depend crucially on the ecological costs of grouping. Thus, friendly social relationships among female bonobos are thought to be possible because their parties are relatively stable, with individuals rarely forced to be solitary. Equivalently friendly relationships among female chimpanzees, on the other hand, are prohibited because parties are regularly forced to fragment, as a result of feeding competition when fruits are scarce (Chapman *et al.,* 1994). Over evolutionary time, such differences have led to differences in species psychology (e.g. Wrangham, 1993).

Resource-based sociality is the only framework so far proposed to explain the ultimate sources of behavioral differences between the two species. To explore one component of it, this chapter examines feeding competition. We begin by considering general principles of grouping in the great apes.

GROUP SIZE AND FEEDING COMPETITION IN THE GREAT APES

Tree-fruits are generally distributed in discrete food patches (trees, or small groups of trees). Individuals feeding in groups share these resources. When a patch has fewer meals than the number of individuals in the group, the result is 'within-group scramble competition' (WGS), realized as reduced food intake or increased travel distance (to reach extra food sources). All frugivorous primates, such as the great apes, experience WGS. The question is, what affects its intensity?

Janson & Goldsmith (1995) found that group size in frugivorous primates is well predicted by the cost of additional travel imposed by extra companions, i.e. by the steepness of the curve relating extra travel to group size. If the slope of the curve relating extra travel to group size is shallow, groups are large; if it is steep, groups are small. The relative shallowness or steepness of this curve is determined both by the nature of the food supply and by the costs of locomotion. There are therefore two kinds of explanation for differences in the intensity of WGS among frugivores: type of food or costs of locomotion (Wrangham *et al.,* 1993).

Both principles can be seen in fission–fusion societies. First, food abundance varies between seasons. Increased fruit abundance means that the amount of additional travel imposed by extra companions is reduced. Thus, groups can be large when fruit is abundant.

Second, the energetic cost of extra travel differs between the sexes. Interestingly, the variation occurs in different directions in different species. In orangutans, *Pongo pygmaeus,* travel is costlier for adult (large) males than for (smaller) females and female-sized males: the

[45]

large males cannot travel as fast as females or their smaller rivals (van Schaik & van Hooff, Chapter 1). As expected, therefore, the more social individuals are the females and female-sized males. In less sexually dimorphic chimpanzees, by contrast, travel is costlier for mothers than for males: the burden of carrying an infant may be responsible for the lower velocity of females than males (2.3 vs. 2.8 feet per second, Hunt, 1989). Again, as expected, the more social sex is the one with cheaper locomotion, which in the case of chimpanzees is males. In support of this explanation, females without infants are as social as adult males (Goodall, 1986). Thus, sex differences in the cost of locomotion may explain why male-bonding is possible in chimpanzees, without resort to other factors such as differential benefits to males and females.

THE CHIMPANZEE–BONOBO PROBLEM

Differences in both locomotion and food are therefore candidates for explaining differences in social ecology. However, relevant locomotor aspects have barely been studied. Bonobo and chimpanzee locomotor behavior is detectably different, with bonobos traveling in trees more often and using more agile movements (Doran, Chapter 16). Possibly, bonobos have a lower cost of locomotion than chimpanzees (that is, less time or energy to travel a given distance). If so, this could explain differences in the intensity of WGS, and therefore in group size and stability, between bonobos and chimpanzees. This hypothesis, which has not previously been considered, needs to be tested by appropriate study of the costs of locomotion and their relation to group size.

Attention to date, however, has focused on the food supply, which presents an obvious point of comparison because the forests inhabited by chimpanzees and bonobos are similar and geographically close, especially in the equatorial regions. The two species are allopatric, separated merely by the Zaïre River, a boundary which curls across the equator in such a way that the latitudinal (as well as longitudinal) distribution of bonobos (latitudes 2° N to 4° S) is entirely overlapped by that of chimpanzees (latitudes 13° N to 7° S). As a result, the range of climates and habitats experienced by bonobos is merely a subset of those experienced by chimpanzees. Two kinds of ecological influence have been proposed.

First are habitat differences resulting from plant structure and dynamics. For example, fruit trees used by bonobos could be larger, as proposed by White (1989). However, the only study of tree measurements found the size and density of fruit trees used by the two species to be essentially identical (Kibale vs. Lomako; Chapman *et al.*, 1994). Alternatively, Malenky (1990) suggested that seasonal variance in fruit production may be less for bonobos. But seasonality in lowland forest is strongly related to latitude (van Schaik *et al.*, 1993), so fruiting seasonality is unlikely to differ consistently between the two species. There is currently no evidence of any consistent differences in plant structure and dynamics between the forests occupied by chimpanzees and bonobos.

Second is a difference in food availability resulting from feeding competition with gorillas (*Gorilla gorilla*). Like chimpanzees, gorillas live only north of the Zaïre River. They are therefore sympatric with chimpanzees throughout most of their range, but never with bonobos. The relevant food is the leaves and stems of terrestrial herb vegetation (THV), a component of the diet of all three species of African ape.

Gorillas, restricted to a diet of THV, experience very low levels of within-group feeding competition (Watts, 1994; Janson & Goldsmith, 1995). Their only other major dietary component is fruits, similar to those eaten by chimpanzees. Since gorillas live in stable groups, their dependence on THV is presumably the critical difference reducing feeding competition for gorillas compared with chimpanzees.

Because THV is invariably prominent in gorilla diets, gorilla foraging has been hypothesized to reduce the availability of THV for sympatric chimpanzees as compared with bonobos (Wrangham, 1986). Accordingly, the reason why bonobos have more stable groups than chimpanzees may be their ability to eat more THV than chimpanzees.

Two kinds of evidence support this THV hypothesis. The only direct comparison between chimpanzee and bonobo diets suggests that Lomako bonobos eat significantly more THV than do Kibale chimpanzees (Malenky & Wrangham, 1994). At some bonobo and chimpanzee sites, THV is known to be eaten more when arboreal fruit is scarce (Wamba: Kano & Mulavwa, 1984; Ndoki: Kuroda *et al.*, Chapter 6; Lopé: Tutin *et al.*,

1991; Kibale: Wrangham *et al.*, 1991; Kahuzi-Biega: Yamagiwa *et al.*, Chapter 7). Thus, there is preliminary evidence that bonobos eat more THV than do chimpanzees and that it can be a fallback food when competition is relaxed. Whether THV provides enough food to be important for bonobo social ecology, however, is still uncertain (Malenky & Stiles, 1991; Malenky *et al.*, 1994).

WHY DON'T KANYAWARA FEMALES TRAVEL IN PARTIES?

If THV is indeed responsible for allowing bonobo parties to be more stable and larger than those of chimpanzees, it can be expected to have similar effects when different populations of chimpanzees are compared. For example, the most stable chimpanzee parties reported from the wild appear to be those at Bossou, where there are no gorillas, and where THV seems particularly abundant and important in the chimpanzee diet (Sugiyama & Koman, 1992). However, recent comparisons of Gombe and Kanyawara chimpanzees indicate that although the two populations differ in the availability of THV, they do not show the expected differences in grouping patterns.

Comparison of these two populations is instructive because their social organization is generally similar, despite ecological differences. For example, in a preliminary description of social relationships in Kanyawara, chimpanzees were found to be similar to Gombe in having strong male–male friendships and relatively antisocial females (Wrangham *et al.*, 1992). These patterns have been robustly confirmed by subsequent observations. Additional similarities include systematic intercommunity migration by adolescent females, long-term intracommunity alliances used in dominance struggles between males and territories defended through calls and violence, including lethal aggression (Wrangham *et al.*, in prep.).

Gombe chimpanzees, on the one hand, conform to expectation. During fruit-poor seasons in Gombe, chimpanzees have very little THV and no consistent fall-back foods (personal observation). Consistent with the lack of fall-back foods, fruit scarcity leads to relatively intense seasonal food stress (apparently responsible for weight loss, ill health, increased mortality and reduced reproductive effort; Goodall, 1986; Wallis, 1992, 1995). During periods of apparent fruit scarcity, Gombe chim-

panzees travel in small parties and females especially spend much time alone (Wrangham, 1977).

Kanyawara chimpanzees, on the other hand, have relatively abundant THV compared with Gombe and they eat it when fruit is scarce (Wrangham *et al.*, 1991). According to the THV hypothesis, this means that Kanyawara chimpanzees should be able to maintain relatively large parties during periods of fruit scarcity, including having females that commonly travel together, as happens at other chimpanzee sites (whether or not it is THV that allows them to do so, e.g. Bossou, Mahale, Taï; Boesch, Chapter 8). However, Kanyawara's females do not do so. Instead, Kanyawara party size is consistently correlated with phenological measures of fruit availability (1988–89: Wrangham *et al.*, 1992; 1989–92: Chapman *et al.*, 1995); and parties become small during fruit-poor seasons, with the result that mean party size is very similar in Kanyawara and Gombe (Kanyawara: 5.6 (1984–85), 6.1 (1988–89), 5.1 (1989–1992); Wrangham *et al.*, 1992; Chapman *et al.*, 1995; Gombe: 4–5, Goodall 1986; numbers exclude dependent offspring). Furthermore, adult females in Kanyawara spend about as much time alone (with their dependent offspring) as do their Gombe counterparts (70% in Kanyawara vs. 65% in Gombe; Wrangham *et al.*, 1992). Thus, the presence of THV in Kibale does not cause parties to be large or relatively stable and therefore does not have the social effects expected of it.

TOWARD A SOLUTION

Kanyawara chimpanzees accordingly present a problem for the THV hypothesis. Below, we examine data relevant to finding a solution. We ask whether seasonal variation in fruit availability is ecologically and demographically significant, and what drives variation in party size. We conclude that Kanyawara chimpanzees search for fruits even when they are scarce, so that THV is never an important influence on feeding and grouping.

STUDY SITE AND METHODS

Kanyawara's chimpanzees have been studied irregularly since 1979, when they were observed briefly (Ghiglieri, 1984). From 1983 to 1985 they were studied by Isabirye-Basuta (1988, 1989) for 30 months. Since 1987, attempts at observation have been continuous. Habituation has increased steadily. For example, by 1988 parties could be watched on the ground and by 1990 they tolerated observers walking about 15 m behind.

Chapman & Wrangham (1993) described the study site and population. Here we note aspects relevant to understanding some of the more important constraints and biases on observation. Briefly, Kanyawara contains medium-altitude forest (1470–1750 m) bordered by subsistence agriculture. The forest has a history of substantial disturbance, so that only about 30% of the Kanyawara community's range is relatively undisturbed. Other sectors include forest that was logged in the 1960s either lightly (*c.* 20% of the community range, depending on total area measured) or heavily (*c.* 20%), and plantations of exotic trees such as pine (*Pinus caribbaea*) established on hill-top grasslands in the 1960s (*c.* 15%). The balance of the Kanyawara community's range is composed of undisturbed papyrus (*Cyperus papyrus*) swamps (*c.* 3%) and occupied villages, composed mostly of farm bush and fields in various states of regeneration. The population has never been deliberately provisioned, but during periods of fruit scarcity Kanyawara chimpanzees regularly raid village crops such as banana stems. Observation of crop-raiding is difficult.

Since December 1987 attempts to contact and follow chimpanzees have been made about six days per week, normally by one or two teams, each with two observers. Despite the rather constant search effort, the number of observation hours per month has varied widely (mean hours per month 83.6, standard deviation 53.7, range 4.7–219.7, $N = 41$ months, January 1991–May 1994).

Behavioral data analyzed in this paper come from two sources. First, party composition data were compiled by ourselves, Ugandan field assistants or other observers. They consist of records every 15 minutes of: all individuals present or believed to be present (i.e. if seen before and after that 15-minute point without any sign of having left in the interval); map location; and the food being eaten by most of the party. Second, RWW collected 10-minute focal data from February to June 1993. These consisted of 5 scans at 2 minute intervals of a focal individual, recording activity, nearest neighbors and party composition. To reduce dependence in the data, no individual was scanned more frequently than every 30 minutes.

Phenological data were recorded every two weeks from a 12 km trail (starting December 1987). Fecal analysis was done on 1696 dungs over 78 months, December 1987–May 1994 (mean per month = 21.7, SD = 18.7). For phenological and dung analysis methods, see Wrangham *et al.* (1991).

THV is defined by its distribution, rather than in terms of its floral composition or its plant parts. The two key features are its terrestriality and the fact that it occurs in patches sufficiently large to allow individuals to co-feed with minimal within-group competition. Although this definition is frustratingly vague as far as patch size is concerned, THV is a useful concept because it describes a terrestrial component of the diet that is important for all the African apes (e.g. Kuroda, 1979). The particular foods that constitute THV vary widely and routinely include several types during the same feeding bout. Four plant families are especially prominent: Marantaceae, Zingiberaceae, Gramineae and Acanthaceae. The parts eaten include pith (especially of monocotyledonous stems), leaf-shoots, leaves and fruits (especially Zingiberaceae). THV is not strictly 'herbaceous,' because chimpanzees eat items from some woody plants that are found in combination with herbs, such as a shrub-fig (*Ficus asperifolia (urcelolaris)*; Wrangham *et al.*, 1991). This means that THV is not an ideal term, but we use it here in acknowledgement of current practice.

Party size refers to total number of chimpanzees present, excluding dependent offspring (see Wrangham *et al.*, 1992). Computer data entry has not yet been finalized for some years, so different periods are used for different analyses. Statistical tests are 2-tailed throughout.

SOCIAL ECOLOGY IN KANYAWARA

Were fruit shortages important in Kanyawara?

Major fruits were produced irregularly

As in all other wild chimpanzee and bonobo populations, arboreal fruits dominated the diet of Kanyawara chimpanzees (monthly average of 79% of feeding time from 1991 to 1993; see Table 4.1). The potential importance of food scarcity was therefore expected to depend on the availability of arboreal fruits.

Table 4.1. *Monthly feeding time on major food types*

	1991	1992	1993	1991–93
Ficus natalensis	28.4	10.4	11.7	16.8
Mimusops bagshawei	13.6	15.1	10.4	13.0
Ficus exasperata	5.7	11.9	15.7	11.1
Pseudospondias microcarpa	11.4	12.4	0.6	8.2
Ficus sansibarica	3.4	5.7	12.7	7.3
Uvariopsis congensis	12.3	0.1	8.3	6.9
Ficus saussureana	5.3	4.7	4.4	4.8
7 top fruit species (total of the above)	80.1	60.4	63.8	68.1
24 other arboreal fruit species	5.4	16.6	10.7	10.9
Total arboreal fruit	85.5	77.0	74.5	79.0
THV	12.3	19.9	18.4	16.9
Arboreal leaf	1.3	1.1	5.3	2.6
Meat	0.8	1.2	0.7	0.9
Other (wood, seed, flower, bark, honey, etc.)	0.1	0.8	1.1	0.6
N (15-minute points observed)	3109.0	3242.0	5459.0	
Mean % time feeding per month	59.2	53.9	49.8	54.3

Note: Cells show mean monthly percentage of feeding records for each food type. Each year represents 12 months of observation.

Major fruit species were defined as those that accounted for > 75% of time spent eating fruit. From 1991 to 1993 there were seven such species (Table 4.1). Three produced succulent drupes (*Mimusops bagshawei* (Sapotaceae), *Uvariopsis congensis* (Annonaceae) and *Pseudospondias microcarpa* (Anacardiaceae)). The remaining four were figs (*Ficus natalensis*, *Ficus exasperata*, *Ficus sansibarica* (*brachylepis*) and *Ficus saussureana* (*dawei*) (Moraceae)). These seven species accounted for between 78% and 94% of total time spent eating fruits (1992 and 1991 respectively), or 60–80% of total feeding time.

Because the diet was dominated by these seven species, their fruiting patterns were critical to overall levels of fruit availability (Fig. 4.1). Fruit production in drupes was tightly synchronized within species, but fruiting was unpredictable. Thus drupe species sometimes failed to fruit at all within a year, whereas figs invariably fruited each year. The result was that drupes were less often available than figs. From December 1987 to September 1993, there were on average 4.5 months per year with none of the drupes fruiting.

By contrast, there were only 1.2 months per year with none of the figs fruiting and only 0.7 months per year with none of the seven major food species in fruit. But figs

Fig. 4.1. Adolescent male chimpanzee eats figs at Mahale. (Photo by T. Nishida)

were eaten less than drupes when drupes were available. Across months, there was a strong negative correlation between drupe-eating and fig-eating (major fruits: Spearman $r_s = -0.75$, $N = 41$, $P < 0.001$; all non-fig vs. fig fruits: $r_s = -0.89$, $N = 41$, $P < 0.001$; data from January

1991–May 1994). Since figs were relatively continuously available, this implies that figs were a fallback food.

Supporting the concept of figs as a fallback food, feeding time was generally more closely related to fruit availability for drupes than figs (percent time being eaten vs. percent trees on the phenology trail with ripe fruit: *M. bagshawei*, Spearman $r_s = 0.84$, $p < 0.001$; *P. microcarpa*, $r_s = 0.42$, $p < 0.05$; *U. congensis*, $r_s = 0.58$, $p < 0.01$; *F. sansibarica*, $r_s = -0.05$, ns; *F. saussureana*, $r_s = 0.35$, ns; *F. exasperata*, $r_s = 0.50$, $p < 0.01$; *F. natalensis*, $r_s = 0.37$, ns; $N = 33$ months, January 1991–September 1993).

Preference for drupes over figs has been found in other studies of apes (Leighton, 1993; Kuroda *et al.*, Chapter 6). Kanyawara's figs tend to have little sugar and therefore generally provide less digestible carbohydrate than do drupes (Conklin & Wrangham, 1994).

In sum, abundance of major drupes varied greatly between months, with chimpanzees relying on figs when drupes were scarce.

Reproductive rate in Kanyawara was low compared with other sites

The ultimate importance of total fruit scarcity is a negative impact on the chimpanzees' reproductive rate. Kanyawara's reproductive rate seems low when compared with other chimpanzee populations: for 11 Kanyawara females for whom minimum interbirth interval has been seen or inferred, the shortest interval between surviving offspring has been 5.5 years. The mean IBI, including minimum IBIs for females who have not yet given birth, was 7.2 years, SD 1.7 years. This is longer than at other sites, especially since data from other sites do not include females waiting to conceive (Bossou 4.4 years, Sugiyama, 1989; Gombe 5.5 years, Goodall, 1986; Mahale 6.0 years, Nishida *et al.*, 1990).

On the other hand, there was no indication that fruit scarcity was associated with ill health or mortality. No widespread bouts of diarrhea occurred. Respiratory infections were rare. Parasitic nematodes and protozoa occurred at low frequency (monitored over five years, Reid & Wrangham, in preparation). One trematode outbreak occurred for a seven-month period, but without any obvious relationship to fruit availability (Wrangham, 1995). No infants died. Kanyawara chimpanzees are relatively large-bodied (*c.* 40–50 kg, Kerbis Peterhans *et al.*, 1993) and, unlike Gombe chimpanzees, visual assessment

did not suggest seasonal weight loss. They never lost their hair sheen. This is in contrast to Gombe, where individuals can look thin and dull-haired during periods of fruit scarcity (personal observation).

We tentatively interpret this pair of results as indicating that the main fallback foods, figs and THV (plus agricultural crops such as banana stems), are adequate for the physiological maintenance of chimpanzees, but provide a poor basis for reproduction. More data are needed to test this conclusion and to establish the demographic and growth parameters of this population. However, the evidence from reproductive rates clearly suggests that the overall level of food abundance was low compared to other chimpanzee sites and that periods of drupe scarcity were ecologically important.

Fruit patches were highly attractive to chimpanzees

The major drupes were produced in groves of locally high density. Figure 4.2(*a*) gives an example for the eating of *Mimusops bagshawei* fruits during 1993–94, showing that it was consumed in a limited portion of the community range compared with THV (cf. Fig. 4.2(*b*)). Such fruiting groves occurred in areas that were visited routinely by Kanyawara chimpanzees between peaks of drupe production, because they found other foods there also. Therefore, we have long-term records of which individuals use the areas.

The community had eight resident adult females, compared with a conservative count of 12 peripheral adult females who shifted their core areas for a week or more at a time, sometimes within their own community's range and sometimes to another community's range. During peaks of drupe production, there were more sightings both of familiar-but-rarely-seen females or mother–offspring units, and of completely unfamiliar females. The behavior of males suggested that these stranger females visited from neighboring communities: mothers were the object of great social interest by males and offspring were twice subject to apparent attempts at infanticide (February 1993, February 1994). This suggests that females visited fruiting groves both from distant parts of the Kanyawara community (e.g. mother *EK*, Fig. 4.3(*c*)), and from other communities (e.g. mother *PU*, Fig. 4.3(*b*)), in contrast to the even use of the range by central females (e.g. mother *LP*, Fig. 4.3(*a*)).

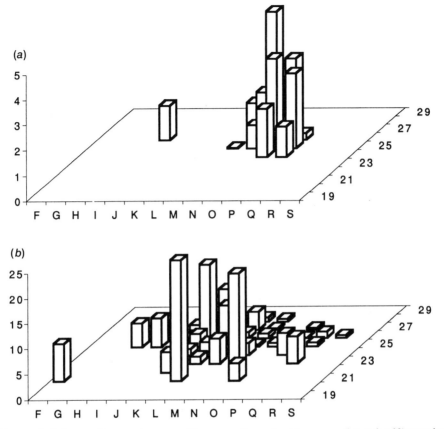

Fig. 4.2. Grid maps of relative use of community range at Kany-awara: feeding sites. Data from January 1993 to May 1994, $N = 7172 \times 15$-minute points. Grid represents an area of 7 km north-south (F to S), and 5.5 km east–west (19 to 29): each square is 0.25 km². 'Relative use' shows the proportion of obser-vations recorded for a given food, divided by the proportion of all observations recorded in that grid square. (a) Relative use of range by chimpanzees when eating *Mimusops bagshawei* fruits, showing a patchy distribution centred in the south and east. $N = 547 \times 15$-minute points. (b) Relative use of range by chim-panzees when eating THV. THV-feeding was more widely spread than when eating *M. bagshawei* fruits, but was still patchy. The western concentration of data occurred near the forest edge. $N = 300 \times 15$-minute points.

An alternative explanation for the occasional sightings of peripheral females is that they were in the process of migrating between communities, but immigration by adolescent females occurred rapidly. Thus, we recorded six newly arrived adolescent females. After their initial sightings, five of these females were seen as often as cen-tral females (70–100% of months, $N = 2$–41 months, mean 24.8, compared with 44–100% for resident adult females). Only the sixth (*TW*) behaved like a peripheral female by visiting for two days only.

In sum, variation in the spatial and temporal pro-duction of fruits seems very important, because it appeared to have a major impact on chimpanzee social relationships and use of space.

Was THV density sufficient to provide food for chimpanzees?

THV species were abundant
Stem densities of THV species eaten at Kanyawara have been shown to be significantly lower than at Lomako, and appear to be lower than at Ndoki (Malenky *et al.*, 1994). However, the data on which these figures are

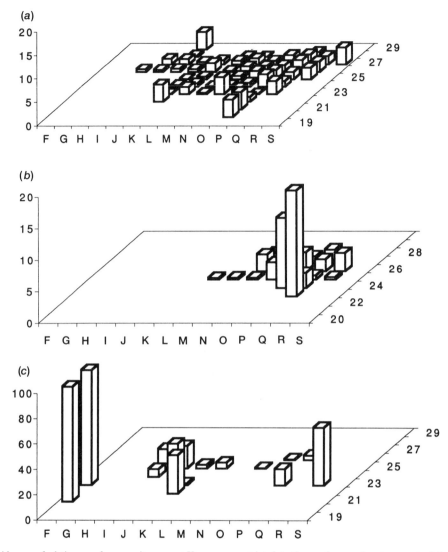

Fig. 4.3. Grid maps of relative use of community range at Kany-awara: individual occupancy. Data source and analysis are as in Fig. 4.2. (*a*) Relative range use by LP, a resident central mother. Her time was spread evenly over the central and southern part of the range. $N = 1948 \times 15$-minute points. (*b*) Relative range use by PU, a peripheral southern mother. Her visits were restricted to the southern and eastern part of the range. $N = 332 \times 15$-minute points. (*c*) Relative range use by EK, a resident northern mother. Map shows that EK made an occasional journey to the southern part of the range. $N = 79 \times 15$-minute points.

based came from natural forest areas of Kanyawara, and excluded the plantations of exotic trees (such as pines) and the swamps. Both plantations and swamps were highly productive of THV foods. Pine plantations were routinely visited in the late afternoon by chimpanzees who foraged there for the relatively abundant herbs. Swamps in Kanyawara are dominated by papyrus, a species whose pith is eaten by chimpanzees during periods of extreme food scarcity, which grows all year and has the highest growth rate known for any herb

(Thompson, 1976). Papyrus pith is not a preferred food but it is enormously abundant.

A second type of evidence that THV was abundant came from how easily chimpanzees found it. When drupes were scarce, chimpanzees sometimes visited a particular fig tree daily for a week or more. Typically no other fruits were eaten within several hundred meters of the target fig tree, apparently because none occurred, whereas THV food-patches were found every day at high density within a few minutes' walk from the fig-tree. Even if no THV happened to be nearby, chimpanzees could find THV patches predictably, for example in the ecotone areas between exotic plantations and natural forest, or in large gaps in logged areas.

THV was eaten all year

Previously, we found that the presence of THV remains (fibrous strands) in the dung was predicted by fruit scarcity (when variation in rainfall was statistically controlled, Wrangham et al., 1991). This suggests that the percentage of feeding bouts on THV is negatively correlated with the percentage on fruits. When non-fig and fig fruits were combined into a single 'fruit' category, there was indeed a significant negative correlation between feeding time on THV and feeding time on fruits (Spearman $r_s = -0.94$, $N = 41$ months, $p < 0.001$; January 1991–May 1994). When only non-fig fruit are considered, however, the correlation with THV was not significant ($r_s = -0.31$, $N = 41$ months; partial correlation using log-transformed data showed neither non-figs nor figs were correlated with THV: non-fig fruits, partial $r_s = -0.09$; figs, partial $r_s = 0.02$). This suggests that THV was only a secondary fallback food, after figs.

THV was the most important component of the diet after arboreal fruits, with a monthly average of 17% of feeding points (standard deviation 12%, N = 41 months, 1991–94; see Table 4.1). Typically it was eaten late in the day (correlation between hour of day and percentage of feeding time spent eating THV, Spearman $r_s = 0.96$, $N = 14$ hours, $p < 0.01$ ($N = 2213$ feeding points, 1993–94). This was probably partly the result of fruit shortage later in the day. However, even during periods of drupe superabundance, chimpanzees often left fruit-rich trees to eat THV in the evening. THV therefore appeared to have a nutritional role beyond being a fallback food. It is rich in fermentable fiber (Conklin & Wrangham, 1994)

and may provide a more efficient use of overnight inactivity than the more rapidly digested nutrients in fruits.

There was no indication that THV induced feeding competition. Individuals fed close to each other. For example, no aggression was observed while eating THV, even though individuals fed closer to each other when eating THV than when eating non-fig fruits (mean number of other individuals within 5 m = 0.8, $N = 152$ scans for THV, vs. 0.2, $N = 291$ scans for fruits; Wilcoxon $T = 0$, $N = 7$ individuals each with > 5 data-points as focal when eating each food type, $p < 0.05$).

Did THV buffer party size?

Party sizes while eating non-fig fruits, fig fruits and THV tended to rise and fall in parallel, all influenced in the same way by overall monthly changes in the grouping pattern. Thus, inter-monthly correlations between sizes of party when eating each of the three main food types were positive and significant. For THV and figs, Spearman $r_s = 0.68$, $N = 33$, $p < 0.001$; for THV and non-fig fruits, $r_s = 0.64$, $N = 34$ months, $p < 0.001$; for figs and non-fig fruits, $r_s = 0.51$, $N = 30$, $p < 0.05$ (data from 1991–93). This implies that fruit abundance drove party size, regardless of the particular food being eaten at a given time.

Comparisons of party size on days when two food types were recorded support this conclusion. Thus, within days, no differences were found in the size of parties eating non-figs, figs or THV (non-figs, figs and THV compared within days from 1991–92; Wilcoxon $z \leqslant 0.30$, $N \leqslant 49$, ns). Similarly, Malenky & Wrangham (1994) found that parties did not increase in size when moving from fruit trees to THV.

The only evidence for THV buffering party size was that mean party size within months was slightly higher when eating THV than when eating figs (7.3 vs. 6.3, Wilcoxon $z = 2.30$, $N = 33$ months, $p = 0.05$; 1991–93).

Arboreal fig fruits were a dominant component of the diet in many months, averaging 42.5% of monthly feeding time over a three-year period (1991–93). In months when more figs were eaten, parties were smaller (mean monthly party size vs. percent feeding records on arboreal fig fruits: Spearman $r_s = -0.39$, $N = 41$ months, $p < 0.05$, 1991–94), whereas when more non-fig fruits were eaten, they were larger ($r_s = 0.35$, $N = 41$, $p < 0.05$).

Correlations between party size and feeding time on arboreal leaves or THV, by contrast, were not significant ($r_s = 0.29$, leaf; 0.22, THV; $N = 41$, ns). Thus, more fig-eating was associated with smaller parties. This is presumably because fruits were generally scarce at those times.

DISCUSSION

THV did not ameliorate feeding competition for Kanyawara chimpanzees

The Kanyawara population is not representative of chimpanzees in general. For example, no other population of chimpanzees or bonobos has been described with so few fruit species accounting for the majority of feeding time. Similarly, movements by peripheral females into communities with temporarily abundant fruit patches, which indicate the importance in Kanyawara of a heterogeneous fruit supply, appear to be uncommon elsewhere. Furthermore, the relationships between diet and life-history variables in Kanyawara chimpanzees appear unusual. On the one hand, low mortality and good body condition suggest that the population had a good food supply. On the other hand, frequent periods of non-fig fruit shortage and long interbirth intervals suggest that nutrition was poor. Nevertheless, even if Kanyawara has ecological idiosyncracies, the relationships between food supply and grouping should follow general rules. We therefore use these rules to address the THV hypothesis.

The effect of THV as an ecological (and social) buffer was originally expected to depend on its abundance, without regard to quality (Wrangham, 1986). Our data, however, show that although THV was abundant in Kanyawara, drupe scarcity led chimpanzees to forage on figs, though rare, in preference to THV. The result was that parties were small. This contradicts the prediction of the THV hypothesis, by showing that even when THV was present, the fruit component of the diet continued to drive party size. THV thus had little effect on the intensity of feeding competition.

Importance of THV quality

Our data require that the THV hypothesis be rejected in its original form, because Kanyawara chimpanzees preferred fruit to THV. But preliminary data show that THV eaten by bonobos has higher nutritional quality than that eaten by Kanyawara chimpanzees (Malenky & Stiles, 1991; Wrangham *et al.*, 1991), so this problem may not apply to bonobos.

Accordingly, and in line with the distinction made by Kuroda *et al.* (Chapter 6), we suggest a classification into high-quality (H–THV) or low-quality (L–THV). H–THV is protein-rich, is of sufficient nutritional value to allow growth and reproduction, is more preferred than typical fig fruits, and typically appears to occur at low density requiring 'feed-as-you-go' foraging: examples include the shoots of *Haumania* spp. and the pith of *Megaphrynium* spp., which occur at low density in Wamba, Lomako and elsewhere (Kano & Mulavwa, 1984; Malenky & Wrangham, 1994; Kuroda, Chapter 6). L–THV, in contrast, is cellulose-rich, is of insufficient quality to allow growth and reproduction by chimpanzees and bonobos, is less preferred than most fig fruits, and may occur in high-density patches: a clear example is papyrus pith, occurring in dense patches in Kanyawara but eaten only *in extremis*. The revised hypothesis substitutes H–THV for THV. Thus, it proposes that the evolution of differences in the social ecology of chimpanzees and bonobos has depended on bonobos having greater access to H–THV, that H–THV is an important fallback food where it occurs in sufficient abundance and that it allows for relatively stable foraging parties by reducing the intensity of WGS.

The revised hypothesis has the merit of distinguishing more clearly between the diets of *Gorilla* and *Pan*. It suggests that L–THV is an adequate diet for *Gorilla* (as seen in high-altitude gorillas, e.g. Yamagiwa *et al.*, Chapter 7) but not for reproduction by *Pan*. It conforms to the observation that H–THV appears important in the diet of bonobos (Kano & Mulavwa, 1984; White, Chapter 3), and that there are important distinctions in the way that gorillas, chimpanzees and bonobos forage according to THV quality (Kuroda *et al.*, Chapter 6).

Two important questions concern differences across species and within species. First, how does the use of H–THV differ between bonobos and those chimpanzees that occur sympatrically with gorillas in bonobo-like habitats, such as in Lopé and Ndoki (Tutin, Chapter 5; Kuroda, Chapter 6). In particular, does competition between gorillas and chimpanzees limit the availability of H–THV for chimpanzees?

Second, to what extent can intraspecific variation be explained by variation in H–THV availability? In the case of Kanyawara, there is essentially no H-THV. We hypothesize that Kanyawara chimpanzees are therefore forced to search for fruits even when few are available, and this leads them to disperse into small parties. This pattern appears likely to explain the fact that Kanyawara chimpanzees conform to the classic type of social organization for *Pan troglodytes schweinfurthii*. Comparisons between Lomako and Wamba, or between Gombe and Mahale, will be helpful because these cases involve sites with similar climate but different party sizes.

Such tests will not explain all differences within a species. For example, Taï chimpanzees form large and relatively stable parties, but eat little THV (C. Boesch, personal communication). This emphasizes that there are several potential sources of explanation for species differences in grouping.

The THV hypothesis has the merit, however, of being the only one to date that also incorporates an explanation for the evolutionary differentiation of bonobos and chimpanzees. By showing that L–THV has little social effect in chimpanzees, our present data have sharpened the search for the origins of the bonobo–chimpanzee differences. The absence of gorillas on the south side of the Zaïre River certainly means that a THV-eating niche has been left vacant, so the question now is: do bonobos eat enough H–THV to account for their reduced feeding competition compared to chimpanzees? If not, a renewed focus on locomotor differences between bonobos and chimpanzees may be required to explain their social differences.

ACKNOWLEDGMENTS

This paper was originally prepared for the Wenner-Gren conference 'The Great Apes Revisited,' held November 12–19, 1994 in Cabo San Lucas, Baja, Mexico. We thank the Government of Uganda, especially the National Research Council, Forestry Department and National Parks Board, for permission to work in the Kibale National Park. Facilities were provided by Makerere University Biological Field Station and the Wildlife Conservation Society. The Department of Zoology, Makerere University, assisted at all times. Acknowledgement for funding is due to the National Science Foundation (BNS-8704458), National Geographic Society (3603–87 and 4348–90), Leakey Foundation, USAID and MacArthur Foundation. Assistance was provided by L. Barrett, J. Basigara, B. Beerlage, A. Berle, J. Byaruhanga, L. Chapman, N. L. Conklin, K. Clement, the late G. Etot, B. Gault, J. Gradowski, M. Hauser, K. Hunt, G. Isabirye-Basuta, the late G. Kagaba, E. Karwani, C. Muruuli, P. Novelli-Arcadi, the late J. Obua, E. Ross, E. Tinkasimire, P. Tuhairwe, P. Weeks and R. G. Wrangham. RWW thanks T. Kano for his hosting a fascinating and informative visit to Wamba. B. Beerlage and the editors provided helpful comments.

REFERENCES

Chapman, C. A. & Wrangham, R. W. 1993. Range use of the forest chimpanzees of Kibale: implications for the understanding of chimpanzee social organization. *American Journal of Primatology*, **31**, 263–73.

Chapman, C. A., White, F. J. & Wrangham, R. W. 1993. Defining subgroup size in fission–fusion societies. *Folia Primatologica*, **61**, 31–4.

Chapman, C. A., White, F. J. & Wrangham, R. W. 1994. Party size in chimpanzees and bonobos. In: *Chimpanzee Cultures*, ed. R.W. Wrangham, W.C. McGrew, F.B.M. de Waal & P.G. Heltne, pp. 41–58. Cambridge, MA: Harvard University Press.

Chapman, C. A., Wrangham, R. W. & Chapman, L. J. 1995. Ecological constraints on group size: an analysis of spider monkey and chimpanzee subgroups. *Behavioral Ecology and Sociobiology*, **36**, 69–70.

Conklin, N. L. & Wrangham, R. W. 1994. The value of figs to a hind-gut fermenting frugivore: a nutritional analysis. *Biochemical Ecology and Systematics*, **22**, 137–51.

Ghiglieri, M. P. 1984. *The Chimpanzees of the Kibale Forest: A Field Study of Ecology and Social Structure*. New York: Columbia University Press.

Goodall, J. 1986. *The Chimpanzees of Gombe: Patterns of Behavior*. Cambridge, MA: Harvard University Press.

Hunt, K. D. 1989. *Positional Behavior in Pan troglodytes at the Mahale Mountains and the Gombe Stream National Parks, Tanzania*. Ann Arbor, MI: University Microfilms.

Isabirye-Basuta, G. 1988. Food competition among individuals in a free-ranging chimpanzee community in Kibale Forest, Uganda. *Behaviour*, **105**, 135–47.

Isabirye-Basuta, G. 1989. Feeding ecology of chimpanzees in the Kibale Forest, Uganda. In: *Understanding Chimpanzees*, ed. P. G. Heltne & L. A. Marquardt,

pp. 116–27. Cambridge, MA: Harvard University
Press.

Janson, C. H. & Goldsmith, M. L. 1995. Predicting group
size in primates: foraging costs and predation risks.
Behavioral Ecology, 6, 326–36.

Kano, T. & Mulavwa, M. 1984. Feeding ecology of the
pygmy chimpanzees (*Pan paniscus*) of Wamba. In: *The
Pygmy Chimpanzee: Evolutionary Biology and
Behavior*, ed. R. L. Susman, pp. 233–74. New York:
Plenum Press.

Kerbis Peterhans, J., Wrangham, R. W., Carter, M. L. &
Hauser, M. D. 1993. A contribution to tropical rain
forest taphonomy: retrieval and documentation of
chimpanzee remains from Kibale Forest, Uganda.
Journal of Human Evolution, 25, 485–514.

Kuroda, S. 1979. Grouping of the pygmy chimpanzee.
Primates, 20, 161–83.

Leighton, M. 1993. Modeling dietary selectivity by
Bornean orangutans: evidence for integration of
multiple criteria in fruit selection. *International
Journal of Primatology*, 14, 257–313.

Malenky, R. K. 1990. *Ecological Factors Affecting Food
Choice and Social Organization in* Pan paniscus. Ann
Arbor, MI: University Microfilms.

Malenky, R. K. & Stiles, E. W. 1991. Distribution of
terrestrial herbaceous vegetation and its consumption
by *Pan paniscus* in the Lomako Forest, Zaïre.
American Journal of Primatology, 23, 153–69.

Malenky, R. K. & Wrangham, R. W. 1994. A quantitative
comparison of terrestrial herbaceous food
consumption by *Pan paniscus* in the Lomako Forest,
Zaïre, and *Pan troglodytes* in the Kibale Forest,
Uganda. *American Journal of Primatology*, 32, 1–12.

Malenky, R. K., Kuroda, S., Vineberg, E. O. &
Wrangham, R. W. 1994. The significance of terrestrial
herbaceous foods for bonobos, chimpanzees and
gorillas. In: *Chimpanzee Cultures*, ed. R. W.
Wrangham, W. C. McGrew, F. B. M. de Waal & P.
G. Heltne, pp. 59–75. Cambridge, MA: Harvard
University Press.

Nishida, T., Takasaki, H. & Takahata, Y. 1990.
Demography and reproductive profiles. In: *The
Chimpanzees of the Mahale Mountains: Sexual and Life
History Strategies*, ed. T. Nishida, pp. 63–98. Tokyo:
University of Tokyo Press.

van Schaik, C. P., Terborgh, J. W. & Wright, S. J. 1993.
The phenology of tropical forests: adaptive
significance and consequences for primary consumers.
Annual Review of Ecology and Systematics, 24, 353–77.

Sugiyama, Y. 1989. Population dynamics of chimpanzees

at Bossou, Guinea. In: *Understanding Chimpanzees*, ed.
P. G. Heltne & L. A. Marquardt, pp. 134–45.
Cambridge, MA: Harvard University Press.

Sugiyama, Y. & Koman, J. 1992. The flora of Bossou: its
utilization by chimpanzees and humans. *African Study
Monographs*, 13, 127–69.

Thompson, K. 1976. Swamp development in the head
waters of the White Nile. In: *The Nile, Biology of an
Ancient River*, ed. J. Rzoska, pp. 177–95. The Hague:
Dr. W. Junk.

Tutin, C. E. G., Fernandez, M., Rogers, M. E.,
Williamson, E. A. & McGrew, W. C. 1991. Foraging
profiles of sympatric lowland gorillas and chimpanzees
in the Lopé Reserve, Gabon. *Philosophical
Transactions of the Royal Society of London, Series B*,
334, 179–86.

Wallis, J. 1992. Socioenvironmental effects on timing of
first postpartum cycles in chimpanzees. In: *Topics in
Primatology, Vol. 1, Human Origins*, ed. T. Nishida,
W. C. McGrew, P. Marler, M. Pickford & F. B. M.
de Waal, pp. 119–30. Tokyo: University of Tokyo
Press.

Wallis, J. 1995. Seasonal influence on reproduction in
chimpanzees of Gombe National Park. *International
Journal of Primatology*, 16, 435–51.

Watts, D. P. 1994. Agonistic relationships between female
mountain gorillas (*Gorilla gorilla beringei*). *Behavioral
Ecology and Sociobiology*, 34, 347–58.

White, F. J. 1989. Ecological correlates of pygmy
chimpanzee social structure. In: *Comparative
Socioecology: Behavioral Ecology of Humans and Other
Mammals*, ed. V. Standen & R. A. Foley, pp. 151–64.
Oxford: Blackwell Scientific Publications.

Wrangham, R. W. 1977. Feeding behaviour of
chimpanzees in Gombe National Park, Tanzania. In:
Primate Ecology, ed. T.H. Clutton-Brock, pp. 504–38.
London: Academic Press.

Wrangham, R. W. 1986. Ecology and social evolution in
two species of chimpanzees. In: *Ecology and Social
Evolution: Birds and Mammals*, ed. D. I. Rubenstein &
R. W. Wrangham, pp. 352–78. Princeton, NJ:
Princeton University Press.

Wrangham, R. W. 1993. The evolution of sexuality in
chimpanzees and bonobos. *Human Nature*, 4, 47–79.

Wrangham, R. W. 1995. Relationship of chimpanzee
leaf-swallowing to a tapeworm infection. *American
Journal of Primatology*, 37, 293–304.

Wrangham, R. W., Conklin, N. L., Chapman, C. A. &
Hunt, K. D. 1991. The significance of fibrous foods
for Kibale Forest chimpanzees. *Philosophical*

Transactions of the Royal Society of London, Series B, **334**, 171–8.

Wrangham, R. W., Clark, A. P. & Isabirye-Basuta, G. 1992. Female social relationships and social organization of Kibale Forest chimpanzees. In: *Topics in Primatology Vol. 1, Human Origins,* ed. T. Nishida, W. C. McGrew, P. Marler, M. Pickford & F. B. M.

de Waal, pp. 81–98. Tokyo: University of Tokyo Press.

Wrangham, R. W., Gittleman, J. L. & Chapman, C. A. 1993. Constraints on group size in primates and carnivores: population density and day-range as assays of exploitation competition. *Behavioral Ecology and Sociobiology,* **32**, 199–209.

5 • Ranging and social structure of lowland gorillas in the Lopé Reserve, Gabon

CAROLINE E. G. TUTIN

INTRODUCTION

At the time of the first Wenner-Gren conference on 'The Great Apes' in 1974, western lowland gorillas had been little studied in the wild. Apart from the pioneering field study of Jones & Sabater Pi (1971), knowledge of gorilla ecology in central west Africa came from the anecdotes of explorers and big game hunters. Jones & Sabater Pi, working in Equatorial Guinea (Rio Muni), found that gorillas lived in smaller groups and ate more fruit than the mountain gorillas of the Virunga Volcanoes (Schaller, 1963; Fossey & Harcourt, 1977). Jones & Sabater Pi conducted their study in secondary and regenerating forests, and lowland gorillas were believed to be restricted to disturbed forest habitats with dense herbaceous growth and not to occur in primary forest (Groves, 1971; Jones & Sabater Pi, 1971). This assumption was not challenged until a census of apes in Gabon showed gorillas to occur throughout the forested parts of the country although, at that time, most of these forests were primary, i.e. 'untouched by human hand' (Tutin & Fernandez, 1984). The widespread presence of gorillas has now been confirmed in tropical forests, seasonally inundated forests and even swamp forests throughout central west Africa (Carroll, 1988; Fay, 1989; Fay et al., 1989; Blake, 1993).

In these botanically diverse tropical forests, gorillas eat large amounts and a diverse array of succulent fruit (Williamson et al., 1990; Nishihara, 1992; Tutin & Fernandez, 1993a; Remis, 1994; Kuroda et al., Chapter 6). The same is true of geographically distinct gorilla populations in lowland tropical forests in eastern Zaïre (Yamagiwa et al., 1994), suggesting that ripe fruit are eaten whenever available (see also Watts, Chapter 2). Habitat-dependent dietary differences are striking and

consistent, regardless of subspecies, making it more appropriate to divide gorillas into 'frugivorous' (lowland primary forest) and 'folivorous' (montane habitat) populations (Watts, 1990). This distinction reflects differential availability of fleshy fruit, but frugivorous gorillas also eat piths and leaves and, during periods of fruit scarcity, these vegetative items become keystone foods [sensu Terborgh, 1986] (Rogers et al., 1988; Tutin et al., 1991; White et al., 1995; Yamagiwa et al., Chapter 7). At Lopé, gorillas eat fruit of at least 97 species, but seasonal variation in diet is pronounced, as are differences between years (Tutin et al., 1991; Tutin & Fernandez, 1993a). This is because most plant species reproduce seasonally, but not always annually, resulting in an ever-changing array of potential fruit foods and in great variations in fruit abundance. In contrast, the diet of gorillas living in montane habitats is less diverse and is composed mainly of stems and leaves, the availability of which varies little over time (Schaller, 1963; Fossey & Harcourt, 1977; Vedder, 1984; Watts, 1984). Thus, for simplicity, the term 'forest' gorillas is used here to refer to populations that have diverse diets including much fruit, and 'montane' for those with diets dominated year-round by vegetative plant parts.

Preference for fruit might be expected to affect the social structure and ranging patterns of forest gorillas as availability of fruit is often limited (Fig. 5.1). It has been proposed that the smaller group sizes of forest, compared with montane, gorillas are related to dietary differences, as competition among group members for food will be greater for fruit than for non-fruit foods (Wrangham, 1979; Harcourt et al., 1981). However, an analysis of gorilla group size at Lopé (Tutin et al., 1992) showed the average group size to differ little from published figures

Fig. 5.1. Adult male gorilla eats fruit at the top of an emergent tree at Lopé (centre, top of picture). (Photo by C. Tutin)

for gorillas in montane habitats (Murnyak, 1981; Harcourt *et al.*, 1981). Data from Equatorial Guinea (Harcourt *et al.*, 1981; Jones & Sabater Pi, 1971) indicated generally small group sizes (averaging five) for gorillas in secondary forests close to human settlements. As plant diversity is lower, and as fewer species produce succulent fruit in secondary forests than in primary forests, the small group size found in Equatorial Guinea was probably a result of hunting pressure. However, a small average group size of 7.5 ($N = 6$ groups) was recorded also at the Nouabalé-Ndoki Reserve in northern Congo, an area of primary forest where hunting pressure is low (Mitani *et al.*, 1993; Olejniczak, 1994).

Forest gorillas have proven difficult to habituate even after ten consecutive years of research at Lopé, so our data on ranging and social structure come from tracking and nest sites as well as from observation. Day ranges, both of gorilla groups and solitary males, are longer in lowland forests than in montane habitats (Tutin *et al.*, 1992; Remis, 1994; Yamagiwa & Mwanza, 1994;

Yamagiwa *et al.*, 1994). Home-range size also appears to be larger although few data have been published (Tutin *et al.*, 1992; Remis, 1994). Both of these differences are likely to be related to the fruit component of the diet as fruit crops are usually spatially dispersed and temporally ephemeral. Ranging over a large area brings gorillas to fruit crops more frequently than would be the case if they traveled less. The ranges of neighboring gorilla groups at Lopé overlap so extensively that no group appears to have exclusive access to any area (Tutin *et al.*, 1992).

Here, we present an analysis of ranging and social structure of the Lopé gorillas based on traditional field methods. We focus on Porthos' group, which was followed from 1984 to its extinction in 1993 when the silverback male died (Tutin & Fernandez, 1993b). Incomplete habituation and the difficulties of tracking frugivorous gorillas have limited the amount of data we have been able to amass on ranging patterns. Since 1990 we have collected hairs from gorillas' nests, and DNA–PCR techniques should produce 'fingerprints' of bands allowing consistent identification of individuals (Wickings, 1993). The genetic information from these hairs will provide crucial additional data on the ranging and social structure of forest gorillas at Lopé. Findings from other studies of forest gorillas, notably Bai Hokou, Central African Republic (Remis, 1993, 1994) and Nouabalé-Ndoki, Congo (Mitani *et al.*, 1993; Olejniczak, 1994; Kuroda *et al.*, Chapter 6) are included in an attempt to synthesise what is known, suspected, or remains obscure about the social organization of gorillas in tropical forest habitats.

STUDY SITE AND METHODS

The Lopé Reserve covers 5000 km^2, most of which is tropical rain forest, but a zone of savanna and gallery forest extends about 10 km south of the Ogooué River, which forms the northern boundary of the Reserve. The study area covers about 50 km^2, immediately south of the savanna zone (0°10′ S; 11°35′ E). Gorillas occur at a mean density of 1.0 km^{-2} and are, at present, well protected from hunting.

The vegetation, topography and climate of the study area are described by Tutin *et al.* (1991, 1994) and White *et al.* (1995). Field procedures involve searching through the forest for gorillas, or noting indirect signs of their presence and activities. When gorillas are located, we try to observe and to follow their movements for as long as possible. We record groups of nests and map routes of travel by following trail signs to supplement the observational data.

As gorillas (*Gorilla gorilla gorilla*) and chimpanzees (*Pan troglodytes troglodytes*) are sympatric and eat many of the same foods (Tutin & Fernandez, 1993a), great care must be taken to identify the species of ape responsible for feeding remains, feces and nests. Gorillas' overnight-nest groups and feces can usually be distinguished with certainty from that of chimpanzees if they are fresh, as it is rare for a complete group of gorillas to nest in trees (Tutin *et al.*, 1995), and gorilla feces are usually lobed and of firmer consistency than those of chimpanzees. However, the two species process many foods in the same way (Tutin & Fernandez, 1994), so feeding and day resting sites cannot always be distinguished unless fresh feces are also present.

Gorilla nest sites are rich sources of data, as the number of nests gives an indication of the number of weaned individuals in the group. However, a large data set is needed, as, at Lopé, only at one-third of fresh nest sites did the number of nests correspond to the number of weaned individuals in the group (Tutin *et al.*, 1995). Remis (1993) found that at Bai Hokou in the Dzanga-Sangha Reserve, Central African Republic, 47% of gorillas did not construct night nests but instead slept on bare ground. This reinforces the need for great caution in making deductions about the size, composition or dynamics of forest gorilla groups from nest sites.

The locations of contacts, nest sites and trails of gorillas were mapped onto large scale maps (1 : 10 000). For analysis of ranging, the data were transcribed onto copies of topographic maps (1 : 50 000, IGN Paris) onto which a 100 by 100 m grid system was superimposed.

The amount of ripe fruit of 62 species that are important in apes' diets was recorded monthly by monitoring ten individual plants of each species. The quantity of leaves, flowers and ripe and unripe fruit was noted on a 10-point scale from 0 (none) to 10 (maximum crop). A total ripe fruit score was calculated each month for each species by dividing the summed score by the number of monitored trees. 'Poor Fruit Months' were defined as those when six or fewer species in the phe-

nology sample had total fruit scores ≤5; all other months were considered as 'Good Fruit Months'.

RESULTS

Group size

Determination of mean (or median) group size is not easy, as opportunities when complete group counts can be made unambiguously by observation are rare in tropical forest habitats. The size of each gorilla group changes over time through births, deaths and migration, yet statistically each group should only contribute a single datum to a calculation of average group size within a local population. Estimations of group size can be made from fresh nest sites, but analysis of 137 sites of Porthos' group showed that only 34% contained the number of nests corresponding to the number of weaned individuals. When there were eight weaned individuals in Porthos' group, the number of nests found at fresh nest sites ranged from 3–16 (Tutin *et al.*, 1995). Another problem of using indirect group counts obtained from nest sites is that of establishing the identity of the group of gorillas. This is not always possible by observation, and no assumptions should be made either from the number of nests (see above) or from the location of the nest site as home ranges overlap so extensively (Tutin *et al.*, 1992). Analysis of our data from Lopé gave a median group size of ten members for eight groups (range 4–16) (Tutin *et al.*, 1992), but doubt remains about the independence of these data points, collected over a 6.5-year period from groups that could not always be identified with certainty. Analysis of DNA from hairs will allow accurate and objective group counts and thus the calculation of average group sizes in the population from independent counts, as the identity of the group and of individual members, can be established.

Porthos' group was first recognized in 1984 and over ten years, the number of members varied from a high of 11 (1985 and 1987) to a low of 5 (1993) (see Table 5.1). Porthos was the only fully adult male in the group for most of the 10-year period but two of his sons matured in the group and left when they were young silverbacks. The group contained at most four adult females. Porthos was very old when he died as a result of the massive infection of wounds inflicted by another male gorilla in a fight (Tutin & Fernandez, 1993b). The fate of the four remaining members is not known. Data from a second group, St Exupéry's, which we followed for six years, showed similar fluctuations of membership from 9 to 14.

The composition of the main study group at Bai Hokou varied from 12–15 weaned individuals over 27 months and included at least two silverbacked males (Remis, 1994).

Day ranges

A combination of tracking between successive night nests and observation allowed measurement of 80 complete day ranges of gorilla groups at Lopé between 1986 and 1994. The average distance traveled in a day was 1105 ± 553 m (range 220–2790). At least five gorilla groups contributed to the data set, but day ranges of Porthos' group were measured 42 times. Comparison of day ranges of Porthos' group and the other groups showed no consistent difference in daily travel distance (Mann Whitney U test, $N_1 = 38$, $N_2 = 42$, $z = 0.01$, 2-tailed, ns).

A comparison of day-range lengths during Good Fruit Months and Poor Fruit Months (see Methods), showed that gorilla groups travelled further when fruit was abundant than when it was scarce (Mann Whitney U test, $N_1 = 32$, $N_2 = 48$, $Z = 4.37$, $p < 0.001$). This was confirmed with a comparison of the data from Porthos' group, which travelled on average 1266 m when fruit was abundant, compared with 749 m when fruit was scarce ($N_1 = 16$, $N_2 = 26$, $z = 3.02$, $p < 0.005$). There was no significant difference between the lengths of 16 day ranges of Porthos' group measured between 1988 and 1990 and 26 day-ranges between 1991 and 1993 despite a slight change in group size (Table 5.1). The data on day range lengths are summarized in Table 5.2.

Day ranges were measured by transcribing direct and indirect observations onto the 1:10 000 maps and then measuring the route taken between successive nest sites. Following gorilla trail over a complete day's range was difficult and time-consuming (usually requiring 2–3 days), and it was impossible to measure the distance covered (with either a topofil or by counting paces) during tracking, because of the frequent need to backtrack. Only twice did we return to measure with topofil a complete day's range that had been established by

tracking. On these two occasions, the distances measured on the ground were 18 and 28% greater than those measured on the map. Even on large-scale maps it is impossible to mark precisely circuitous routes and the rugged nature of the terrain compounded the error. In addition, the difficulties of tracking gorillas in tropical forests mean that we were more likely to follow successfully complete day ranges when gorillas traveled little than when they moved great distances to fruiting trees. This suggests that our indirect measurements of day-range length are under-estimates by about 20–25%.

Remis (1994) estimated the average day range of a group of 13 gorillas at Bai Hokou to be 1600 m but believed this to be an underestimate for the same reasons.

Home ranges

Measuring home ranges presents a number of additional problems (Craig, 1981; Anderson, 1982) that are related to identifying correctly the gorilla group responsible for the trails or nests. In a preliminary analysis of gorilla home-range size at Lopé (Tutin et al., 1992) we noted that the data from observations, nest groups and trails represented only 11% of days. From these data, home-range sizes of three gorilla groups during one year were 7–14 km^2, but overlap between the three groups' ranges was extensive and the range of one group was overlapped completely by the other two.

All of our observations of Porthos' group and their nest-sites found in ten years (1984–93) were located in a minimum convex polygon (Southwood, 1966) of 21.7 km^2. We cannot say from our limited data whether or not the whole of this area was visited each year but we believe not. Our knowledge of Porthos' group's range increased steadily over time, as we were able to follow their movements with greater success due both to their increasing tolerance of our presence and to our accumulated knowledge of the terrain, particularly the location of fruiting trees. The group spent most of the time in an area of about 10 km^2 and this was the zone in which we concentrated our searches. We were able to document five nomadic movements that took the group out of their main range (or core area) for periods ranging from 1–8 weeks. (A sixth major move that will extend the total home range of the group to about 25 km^2 is suspected

Table 5.1. *Composition of Porthos' group, 1984 to 1993 at Lopé*

	1984	1985	1986	1987	1988	1989	1990	1991	1992	1993	Comments
Porthos Silverback (c.1950)	xxxxxx	xxxxxx	xxxxxx	xxxxxx	xxxxxx	xxxxxx	xxxxxx	xxxxxx	xxxxxx	xx	Died 23 April 1993
Unnamed 1 Adult female	xxxxxx	xxxxxx	?????	04/87							
Unnamed 2 Adult female	xxxxxx	xxxxxx	xxxxxx	xx????	08/88						
Esmeralda Adult female	xxxxxx	xxxxxx	xxxxxx	xxxxxx	xxxxxx	xxxxxx	xxxxxx	xxxxxx	08/92		
Madame Bovary Adult female	xxxxxx	xxxxxx	xxxxxx	xxxxxx	xxxxxx	xxxxxx	xxxxxx	xxxxxx	xxxxxx	xx	
DArtagnan Male (c.1975/76)	xxxxxx	xxxxxx	xxxxxx	xxxxxx	xxxxxx	xxxxxx	x-x-x	01/91			Sporadic presence 1990
Aramis Male (c.1978/79)	xxxxxx	xxxxxx	xxxxxx	xxxxxx	xxxxxx	xxxxxx	xxxxxx	xxxxxx	05/92		Sporadic presence during early 1992
Visage Rose Female (c.1980/81)	xxxxxx	xxxxxx	xxxxxx	xx????	05/88						Observed in St. Exupéry group in August 1989
Dois Rose Female (c.1981/82)	xxxxxx	xxxxxx	xxxxxx	xxxxxx	08/88						
Ignace Male (c.1983)	xxxxxx	xxxxxx	xxxxxx	xxxxxx	xxxxxx	xxxxxx	xxxxxx	xxxxxx	xxxxxx	xx	
Sapho ? Male (c.1985)		xxxxxx	xxxxxx	xxxxxx	xxxxxx	xxxxxx	xxxxxx	xxxxxx	xxxxxx	xx	
Papillon Male (c.1987)				xxxxxx	xxxxxx	xxxxxx	xxxxxx	xxxxxx	xxxxxx	xx	
Unnamed Infant (09/91)									xx	xxx	
Unnamed Infant (01/92)									xx		
Unnamed ? Female (c.1982)									x		Brief presence in group May 1992
Total for most of year	10	11	10	10	10	8	8	7	7	5	

KEY: x, presence in group during 2-month period; date, last month of confirmed presence; (date), year of birth, known or estimated.

Table 5.2. *Day range lengths of gorilla groups at Lopé*

Group	Season	N	Mean day range (M)	Standard error	Range (m)
All	All	80	1105	553	220–2790
	Good fruit	48	1316	592	420–2790
	Bad fruit	32	789	277	220–1650
Porthos	All	42	1138	556	430–2790
	Good fruit	26	1327	613	430–2790
	Bad fruit	16	830	237	430–1260

but awaits confirmation by identification of hairs from nest sites.) Other temporary range shifts are likely, as the group sometimes was not found in the core area for periods of up to three months. Four of the six range shifts took the group to exceptionally high concentrations of ripe fruit, and two occurred when ripe fruit was atypically scarce due to crop failures. On the latter occasions the group moved to habitat-types where other foods were locally available. One of these, in February–March 1993, was to the highest part of the study area (680 m), where they fed on seeds of two tree species that grow only at altitudes above 450 m (White, 1992).

Overlap of Porthos' home range by other gorillas was complete, although in the northern savanna and gallery forest zone only lone males were encountered. Other groups of gorillas are believed to have home ranges of similar sizes, and their sporadic appearances in Porthos' central range suggest that temporary range shifts are a general phenomenon. In addition to the temporary range shifts, more permanent shifts occur. From 1984–89, Porthos' group was not our focal study group but they were encountered regularly in the western part of the home range of St Exupéry's group. During 1987–89, following a change of silverback male, St Exupéry's group ranged further and further east and only rarely visited the center of their old range. By 1989, Porthos' group had shifted east to occupy the area largely vacated by St Exupéry's group. By 1992, Porthos' group rarely visited the western area that had previously been the center of their range. These temporary and more permanent range shifts add to the difficulties of studying gorillas in tropical forests.

Remis (1994) found the home range of her study group (C) at Bai Hokou to encompass at least 23 km^2 over 27 months, and she recorded similar patterns of nomadic shifts related to seasonal variation in fruit avail-

ability. The ranges of at least three neighboring groups partially overlapped the range of Group C. At Nouabalé-Ndoki, range overlap also must be extensive as at least six groups visited the large swamp watched by Olejniczak (1994). The importance of swamps for gorillas in northern Congo is underlined by Kuroda et al. (Chapter 6) who report that regular visits are made year-round. The aquatic plants in the swamps provide an abundant supply of staple herbaceous foods for gorillas and may be the key factor permitting high population densities of up to four individuals km^{-2} (Fay, 1989; Mitani et al., 1993) compared with areas like Lopé that lack such swamps.

Intergroup encounters

Given extensive or complete overlap of home ranges of neighboring groups, the potential for intergroup encounters is great. However, in almost 11 years of observations at Lopé (October 1983–May 1994) only 40 encounters between two groups or between a group and a lone male have been seen. Of these, 22 involved two groups, 16 a group and a lone male, and twice we could not be sure whether the group under observation was interacting with another group or with a lone male. The encounters occurred at distances of 50–300 m between groups and lasted an average of 54 minutes (range 1–186). We never witnessed fighting, but it evidently did occur as Porthos died after an attack in which he received 16 deep puncture wounds, presumably from another adult male. On two other occasions small quantities of blood were found: once, evidence of smashed vegetation and tracks indicated that a fight had occurred during the night between a small group of gorillas and a lone silverback who was seen alone and not visibly injured early the next morning; in the other case, blood was found after a prolonged nocturnal vocal exchange. Evidence found the next day indicated that a lone male and a small group of gorillas (two groups of nests were found separated by about 250 m) were involved, but we could not rule out an attack by a leopard as the cause of the alarm and distress vocalizations heard from 19.20 to 22.26 hours. The episode occurred in the center of Porthos' range in May 1993, a month after the silverback's death, and may have involved the three young males and the adult female who survived him.

Nine of the intergroup encounters between groups and lone males involved lone males silently tracking groups and being discovered by researchers doing the same thing. Indirect evidence indicated that on several occasions lone males tracked groups in this way for several consecutive days. The other interactions between groups and lone males occurred close to large fruiting trees and included exchanges of vocalizations or chest beats. Most interactions between two groups occurred close to concentrations of ripe fruit. Fourteen of 22 encounters included auditory exchanges. In the other cases, the silent group was under observation when vocalizations or chest beats from a second group were heard. Thus, the true frequency of intergroup encounters is higher than recorded here, as instances when an undetected group heard sounds from a group under observation could not be detected by observers.

In October–November 1989, four groups of gorillas were attracted to a small area (of about 1500×500 m) where fruiting trees of three species were concentrated on the forest's edge. Seven intergroup encounters were seen: for six, the identity of both groups was known and for the seventh, only one group was identified. Interactions occurred between three of the six possible dyads and Porthos' group was involved in six of the seven interactions. In three cases there was an exchange of vocalizations and chest beats, most, if not all, given by the silverbacks and in the four other cases, only one group vocalized. It was rarely possible to follow the movements of both groups at the end of an encounter but we know that after at least five of the six interactions involving Porthos, his group stayed in the area and fed in the fruiting trees while the other group moved away. This suggests that Porthos' group was dominant in this area, which was central in their home range. Another example of several gorilla groups being attracted to the same place occurred in 1988 when a huge *Gambeya*

africana, a rare tree species in the study area, had a large fruit crop. We watched the tree from a hide for 22 days and gorillas visited it 11 times. Four groups and one lone male came to feed but only one intergroup encounter was witnessed. The outcome of this encounter, and the behavior shown by gorillas approaching the tree, indicated that St Exupéry's group was dominant in the area at that time. St Exupéry's group made five visits to the *Gambeya* and always arrived silently. The other groups, and the lone male, either vocalized or gave chest beats while some distance (>100 m) away, suggesting that they were seeking information about the presence of other gorillas before approaching the tree. In the case of the intergroup encounter, the silverback St Exupéry, having fed with his group in the tree an hour earlier, rushed back and sat below the *Gambeya* giving chest beats in response to those of the other group. The exchange lasted for two hours before St Exupéry left to rejoin his group and the second group was able to enter the *Gambeya* and feed.

There were indications of tolerance between certain groups, particularly Porthos' and Petit Prince's. Petit Prince led a small group with only four members when first recognized in late 1988. Two infants were born in 1989 and 1990 and another gorilla joined the group before they disappeared from the main study area in 1991. Several times Porthos' and Petit Prince's groups were known to have nested in proximity and to have spent long periods within auditory range. One encounter between these groups occurred at the end of a very wet day when Petit Prince's group were resting in a savanna 80 m from the forest edge (Fig. 5.2). Porthos arrived on the edge of the forest and also rested, giving 12 chest beats over a period of an hour to which Petit Prince did not respond. The encounter ended at darkness when both groups re-entered the forest and nested 200 m apart. Perhaps this tolerance was because the silverbacks were related: given their estimated ages, it is possible that Petit Prince was Porthos' son and analysis of hair derived DNA should provide information about relatedness.

Social organization

Social organization of lowland gorillas at Lopé appears to resemble that of the well-studied mountain gorillas of Rwanda with cohesive groups led by one or more adult

Fig. 5.2. Adult male gorilla ventures out of forest onto savanna at Lopé. (Photo by C. Tutin)

males (Harcourt, 1979; Yamagiwa, 1987; Watts, 1994). The two groups that we have been able to follow and observe regularly were both led by a single silverback male (Porthos and St Exupéry), although for some of the time two silverbacks were in the group, with the second one being a maturing son who later emigrated.

Two cases of emigration by young adult males from Porthos' group were documented. Both of these males left the group when estimated to be aged 14–15 years (see Table 5.1). The older male, D'Artagnan, left gradually with observations over a year confirming his presence at times and his absence at others. The second male, Aramis, left abruptly and his departure coincided with the disappearance of a young adult (presumed to have been female) who joined the group briefly in May 1992. It is tempting to speculate that Aramis succeeded in taking this young female away from the group led by his aging father. Two females born in Porthos' group disappeared when aged between six and eight years and one was subsequently seen in St Exupéry's group. Three adult females disappeared between 1987 and 1992, but it is not known whether they emigrated or died.

D'Artagnan and Aramis were both young silverbacks when they left Porthos' group, and it is likely that in some cases of groups with more than one silverback, the males are father and son. Groups with two and three silverbacks have been seen, but the limitations of our

data on all but Porthos' group preclude firm conclusions about gorilla social structure at Lopé, which must await the results of DNA analyses. The three silverbacked males observed in one group were of visibly different ages: one was very old and obviously handicapped by arthritis, and one was a young adult male who has since been seen alone, while the third was in his prime. At Lopé, groups with more than one silverback tend to be larger than those with a single adult male, but we cannot yet say whether or not this reflects a genuine dichotomy in social structure, with small one-male groups and larger multi-male groups led by males who are not father and son. The alternative is that one-male groups are the basic unit of social organization but, as a group matures, its size increases through births and immigrations, with maturing sons being present as young silverbacks (see also Yamagiwa, 1987). Group C at Bai Hokou contained three silverbacks at the end of Remis' study period, due to maturation of a blackbacked male, but the relationship between the two permanent silverbacks was unknown (Remis, 1994).

Some authors have suggested that forest gorillas differ fundamentally from montane gorillas by having flexible social units that show fission–fusion organization similar to that of chimpanzees (Jones & Sabater Pi, 1971; Mitani, 1992; Remis, 1994). We have seen no evidence of this at Lopé, although it is clear that for forest gorillas the group's spread in space is sometimes very large. It is very hard to quantify group spread systematically, especially when it is large, but we have sometimes been able to confirm spreads of over 500 m and twice, subgroups of gorillas spent 1–3 hours at distances of more than 500 m from other group members. Most overnight nest groups are spatially compact, rarely covering an area of more than 300 m^2, and we have never documented a case where subgroups did not rejoin to nest. However Remis (1994) reported 34 occasions (19% of days on which the group was contacted) when her focal group of 13 gorillas, with two silverback males, split into two subgroups and twice, nest groups were found separated by up to 1 km, indicating that they had slept apart with a silverback in each subgroup.

At Lopé, gorillas harvest most of their foods arboreally and 35% of their night nests are built in trees (Tutin & Fernandez, 1993a; Tutin et al., 1995). This contrasts with mountain gorillas who harvest 90% of their food terrestrially and rarely nest above the ground

(Watts, 1984). Trees offer a third dimension for inter-individual spacing, but observational conditions in tropical forest rarely permit data to be collected on the positions of all members of a group. For example, in 125 observations of Porthos' group between 1989 and 1992 that lasted for at least 15 minutes, all group members were seen only eight times. A greater proportion of the group were visible when at least some were in trees (average 56% of the group) than when all were on the ground (40%). This result is not unexpected as, in much of the forest at Lopé, visibility at ground level is restricted by dense herbaceous vegetation. This has been a major handicap in habituation and also makes it difficult to collect unbiased data on inter-individual distances, group spread and even group size.

Aggressive interactions among group members were rarely seen at Lopé, and only three clear cases of displacement or aggression were seen in feeding contexts. In contrast, we often heard gorillas running toward fruiting trees giving excited vocalizations, and this may reflect 'first-come-first-served', scramble competition for the limited number of feeding places in smaller fruiting trees. In general, the large group spread and low proportion of group members visible at any one time suggest that within-species feeding competition influences the behavior of the frugivorous Lopé gorillas. Two facets of gorilla feeding ecology at Lopé also lend support to the idea that conspecific feeding competition can be important. First, compared with chimpanzees, gorillas at Lopé spend little time processing food before ingestion: they do not wadge foods and they swallow larger seeds (Tutin & Fernandez, 1994). These differences cannot be explained entirely by differences between the ape species in body weight, as even infant gorillas swallow larger seeds than do adult chimpanzees. Preingestion food processing, to separate digestible flesh from indigestible fruit skin and seeds, reduces the rate at which food is acquired. Swallowing fruit whole with minimal processing (although piercing of fruit skin is essential if any nutritive value is to be obtained, as most fruit skins are impervious to digestion) means that gorillas' rate of feeding on fruit is faster than that of chimpanzees. Gorillas at Lopé do spend time on preingestion processing of abundant vegetative foods such as *Aframomum* pith and Marantaceae young leaves, so their non-selective feeding on scarcer fruit might be a result of higher levels of intra-specific competition. Second, in Good Fruit Months, gorillas have a more diverse fruit diet than do sympatric chimpanzees both at Lopé (Tutin *et al.*, 1991) and Noua-balé-Ndoki (Kuroda *et al.*, Chapter 6). This could reflect less developed preferences of gorillas for particular species of fruit, compared with chimpanzees who eat fruit such as *Dialium* and *Celtis* exclusively when these are available. However, an alternative explanation is that being a member of a closed group of constant size means that gorillas are not always able to gain access to the preferred fruit, because of limited number of feeding sites in all but large fruit trees, while chimpanzee fission–fusion social organization allows feeding group sizes to fluctuate in response to fruit availability.

DISCUSSION

Field work on lowland gorillas in tropical forest habitats was late starting compared with studies of the other great apes, but consensus on some aspects of their ecology and behavior is beginning to emerge. The frugivorous diet of these gorillas, which contrasts so strongly with the diet of mountain gorillas, is possible because of the availability of succulent fruit of a diverse array of plant species in tropical rain forests. In these habitats, 70–90% of plant species have fruit adapted to dispersion by animals (Alexandre, 1980; Hall & Swaine, 1981; Foster, 1982; White, 1994) providing a wide niche for frugivores (Gautier-Hion, 1983). It is difficult to draw conclusions about any general relationship between group size and diet, as other factors, particularly human hunting but perhaps also the biomass of sympatric frugivores and the presence of predators, vary between habitats and confound the picture. However, the very large groups of gorillas that can occur in montane habitats have not yet been seen in tropical forests. The relatively large home ranges and long day ranges of gorillas in these forests seem to be related to diet, as day ranges were longer when fruit was abundant than when it was scarce (Table 5.2).

Home ranges of gorilla groups in lowland forests overlap extensively, and it seems that no group has exclusive access to any area. Use of home ranges is not regular, and both temporary and more permanent shifts in ranging occur. Temporary range shifts were related to the non-random dispersion of preferred foods (usually fruit), and particularly attractive concentrations of fruit sometimes brought as many as four gorilla groups into the same area. The home ranges of gorilla groups in montane habitats also overlap extensively and shifts in range use occur, but most of these have been attributed not to food but to mating tactics, specifically to attempts to acquire females or to protect unweaned offspring from infanticide (Watts, 1994).

Intergroup interactions were observed at Lopé, but the true frequency was underestimated as only half of those recorded included an exchange of sounds. Lone male gorillas tracked feeding trails of groups, and it seems likely that trail signs and nest sites provide information to both lone males and to groups about the presence, activity and identity of conspecifics. The small number of observed intergroup encounters probably represents the 'tip of the iceberg', and the influence of other groups on ranging and other behavior cannot yet be interpreted. Intergroup encounters among mountain gorillas are related to acquisition of females and not to defense of home ranges or resources (Fossey, 1983; Yamagiwa, 1987; Sicotte, 1993; Watts, Chapter 2). At Lopé, the majority of encounters between groups were related to access to, or defense of, food resources. We obtained evidence of female transfer, and it is possible that this occurs when groups are drawn into close proximity by concentrations of fruit. In contrast to the intergroup encounters of mountain gorillas, where close-range displays were common and fighting between silverbacks is not unusual (74% and 17% respectively of 58 encounters; Sicotte, 1993), encounters at Lopé were generally peaceful with groups remaining at least 50 m apart. None of the 40 inter-group encounters witnessed involved close range displays, but three times we found indirect evidence of fights. This gives a maximum of 3 of 43 interactions (7%) escalating to fighting and includes the fatal attack on Porthos. It seems likely that forest gorillas do not fight over access to food, even rare succulent fruit, but that, as is the case for montane gorillas, violent interactions occur over acquisition or defense of females. An extreme case of tolerance between groups of forest gorillas comes from Nouabalé-Ndoki in northern Congo where up to 31 gorillas (4 groups and 1 lone male) were seen at once in a large swamp. Intergroup interactions included play between immatures and displays, but no fighting was seen (Olejniczak, 1994; Fay, personal communication).

Questions about intra-group competition for food, whether or not it exists and, if so, whether it applies in all feeding situations or only for preferred fruit, merit close attention. The large group spread and the suggestion of scramble competition to gain access to smaller fruiting trees found at Lopé and at Bai Hokou (Remis, 1994; Goldsmith, personal communication) suggest that competition between group members for food (at least for preferred fruits) may be a cost of gorilla sociality. Our observations and those of Remis (1994) indicate that the number of feeding places in trees, where lowland gorillas harvest most of their food, are limited by the space available on branches capable of supporting a gorilla's weight. Gorillas appear able to assess the number of feeding sites in a tree and do not climb it unless they are sure to be able to feed. The rarity of fights, displays and displacement in trees may reflect an awareness of the danger of falling, and if this is the case, manifestations of competition will be related to access to trees and be difficult to observe. Some aspects of gorilla feeding ecology compared with that of sympatric chimpanzees, such as higher diversity of fruit per fecal sample during Good Fruit Months in gorillas and less food processing of fruit than of vegetative foods by gorillas, also suggest that within-species competition for fruit is important.

The details of social behavior of gorillas in tropical rain forest habitats remain enigmatic, and although at all sites the average group size is seven to ten and one-male groups predominate, larger groups with more than one silverback occur. At two sites (Bai Hokou and Nouabalé-Ndoki) evidence exists of fission–fusion in groups with more than one silverback with subgroups nesting apart on occasions (Mitani, 1992; Remis, 1994). Such behavior has not been seen at Lopé, although group spread can be so great during feeding that temporary fission would be recorded by most definitions of subgroups or parties (see Boesch, Chapter 8). It is not yet clear whether or not local differences in social organization exist, or if fission-fusion is rare, occurring only in large multi-male groups and perhaps only prior to a permanent split (see also Remis, 1994). These observations and the absence of very large groups (of > 16) in lowland forests, suggest that maximum group size for forest gorillas is constrained stringently by within-group competition for fruit. If so, then a negative correlation between group size and degree of frugivory may emerge across gorilla populations, if accurate and independent group counts can be obtained by use of hair derived DNA. Social behavior of gorillas is subtle and occurs at the low frequencies typical of folivorous primates. It may take several more years before the elusive forest gorillas reveal their secrets, but there are reasons for optimism: several long-term studies are underway and new ones are beginning or planned, and, as well as steadily amassing data, all are contributing actively to conservation (e.g. Aveling, 1993). Because of the difficulties of conducting behavioral studies, the emphasis has been on ecology and on understanding the gorillas' role within tropical forest ecosystems, and conscious efforts have been made to standardize methods and to share findings between sites (Kuroda & Tutin, 1993; Mitani *et al.*, 1994). Although many questions remain, the future is bright, as new techniques such as DNA 'fingerprinting' will provide some answers, and the new conservation initiatives linked with all of the long-term research sites should help to ensure the gorillas' future in the Central African Republic, Congo, Gabon and Zaïre.

ACKNOWLEDGMENTS

Thanks go first and foremost to the Centre International de Recherches Médicales de Franceville for providing core funding for research at Lopé since 1983. Important grants from the L. S. B. Leakey Foundation and the World Wide Fund for Nature are also gratefully acknowledged. The following colleagues contributed to data collection at Lopé and shared the joys and frustrations of watching, tracking and searching for Porthos (whom we sorely miss) and his friends: Kate Abernethy, Michel Fernandez, Rebecca Ham, Karen McDonald, Patricia Peignot, Liz Rogers, Ben Voysey, Lee White, Jean Wickings, Liz Williamson and especially Richard Parnell. This paper was originally prepared for the Wenner-Gren conference 'The Great Apes Revisited,' held from November 12–19, 1994 in Cabo San Lucas, Baja, Mexico. Warm thanks go: to Sydel Silverman, Laurie Obbink and Mark Mahoney of the Wenner-Gren Foundation; to all the conference participants for much stimulating discussion; and to Bill McGrew, Linda Marchant and Toshisada Nishida for valuable editorial comments.

REFERENCES

Alexandre, D. Y. 1980. Charactére saisonnier de la fructification dans une forêt hygrophile de Côte d'Ivoire. *Revue d'Ecologie (Terre et Vie)*, **34**, 335–50.

Anderson, D. J. 1982. The home range: a new nonparametric estimation technique. *Ecology*, **63**, 103–12.

Aveling, C. 1993. A new conservation project for Central African tropical forests. *Gorilla Conservation News*, **7**, 3–4.

Blake, S. 1993. A reconnaissance survey in the Likouala swamps of northern Congo and its implications for conservation. M.Sc. Thesis, University of Edinburgh.

Carroll, R. W. 1988. Relative density, range extension and conservation potential of the lowland gorilla in the Dzanga–Sangha region of southwestern Central African Republic. *Mammalia*, **52**, 309–23.

Craig, E. J. 1981. A relief-adjusted method for determining home range in mountainous areas. *Journal of Mammalogy*, **62**, 837–9.

Fay, J. M. 1989. Partial completion of a census of the western lowland gorilla (*Gorilla gorilla gorilla*) in southwestern Central African Republic. *Mammalia*, **53**, 203–15.

Fay, J. M., Agnagna, M., Moore, J. & Oko, R. 1989. Gorillas in the Likouala swamp forests of north central Congo. *International Journal of Primatology*, **10**, 477–86.

Fossey, D. 1983. *Gorillas in the Mist*. Boston: Houghton Mifflin.

Fossey, D. & Harcourt, A. H. 1977. Feeding ecology of free ranging mountain gorillas. In: *Primate Ecology*, ed. T. H. Clutton-Brock, pp. 415–47. London: Academic Press.

Foster, R. B. 1982. The seasonal rhythm of fruit fall on Barro Colorado Island. In: *The Ecology of a Tropical Forest: Seasonal Rhythms and Long Term Changes*, ed. E. G. Leigh, A. S. Rand & D. M. Windsor, pp. 151–72. Washington: Smithsonian Institution Press.

Gautier-Hion, A. 1983. Leaf consumption by monkeys in western and eastern Africa: a comparison. *African Journal of Ecology*, **21**, 107–13.

Groves, C. P. 1971. Distribution and place of origin of the gorilla. *Man*, **6**, 44–51.

Hall, J. B. & Swaine, M. D. 1981. *Distribution and Ecology of Vascular Plants in a Tropical Rain Forest:*

Forest Vegetation in Ghana. The Hague: Dr W. Junk.

Harcourt, A. H. 1979. Social relationships between adult male and female mountain gorillas. *Animal Behaviour,* 27, 325–42.

Harcourt, A. H., Fossey, D. & Sabater Pi, J. 1981. Demography of *Gorilla gorilla. Journal of Zoology,* 195, 215–33.

Jones, C. & Sabater Pi, J. 1971. Comparative ecology of *Gorilla gorilla* (Savage & Wyman) and *Pan troglodytes* (Blumenbach) in Rio Muni, West Africa. *Bibliotheca Primatologica,* 13, 1–96.

Kuroda, S. & Tutin, C. E. G. (eds.) 1993. Field studies of African apes in tropical rain forest: methods to increase the scope and accuracy of intersite comparisons. *Tropics,* 2, 187–255.

Mitani, M. 1992. Preliminary results of the studies on wild western lowland gorillas and other sympatric diurnal primates in the Ndoki Forest, northern Congo. In: *Topics in Primatology, Vol. 2, Behavior, Ecology and Conservation,* ed. N. Itoigawa, Y. Sugiyama, G. P. Sackett & R. K. R. Thompson, pp. 215–24. Tokyo: University of Tokyo Press.

Mitani, M., Yamagiwa, J., Oko, R. A., Moutsambolé, J. M., Yumoto, T. & Maruhashi, T. 1993. Approaches in density estimates and reconstruction of social groups of a western lowland gorilla population in the Ndoki Forest, northern Congo. *Tropics,* 2, 219–29.

Mitani, M., Kuroda, S. & Tutin, C. E. G. (eds.) 1994. Floral lists from five study sites of apes in the African tropical forests. *Tropics,* 3, 247–340.

Murnyak, D. F. 1981. Censusing gorillas in Kahuzi-Biega National Park. *Biological Conservation,* 21, 163–76.

Nishihara, T. 1992. A preliminary report on the feeding habits of western lowland gorilla (*Gorilla gorilla gorilla*) in the Ndoki Forest of northern Congo. In: *Topics in Primatology, Vol. 2, Behavior, Ecology and Conservation,* ed. N. Itoigawa, Y. Sugiyama, G. P. Sackett & R. K. R. Thompson, pp. 225–40. Tokyo: University of Tokyo Press.

Olejniczak, C. 1994. Report on pilot study of western lowland gorillas at Mbeli Bai, Nouabalé-Ndoki Reserve, northern Congo. *Gorilla Conservation News,* 8, 9–11.

Remis, M. J. 1993. Nesting behavior of lowland gorillas in the Dzanga-Sangha Reserve, Central African Republic: implications for population estimates and understandings of group dynamics. *Tropics,* 2, 245–55.

Remis, M. J. 1994. Feeding ecology and positional behavior of western lowland gorillas (*Gorilla gorilla gorilla*) in the Central African Republic. Ph.D. dissertation, Yale University.

Rogers, M. E., Williamson, E. A., Tutin, C. E. G. & Fernandez, M. 1988. Effects of the dry season on gorilla diet in Gabon. *Primate Report,* 22, 25–33.

Schaller, G. B. 1963. *The Mountain Gorilla.* Chicago: University of Chicago Press.

Sicotte, P. 1993. Inter-group encounters and female transfer in mountain gorillas: influence of group composition on male behavior. *American Journal of Primatology,* 30, 21–36.

Southwood, T. R. E. 1966. *Ecological Methods.* London: Methuen.

Terborgh, J. 1986. Keystone plant resources in the tropical forest. In: *Conservation Biology: The Science of Scarcity and Diversity,* ed. M.E. Soule, pp. 330–44. Sunderland, MA: Sinauer.

Tutin, C. E. G. & Fernandez, M. 1984. Nationwide census of gorilla and chimpanzee populations in Gabon. *American Journal of Primatology,* 6, 313–36.

Tutin, C. E. G. & Fernandez, M. 1993a. Composition of the diet of chimpanzees and comparisons with that of sympatric lowland gorillas in the Lopé Reserve, Gabon. *American Journal of Primatology,* 30, 195–211.

Tutin, C. E. G. & Fernandez, M. 1993b. Lowland gorillas in the Lopé Gabon: A profile of Porthos' group. *Gorilla Gazette,* 7, 1–2.

Tutin, C. E. G. & Fernandez, M. 1994. Comparison of food processing by sympatric apes in the Lopé Reserve, Gabon. In: *Current Primatology, Vol. 1, Ecology and Evolution,* ed. B. Thierry, J. R. Anderson, J. J. Roeder & N. Herrenschmidt, pp. 29–36. Strasbourg: Université Louis Pasteur.

Tutin, C. E. G., Fernandez, M., Rogers, M. E., Williamson, E. A. & McGrew, W. C. 1991. Foraging profiles of sympatric lowland gorillas and chimpanzees in the Lopé Reserve, Gabon. *Philosophical Transactions of the Royal Society of London, Series B,* 334, 179–186.

Tutin, C. E. G., Fernandez, M., Rogers, M. E. & Williamson, E. A. 1992. A preliminary analysis of the social structure of lowland gorillas in the Lopé Reserve, Gabon. In: *Topics in Primatology, Vol. 2, Behavior, Ecology and Conservation,* ed. N. Itoigawa, Y. Sugiyama, G. P. Sackett & R. K. R. Thompson, pp. 245–54. Tokyo: University of Tokyo Press.

Tutin, C. E. G., White, L. J. T., Williamson, E. A., Fernandez, M. & McPherson, G. 1994. List of plant species identified in the northern part of the Lopé Reserve, Gabon. *Tropics*, **3**, 249–76.

Tutin, C. E. G., Parnell, R. J., White, L. J. T. & Fernandez, M. 1995. Nest building by lowland gorillas in the Lopé Reserve, Gabon: environmental influences and implications for censusing. *International Journal of Primatology*, **16**, 53–76.

Vedder, A. L. 1984. Movement patterns of a group of free-ranging mountain gorillas (*Gorilla gorilla beringei*) and their relation to food availability. *American Journal of Primatology*, **7**, 73–89.

Watts, D. P. 1984. Composition and variability of mountain gorilla diets in the Central Virungas. *American Journal of Primatology*, **7**, 323–56.

Watts, D. P. 1990. Ecology of gorillas and its relation to female transfer in mountain gorillas. *International Journal of Primatology*, **11**, 21–45.

Watts, D. P. 1994. Gorilla mating tactics and habitat use. *Primates*, **35**, 35–47.

White, L. J. T. 1992. Vegetation history and logging disturbance: Effects on rain forest mammals in the Lopé Reserve, Gabon. Ph.D. thesis, University of Edinburgh.

White, L. J. T. 1994. Patterns of fruit-fall phenology in the Lopé Reserve, Gabon. *Journal of Tropical Ecology*, **10**, 289–312.

White, L. J. T., Rogers, M. E., Tutin, C. E. G., Williamson, E. A. & Fernandez, M. 1995. Herbaceous vegetation in different forest types in the Lopé Reserve, Gabon: implications for keystone food availability. *African Journal of Ecology*, **33**, 124–41.

Wickings, E. J. 1993. Hypervariable single and multi-locus DNA polymorphisms for genetic typing of non-human primates. *Primates*, **34**, 323–31.

Williamson, E. A., Tutin, C. E. G., Rogers, M. E. & Fernandez, M. 1990. Composition of the diet of lowland gorillas at Lopé in Gabon. *American Journal of Primatology*, **21**, 265–77.

Wrangham, R. W. 1979. On the evolution of ape social systems. *Social Science Information*, **18**, 335–68.

Yamagiwa, J. 1987. Male life history and the social structure of wild mountain gorillas. In: *Evolution and Coadaptation in Biotic Communities*, ed. S. Kawano, J. H. Connell & T. Hidaka, pp. 31–51. Tokyo: University of Tokyo Press.

Yamagiwa, J. & Mwanza, N. 1994. Day journey length and daily diet of solitary male gorillas in lowland and highland habitats. *International Journal of Primatology*, **15**, 207–24.

Yamagiwa, J., Mwanza, N., Yumoto, T. & Maruhashi, T. 1994. Seasonal change in the composition of the diet of eastern lowland gorillas. *Primates*, **35**, 1–14.

6 • Sympatric chimpanzees and gorillas in the Ndoki Forest, Congo

SUEHISA KURODA, TOMOAKI NISHIHARA, SIGERU SUZUKI AND RUFIN A. OKO

INTRODUCTION

After the pioneering study of sympatric western lowland gorillas (*Gorilla gorilla gorilla*) and tschego chimpanzees (*Pan troglodytes troglodytes*) in Rio Muni (Equatorial Guinea) by Jones & Sabater Pi (1971), several study sites were established in the 1980s in tropical forests where both chimpanzees and gorillas occur (western lowland gorillas and tschego chimpanzees – Gabon: Tutin & Fernandez, 1984, 1993; Tutin *et al.*, 1991; Tutin, Chapter 5; Central African Republic: Carroll, 1988; Fay, 1988; Remis, 1993, 1994; Congo: Kuroda, 1992; Nishihara, 1992, 1994; Mitani *et al.*, 1993; eastern lowland gorillas (*Gorilla gorilla graueri*) and long-haired chimpanzees (*Pan troglodytes schweinfurthii*) eastern Zaïre: Yamagiwa *et al.*, 1992, 1994, Chapter 7).

Recent studies at these sites suggest common features of interspecific relationships between gorillas and chimpanzees. Generally, dietary overlap in plant food is great, ranging from about 50% at Kahuzi-Biega (Yamagiwa *et al.*, 1994) to 60–80% at Lopé (Williamson *et al.*, 1990; Tutin & Fernandez, 1993) and Ndoki (Kuroda, 1992). Overlap in fruit consumption is particularly high (70–90%) and habitat use patterns are similar. Thus, they share a similar niche and ecological competition is likely to occur between these two ape species (Tutin & Fernandez, 1985, 1993; Williamson *et al.*, 1990; Kuroda, 1992). Low population densities of gorillas and chimpanzees in Lopé and Kahuzi-Biega might partly be due to this competition (Tutin & Fernandez, 1985; Yamagiwa *et al.*, 1992).

However, overt interspecific competition has not been observed at any of these sites. Instead, competition avoidance is commonly seen. During the fruiting season both species concentrate on fruits, and in the lesser fruiting season both consume a higher proportion of the abundant fibrous foods. This seasonal shift is more marked among gorillas, whereas chimpanzees maintain a high intake of fruits (Williamson *et al.*, 1990; Tutin *et al.*, 1991; Nishihara, 1992; Yamagiwa *et al.*, 1992; Tutin & Fernandez, 1993; Remis, 1994). Thus, substantial niche separation between the two species occurs during the lesser fruiting season. This seasonal niche separation may be related to differences in the social systems of both species (Tutin *et al.*, 1991).

The general features of interspecific relationships between gorillas and chimpanzees in the Ndoki Forest do not differ from those recognized at other sites. However, we have observed some unique phenomena, such as frequent, year-round swamp-use by gorillas (Mitani *et al.*, 1993; Nishihara, 1994) and peaceful co-feeding on figs during the lesser fruiting season by both apes (Suzuki & Nishihara, 1992). The former functions to enlarge niche separation, whereas the latter seems to contradict the theory of competitive exclusion. Ndoki is also unique in that population densities of both apes are extremely high among sites of sympatric gorillas and chimpanzees studied (Nishihara, 1994; Suzuki, unpublished data).

This study will review and examine data on the ecology of gorillas and chimpanzees in the Ndoki Forest to clarify their interspecific relationships.

STUDY SITE AND METHODS

The data presented here were collected from 1989–92, but the main discussion focuses on data collected from August 1991 to November 1992. The main study site covers a tropical forest of about 20 km, and is located in the southwest part of the Nouabalé-Ndoki National Park, Congo. The main camp was set at 2° 20′ N and 16° 19′ 11″ E (see Fig. 6.1). The area around the study site has been free of human activity for the last 50 years, thus apes and other animals are found in high densities (gorilla: 1.9–2.6 individuals/km² by bed census, 2.3–2.6 individuals/km² by direct observation and bed count; Nishihara, 1994; chimpanzee: 2.7 individuals/km² by direct observation; Suzuki, unpublished data). Nine species of monkey are also found in high densities (Mitani, 1992).

The annual cumulative rainfall ranged from 1430 mm (December 1991–November 1992) to 1649 mm (September 1991-August 1992). Using monthly rainfall patterns, the year can be divided into two seasons: rainy season from March to November and dry season from December to February. These seasonal differences reflected the amount of fruit availability and diversity, with the rainy season known as the fruiting season and the dry season known as the lesser fruiting season (Nishihara, 1995; Suzuki *et al.*, 1995). For the estimation of fruit availability and diversity, numbers of individuals and species of fruiting plants were recorded monthly along a 10 m wide transect of 37 km, from December 1991 to November 1992.

The following four vegetation types are distinguished by both structure and species composition (Moutsamboté *et al.*, 1994). First, the mixed-species forest is diverse in vegetation structure and species composition, and occupies the largest area in the study site. Second, the swamp forest occupies 15–20% of the area of intensive study. There are clearings ('marshy grassland' in Mitani *et al.*, 1993) with aquatic herbs along watersides in the swamp forest. Third and fourth, riverine and inland *Gilbertiodendron* forests are dominated by almost-pure stands of *Gilbertiodendron dewevrei*; the former is a closed forest but the latter has plenty of light gaps.

If the species of ape could be confirmed, fresh feces were collected and washed in wire baskets with about 1-mm mesh. Fibrous material, including leaves, seeds and fruit tissue of each plant species, was separated to estimate them by 5% step intervals in volume to whole

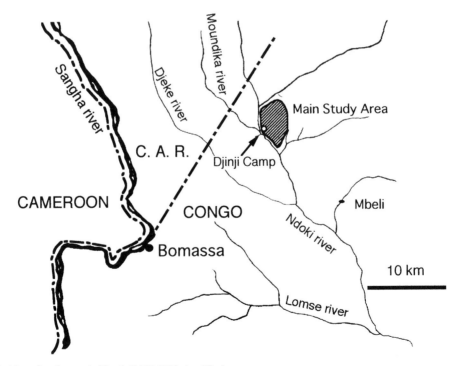

Fig. 6.1. Map of study area in Nouabalé-Ndoki National Park, Congo.

Table 6.1. *Number of plant foods of gorillas and chimpanzees in Ndoki[a]*

	Gorilla(A)	Chimpanzee(B)	Common(C)	C/A(%)	C/B(%)
Number of species	152	108	64	42	59
Number of items	182	114	68	37	60
fruit	115	100	59	51	59
seed	18	3	2	11	17
leaf	29	3	1	3	33
shoot/pith/stem/bark/root	18	6	6	33	33
flower	2	2	0	0	0

[a]Confirmed 1990–94.

material. Thin and small items, such as fragments of shoots and insects and small seeds, were checked both before and after full washing and were identified when possible. Small seeds, such as *Ficus* that may leak through 1-mm mesh, were also included in the analysis by careful washing and estimation. Percent of fibrous material and that of seeds and fruit tissue were used as fiber and fruit scores to ascertain the relative importance of fibrous and fruit foods in the diet. This method is different from that employed in Lopé (Tutin & Fernandez, 1993), but the foliage score used at Lopé is apparently equivalent to the fiber score used here.

Food items of gorillas and chimpanzees were identified by direct observation, fecal analysis and fresh remains when the species of ape could be determined.

RESULTS

Plant food

Dietary overlap

We identified 114 food items for chimpanzees and 182 items for gorillas (see Table 6.1). Table 6.1 compares the plant diet in terms of different food types. Fruit was the most dominant food class for both apes, and thus dietary overlap was largest: 51% for gorilla and 59% for chimpanzee. Overlapping proportions did not change much regardless of the season. All of the 20 dominant fruit foods of both species overlapped with each other (Suzuki & Nishihara, unpublished data), whereas Tutin *et al.* (1991) counted seven important fruit foods of chimpanzees that were not eaten by gorillas at Lopé. In contrast, many food items other than fruits were used only by gorillas, thus the dietary overlap was small for

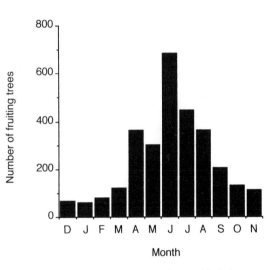

Fig. 6.2. Number of fruiting trees in the monthly fruit census, Dec. 1991–Nov. 1992.

gorillas (13%), whereas it was large for chimpanzees (64%).

Chimpanzees are known to eat the seeds (beans) of Caesalpiniaceae, but not the seeds of fruits. Gorillas also eat the beans of Caesalpiniaceae, including *Gilbertiodendron* and *Dialium* spp. as well as *Dialium* pulp; however, seeds of fruits are more important for them. They frequently crunch and swallow fruit seeds of various species that are gorillas' exclusive food resources (Kuroda, 1992; Nishihara, 1992, 1995). Seeds are also often eaten by gorillas in Lopé (Williamson *et al.*, 1990). Although seeds tend to contain many tannins or toxins that protect them from seed-predation, they are high in protein and lipid foods in general (Rogers *et al.*, 1990).

With a single exception, the young leaves of *Celtis mildbraedii*, leaves were exclusively eaten by gorillas.

The leaves of *C. mildbraedii* taste like okra, suggesting high protein content and were the only foliage food that the chimpanzees ate intensively. Chimpanzees concentrated on *Celtis* leaves when they were available in the lesser fruiting season. In September 1990, chimpanzees ate figs and *Celtis* leaves alternatively over the day's foraging and this pattern lasted over a week. Such intensive feeding on *Celtis* leaves was the primary cause of increased fiber scores in chimpanzee feces during the lesser fruiting season (see Figs. 6.2, 6.3). As noted above, the gorillas also ate these leaves, but they also ate tree bark in the lesser fruiting season. *Celtis mildbraedii* is one of the most dominant species in the mixed-species forest and most individual trees have a large enough canopy to sustain more than ten apes at a time. Thus, its young leaves are super-abundant and provide large food patches when available.

Most of the remaining fibrous food classes were shoots and pith of THV (terrestrial herbaceous vegetation; Wrangham *et al.*, 1991; Wrangham *et al.*, Chapter 4) and roots and stems of AHV (aquatic herbaceous vegetation). Some THV species were shared by both chimpanzees and gorillas, but the overlap was small. Both fecal analysis and feeding remains suggested that chimpanzees' intake of these food classes was small (Suzuki & Nishihara, unpublished data). On the contrary, gorillas ate large amounts of these foods (see Fig. 6.3).

THV and AHV used by gorillas were classified into two categories: first, those eaten throughout the year; and second, those eaten mainly during the lesser fruit-ing season, although these too were available in large amounts throughout the annual cycle.

The first type of herbaceous vegetation is composed of roots and stems of *Hydrocharis* sp. in swamp clearings and shoots of *Haumania danckelmaniana* on dry land, which are relatively easily digested and thus difficult to find in feces (Nishihara, 1995). These roots and stems are rich in minerals and proteins (Calvert, 1985; Nishihara, 1995), which may be why gorillas feed so intensively on them (see Table 6.2). Thus, these can be seen as high-quality THV and AHV because of their nutritional value. Chimpanzees were observed eating small amounts of *Haumania* but not *Hydrocharis* in the swamp. Shoots of *Megaphrynium* are also consumed year-round by gorillas on the west side of the Ndoki River, but this species is rare in the study site. It should be noted that high-quality THV foods are not so abundant and thus are not fallback foods. Shoots of *Haumania* and *Megaphrynium* were distributed widely but sparsely, even in the area where these species are dominant and shoots were fewer during the dry season, i.e. the period of fruit scarcity.

The second type of herbaceous vegetation comprises pith of *Palisota* spp., *Aframomum* spp., *Renealmia* spp., *Costus afer* and other Marantaceae such as *Marantoch-loa* spp. These can be seen as low-quality THV and AHV because of their reduced nutritional value (see Table 6.2). This may be why they are eaten only during the lesser fruiting season, despite their perma-nent abundance. The gorilla feeding pattern on low-quality THV and AHV corresponds to that of foods categorized as fallback foods in other sites (Tutin *et al.*, 1991; Wrangham, *et al.*, Chapter 4).

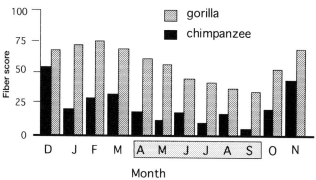

Fig. 6.3. Fiber scores of fecal samples of chimpanzees and gorillas, monthly means, Dec. 1991–Nov. 1992.

Table 6.2. *Nutritional content of major THV and AHV species*

Species	Part eaten	Fe (ppm)	Na (ppm)	Protein	Lipid	Reference
Haumania danckelmaniana	shoot	95	68	35.5	3.7	This study
Hydrocharis sp.	leaf	549	361	19.5	3.3	
	stem	728	256	9.7	1.7	
	root	7280	1080			
Palisota ambigua	pith	54	185	10.3	1.7	Calvert (1985)
Hypselodelphis sp.	shoot	87	115	13.8	3.2	
Megaphrynium macrostachyum	shoot	97	40	10.5	2.4	
Aframomum spp.	stem	92	200	3.8	1.6	

Protein and lipid, by dry matter %. Fe and Na analyzed by K. Kanaya, protein and lipid by R. Azato.

Bark was eaten almost exclusively by gorillas. They fed on the bark of *Celtis mildbraedii* most often, while chimpanzees rarely ate the bark of this species.

Seasonal variation and dietary diversity

According to monthly changes in fiber scores (see Fig. 6.3), gorillas generally ate more fibrous foods than chimpanzees, and both ate more fibrous foods during the lesser fruiting season. This pattern is also seen in Lopé and Hokou (Tutin & Fernandez, 1993; Remis, 1994). Monthly change in fiber scores for gorillas was negatively correlated with fruit availability; this was estimated from the number of fruiting trees and vines within a transect (Spearman rank correlation coefficient, $r_s = -0.74$, $p < 0.05$), as was the case in Lopé (Tutin *et al.*, 1991). For chimpanzees, there was no clear tendency ($r_s = -0.44$, $p > 0.15$).

The mean number of different fruit species found in fecal samples reflects the diversity of the fruit diet (Tutin *et al.*, 1991; Kuroda, 1992; Remis, 1994). Gorillas in Ndoki fed on more fruit species in the fruiting season than in the lesser fruiting season (see Fig. 6.4), a pattern seen at other sites (Tutin *et al.*, 1991; Remis, 1994; Nishihara, 1995). For chimpanzees, however, the number of fruit species found in feces was more constant, so it exceeded that of gorillas in the lesser fruiting season.

Monthly variation in the number of plant species bearing fruits along a transect correlated with the number of fruit species in gorilla feces ($r_s = 0.71$, $p < 0.01$). But for chimpanzees these numbers were not correlated ($r_s = 0.24$, $p > 0.45$), given that the mean

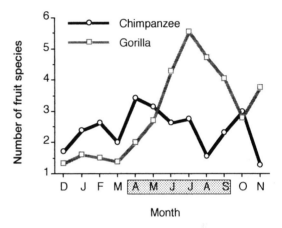

Fig. 6.4. Number of different fruit species per fecal sample of chimpanzees and gorillas, monthly means, Dec. 1991–Nov. 1992.

number of fruit species per chimpanzee fecal sample was relatively constant through the seasons.

Animal food

The food niches of gorillas and chimpanzees are notably separated by the types of animal food that each species selects. The frequency of chimpanzee predation upon mammals and birds in Ndoki seems to be as high as in Mahale and Tai (Kuroda *et al.*, unpublished data), but there has been no evidence of predation on vertebrates by gorillas. In three observations of predation, chimpanzees ate all or most of the prey, including bones. Every fecal sample ($N = 8$) that included vertebrate tissues contained bone fragments. Predation apparently provides chimpanzees with many minerals as well as with protein.

Insect-eating by gorillas and chimpanzees occurs throughout the year and termites are their main prey. Their termite-eating habits differ in terms of prey species and feeding technique (Kuroda, 1992; Nishihara, 1992; Suzuki *et al.*, 1995). Gorillas usually fed on *Cubitermes heghi* by breaking open their nests by hand, but it seemed that gorillas ate only small amounts of termites, as they comprised far less than 1% of the undigested materials in their feces. The percentage of feces containing termites during the dry season (January to March) was much lower than that during the rainy season (Suzuki *et al.*, 1995). We have no evidence of gorillas in the main study area eating *Macrotermes muelleri*, but the population on the west side of the Ndoki River is known to eat them intensively when termites swarm (Kuroda, unpublished data).

Chimpanzees use a tool set of perforating stick and fishing brush to get *Macrotermes muelleri* termites from their underground nests. Although these termites showed a marked difference in foraging activity between the dry and rainy seasons, such seasonal variation was not detected in the occurrence of termites in chimpanzee feces. This may be the result of the perforating technique that allows chimpanzees to obtain termites from underground anytime (Suzuki *et al.*, 1995). Feces containing *Macrotermes* accounted for 50% of all fecal samples ($N = 214$) collected during the main study period. Sometimes feces contained *Macrotermes* at levels of more than 5% in volume of whole undigested material. Another important prey insect for chimpanzees is the driver ant (*Dorylus* sp.), and occasionally these ants comprised more than 20% of washed feces.

Weaver ants (*Oecophylla* sp.), which are commonly eaten by gorillas and chimpanzees in Lopé (Tutin & Fernandez, 1992), were rarely found in gorilla feces (Nishihara, unpublished data) and never in chimpanzee feces in Ndoki. Earthworms were eaten only by gorillas (Nishihara & Kuroda, 1991); this is also seen in eastern lowland gorillas in Kahuzi-Biega (Yamagiwa *et al.*, 1992).

Swamp use by gorillas

The importance of swamp vegetation in primate ecology, including that of western lowland gorillas, has been shown by several researchers (Oates, 1978; Uehara, 1988; Williamson, 1988; Fay *et al.*, 1989; Fay & Agnagna, 1992). Gorillas in Ndoki frequently foraged on swamp vegetation, especially to feed on roots of *Hydrocharis* sp. (Mitani *et al.*, 1993; Nishihara, 1995). Most swamp observations of gorillas were made when they were feeding on *Hydrocharis* and this occurred regularly through the seasons (Nishihara, 1995).

Other AHV, including *Aframomum* spp. and fruits, for example of *Nauclea* spp., *Mytragia* sp. and *Landolphia* spp., are also eaten by gorillas in the swamp, but chimpanzees only rarely entered into the swamp to forage (Mitani *et al.*, 1993; Suzuki & Nishihara, unpublished data). We have no evidence that chimpanzees eat *Hydrocharis*. This species is found in large patches in deep swamps and is impossible to feed on without sitting in the water, a habit that seems to be avoided by these chimpanzees. Thus, swamp vegetation, occupying 15–20% of the study area, is used almost exclusively by gorillas.

Co-feeding during lesser fruiting season

During the lesser fruiting season, gorillas depended upon fibrous foods of THV, AHV and *Celtis mildbraedii* (see Fig. 6.3). However, they also ate large amounts of fruits of *Duboscia macrocarpa*, *Ficus* spp. and *Drypetes* spp. Chimpanzees increased their dependency on fruits of *Ficus* and *Duboscia*, which were available year-round but not eaten much during the fruiting season (Suzuki & Nishihara, unpublished data). This pattern was observed at Lopé, and these fruit species were classified as keystone or fallback foods (Tutin & Fernandez, 1993). Thus, fig intake increased in relative importance for both species of ape in the lesser fruiting season. As well as *Ficus* spp., five more fruit species, including *Duboscia macrocarpa* and *Drypetes* spp., were shared by these apes during the lowest fruiting months, though the total fruit score of gorillas was much lower than that of chimpanzees (see Fig. 6.4). This indicates that these apes share important fruit foods even during periods of fruit scarcity.

Co-feeding on figs by gorillas and chimpanzees was seen four times during the lesser fruiting season at Ndoki (Suzuki & Nishihara, 1992). M. Fay (personal communication) also observed co-feeding on figs at

Ndakan, 20 km west of our main study site. In three cases of co-feeding in Ndoki, gorillas came to the fig trees after chimpanzees. In another case, although a female gorilla came earlier than the chimpanzees, she retreated when she saw the observer and came back after the chimpanzees arrived in the fig tree, as if she was encouraged by the chimpanzees' presence. This may have also shown her preference for figs. In all cases, both species approached each other to within 7–8 m, but neither overt interaction nor apparent sign of competition over figs was observed. There were some signs that chimpanzees were a bit frightened, but the chimpanzees did not flee from the gorillas (Suzuki & Nishihara, 1992).

In all cases of co-feeding at Ndoki, differences in feeding patterns between the apes were clearly distinguished (Suzuki & Nishihara, unpublished data). Chimpanzees at fig trees seemed to concentrate on eating figs and repeatedly visited the same fig trees, but gorillas did not. Only 38% of gorilla feces ($N = 55$) collected around fig trees contained fig tissue. In two cases that occurred over two consecutive days, the numbers of gorillas seen in a tree ($N = 4$, $N = 4$) differed from the numbers of bed counts ($N = 25$, $N = 30$). It appeared that the gorilla group had split and that many individuals foraged in other places, despite the fact that trees were large enough (as the canopies of all these figs were about 30 m in diameter) to provide many group members with feeding places. Chimpanzee feces, being fragmented after being dropped from the canopy, contained figs only, while the fig tissue contained in gorilla feces, if any, accounted for less than 50% of total matter. This suggests that gorillas did not have as 'strong' a preference for fig-eating as chimpanzees did. This tendency was also reflected in the different lengths of time the two ape species remained in the fig trees during the observers' 'hide and wait' observations (Suzuki & Nishihara, unpublished data); chimpanzees were seen in fig trees for an average of 107 minutes ($N = 13$), whereas gorillas stayed an average of only 19 minutes ($N = 7$).

Other sightings of close proximity between the two species of ape were few, but in all cases we saw no signs of competition. In one case, a mother–offspring group of chimpanzees was observed feeding on *Dialium* fruits up in the branches, 20 m off the ground, while a gorilla group rested and played underneath. It was at the end of *Dialium* season when both apes concentrated on these fruits. A large broken branch in the same tree suggested that gorillas had also fed on it within one or two days. When a silverback roared, some chimpanzees glanced down but did nothing more. In another case, when 'wrah' calls of chimpanzees were heard 50 m away, a juvenile gorilla near the observer ran towards them but soon came back. During a co-feeding episode, chimpanzees and gorillas made bed sites of 10 and 30 beds, respectively, at distances of less than 50 m apart, both near the fig tree. The distance was short enough for them to know of each other's presence. During the ensuing days, the gorillas and chimpanzees climbed up into the same fig tree. They ate and otherwise acted in their own interest without interfering with each other.

DISCUSSION

General features

The diets of gorillas and chimpanzees at Ndoki are composed of various plant and animal foods, like those at Lopé (Williamson et al., 1990; Tutin & Fernandez, 1993; Tutin et al., 1994). The diet of chimpanzees and its seasonal variation differs little from those reported at other sites (Wrangham, 1977; Nishida & Uehara, 1983; Goodall, 1986; Tutin & Fernandez, 1993). They are essentially frugivorous, complemented by foliage when fruits are scarce and by a regular intake of insects and meat. On the contrary, the variety and flexibility of the gorilla diet far exceeds that of the folivorous mountain gorillas reported in detail by Fossey & Harcourt (1977), Goodall (1977) and Watts (1984). In the fruiting season, Ndoki's lowland gorillas eat a larger variety of fruits than do the chimpanzees, and gorillas regularly eat insects and earthworms. Tutin et al. (1991) described gorillas and chimpanzees at Lopé to be 'generalized, opportunistic frugivor[es]' and this is also true for the Ndoki population.

Dietary overlap between these apes at Ndoki varies according to food class. All major fruit foods are shared by both species, but animal foods are distinct. Most of the foliage foods, including high-quality AHV and THV species that contain high amounts of minerals and protein, are exclusively eaten by gorillas, with the exception of the young leaves of *Celtis mildbraedii*. Since animal food can generally be categorized as protein and mineral food, we can simply say that fruits are shared, whereas protein and mineral foods are not. Thus, the central issue for understanding the relationship between chimpanzees and gorillas concerns their fruit-eating habits.

Fruit-eating opportunists and pursuers

The relative amount of fruits (indicated by 100% fiber score) and the diversity of fruit foods in gorilla feces are correlated with fruit availability and the number of fruiting species, respectively. Such correlations appear when mammals have the following feeding strategy: eat fruits when encountered and do not stay at one feeding place for long. This strategy is characterized by opportunism, low selectivity and low insistence on a particular fruit species. This strategy is feasible if complemented by fibrous foods, in large amounts when necessary.

These correlations were not found in the chimpanzees at Ndoki. This may be caused partly by the small number of fecal samples (214 samples over the main study period). However, observations of chimpanzee fruit-eating, in contrast to gorilla observations, suggest that these correlations, if present, are low. Chimpanzees repeatedly visited the same fruit trees or stayed long and ate a great deal of fruit (Suzuki & Nishihara, unpublished data; Fig. 6.5). During the fruiting season they selected a small number of highly preferred fruit species, such as *Landolphia* spp. and *Dialium* spp. There was a low diversity of different fruit species in their feces. During the lesser fruiting season, selectivity on fruit

Fig. 6.5. Adult male chimpanzee rests between bouts of eating in Ndoki. (Photo by S. Kuroda)

species decreased but still they consistently ate fruits. Species that employ this strategy are characterized as being pursuers, with high selectivity and concentration on fruits found.

The gorillas' feeding strategy on fruits is curious from the viewpoint of optimal foraging theory. That is, they leave fruiting trees when there is still much fruit remaining. This was typically seen in the gorillas' short stays in fig trees during periods of fruit scarcity. Such a feeding pattern is also seen among eastern lowland gorillas during the fruiting season, i.e. gorillas usually leave fruiting trees after short feeding bouts and do not come back for several weeks (Yamagiwa *et al.*, 1994).

Short stays at a fruiting tree were also seen when gorillas ate the fruits of *Dialium* at Ndoki. Though the gorillas strongly preferred this fruit and its seeds, which sometimes occupied up to 60% of gorillas feces during the fruiting season, no lengthy feeding bouts at a *Dialium* tree were seen. Instead, gorillas fissioned into smaller aggregations, spreading out over more than 50 m. We could hear them eating *Dialium* at one tree for about ten minutes before moving on to another. Some members also ate other foods, such as the shoots of *Haumania*. Short stays were not due to small-sized *Dialium* trees, because they not only left many fruits but also did not repeatedly use even the same large trees over several consecutive days. *Dialium pachyphyllum* is a common species at Ndoki and reaches maturity at 10 cm dbh. Although most fruiting trees were between 10–25 cm in diameter, several trees were often distributed within 10 m or so (Kuroda & Nishihara, unpublished data). Thus, although gorillas fed on *Dialium* selectively, their feeding pattern was somewhat opportunistic (Fig. 6.6).

Because of such a feeding strategy, it appears that damage by gorillas' fruit-eating to chimpanzees' food resources is not severe. When gorillas come to a fig tree occupied by chimpanzees, chimpanzees probably know they will not stay long and remain indifferent to their presence while continuing to eat.

Feeding patterns on THV and fruits of gorillas

Why do gorillas feed on fruits so opportunistically? This feeding pattern may reduce intragroup competition over fruits because of low insistence on particular fruits. However, avoidance of intragroup competition does not entirely explain this pattern, because it is not solely seen

Fig. 6.6. Adult male gorilla high in a tree at Ndoki. (Photo by S. Kuroda)

in fruit-eating, but is also a general feeding pattern, seen even when gorillas eat low-quality and abundantly distributed THV, such as *Palisota* and *Aframomum*.

The pattern is adaptive when they eat unripe fruits and mature foliage, because they avoid the accumulation of great amounts of particular secondary compounds present in these items. However, this too does not adequately explain the feeding pattern because such items are not their main food.

The opportunistic pattern was also seen when gorillas ate *Hydrocharis*, which may provide them with indispensable minerals. *Hydrocharis* is found in large patches with large amounts of edible parts, but gorillas were not seen to stay at one patch for a long time. Rather, they foraged widely, splitting up into smaller groups. This may be adaptive feeding because when they forage on the plant's roots, they damage the whole plant, and thus by limiting their use of the area the whole patch is not destroyed. However, the distribution of *Hydrocharis* is limited to a swamp along the east side of the Ndoki River and is rare along the west side of the Ndoki River, where gorillas also generally show this feeding pattern.

The most likely food resources that might induce gorillas to show this pattern should be common and essential food types, that is, high-quality THV. Gorillas forage year-round in the mixed-species forest, feeding on

shoots of *Haumania* at Ndoki, and *Megaphrynium* on the west bank of the Ndoki River. Shoots of these species are not large and are distributed widely but sparsely, so that even where these species are dominant, gorillas are forced to split up and move around. They eat some and move, thus requiring a larger space and movement in a certain spread. They must pay attention to the movement of others to keep group cohesion, because gorillas have no vocalization to call to each other when a group forages in such a situation. This limits the amount of time an individual spends at one place if others move on. Eastern lowland gorillas also show the same feeding pattern on THV, small vines and fruits (Yamagiwa, personal communication).

In such situations, direct competition over food seems to occur rarely. If gorillas feed on fruits in the same pattern, it will function to keep intragroup competition low, and additionally it reduces the chance of direct interspecific competition with chimpanzees, as discussed above. It is also likely that gorilla intergroup competition is reduced by this pattern. However, direct intergroup competition over fruiting trees has been observed in Lopé (Tutin, Chapter 5). At Bai Mbeli, near our study site, several gorilla groups were reported to intermingle when they ate AHV (Olejniczak, 1994). These observations from Bai Mbeli support our hypothesis, but more data are needed to clarify this issue.

Gorillas at Ndoki and at Hokou have been reported to split up during foraging. This has been interpreted as an adaptation to fruit-eating (Mitani *et al.*, 1993; Remis, 1994). However, it may be better to explain this from the perspective of adaptation to high-quality HTV, because the degree of foraging group flexibility is much more restricted than the fission and fusion of chimpanzees. Furthermore, gorillas' opportunistic fruit-eating is not analogous to chimpanzees' persistent fruit-eating, which requires a fission and fusion of foraging groups to regulate intraspecific competition. Western lowland gorillas have a different feeding strategy that ought not to be simplified as being a quasi-chimpanzee strategy, even though they feed on many fruits.

ACKNOWLEDGMENTS

This paper was originally prepared for the Wenner-Gren conference 'The Great Apes Revisited,' held November 12–19, 1994 in Cabo San Lucas, Baja, Mexico. This study was financed by a grant from the Monbusho International Scientific Research Program (No. 02041046 to S. Kuroda). The Ministry of Science and Technology of the Republic du Congo permitted us to study in the Ndoki Forest. Drs J. Dinga-Reassi, A. Dinga-Makanda and M. Onanga supported our research in Congo. Drs J. M. Moutsamboté and M. Mitani cooperated with us in the field and exchanged useful information. Dr M. Fay and Mr M. Hatchwell of the Wildlife Conservation Society kindly helped and cooperated with us. To these people and institutions, we make grateful acknowledgment.

REFERENCES

Calvert, J. J. 1985. Food selection by western lowland gorillas (*G. g. gorilla*) in relation to food chemistry. *Oecologia*, **65**, 236–46.

Carroll, R. W. 1988. Relative density, range extension, and conservation potential of the lowland gorilla (*Gorilla gorilla gorilla*) in the Dzanga-Sangha region of southwestern Central African Republic. *Mammalia*, **52**, 309–23.

Fay, M. J. 1988. Partial completion of a census of the western lowland gorilla (*Gorilla gorilla gorilla*, Savage & Wyman) in southwestern Central African Republic. *Mammalia*, **52**, 203–15.

Fay, M. J. & Agnagna, M. 1992. Census of *Gorillas* in northern Republic of Congo. *American Journal of Primatology*, **27**, 275–84.

Fay, M. J., Agnagna, M., Moore, J. & Oko, R. 1989. Gorillas in the Likouala swamp forest of north central Congo. *International Journal of Primatology*, **10**, 477–86.

Fossey, D. & Harcourt, A. H. 1977. Feeding ecology of free-ranging mountain gorilla (*Gorilla gorilla beringei*). In: *Primate Ecology: Studies of Feeding and Ranging Behaviour in Lemurs, Monkeys and Apes*, ed. T. H. Clutton-Brock, pp. 415–47. London: Academic Press.

Goodall, A. G. 1977. Feeding and ranging behaviour of a mountain gorilla group (*Gorilla gorilla beringei*) in the Tshibinda-Kahuzi region (Zaïre). In: *Primate Ecology: Studies of Feeding and Ranging Behaviour in Lemurs, Monkeys and Apes*, ed. T. H. Clutton-Brock, pp. 449–79. London: Academic Press.

Goodall, J. 1986. *The Chimpanzees of Gombe: Patterns of Behavior*. Cambridge, MA: Harvard University Press.

Jones, C. & Sabater Pi, J. 1971. Comparative ecology of *Gorilla gorilla* (Savage & Wyman) and *Pan troglodytes* (Blumenbuch) in Rio Muni, West Africa. *Bibliotheca Primatologica*, **13**, 1–96.

Kuroda, S. 1992. Ecological interspecies relationships between gorillas and chimpanzees in the Ndoki-Nouabalé reserve, northern Congo. In: *Topics in Primatology, Vol. 2, Behavior, Ecology, and Conservation*, ed. N. Itoigawa, Y. Sugiyama, G. P. Sackett & R. K. R. Thompson, pp. 385–94. Tokyo: University of Tokyo Press.

Mitani, M. 1992. Preliminary results of the studies on wild western lowland gorillas and other sympatric diurnal primates in the Ndoki Forest, northern Congo. In: *Topics In Primatology, Vol. 2, Behavior, Ecology and Conservation*, ed. N. Itoigawa, Y. Sugiyama, G. P. Sackett & R. K. R. Thompson, pp. 215–24. Tokyo: University of Tokyo Press.

Mitani, M., Yamagiwa, J., Oko, R. A., Moutsamboté, J. M., Yumoto, T. & Maruhashi, T. 1993. Approaches in density estimates and reconstruction of social groups in the Ndoki forest, northern Congo. *Tropics*, **2**, 219–29.

Moutsamboté, J. M., Yumoto, T., Mitani. M., Nishihara, T., Suzuki, S. & Kuroda, S. 1994. Vegetation and list of plant species identified in the Nouabalé-Ndoki Forest, Congo. *Tropics*, **3**, 277–94.

Nishida, T. & Uehara, S. 1983. Natural diet of chimpanzees (*Pan troglodytes schweinfurthii*): long-term record from the Mahale Mountain, Tanzania. *African Study Monographs*, **3**, 109–30.

Nishihara, T. 1992. A preliminary report on the feeding habits of western lowland gorilla (*Gorilla gorilla gorilla*) in the Ndoki Forest, northern Congo. In: *Topics In Primatology, Vol. 2, Behavior, Ecology and Conservation*, ed. N. Itoigawa, Y. Sugiyama, G. P. Sackett & R. K. R. Thompson, pp. 225–40. Tokyo: University of Tokyo Press.

Nishihara, T. 1994. Population density and group organization of gorillas (*Gorilla gorilla gorilla*) in the Nouabalé-Ndoki National Park, Congo. *Afurika Kenkyu/African Area Studies*, **44**, 29–45 (in Japanese with English abstract).

Nishihara, T. 1995. Feeding ecology of western lowland gorillas in the Nouabalé-Ndoki National Park, northern Congo. *Primates*, **36**, 151–68.

Nishihara, T. & Kuroda, S. 1991. Soil scratching behavior by western lowland gorillas. *Folia Primatologica*, **57**, 48–51.

Oates, J. F. 1978. Water-plant and soil consumption by guereza monkeys (*Colobus guereza*): a relationship with minerals and toxins in the diet? *Biotropica*, **10**, 241–53.

Olejniczak, C. 1994. Report on a pilot study of western lowland gorillas at Mbeli Bai, Nouabalé-Ndoki Reserve, northern Congo. *Gorilla Conservation News*, **8**, 9–11.

Remis, M. J. 1993. Nesting behavior of lowland gorillas in the Dzanga-Sangha Reserve, Central African Republic: Implications for population estimates and understanding of group dynamics. *Tropics*, **2**, 245–55.

Remis, M. J. 1994. Feeding ecology and positional behavior of western lowland gorillas (*Gorilla gorilla gorilla*) in the Central African Republic. Ph.D. dissertation, Yale University.

Rogers, M. E., Maisels, F., Williamson, E. A., Fernandez, M. & Tutin, C. E. G. 1990. Gorilla diet in the Lopé Reserve, Gabon: A nutritional analysis. *Oecologia*, **84**, 326–39.

Suzuki, S. & Nishihara, T. 1992. Feeding strategies of sympatric gorillas and chimpanzees in the Ndoki-Nouabalé Forest, with special reference to co-feeding behaviour by both species. *Abstracts of the XIVth Congress of the International Primatological Society*, Strasbourg, France.

Suzuki, S., Kuroda, S. & Nishihara, T. 1995. Tool-set for termite fishing by chimpanzees in the Ndoki Forest, Congo. *Behaviour*, **132**, 219–35.

Tutin, C. E. G. & Fernandez, M. 1984. Nationwide census of gorilla (*Gorilla g. gorilla*) and chimpanzee (*Pan t. troglodytes*) populations in Gabon. *American Journal of Primatology*, **6**, 313–36.

Tutin, C. E. G. & Fernandez, M. 1985. Foods consumed by sympatric populations of *Gorilla g. gorilla* and *Pan t. troglodytes* in Gabon: some preliminary data. *International Journal of Primatology*, **6**, 27–43.

Tutin, C. E. G. & Fernandez, M. 1992. Insect-eating by sympatric lowland gorillas (*Gorilla g. gorilla*) and chimpanzees (*Pan t. troglodytes*) in the Lopé Reserve, Gabon. *American Journal of Primatology*, **28**, 29–40.

Tutin, C. E. G. & Fernandez, M. 1993. Composition of the diet of chimpanzees and comparisons with that of sympatric lowland gorillas in the Lopé Reserve, Gabon. *American Journal of Primatology*, **30**, 195–211.

Tutin, C. E. G., Fernandez, M., Rogers, M. E., Williamson, E. A. & McGrew, W. C. 1991. Foraging profiles of sympatric lowland gorillas and chimpanzees in the Lopé Reserve, Gabon. *Philosophical Transactions of the Royal Society of London, Series B*, **334**, 179–87.

Tutin, C. E. G., White, L. J. T., Williamson, E. A., Fernandez, M. & McPherson, G. 1994. List of plant species identified in the northern part of the Lope Reserve, Gabon. *Tropics* **3**, 249–76.

Uehara, S. 1988. Utilization patterns of a marsh grassland within the tropical rain forest by the bonobos (*Pan paniscus*) of Yalosidi, Republic of Zaïre. *Primates*, **31**, 311–22.

Watts, D. P. 1984. Composition and variability of mountain gorilla diets in the central Virungas. *American Journal of Primatology*, **7**, 323–56.

Williamson, E. A. 1988. Behavioural ecology of western lowland gorillas in Gabon. Ph.D. thesis, University of Stirling.

Williamson, E. A., Tutin, C. E. G., Rogers, M. E. & Fernandez, M. 1990. Composition of the diet of lowland gorillas at Lopé in Gabon. *American Journal of Primatology*, **21**, 265–77.

Wrangham, R. W. 1977. Feeding behavior of chimpanzees in Gombe National Park, Tanzania. In: *Primate Ecology: Studies of Feeding and Ranging Behaviour in Lemurs, Monkeys, and Apes*, ed. T. H. Clutton-Brock, pp. 503–38. London: Academic Press.

Wrangham, R. W., Conklin, N. L., Chapman, C. A. & Hunt, K. D. 1991. The significance of fibrous foods for Kibale Forest chimpanzees. *Philosphoical Transactions of the Royal Society of London, Series B*, **334**, 171–8.

Yamagiwa, J., Mwanza, N., Spangenberg, A., Maruhasi, T., Yumoto, T., Fisher, A., Steinhauser, B. B. & Refisch, J. 1992. Population density and ranging pattern of chimpanzees in Kahuzi-Biega National Park, Zaïre: A comparison with a sympatric population of gorillas. *African Study Monographs*, **13**, 217–30.

Yamagiwa, J., Mwanza, N., Yumoto, T. & Maruhasi, T. 1994. Seasonal changes in the composition of the diet of eastern lowland gorillas. *Primates*, **35**, 1–14.

7 • Dietary and ranging overlap in sympatric gorillas and chimpanzees in Kahuzi-Biega National Park, Zaïre

JUICHI YAMAGIWA, TAMAKI MARUHASHI, TAKAKAZU YUMOTO
AND NDUNDA MWANZA

INTRODUCTION

Most primates that live in tropical forests form sympatric populations with other primate species. In some cases they avoid contact or interact aggressively with other primate species, while in other cases they forage, groom, play or mate with others, forming mixed-species groups (Struhsaker, 1978; Gautier-Hion, 1980; Terborgh, 1983). Since these interactions between species are difficult to see in the wild, details of interference competition among sympatric primates are sketchy. However, at least ecological (exploitation) competition can be estimated when two sympatric species show similarities in resource use (Waser, 1987).

Gorillas and chimpanzees live sympatrically and with other primate species in wide areas of Equatorial Africa. Both apes use all strata of the tropical forest and range in various types of vegetation. Unlike *Cercopithecus*, *Cercocebus* or *Colobus* monkeys, these apes do not form mixed-species groups. Because of their larger body size, both apes may dominate the other primate species and be able to exclude them from food resources, even when niche overlap occurs extensively with them. Chimpanzees usually chase monkeys at aggressive encounters and occasionally prey on them (McGrew *et al.*, 1979; Nishida *et al.*, 1983; Goodall, 1986; Sugiyama & Koman, 1987; Boesch & Boesch, 1989), although some arboreal monkeys (such as *Colobus badius*) may aggressively mob and displace chimpanzees (Struhsaker, 1981; Ghiglieri, 1984). Gorillas and chimpanzees are genetically close and show similarities in morphological features such as craniodental form (Sarich & Cronin, 1977; Shea, 1983), yet they show remarkable differences in ecological features and social organization. Gorillas are regarded as folivores and usually form cohesive groups consisting of one male

and several females (Schaller, 1963; Fossey & Harcourt, 1977; Watts, 1984). Chimpanzees usually eat fruits and insects and form flexible (fission-and-fusion) groups based on male–male associations (Goodall, 1968; Nishida, 1979). These studies conducted in allopatric populations of gorillas and chimpanzees suggest a small niche overlap between them.

However, recent studies on sympatric populations of gorillas and chimpanzees in lowland tropical forests report extensive overlap in ranging and in composition of the diet. Both western and eastern lowland gorillas consume many kinds of fruits and regularly eat insects (Tutin & Fernandez, 1985, 1992; Yamagiwa *et al.*, 1991, 1994; Nishihara, 1992). Gorillas often climb high in trees for foraging and nesting, and range in various types of vegetation like chimpanzees (Mwanza *et al.*, 1992; Yamagiwa *et al.*, 1992a; Tutin & Fernandez, 1993). Such similarities in resource use suggest the existence of interspecific competition between gorillas and chimpanzees in lowland forests.

Interspecific competition increases when and where the main foods are scarce. Dietary overlap between sympatric primate species decreases during periods of food scarcity, and niche divergence is conspicuous during the dry season (Hladik, 1978; Gautier-Hion, 1980; Terborgh, 1983).

The present study shows ranging and dietary overlap between semi-habituated gorillas and chimpanzees of the Kahuzi-Biega National Park, Zaïre. Data presented here were mainly collected in a dry season when both kinds of ape concentrated their ranging in mixed primary and secondary forest. Their foraging patterns and the composition of their diets are described on a daily basis, and niche divergence between them is discussed in relation to interference and exploitation competition.

EASTERN LOWLAND GORILLAS: BROAD ECOLOGICAL NICHE AND SYMPATRY WITH CHIMPANZEES

The eastern lowland gorilla (*Gorilla gorilla graueri*) has been classified as a third subspecies of *Gorilla gorilla* (Corbet, 1967; Groves, 1967). Recent analyses on mitochondrial DNA D-loop and COII (cytochrome oxidase subunit II) sequences show distinct differences between the three subspecies (Garner & Ryder, 1992; Ruvolo, 1994) and suggest the possibility that the western and eastern populations might have evolved into separate species (Ruvolo *et al.*, 1994).

Eastern lowland gorillas appear to be intermediate between the other two subspecies in morphological and ecological features (Goodall & Groves, 1977; Yamagiwa *et al.*, 1994). Mountain gorillas (*Gorilla gorilla beringei*) that live in the higher montane forest of the Virunga Volcanoes are regarded as relatively specialized, terrestrial folivores. Western lowland gorillas (*Gorilla gorilla gorilla*) that live in the lowland tropical forest have a frugivorous diet and regularly eat social insects, such as termites and ants. Eastern lowland gorillas inhabit both highland and lowland tropical forests. A comparison of dietary composition indicates that the eastern lowland gorillas that range in the highlands show similarity to mountain gorillas; the diet of eastern lowland gorillas that range in the lowlands is similar to western lowland gorillas (see Fig. 7.1). More than 90% of plant foods do not overlap between the highland and the lowland habitats of eastern lowland gorillas and are not available in both places. Dietary breadth and frugivory of gorillas may reflect the intraspecific flexibility shaped by environmental constraints in their wide range of habitats in Equatorial Africa.

Eastern lowland gorillas occur in small isolated populations scattered over eastern Zaïre (Emlen & Schaller, 1960; Goodall & Groves, 1977; Mwanza & Yamagiwa, 1989; Wilson & Catsis, 1990; Hall & Wathaut, 1992). The Kahuzi-Biega National Park is one of the rare areas where gorillas are continuously distributed from lowland to highland tropical forests (Merz, 1989). The Park is located west of Lake Kivu and covers an area of 6000 km^2 at an altitude of 600–3308 m.

We have studied sympatric populations of gorillas and chimpanzees in the Itebero region (lowland) at an altitude of 600–1300 m from 1987 to 1991 and in the Kahuzi region (highland) at an altitude of 1800–3300 m from 1990 onwards. The two study regions are separated by about 100 km but are interconnected by a corridor of forest.

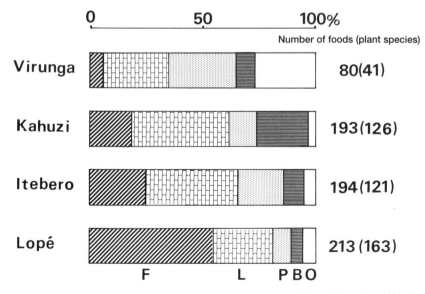

Fig. 7.1. Percentage of food items in the plant food species eaten by gorillas in four study areas. Virunga: *Gorilla gorilla beringei*, Watts, 1984; Kahuzi: *Gorilla gorilla graueri*, this study; Itebero: *Gorilla gorilla graueri*, Yamagiwa *et al.*, 1994; Lopé: *Gorilla gorilla gorilla*, Tutin & Fernandez, 1993. F, fruit or seed; L, leaf; P, pith, stem, twig; B, bark; O, others, such as flower, root, frond, etc.

Gorilla and chimpanzee ranges overlapped extensively in both regions, although gorillas ranged over wider areas than did chimpanzees (Yamagiwa *et al.*, 1992a, b). Gorillas constantly used all types of vegetation to forage, travel and rest, while chimpanzees rarely ranged in swamp and open vegetation.

A higher degree of dietary overlap between gorillas and chimpanzees was found in the lowland than in the highland habitats, but the overlap of fruit species was extensive in both habitats (Yamagiwa *et al.*, 1992a). Both consumed more kinds of fruits in the lowlands than in the highlands. However, differences in dietary composition between habitats were larger for chimpanzees than gorillas. Chimpanzees ate many kinds of fibrous plant parts when fruit was scarce in the highland habitat. These observations suggest that the highland habitat may be testing for frugivorous chimpanzees and predict higher exploitation competition between gorillas and chimpanzees in this habitat.

The population censuses of 1987 and 1990 estimated the apes' densities to be 0.27–0.32 gorillas/km^2 and 0.27–0.33 chimpanzees/km^2 in the Itebero region, and 0.43–0.47 gorillas/km^2 and 0.13 chimpanzees/km^2 in the Kahuzi region (Yamagiwa *et al.*, 1989, 1992b, 1993).

METHODS

The Kahuzi region covers 600 km^2 and is made up of bamboo forest (*Arundinaria alpina*) (37%), primary montane forest (28%) in the western and northern parts of the park, secondary montane forest (20%) in the eastern part, *Cyperus* swamp (7%) and other vegetation (8%), as described by A. Goodall (1977) and by Murnyak (1981). Dominant tree species in each vegetation type are *Podocarpus* sp., *Ficus* spp. and *Symphonia globulifera* in primary forest; *Hagenia abyssinica*, *Myrianthus holstii* and *Vernonia* spp. in secondary forest; *Hypericum lanceolatum* and *Rapanea pulchra* in *Cyperus latifolius* swamp (Casimir, 1975; A. Goodall, 1977).

The rainfall records of the meteorological station near Mt Kahuzi show an annual pattern that can be divided into four seasons: long rainy season, March–June; long dry season, June–September; short rainy season, September–December; and short dry season, December–March.

The 1990 census revealed 25 gorilla groups, 9 solitary male gorillas and 3 unit-groups of chimpanzees in the Kahuzi region (Yamagiwa *et al.*, 1992b, 1993). Since 1991 we have concentrated on the main study area around Tshibati; this comprises 30 km^2 along the western border of the park at an altitude of 2050–2350 m (see Fig. 7.2).

The census of the study reported here was made during the 1991 long dry season (July–September). Before starting, we found four groups of gorillas and a single unit-group of chimpanzees in the study area (see Table 7.1). Both ape populations were semi-habituated and occasionally tolerated the presence of human observers when we stayed at a distance of 20–50 m. Since both species of apes usually moved on the ground, we tried to find their fresh trails (up to 1 day old) and to follow them to the night bed-sites. When we encountered the apes, we carefully approached to watch their behavior but kept our distance so as not to disturb their activities. Party size was noted when we found apes feeding in trees. If gorillas or chimpanzees fled, we did not approach them again for at least an hour. We made 39 contacts (totaling 28 h 18 min) with gorillas and 47 contacts (31 h 35 min) with chimpanzees.

Fresh feeding signs and fecal samples were collected on the apes' trails. Since the size of dung segments is correlated with age in gorillas (Schaller, 1963), we measured the diameter of their fresh feces left on trails and in or beside beds. Beds of gorillas and chimpanzees were distinguished by shape and odor of feces that remained in and around them; gorilla feces form a ball, while chimpanzee feces are cylindrical. Ground beds with large amounts of feces were regarded as those of gorillas. In order to avoid overlapping of samples, we collected only one fecal sample from each night bed and a few fecal samples on the trails per day. We assigned the gorilla feces to four age-sex classes by size: infant, less than 3 cm in diameter; immature, 3–5 cm; subadult and adult, 5–7 cm; silverback male, more than 7 cm. The fecal samples of chimpanzees were assigned to three age classes: infant, less than 2 cm in diameter; immature, 2–4 cm; subadult and adult, more than 4 cm.

Feces collected in the field were sluiced in 1–mm mesh sieves, dried in sun-light and stored in plastic bags. The contents of each sample were divided into seeds, fruit skins, fiber, leaves, fragments of insects and other matter. Large seeds were counted. Small seeds, fiber and

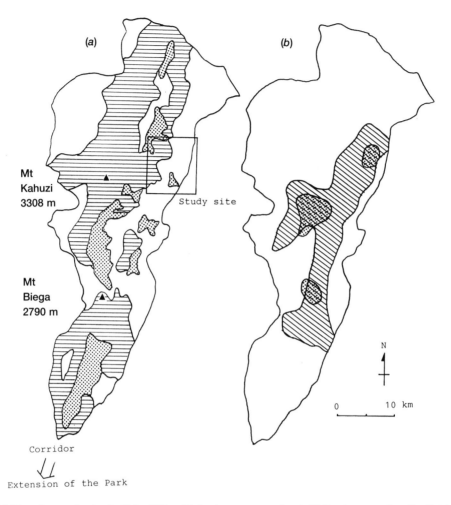

Fig. 7.2. (*a*) Map of vegetation in the Kahuzi-Biega National Park. Dotted area, *Cyperus* swamp; horizontal lined area, bamboo forests; white area within the park, primary or secondary montane forest. (*b*) Ranging areas of gorillas (hatched area) and chimpanzees (dotted area) during the bamboo season (September–November) in 1990.

leaves were rated as being abundant, common or rare. Samples of seeds and insects were preserved in 10% ethanol. Plant specimens were identified at the National Botanical Garden in Brussels.

RESULTS

Group composition and ranging

From direct observations and bed counts, we confirmed the age–sex composition of four gorilla groups and one chimpanzee unit-group (see Table 7.1). Gorillas usually formed cohesive groups, and almost all members of each group were counted during observations or bed counts. On the other hand, chimpanzees usually formed small parties and occasionally moved alone. Although the unit-group contained 22 individuals based on individual identifications, the mean party size was 4.5 chimpanzees ($N = 37$, range: 1–19) from direct observations. The mean number of bed-makers at a night bed site was 2.7 chimpanzees ($N = 43$, range: 1–9). Solitary beds were those found more than 50 m away from other beds. The largest party and the most beds found were in fig (*Ficus* spp.) trees.

Table 7.1. *Age-sex composition of the study groups at Kahuzi-Biega*

(a) Gorillas

	Silverback male over 13 yrs.	Blackback male 9–12 yrs.	Adult female over 9 yrs.	Subadult 7–8 yrs.	Immature 4–6 yrs.	Infant 0–3 yrs.	Total
G1	1		2	1	1		5
G2	1		4	2	3	2	12
G3	1		5	2	6	3	17
G4	2	1	6	4	4	3	20

(b) Chimpanzees

	Adult male over 15 yrs.	Young male 9–15 yrs.	Adult female over 12 yrs.	Young female 9–12 yrs.	Subadult 7–8 yrs.	Immature 5–6 yrs.	Infant 0–4 yrs.	Total
C1	5	3	3	3	2	4	2	22

Eleven of 14 solitary chimpanzees observed traveling were adult females and five of eight dyadic parties consisted of mothers with immatures or infants. The parties formed by more than ten chimpanzees included at least three adult males. These observations reflect the fission-and-fusion nature of the chimpanzee unit-group, the solitary nature of female chimpanzees in the Kahuzi region, and mirror reports for chimpanzees in other areas (Goodall, 1965; Reynolds & Reynolds, 1965; Nishida, 1968; Baldwin *et al.*, 1982; Boesch, Chapter 8).

The ranging areas of the study groups of gorillas overlapped with those of other non-habituated groups of gorillas in the north and south, while the study group of chimpanzees did not contact other groups of chimpanzees during the study period. No trail signs of chimpanzees were found between the study group of chimpanzees and its nearest neighbors; that chimpanzee group lived about 10 km away from the study site. On the other hand, the ranging areas of the four gorilla groups and the unit-group of chimpanzees extensively overlapped in the study area (see Fig. 7.3). Each gorilla group's range overlapped with those of at least two other gorilla groups, and the chimpanzee unit-group did so with all four gorilla groups. The size of the ranging area of gorilla groups (2.7–6.5 km²) was similar to that of chimpanzees (5.3 km²) during the study period. The mean distance between bed sites on consecutive nights for chimpanzees (1365 m, $N = 11$, range: 485–3037 m) was longer (Mann-Whitney U test: $z = 2.15$, $p = 0.03$, 2-tailed) than that of gorilla groups (813 m, $N = 31$, range: 154–2280 m).

In contrast, types of vegetation visited differed between the two species of ape. The home range of each gorilla group included all types of vegetation, and each group visited the bamboo forest and *Cyperus* swamp at least twice for feeding and bed-building. The home range of the chimpanzees also included the swamp, but it is likely from their trail signs that they did not visit the bamboo forest and only passed through the swamp, without feeding.

Gorilla groups tended to use a small area for a few days and then to travel relatively long distances to another area. They usually traveled slowly together, eating ground foliage and occasionally visiting fruiting trees, such as *Myrianthus* spp. or *Syzygium* spp. Several group members climbed these trees to eat the fruits, while other members fed on bark or leaves in trees, or collected fallen fruits off the ground. They stayed around the fruiting trees for only 10–30 minutes and did not visit the same tree again for 1–2 weeks. Chimpanzees also traveled on the ground, but they moved more rapidly than gorillas between fruiting trees, and only occasionally ate ground foliage. Each morning a few chimpanzees appeared in particular fig trees, made hooting vocalizations, ate fruits and were joined by others. They sometimes stayed in the trees for hours, feeding and resting, and visited the same trees more than twice a day. Moreover, they often

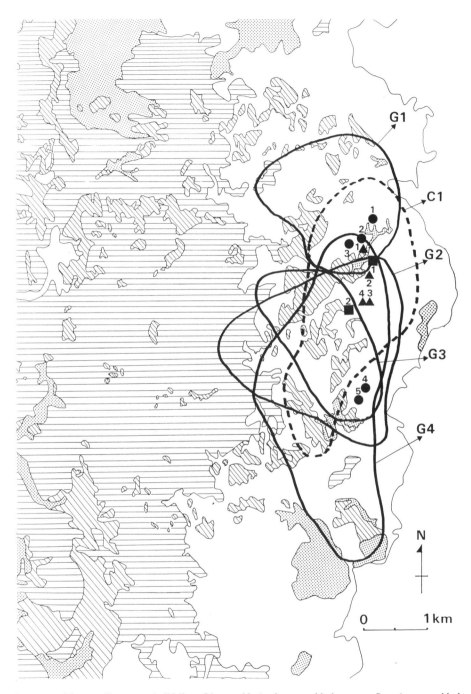

Fig. 7.3. Ranging areas of four gorilla groups (solid line: G1, G2, G3 and G4) and one chimpanzee unit-group (dotted line: C1) during the dry season (July–September) in 1991. Fruiting trees marked and periodically visited by observers: black circle, *Myrianthus* spp.; black square, *Syzygium* spp.; black triangle, *Ficus* spp. Dotted area, *Cyperus* swamp; horizontal lined area, bamboo forests; hatched area, primary forests; white area, secondary forests.

Table 7.2. *Number of species per plant form and per plant part eaten by gorillas and chimpanzees during the long dry season in the Kahuzi region*

	Gorillas only	Number of species eaten by both gorillas and chimpanzees	Chimpanzees only
Plant form			
Tree	3	12	10
Shrub	7	10	7
Vine	12	8	6
Herb	5	8	2
Others	5	9	3
Total	32	47	28
Plant part			
Fruit/Seed	0	20	18
Leaf	34	17	14
Pith/Stem/Twig	12	11	8
Bark	24	5	1
Others	4	2	3
Total	74	55	44

Note: Others: (Plant form), Ensete banana, epiphyte, fern, grass, mushroom, lichens, tree fern; (Plant part), flower, frond, gum, petiole, root, rotten wood.

visited the same fruiting trees for weeks; one fig tree was revisited daily for 17 days, as was a *Syzygium* tree for eight days. In general, gorillas are terrestrial folivores and opportunistic frugivores, while chimpanzees are consistent frugivores and persist in foraging on particular fruit resources.

Composition of diet

During the study period gorillas ate 129 types of food-item from at least 79 plant species, and chimpanzees ate 99 types of food-item from at least 75 plant species (see Table 7.2). More than half of the plant species were eaten by both apes (59% for plant species eaten by gorillas and 63% for plant species eaten by chimpanzees). About half of their food-plants were eaten by both apes (43% of gorilla foods and 56% of chimpanzee foods).

However, dietary overlap varied across food items. Vine species (plant forms) and leaf and bark species (plant parts) overlapped less between gorillas and chimpanzees than other food choices. Overlap in fruit or seed species was complete (100%) for gorillas while only partial (53%) for chimpanzees. For bark species, little overlap (17%) was found for gorillas, while it was extensive

(83%) for chimpanzees. In general, gorillas tended to eat many leaf and bark species that were not consumed by chimpanzees, and chimpanzees tended to eat many fruit or seed species that were not consumed by gorillas.

Gorillas tended to eat different parts of plant species. From ten (13%) plant species eaten, gorillas ate three or more different parts. They ate four different parts of a shrub (*Galiniera coffeoides*: fruit, leaf, pith and bark) and of two epiphyte species (*Englerina woodfordioides* and *Tapinanthus constrictiflorus*: fruit, leaf, bark and flower). Chimpanzees ate three parts from only four (5%) plant species. Gorillas ate a single part from 42 (55%) plant species and chimpanzees ate a single part from 53 (72%) species. Gorillas ate only leaves of 24 species and chimpanzees ate only fruits of 27 species.

Animal foods were rarely the same for gorillas and chimpanzees. During the study period, we collected 256 fresh gorilla feces and 394 fresh chimpanzee feces. Fecal analysis found fragments of insects (ants: *Crematogaster* sp.) in only two (0.1%) fecal samples of gorillas and remains of larval skins (possibly of a butterfly or moth) in only one (0.04%) fecal sample. In contrast, about 30% of chimpanzee fecal samples included fragments of insects: 10% had ants, 7% had bees, 5% had larval skins

Table 7.3. *Mean number of food species eaten by gorillas and chimpanzees per day*

(a) Plant form

| | No. of samples | No. of food species eaten | | | | | | |
		Tree	Shrub	Vine	Herb	Others	Animal	Total
G	12	4.8 (2–8)	1.3 (0–3)	5.9 (3–10)	1.0 (0–3)	0.6 (0–3)	0	13.3 (8–17)
C	11	10.6 (9–13)	0.5 (0–2)	2.5 (1–5)	0.9 (0–3)	4.2 (2–6)	2.2 (0–4)	20.8 (18–26)

(b) Plant part

| | No. of samples | No. of food species eaten | | | | | | |
		Fruit	Leaf	Pith	Bark	Others	Animal	Total
G	12	4.6 (3–8)	6.3 (3–11)	1.9 (0–4)	2.0 (1–5)	0.2 (0–1)	0	14.9 (9–21)
C	11	12.5 (10–16)	3.9 (2–6)	1.8 (0–5)	0	1.3 (0–2)	2.2 (0–4)	21.6 (18–27)

Note: G, gorilla; C, chimpanzee. Other: (Plant form), Ensete banana, epiphyte, fern, grass, lichens, mushroom, tree fern; (Plant part), flower, frond, gum, petiole, root, rotten wood. Animal: monkey (*Cercopithecus*), ant, bee, larva, etc.

and 14% had other insects, mostly beetles. Seven (2%) fecal samples of chimpanzees had hairs, teeth or cracked bones of *Cercopithecus mitis*.

Daily diet and frequency of fruit consumption

Fecal analysis may not indicate total dietary breadth or amount of food consumed, because gorillas usually swallowed all food items while chimpanzees sometimes chewed seeds, fruit skins or fibers and then discarded them as a wadge. Therefore, we combined data from fecal samples with those from feeding remains on apes' fresh trails, in order to compare their dietary breadth on a daily basis. These additional data came from day-long follows of fresh gorilla trails on 12 days and similar follows of chimpanzees on 11 days (see Table 7.3). The results show a wider food breadth in the daily menu of chimpanzees than that of gorillas, although the results may be biased by the differences in group or party size and composition between them. Gorillas used more kinds of vine for foods and chimpanzees used more kinds of tree, epiphyte and fern. Gorillas tended to eat more kinds of leaf and bark, while chimpanzees tended to eat more kinds of fruit, flower and root. Neither animal foods by gorillas, nor bark by chimpanzees were eaten during these sampled days.

Both gorillas and chimpanzees ate ripe fruits over the whole study period. Remains of at least one species of fruit were found in 96.5% of gorilla feces and in 99.7% of chimpanzee feces. However, the mean number of different fruit species per gorilla fecal sample was smaller than per chimpanzee fecal sample (see Table 7.4). Fecal samples of gorillas included less fruit and more leaves or fibers than those of chimpanzees.

Fruit remains of *Myrianthus* spp., *Syzygium* spp., *Ficus* spp. and *Bridelia bridelifolia* were often found in fecal samples of both gorillas and chimpanzees. Fruit remains of *Myrianthus* spp. were found in 83% of gorilla feces and in 52% of chimpanzee feces. However, the frequency of fruit consumption by both gorillas and chimpanzees varied over 10-day periods (see Fig. 7.4). For both apes, it was high in August but declined in September when availability of fruit dropped. For chimpanzees, the frequencies of eating *Syzygium* spp. and *Bridelia bridelifolia* were low during the study period, while those for gorillas sometimes increased after mid-August. The frequency of *Ficus* spp. in the diet was low for gorillas, but for chimpanzees it was always high during the study period. These data may reflect different preferences for particular fruit species in gorillas and chimpanzees. When available, fruits of *Myrianthus* spp. were eaten by both apes. However, since the seeds of *Syzygium* spp. were usually chewed and spat out by chimpanzees, the frequency of seeds in their feces may not represent the actual frequency of their eating; from direct observations, chimpanzees often ate the fruits of *Syzyg-*

Table 7.4. *Mean number of fruit or seed species and relative amount of fruit remains per fecal sample of gorillas and chimpanzees*

	No. of samples	No. of species Mean (range)	Relative amount of remains			
			−	+	++	+++
Gorillas	256	1.7 (0–5)	4%	35%	41%	20%
Chimpanzees	394	3.1 (0–7)	0	7	41	51

Note: Relative amount: −, none; +, rare; ++, common; +++, abundant.

ium spp. The four fruit species that were often eaten by both apes for at least a short period may lead to inter-specific competition.

Co-use of fruiting trees by gorillas and chimpanzees

On the basis of fresh feeding remains and feces, particu-lar fruiting trees (5 *Myrianthus* spp., 2 *Syzygium* spp. and 4 *Ficus* spp.) were marked and visited periodically, in order to verify the dates on which gorillas or chimpan-zees fed there (see Fig. 7.3). Gorillas, in cohesive groups, visited these fruiting trees on fewer than 20% of the total days monitored by the observers (see Table 7.5). They did not revisit the same trees more than twice on two consecutive days. In contrast, chimpanzees appeared individually in the fruiting trees and repeated their visits to the same trees for many days. Their fresh feeding signs were found on 11 (34%) days at a *Syzygium* tree and on 21 (60%) days at a *Ficus* tree.

In ten cases, we confirmed that gorillas and chimpan-zees visited the same fruiting tree on the same day. In three cases, we saw encounters at the fruiting trees. Twice at *Myrianthus* trees, chimpanzees approached the place where gorillas of G1 Group were eating fruits in the trees and on the ground. In both cases, gorillas left the trees in less than 20 minutes; this is within their usual range of duration of visits to *Myrianthus* fruiting trees. After a few minutes, one chimpanzee in one case and four chimpanzees in the other case came to the trees and started to eat fruits. In the case of a *Syzygium* tree, gorillas approached the tree in which five chimpanzees (adult male, young male, young female and adult female with an infant) were eating fruits. When chimpanzees noticed the presence of gorillas about 70 m from the tree, they stopped eating, glanced at the gorillas and then

began eating again after a few seconds without making any vocalization. The chimpanzees continued feeding for 13 minutes and then left the tree one by one. During the encounter, the gorillas did not approach the tree but stayed on the ground, forming a line of procession. After the chimpanzees left, the gorillas quickly moved to the tree, climbed up and fed on the fruits.

No antagonism was seen during these three encoun-ters. Although chimpanzees emitted pant-hoots several times and gorillas made cough barks, these sounds might have been made to their conspecifics or to the observers. In this context, the apes need not respond aggressively to each other (Fig. 7.5).

During two encounters between conspecific groups (G3 and G4), adult male gorillas showed aggressive pos-tures and vocalizations (explosive displays with barks and chest-beats), as usually seen in intergroup encounters (Yamagiwa, 1983). In contrast, the gorillas seemed to be calm during the encounters with chimpanzees. These observations suggest that both gorillas and chimpanzees avoid interspecific contacts and may respect the current occupation of fruiting trees by the other species.

DISCUSSION

The densities of gorillas generally have been found to be almost the same as, or slightly lower than, those of sym-patric chimpanzees in lowland tropical forests (Jones & Sabater Pi, 1971; Tutin & Fernandez, 1984, 1993; Yama-giwa *et al.*, 1989). Since gorillas are more than twice the body-size of chimpanzees, their densities should be lower than those of chimpanzees if both apes use the same food resources. In contrast, the density of gorillas is far higher than that of sympatric chimpanzees inhabiting the mon-tane forest of Kahuzi-Biega National Park (Yamagiwa *et al.*, 1992b, 1993). The low density of chimpanzees may

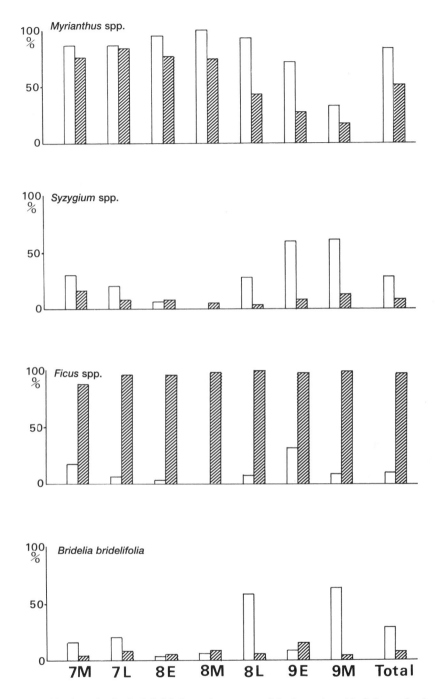

Fig. 7.4. Percentage of fecal samples that included fruit remains by species in each period. Months divided into three periods (E: Early, 1st–10th; M: Middle, 11th–20th; L: Later, 21st–end; 7, July; 8, August; 9, September). Each column shows percent-age of fecal samples with fruit remains in total number of samples collected in each period (white column, gorilla; hatched column, chimpanzees).

Table 7.5. *Number of days when gorillas and chimpanzees visited particular fruiting trees and their encounters at feeding sites*

| | Fruiting tree | | | | | | | | | | |
| | *Myrianthus* spp. | | | | | *Syzygium* spp. | | *Ficus* spp. | | | |
Social unit	1	2	3	4	5	1	2	1	2	3	4
G1	5	1	4			3		3			
G2				1	1	1	1		1		
G3			1	1			4				
G4				2	2		2	1	1	3	1
C1	1	4	4			4	11	4	5	21	6
Number of days observed	28	31	31	10	10	29	32	35	35	35	35
Use in the same day	G1:C1		G1:C1	G3:G4	G3:G4	G3:C1	G1:C1	G2:C1	G4:C1	G4:C1	G4:C1
Encounter observed	G1:C1		G1:C1	G3:G4	G3:G4	G3:C1					

Fig. 7.5. Adult male eastern lowland gorilla at Kahuzi-Beiga. (Photo by J. Yamagiwa)

be a result of their frugivorous diet and the low diversity of fruits in the area. The larger biomass of gorillas compared with sympatric chimpanzees may be due to the distinctive differences in diet between them.

However, the sympatric populations of gorillas and chimpanzees in lowland forests show extensive dietary overlap (Kuroda, 1992; Yamagiwa *et al.*, 1992a; Tutin & Fernandez, 1993). At Lopé Reserve in Gabon, most food types were eaten by both kinds of ape during an 8-year study (Tutin & Fernandez, 1993).

Fruit parts accounted for 55% of gorilla and 76% of chimpanzee plant food items, and the overlap of fruit species between them was greater than that of other food classes (79% for gorillas and 82% for chimpanzees). These results suggest that both species of apes strongly prefer fruits in general and tend to have common dietary features.

In the montane forest of Kahuzi, both kinds of ape also overlap in their diets. Gorillas ate more than half of the types of fruit eaten by the sympatric chimpanzees, and more than 95% of fresh fecal samples of both species included fruit remains. Both species of ape daily consumed one species of herb, two species of pith and more than three species of leaf, on average. However, the percentage of dietary overlap between them was smaller than that of apes in lowland forests. Fruit parts accounted for only 16% of gorilla and 38% of chimpanzee plant food items. Dietary overlap in fruit or seed species was 100% for gorillas but only 53% for chimpanzees. These observations suggest that the major differences in diet between the species may not be qualitative but quantitative, and that the seasonal changes in fruit availability greatly influence ape density.

High diversity of fruit during the dry season in montane forest

Tutin & Fernandez (1993) reported that the diets of gorillas and chimpanzees showed their greatest divergence at Lopé Reserve during the dry season when fruit was scarce. In contrast, both species of ape consumed the most kinds of fruit during the long dry season in the Kahuzi region. In 1993–94, we collected 2055 fecal samples of a gorilla group (G3) and 767 fecal samples of chimpanzees in the study area (mean monthly sample size = 118, range: 16–419). The diversity of fruit was highest for both apes during the long dry season (see Fig. 7.6). After this season, the number of fruit species consumed by gorillas dropped notably and stayed low in the other three seasons. The reduction in fruit species eaten was also seen in fecal samples of chimpanzees, but their diversity of fruit remained higher than that of gorillas in the other three seasons. These observations suggest that gorillas may change food selectivity (from fruit to foliage) after the long dry season while chimpanzees may not. This marked change of feeding strategy seems to be a common feature of eastern lowland gorillas both in highland and in lowland habitats. Gorillas decreased the length of their day-journeys when they consumed fewer kinds of fruit in both the Kahuzi and the Itebero regions (A. Goodall, 1977; Yamagiwa et al., 1992a; Yamagiwa & Mwanza, 1994). This is likely a folivorous strategy that minimizes the expenditure of energy at the expense of food quality (Schoener, 1971; Clutton-Brock, 1977).

Unlike the sympatric populations at Lopé, gorillas and chimpanzees in the Kahuzi region did not show the marked divergence in dietary niche during the period in which fruits were less available (unpublished data). Both apes continued to consume common plant foods and did not increase the range of different food items between the species. Quantitative differences in diet emerged between them: gorillas consumed large amounts of fibrous plant foods, while chimpanzees continued to select fruits and animal foods during the period of fruit scarcity.

Among species of fruit, figs were most often eaten by chimpanzees during this period. At least six species of *Ficus* were found by botanical survey and four of these were eaten by chimpanzees in the study area (Yumoto et al., 1994). Fig fruits are eaten by chimpanzees in various habitats across Africa (Reynolds & Reynolds, 1965; Sabater Pi, 1979; Nishida & Uehara, 1983; Goodall, 1986; Sugiyama & Koman, 1987; McGrew et al., 1988). They are less preferred than non-fig fruits and are often eaten by chimpanzees when non-fig fruits are in short supply (Conklin & Wrangham, 1994; Wrangham et al., Chapter 4). However, as shown in this study, fig seeds were also found in most fecal samples of chimpanzees during the long dry season and may be regarded as their major food over the whole year. Since the montane forest provides

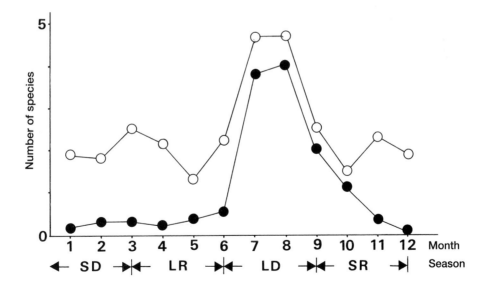

Fig. 7.6. Mean number of fruit species found in fecal samples of gorillas and chimpanzees in each month from 1993 to 1994. SD, short dry season; LR, long rainy season; LD, long dry season; SR, short rainy season (filled circle, gorilla; open circle, chimpanzee).

less diversity of fruits than other habitats over the whole year, fig fruits may not be the fallback food but instead may be the year-round preferred food in the Kahuzi region. Chimpanzees in this region also regularly ate social insects, such as bees and ants, and preyed on mammals. They used sticks for digging up underground nests of stingless bees to eat honey, larvae and nests (Yamagiwa *et al.*, 1988). Fig fruits and animal foods may complement their nutritional requirements under conditions of fruit scarcity.

Competition between gorillas and chimpanzees

Fruit is a limited food resource because of its patchy distribution and seasonal availability. Thus, fruit is the most likely cause of interspecific competition between sympatric primates. The study group of chimpanzees did not overlap with the ranging area of any other unit-group of chimpanzees during the study period. Four study groups of gorillas had overlapping ranging areas and two groups had aggressive encounters around fruiting trees of *Myrianthus* spp. These observations suggest strong intergroup antagonism in range use by chimpanzees and in the use of particular fruiting trees by gorillas. The Kahuzi gorillas also had aggressive encounters in bamboo forests when bamboo shoots were available (Yamagiwa, 1983). Tutin (Chapter 5) suggests that encounters between groups of western lowland gorillas appear to be related primarily to access to, or defense of, food resources, although aggressive encounters may also occur over acquisition or defense of females, as seen in mountain gorillas (Harcourt, 1978; Sicotte, 1993; Watts, 1994, Chapter 2). High intergroup antagonism in both apes, including adult killings and infanticides, are also reported in other areas (Goodall, 1977; Goodall *et al.*, 1979; Kawanaka, 1981; Fossey, 1983, 1984; Nishida *et al.*, 1985; Watts, 1989).

In contrast, interspecific antagonism seems relatively weak. Gorilla and chimpanzee groups had greatly overlapping home ranges, used the same fruiting trees and occasionally met each other at these trees, but no agonistic contacts were seen between gorillas and chimpanzees. These findings suggest that interference competition in the apes over food may be lower between species than between conspecifics.

Different foraging patterns may lower interference competition between the species. Gorillas usually formed cohesive groups, occasionally visiting fruiting trees for a brief time and revisiting them only after long intervals. Chimpanzees moved individually, often visited the same fruiting trees and stayed for a long time while feeding. Gorillas rarely ate fig fruits, while they were avidly eaten by chimpanzees. When interspecific encounters occurred around the fruiting trees of *Myrianthus* spp. and *Syzygium* spp., which were often eaten by both kinds of ape, they remained calm and seemed to avoid confrontation. Gorillas and chimpanzees may recognize the other species' foraging patterns as different from their own and may not regard it as a threat to their feeding activity. Each species of ape tolerated the other's foraging within the same area and seemed to respect the temporary ownership of fruiting trees by the other species. In a sympatric population of gorillas and chimpanzees at Ndoki, Congo, both apes were seen to co-feed on fig fruits in the same tree (Suzuki & Nishihara, 1992; Kuroda *et al.*, Chapter 6). Such co-feeding was not observed in this study. Fruits may constitute the most limited food resources in the Kahuzi region, and gorillas and chimpanzees may have developed mutual avoidance around their preferred fruiting trees to decrease the risk of serious fights (Fig. 7.7).

Fig. 7.7. Adult male eastern lowland gorilla, surrounded by youngsters, at Kahuzi-Biega. (Photo by J. Yamagiwa)

Opportunistic fruit-eating by folivorous gorillas and persistent fruit-eating by frugivorous chimpanzees may have developed from the worldwide changes in climate during the Tertiary and Quaternary periods. The tropical lowland forests, possibly the original habitats of both apes in the Miocene, may have permitted extensive niche convergence of the African great apes. The colder and drier climate in the Pleistocene reduced tropical forests and may have pushed the apes to develop different foraging strategies in refuge forests. Gorillas remained in the moist forests and increased herb and bark eating. This folivorous foraging strategy enabled gorillas to enlarge their range into highland forests. On the other hand, chimpanzees adopted a frugivorous strategy that increased the time spent searching for high-quality foods, such as fruits and animal matter. This study showed that the chimpanzees enlarged their daily food breadths and consumed various animal foods to complement their nutritional requirements under the conditions of fruit scarcity. Chimpanzees remained in the wooded areas, but their frugivorous and omnivorous strategy has enabled them to enlarge their range into arid areas (Nishida, 1976).

These differences in their respective foraging strategies may have resulted in a higher density of gorillas as compared with that of chimpanzees in montane forests. Past interspecific competition may have forced gorillas and chimpanzees to develop different foraging strategies, which have consequently enabled them to coexist sympatrically or allopatrically. Interference competition resulted in a smaller dietary overlap between them in montane forests than in tropical lowland forests, and exploitation competition resulted in a higher density of gorillas than that of chimpanzees in montane forests. This study suggests that interspecific competition may not prevent their sympatry but may still affect their diet and ranging patterns.

ACKNOWLEDGMENTS

This paper was originally prepared for the Wenner-Gren Conference 'The Great Apes Revisited,' held November 12–19, 1994 in Cabo San Lucas, Baja, Mexico. This study was financed by the Monbusho (Ministry of Education, Science and Culture, Japan) International Scientific Research Program (Nos. 630430810, 01041089) in cooperation with C.R.S.N. (Centre de Recherches en Sciences Naturelles) in Zaïre. We thank I.Z.C.N. (Institut Zairois pour Conservation de la Nature) for granting permission to carry out research in their National Park; Dr N. Zana, Dr B. Baluku, Dr M. M. Mankoto, Mr M. O. Mankoto, Dr B. Steinhauer-Burkart, Mr M. Mashagiro and Mr M. M. A. Wathaut for their administrative help and hospitality; and Mr B. Muphumbuko, Mr K. Ndumbo for their technical help. We are also greatly indebted to all guides, guards and field assistants at the Kahuzi-Biega National Park for their technical help and hospitality throughout the field work.

REFERENCES

Baldwin, P. J., McGrew, W. C. & Tutin, C. E. G. 1982. Wide-ranging chimpanzees at Mt. Assirik, Senegal. *International Journal of Primatology*, **3**, 367–85.

Boesch, C. & Boesch, H. 1989. Hunting behavior of wild chimpanzees in the Tai National Park. *American Journal of Physical Anthropology*, **78**, 547–73.

Casimir, M. J. 1975. Feeding ecology and nutrition of an eastern gorilla group in the Mt. Kahuzi region (Republíque du Zaïre). *Folia Primatologica*, **24**, 1–36.

Clutton-Brock, T. H. 1977. Some aspects of intraspecific variation in feeding and ranging behaviour in primates. In: *Primate Ecology*, ed. T. H. Clutton-Brock, pp. 539–56. London: Academic Press.

Conklin, N. L. & Wrangham, R. W. 1994. The value of figs to a hind-gut fermenting frugivore: a nutritional analysis. *Biochemical Ecology and Systematics*, **22**, 137–51.

Corbet, G. B. 1967. The nomenclature of the eastern lowland gorilla. *Nature*, **215**, 1171–2.

Emlen, J. T. & Schaller, G. B. 1960. Distribution and status of the mountain gorilla (*Gorilla gorilla beringei*). *Zoologica*, **45**, 41–52.

Fossey, D. 1983. *Gorillas in the Mist*. Boston: Houghton Mifflin.

Fossey, D. 1984. Infanticide in mountain gorilla (*Gorilla gorilla beringei*) with comparative notes on chimpanzees. In: *Infanticide: Comparative and Evolutionary Perspectives*, ed. G. Hausfater & S. B. Hrdy, pp. 217–36. New York: Aldine, Hawthorne.

Fossey, D. & Harcourt, A. H. 1977. Feeding ecology of free-ranging mountain gorilla (*Gorilla gorilla beringei*). In: *Primate Ecology*, ed. T. H. Clutton-Brock, pp. 415–47. New York: Academic Press.

Garner, K. J. & Ryder, O. A. 1992. Some applications of

PCR to studies in wildlife genetics. *Symposia of the Zoological Society of London*, **64**, 167–81.

Gautier-Hion, A. 1980. Seasonal variations of diet related to species and sex in a community of *Cercopithecus* monkeys. *Journal of Animal Ecology*, **49**, 237–69.

Ghiglieri, M. P. 1984. *The Chimpanzees of Kibale Forest: A Field Study of Ecology and Social Structure*. New York: Columbia University Press.

Goodall, A. G. 1977. Feeding and ranging behavior of a mountain gorilla group (*Gorilla gorilla beringei*) in the Tshibinda-Kahuzi region (Zaïre). In: *Primate Ecology*, ed. T. H. Clutton-Brock, pp. 450–79. New York: Academic Press.

Goodall, A. G. & Groves, C. P. 1977. The conservation of the eastern gorillas. In: *Primate Conservation*, ed. HSH Prince Rainier III & G. H. Bourne, pp. 599–637. New York: Academic Press.

Goodall, J. 1965. Chimpanzees of the Gombe Stream Reserve. In: *Primate Behavior*, ed. I. DeVore, pp. 425–37. New York: Holt, Rinehart and Winston.

Goodall, J. 1968. Behaviour of free-living chimpanzees of the Gombe Stream Reserve. *Animal Behaviour Monographs*, **1**, 161–311.

Goodall, J. 1977. Infant killing and cannibalism in free-living chimpanzees. *Folia Primatologica*, **28**, 259–82.

Goodall, J. 1986. *The Chimpanzees of Gombe: Patterns of Behavior*. Cambridge, MA: Harvard University Press.

Goodall, J., Bandora, A., Bergmann, E., Busse, C., Matama, H., Mpongo, E., Pierce, A. & Riss, D. 1979. Intercommunity interactions in the chimpanzee population of the Gombe National Park. In: *The Great Apes*, ed. D. A. Hamburg & E. R. McCown, pp. 13–53. Menlo Park, CA: Benjamin/Cummings.

Groves, C. P. 1967. Ecology and taxonomy of the gorilla. *Nature*, **213**, 890–3.

Hall, J. S. & Wathaut, W. M. 1992. A preliminary survey of the eastern lowland gorilla. WCI (Wildlife Conservation International)/IZCN (Institut Zairois pour Conservation de la Nature) Report 1992.

Harcourt, A. H. 1978. Strategies of emigration and transfer by primates, with particular reference to gorillas. *Zeitschrift für Tierpsychologie*, **48**, 401–20.

Hladik, C. M. 1978. Adaptive strategies of primates in relation to leaf-eating. In: *The Ecology of Arboreal Folivores*, ed. G. G. Montgomery, pp. 51–71. Washington, DC: Smithsonian Institution Press.

Jones, C. & Sabater Pi, J. 1971. Comparative ecology of *Gorilla gorilla* (Savage and Wyman) and *Pan troglodytes* (Blumenbach) in Río Muni, West Africa. *Bibliotheca Primatologica*, **13**, 1–96.

Kawanaka, K. 1981. Infanticide and cannibalism in chimpanzees – with special reference to the newly observed case in the Mahale Mountains. *African Study Monographs*, **1**, 69–99.

Kuroda, S. 1992. Ecological interspecies relationships between gorillas and chimpanzees in the Ndoki-Nouabalé Reserve, Northern Congo. In: *Topics in Primatology, Vol. 2, Behavior, Ecology, and Conservation*, ed. N. Itoigawa, Y. Sugiyama, G. P. Sackett & R. K. R. Thompson, pp. 385–94. Tokyo: University of Tokyo Press.

McGrew, W. C., Tutin, C. E. G. & Baldwin, P. J. 1979. New data on meat-eating by wild chimpanzees. *Current Anthropology*, **20**, 238–9.

McGrew, W. C., Baldwin, P. J., & Tutin, C. E. G. 1988. Diet of wild chimpanzees (*Pan troglodytes verus*) at Mt Assirik, Senegal: I. Composition. *American Journal of Primatology*, **16**, 213–26.

Merz, G. 1989. Etat actuel des elephants et des gorilles dans le parc national de Kahuzi-Biega, Zaïre. Rapport dans le cadre du projet 'IZCN – Conservation de la nature integrée au Zaïre – est'. Unpublished manuscript submitted to IZCN (Institut Zairois pour Conservation de la Nature).

Murnyak, D. F. 1981. Censusing the gorillas in Kahuzi-Biega National Park. *Biological Conservation*, **21**, 163–76.

Mwanza N. & Yamagiwa, J. 1989. A note on the distribution of primates between the Zaïre-Lualaba River and the African Rift Valley. *Grant-in-Aid for Overseas Scientific Research Report: Interspecies Relationships of Primates in the Tropical and Montane Forests* 1, pp. 5–10.

Mwanza, N., Yamagiwa, J., Yumoto, T. & Maruhashi, T. 1992. Distribution and range utilization of eastern lowland gorillas. In: *Topics in Primatology, Vol. 2, Behavior, Ecology, and Conservation*, ed. N. Itoigawa, Y. Sugiyama, G. P. Sackett & R. K. R. Thompson, pp. 283–300. Tokyo: University of Tokyo Press.

Nishida, T. 1968. The social group of wild chimpanzees in the Mahali Mountains. *Primates*, **9**, 167–224.

Nishida, T. 1976. The bark-eating habits in primates, with special reference to their status in the diet of wild chimpanzees. *Folia Primatologica*, **25**, 277–87.

Nishida, T. 1979. The social structure of chimpanzees of the Mahale Mountains. In: *The Great Apes*, ed. D. A. Hamburg & E. R. McCown, pp. 73–121. Menlo Park, CA: Benjamin/Cummings.

Nishida, T. & Uehara, S. 1983. Natural diet of chimpanzees (*Pan troglodytes schweinfurthii*): Long-term record from the Mahale Mountains, Tanzania. *African Study Monographs*, **3**, 109–30.

Nishida, T., Uehara, S. & Nyundo, R. 1983. Predatory behavior among wild chimpanzees of the Mahale Mountains. *Primates*, **20**, 1–20.

Nishida, T., Hiraiwa-Hasegawa, M., Hasegawa, T. & Takahata, Y. 1985. Group extinction and female transfer in wild chimpanzees in the Mahale Mountains. *Zeitschrift für Tierpsychologie*, **67**, 284–301.

Nishihara, T. 1992. A preliminary report on the feeding habits of western lowland gorillas (*Gorilla gorilla gorilla*) in the Ndoki Forest, northern Congo. In: *Topics in Primatology, Vol. 2, Behavior, Ecology, and Conservation*, ed. N. Itoigawa, Y. Sugiyama, G. P. Sackett & R. K. R. Thompson, pp. 225–40. Tokyo: University of Tokyo Press.

Reynolds, V. & Reynolds, F. 1965. Chimpanzees of the Budongo Forest. In: *Primate Behavior*, ed. I. DeVore, pp. 368–424. New York: Holt, Rinehart and Winston.

Ruvolo, M. 1994. Molecular evolutionary processes and conflicting gene trees: the hominoid case. *American Journal of Physical Anthropology*, **94**, 89–113.

Ruvolo, M., Pan, D., Zehr, S., Goldberg, T., Disotell, T. R. & von Dornum, M. 1994. Gene trees and hominoid phylogeny. *Proceedings of the National Academy of Sciences, USA*, **91**, 8900–4.

Sabater Pi, J. 1979. Feeding behaviour and diet of chimpanzees (*Pan troglodytes troglodytes*) in the Okorobikó Mountains of Rio Muni (West Africa). *Zeitschrift für Tierpsychologie*, **50**, 265–81.

Sarich, V. M. & Cronin, J. E. 1977. Generation length and rates of hominoid molecular evolution. *Nature*, **269**, 354.

Schaller, G. B. 1963. *The Mountain Gorilla: Ecology and Behavior*. Chicago: University of Chicago Press.

Schoener, T. W. 1971. Theory of feeding strategies. *Annual Review of Ecology and Systematics*, **2**, 369–404.

Shea, B. T. 1983. Size and diet in the evolution of African ape craniodental form. *Folia Primatologica*, **40**, 32–68.

Sicotte, P. 1993. Inter-group encounters and female transfer in mountain gorillas: Influence of group composition on male behavior. *American Journal of Primatology*, **30**, 21–36.

Struhsaker, T. T. 1978. Food habits of five monkey species in the Kibale Forest, Uganda. In: *Recent Advances in Primatology, Vol. 1*, ed. D. J. Chivers & J. Herbert, pp. 225–48. New York: Academic Press.

Struhsaker, T. T. 1981. Polyspecific association among tropical rainforest primates. *Zeitschrift für Tierpsychologie*, **57**, 268–304.

Sugiyama, Y. & Koman, J. 1987. A preliminary list of chimpanzees' alimentation at Bossou, Guinea. *Primates*, **28**, 133–47.

Suzuki, S. & Nishihara, T. 1992. Feeding strategies of sympatric gorillas and chimpanzees in the Ndoki-Nouabalé forest, with special reference to co-feeding behavior by both species. *Abstracts of the XIVth Congress of the International Primatological Society*, Strasbourg, France.

Terborgh, J. 1983. *Five New World Primates: A Study in Comparative Ecology*. Princeton, NJ: Princeton University Press.

Tutin, C. E. G. & Fernandez, M. 1984. Nationwide census of gorilla (*Gorilla g. gorilla*) and chimpanzee (*Pan t. troglodytes*) populations in Gabon. *American Journal of Primatology*, **6**, 313–36.

Tutin, C. E. G. & Fernandez, M. 1985. Food consumed by sympatric populations of *Gorilla g. gorilla* and *Pan t. troglodytes* in Gabon: some preliminary data. *International Journal of Primatology*, **6**, 27–43.

Tutin, C. E. G. & Fernandez, M. 1992. Insect-eating by sympatric lowland gorillas (*Gorilla g. gorilla*) and chimpanzees (*Pan t. troglodytes*) in the Lopé Reserve, Gabon. *American Journal of Primatology*, **28**, 29–40.

Tutin, C. E. G. & Fernandez, M. 1993. Composition of the diet of chimpanzees and comparisons with that of sympatric lowland gorillas in the Lopé Reserve, Gabon. *American Journal of Primatology*, **30**, 195–211.

Waser, P. M. 1987. Interactions among primate species. In: *Primate Societies*, ed. B. B. Smuts, D. L. Cheney, R. M. Seyfarth, R. W. Wrangham & T. T. Struhsaker, pp. 210–26. Chicago: University of Chicago Press.

Watts, D. P. 1984. Composition and variability of mountain gorilla diets in the central Virungas. *American Journal of Primatology*, **7**, 323–56.

Watts, D. P. 1989. Infanticide in mountain gorillas: new cases and a reconsideration of the evidence. *Ethology*, **81**, 1–18.

Watts, D. P. 1994. Gorilla mating tactics and habitat use. *Primates*, **35**, 35–47.

Wilson, J. R. & Catsis, M. C. 1990. A preliminary survey of the forests of the 'Itombwe' mountains and the Kahuzi-Biega National Park Extension, east Zaïre, July-September 1989. Project report for WWF project 3902.

Yamagiwa, J. 1983. Diachronic changes in tho eastern

lowland gorilla groups (*Gorilla gorilla graueri*) in the Mt Kahuzi region, Zaïre. *Primates*, **24**, 174–83.

Yamagiwa, J. & Mwanza, N. 1994. Day-journey length and daily diet of solitary male gorillas in lowland and highland habitats. *International Journal of Primatology*, **15**, 207–24.

Yamagiwa, J., Yumoto, T., Mwanza, N. & Maruhashi, T. 1988. Evidence of tool-use by chimpanzees (*Pan troglodytes schweinfurthii*) for digging out a bee-nest in the Kahuzi-Biega National Park, Zaïre. *Primates*, **29**, 405–11.

Yamagiwa, J., Maruhashi, T., Yumoto, T. & Mwanza, N. 1989. A preliminary survey on sympatric populations of *Gorilla g. graueri* and *Pan t. schweinfurthii* in eastern Zaïre. *Grant-in-Aid for Overseas Scientific Research Report: Interspecies Relationships of Primates in the Tropical and Montane Forests* 1, pp. 23–40. Inuyama: Kyoto University.

Yamagiwa, J., Mwanza, N., Yumoto, T. & Maruhashi, T. 1991. Ant eating by eastern lowland gorillas. *Primates*, **32**, 247–53.

Yamagiwa, J., Mwanza, N., Yumoto, T. & Maruhashi, T. 1992a. Travel distances and food habits of eastern lowland gorillas: a comparative analysis. In: *Topics in Primatology, Vol. 2, Behavior, Ecology, and Conservation*, ed. N. Itoigawa, Y. Sugiyama, G. P. Sackett & R. K. R. Thompson, pp. 267–81. Tokyo: University of Tokyo Press.

Yamagiwa, J., Mwanza, N., Spangenberg, A., Maruhashi, T., Yumoto, T., Fischer, A., Steinhauer, B. B. & Fischer, J. 1992b. Population density and ranging pattern of chimpanzees in Kahuzi-Biega National Park, Zaïre: A comparison with a sympatric population of gorillas. *African Study Monographs*, **13**, 217–30.

Yamagiwa, J., Mwanza, N., Spangenberg, A., Maruhashi, T., Yumoto, T., Fischer A. & Steinhauer, B. B. 1993. A census of the eastern lowland gorillas *Gorilla gorilla graueri* in Kahuzi-Biega National Park with reference to mountain gorillas *G. g. beringei* in the Virunga Region, Zaïre. *Biological Conservation*, **64**, 83–9.

Yamagiwa, J., Mwanza, N., Yumoto, T. & Maruhashi, T. 1994. Seasonal change in the composition of the diet of eastern lowland gorillas. *Primates*, **35**, 1–14.

Yumoto, T., Yamagiwa, J., Mwanza, N. & Maruhashi, T. 1994. List of plant species identified in Kahuzi-Biega National Park, Zaïre. *Tropics*, **3**, 295–308.

- PART III

Social relations

8 • Social grouping in Taï chimpanzees

CHRISTOPHE BOESCH

INTRODUCTION

Why do animals live in groups? Generally, animals are thought to live in groups in order to increase their survival and reproductive success. But group living incurs costs, mainly in terms of higher intraspecific competition and higher risks of parasite transmission. Therefore, the discussion has focused around the benefits that could make possible the evolution of sociality. Three main benefits have been proposed; group living decreases predator pressure (van Schaik, 1983; Dunbar, 1988), allows better exploitation and protection of food resources (Wrangham, 1980; Krebs & Davies, 1993), or results from the need for cooperative behavior, like hunting or communal breeding (Caraco & Wolf, 1975; Emlen, 1991). For more specific situations, other factors have been proposed such as rearing strategies (e.g. lions; Packer *et al.*, 1990). In most cases, observers study animals that already live in groups, making it difficult to investigate the factors that were originally responsible for the appearance of social groups. However, we can study the factors that affect the social groups that we observe in the wild.

Predation pressure has been shown to have a large impact on group size in many animals including species of fish, birds, ungulates and primates (Pulliam & Caraco, 1984; Kummer *et al.*, 1985; Cheney & Wrangham, 1987; Dunbar, 1988; Krebs & Davies, 1993). Similarly, seasonal variation in food supply has been shown to have a marked influence in the fragmentation of hamadryas, olive and gelada baboons and of spider monkeys (Kummer, 1971; Dunbar, 1988; Chapman *et al.*, 1995). However, it is generally thought that predation pressure plays a more important role, because if

you miss one meal, you can still eat at another time, but if you fail once to escape from a predator, you are dead (the so-called 'run for your life or run for a meal' principle).

All chimpanzee populations have been reported to live in a 'fission–fusion' social structure (Nishida, 1968; Kummer, 1971; Goodall, 1986). A 'community' includes all individuals that are regularly seen over months in the temporary subgroups, called 'parties'. The community is equivalent to the 'unit-group,' as used by researchers of the Mahale and Bossou chimpanzees (Nishida, 1968; Sugiyama & Koman, 1979). The fission–fusion system also has been observed in eight other species of large-bodied primates and is thought to allow greater flexibility in exploiting resource patches of different sizes in a species free from predation (gelada: Dunbar, 1988; hamadryas: Kummer, 1971; spider monkey: Terborgh & Janson, 1986; Symington, 1990). Proving that a species is free of predation is very difficult, first, because predation is a rare event and has to be so for the prey species to survive (Boesch, 1991). Second, the presence of human observers may directly interfere with predation when the prey is habituated to humans but the predator is not. For example, in a study of vervet monkeys, disappearances due to leopards were 3.6 times more frequent when human observers were absent than when the observers were following the monkeys (Isbell & Young, 1993). In Taï, Ivory Coast, recent observations on chimpanzees showed that predation by leopards was the main cause of mortality, at least during certain periods of time (Boesch, 1991). In Mahale, chimpanzees were preyed upon by lions and this led to a three-fold increase in the rate of disappearance of the chimpanzees (Tsukahara, 1993). Both Bossou and Assirik chimpanzees seem to react to higher predation pressure by forming larger parties (Tutin *et al.*, 1983; Sakura, 1994). Thus, species with a fission-fusion system suffer from predation and the net advantage of this system needs more investigation.

Many attempts have been made to single out the factor responsible for the social organization of chimpanzees. After analyzing two years of data from Gombe, Goodall (1986) proposed sex to be the single most significant factor in shaping grouping patterns. Wrangham (1986) proposed that greater dependence upon terrestrial herbacious vegetation (THV) by bonobos explained their larger party sizes relative to party size in chimpanzees. Subsequent data showed that neither population relied much on THV (Malenky & Stiles, 1991; Wrangham *et al.*, Chapter 4). Then, it was proposed that the size of food patches and their distribution could be the factor explaining such differences (White & Wrangham, 1988). Comparison between Lomako bonobos and Kibale chimpanzees did not support this hypothesis (Chapman *et al.*, 1994). Finally, it has been proposed that *Pan* social organization reflects an adaptation to extended periods of food shortage in which individuals are forced to be in small parties (Malenky, 1990, in Chapman *et al.*, 1994). Food shortage is supposed to be less of a problem in forest environments and thus chimpanzees living in such environments are expected to occur in larger parties.

However, to survive and to reproduce, animals have to solve many problems at once. Thus, we should expect that the solutions they adopt will reflect a balance between many needs. In this chapter, I shall compare the social grouping of the Taï chimpanzees with those of other chimpanzee populations and analyze the different factors that affect party size. However, in the present analysis, I shall exclude the effect of predation on party size due to the impossibility of controlling for its effect by having periods with different intensities of predation pressures. Therefore, I assume predation to have a constant effect on party size.

METHODS

In a fission–fusion system, party size and composition change often and many different methods have been used to study this fluctuation. Several problems emerge from these various methods. First, varying degrees of habituation of different chimpanzee populations have forced observers to use opportunistic sampling of parties, e.g. making a count when a party was seen in good conditions of visibility (Goodall, 1968; Nishida, 1968); 10-minute focal subject samples (Wrangham *et al.*, 1992); 1-hour samples (Doran, 1989); and multiple hour follows of target individuals engaged in many activities and different party sizes (Kawanaka, 1990; Takahata, 1990; this study). Such different methods may produce different results. In Taï, comparison of a continuous recording versus one based on 1-hour interval sampling on the same days, with the same target, produced significantly divergent results for both party size and party type ($p < 0.001$). The second method underestimated small party sizes and overestimated mixed party types. Second, individual identification of all community members is important to guarantee accurate counts of party size. Third, the definition of what constitutes a party varies between observers, from all individuals in sight of one another, to all individuals in auditory contact, or within a certain distance of one another. Both the habituation level and the visibility conditions within a site influence what definition may be used.

At Taï, we recorded all individuals within sight (judged from the observer's point of view) as a party and continuously recorded any changes occurring in the party containing the target individual. In 1987–9, when the data were collected, all individuals of the community were fully identified. In traveling, chimpanzees normally progress in a line with the observer at the rear. I checked the party composition whenever possible and if a change occurred I entered it at the minute following the previous count. Joining and leaving parties were given an arbitrary one minute of duration. I followed individuals for hours or days and changed targets according to our interest in specific behavior, e.g. hunting in the wet season and nut-cracking in the dry season. Previously, I analyzed only initial party counts plus new ones for each party change (Boesch, 1991; used also by White, 1988; Chapman *et al.*, 1994), but for the present analysis I include the time a given party remains constant. Smaller parties have a shorter duration than larger ones (see Table 8.7), so excluding the duration overestimates the importance of small parties. This method gives a precise image of the parties that are observed in a population, as well as the duration of each party size.

Table 8.1. *Mean duration of parties in different chimpanzee populations*

Population	Duration of parties (minutes)
Chimpanzee	
Bossou	126
Gombe	69[a]
Taï	17
Bonobo	
Lomako	102
Wamba	86

Source: Bossou: Sugiyama, 1988; Gombe: Halperin, 1979 ([a]method used not directly comparable to others); Lomako: White, 1989; Wamba: Kuroda, 1979.

RESULTS

I first present a general description of the social structure of the Taï chimpanzees in order to compare it with information from other chimpanzee populations. Then, I proceed to a detailed analysis of the grouping patterns, in order to find out which factors influence party size in the Taï chimpanzees.

Party size

The fission–fusion nature of the Taï chimpanzee society is shown in Table 8.1. Parties last, on average, 17 minutes. This high fluidity in parties is observed for all activities, party types and party sizes, despite some variation to be analyzed later. Figure 8.1 shows the frequency of each party size we observed between August 1987 to February 1989. Mean party size was eight individuals and the largest party was 41 chimpanzees. At that time, the community numbered 76 individuals. This reflects the basic structure of the Taï chimpanzee community: high fluidity, party size rarely exceeding more than a third of the community members, and an average party size of eight chimpanzees.

Comparison with other chimpanzee populations shows that average party size is usually around five individuals, which is smaller than what has been observed in Taï chimpanzees (Table 8.2). Bonobo party sizes in two communities (the Rangers in Lomako [White, 1988] and Wamba) tend to be larger than those observed for chimpanzees. This has been proposed by some authors

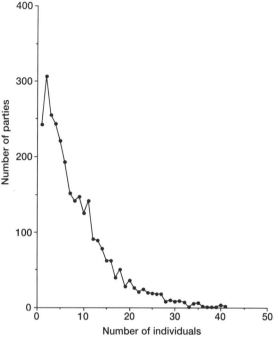

Fig. 8.1. Frequency of different party sizes in Taï chimpanzees from August 1987 to February 1989 ($N = 2912$)

to be species-specific (Kuroda, 1979; Wrangham, 1986; White, 1988; Kano, 1992). However, such a comparison could be influenced by the size of the community, as we might expect that smaller communities form smaller parties. For example, the Bossou community has 20 group members and thus cannot form parties larger than 20 individuals, whereas in the Taï community with 76 individuals, such party sizes are easily possible. Thus, we need to correct for community size if we want to compare different communities. In Table 8.2, I use the *relative party size* – mean party size divided by the community size. Chimpanzee communities have mean relative party sizes between 9 and 21%, whereas in bonobos they are between 22 to 46% (Table 8.2). However, the variation in mean relative party size is primarily explained by the community size: The smaller the community, the larger the mean relative party size (Spearman rank correlation: $r_s = -0.66$, $N = 11$, $p < 0.05$). (This test was done with the largest number of individuals known to be members of each study community as shown in Table 8.2. If I correct for the

Table 8.2. *Party sizes for different chimpanzee and bonobo communities. Community and party size figures include members of all ages, except for party size in Kibale and in Lomako 1993 where infants were excluded. For some studies, community size was not precisely known but only estimated (with an ^{a} in the table), these data must be used with caution. Studies of short duration (less than 12 months) were excluded as data on community size were too speculative – missing data*

| | Mean party size | Number of parties | Community size | Mean relative party size (%) | Proportion of parties containing | | |
					Lone indiv.	<25%	<50%
					of the community members		
Chimpanzee							
Assirik	5.3	267	28^a	19	17%	72%	94%
Bossou	4.0	426	20	20	–	–	–
Gombe	5.6,	498	57	9	13%	95%	99%
Kibale	5.1		37	19	–	–	–
Mahale	6.1	218	29	21	11%	58%	81%
Taï	8.3	2912	76	11	4%	77%	84%
Bonobo							
Lomako 1988							
Blobs	4.3	–	10^a	43	–	–	–
Hedons	7.1	–	22^a	32	–	–	–
Rangers	9.7	–	21^a	46	–	–	–
Lomako 1993	5.8	248	36	22	–	–	–
Wamba							
Kuroka	16.9	147	58^a	29	9%	50%	76%
Furichi		–	35	92	–	–	–

Source: Assirik: Tutin *et al.*, 1983; Bossou: Sakura, 1994; Gombe: Goodall, 1968; Kibale: Clark & Wrangham, 1994; Chapman *et al.*, 1994 (2414 10-minute scans); Mahale: Nishida, 1968; Lomako 1988 (Blobs, Hedons, Rangers): White, 1988; Lomako 1993: Fruth & Hohmann (personal communication), who followed the community previously called the Rangers; Wamba: Kuroda, 1979; Furuichi, 1989.

uncertainties of the estimated communities by including larger figures for the three communities in Lomako and Assirik [using a 50% increase over the known member count from Table 8.2] and 100 members for community B in Wamba [Kuroda, 1979, personal communication], the test remains statistically significant: Spearman rank correlation: $r_s = -0.76$, $N = 11$, $p < 0.02$.) A linear fit between mean relative party size and community size describes 76% of the variance observed (see Fig. 8.2). The bonobo community at Wamba seems to be special, as its mean relative party size increased from 29% in 1975 (Kuroda, 1979) to 92% in 1984 (Furuichi, 1989). The heavy provisioning that started in 1977 may be the most important factor in explaining such a change in social grouping, and as it is a human artefact, we should avoid any generalization from such data. For the other populations, it seems that the smaller a community is the

more individuals are found in parties. In other words, the smaller the community the higher the cohesion between members.

Mahale and Taï chimpanzees and Wamba bonobos tend to have similar proportions of parties containing less than 50% of the community members (see Table 8.2). Gombe chimpanzees fall at the lowest extreme with 95% of the parties having fewer than 25% of the community members, thus presenting a more skewed distribution towards the small end in the distribution of party sizes.

It seems that for small communities of chimpanzees and bonobos, the fission–fusion structure remains but loses much of its flexibility. In addition, parties in smaller communities tend to remain stable for longer periods (see Table 8.1). Data are incomplete for comparing all populations but a trend is apparent. In Wamba, once provisioning was successful, the fission–fusion

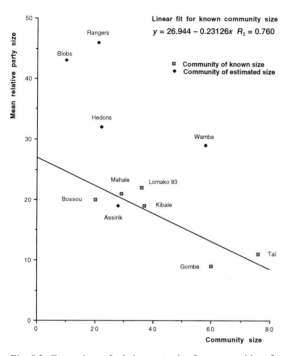

Fig. 8.2. Comparison of relative party size for communities of different sizes in chimpanzees and bonobos

Table 8.3. *Party composition in Taï chimpanzees from August 1987 to February 1989, indicating number of parties recorded and duration of parties*

Party type	Frequency		Duration (minutes)	
	N	%	*N*	%
Mixed	1 799	52	35 773	61
Females	625	18	7 466	13
Males	708	20	12 741	21
Adults	12	0.3	201	0.3
Lone Males	231	6	1 970	3
Females	157	4	1 085	2
Total	3 532		59 236	

Table 8.4. *Adult sex ratio of mixed parties of different sizes in Taï chimpanzees between August 1987 and February 1989. Adult sex ratio of community was 1:4 (male:female)*

Party size	Sex ratio (male/female)
1–10	1.34
11–20	0.90
> 20	0.54

structure tended to disappear, as parties lasted for weeks and averaged 91% of community members (Furuichi, 1989). This tendency for smaller communities to become less fluid indicates that the fission–fussion structure is more than a feeding strategy.

Party type

The second basic aspect of social structure in chimpanzees and bonobos is party composition. Taï chimpanzees are found mainly in mixed parties containing adults of both sexes and their offspring (Table 8.3). Unisexual parties account for only one-third of the observed parties, stressing the high bisexual cohesion of this community. In addition, the very low frequency of adult parties, including males and only those females without dependent offspring, is explained by the very high number of females with babies (on average about 24 of the 27 adult females). Eighty-two percent of the time females were observed in association with males; males associated with females 74% of the time. In a community

containing 27 adult females and 7 adult males, if individuals were randomly associated, we would expect female parties to be the most common ones, followed by mixed parties and male parties to be by far the rarest ones. The data indicate that both sexes associate preferentially with the other sex. In addition, the proportion of females in mixed parties increases as party size increases (see Table 8.4), tending toward the community's adult sex ratio as party size increases. Males seem more actively to seek contact with the opposite sex than do females. In conclusion, Taï chimpanzees live in a bisexually bonded social system, in which both females and males associate together for most of their time.

When party type is compared for chimpanzee communities, it is apparent that mixed parties are most frequently observed (see Table 8.5). The only exception is Bossou, where mother parties are most frequent. The presence of only one adult male for 10 to 12 females may explain the low frequency of mixed parties in this population. Taï chimpanzees have the same proportion of mixed parties as the Lomako bonobos but more than

Table 8.5. *Party types seen in different chimpanzee and bonobo communities. The adult sex ratio (males:females) is indicated for each population when available. In some studies (ª), community size estimated. Data for all party types were not presented in each study and some included unknown parties (–, data missing)*

Site	Adult sex ratio	Mixed	Adults	Males	Mothers	Lone individual
Chimpanzee						
Assirik[a]	–	58	3	2	18	19
Bossou	0.12	42	–	–	49	–
Gombe	0.80	30	18	10	24	18
Kibale	0.66	52	–	–	–	28
Mahale	0.66	52	4	11	13	21
Taï	0.25	52	1	20	18	10
Bonobo						
Lomako[a]	0.62	68	8	5	5	14
Wamba	0.66	74	3	3	5	6

Source: Assirik: Tutin *et al.*, 1983; Bossou: Sakura, 1994; Gombe: Goodall, 1968; Kibale: Chapman *et al.*, 1993; Clark & Wrangham, 1994; Mahale: Nishida, 1968; Lomako: White, 1988; Wamba: Kuroda, 1979.

Gombe and Mahale chimpanzees. As the Bossou chimpanzees' case suggests, the adult sex ratio influences party types; the lower the ratio the more actively males have to look for females if they want to form mixed parties. The adult sex ratio is low in Bossou and Taï chimpanzees, whereas all other populations have one equal or larger to 2 males : 3 females (see Table 8.5). Therefore, Taï male chimpanzees most actively seek contact with the other sex, Lomako male bonobos and Kibale and Mahale males are intermediate and Gombe males seek the least contact with the opposite sex.

In conclusion, the bisexually bonded social system proposed for the Taï chimpanzees (Boesch, 1991) seems to be supported by the present analysis and is rather different from the more male bonded group observed in Gombe chimpanzees. Bonobos, in the present comparison, are less bisexually bonded in comparison to Taï chimpanzees, showing the difficulty of interspecific comparisons, as habitat and demographic differences interact with the grouping patterns (Kuroda, 1979; White, 1988; Kano, 1992).

Factors affecting party size in Taï chimpanzees

Here, I compare two different seasons over a 2-year period (August 1987 to February 1989); the wet season from August to October, which includes the big rainy season, and the long dry season from January to February–March. The rainy season is characterized by: almost daily heavy rainfalls, the major hunting season of the year, a progressive increase in general food availability, and low temperatures. Since 1983, the dry season has become very dry with at least one month without a drop of rain. High temperature prevails unless the 'Harmattan,' a cold desert wind, reaches the forest. The Harmattan may persist at most for two weeks. The dry season is also characterized by a general high fruit availability including the *Coula* trees.

Independence of data-points is required for statistical analysis. This is difficult to obtain from such a set of data, but in order to improve it, only one data point per activity was used in counting the largest party size reached for a given activity. For example, a target individual joined a party in a tree, followed by three other parties joining them; the target left the tree preceded by the departure of two other parties. In this example, I used only one party; the duration was the total time the target was in the tree and party size was the largest count in that tree. For feeding, resting and meat-eating parties, this procedure strongly reduces the sample size. In addition, all joining and leaving parties were excluded from the analysis, as they directly contributed to the party they joined or resulted from the party they left. In this way, sample size was reduced from 3464 to 1335 parties.

Party size in Taï chimpanzees is influenced by the different types of activities being performed (Table 8.6).

Table 8.6. *Mean party size (with standard error) of Taï chimpanzees when engaged in different activities from August 1987 to February 1989. Meat-eating parties are larger than resting parties, which in turn are larger than feeding and travelling parties (Student's t test; p < 0.05). Meat-eating parties last longer than both resting and feeding parties, which also last longer than travelling parties (Student's t test; p < 0.05)*

Activity	Number of parties	Party size	Duration (minutes)
Travel	487	9.0 ± 0.33	21.1 ± 1.3
Unsuccessful hunt	36	10.2 ± 0.33	24.8 ± 4.5
Feed	489	10.1 ± 0.33	41.6 ± 1.3
Rest	281	12.2 ± 0.43	40.4 ± 1.7
Meat	42	20.8 ± 1.12	106.0 ± 4.6

Table 8.7. *Factors influencing party size in Taï chimpanzees, controlling for effects of year, season, activity and duration for wet and dry seasons, 1987 to 1989 (ANOVA test: Proc GLM; SAS, 1985). Interaction effects also shown (×). Only results with p-values less than 0.05 are shown, all other comparisons were not statistically significant*

Source	df	F-value	P
Duration	20	3.30	0.0001
Activity	4	5.54	0.0002
Year × season	1	27.20	0.0001
Season × activity	4	6.30	0.0001
Duration × activity	56	1.66	0.002
Season × duration × activity	33	1.56	0.02

Meat-eating parties are clearly the largest and have the longest duration. This shows the potential importance of this activity, in that chimpanzees spent about 10% of their time in meat-eating. Resting parties are the second largest in party size, while feeding parties are of longer duration than travelling parties.

However, party size seems to be affected by additional factors. To sort out the influence of: the activity of the chimpanzees, the duration of the party, the season of the year, and the year on the party size, we performed a multiple ANOVA test (Proc GLM; SAS, 1985) that controls for the possible interactions between the different parameters (see Table 8.7). When corrected for interactions, year and season had no influence on party size, whereas the influence of duration of the party and the activity were highly significant. As already men-

tioned, the larger the party the longer it remains stable. Meat-eating parties are the largest of all and remain so all year round ($F = 5.54$, df = 4, $p < 0.0001$). Some interactions have even larger effects on party size.

First, the seasonal variations in party size may change between years: in 1987–88, the parties were smaller in the dry season than in the wet season, whereas the opposite was true in the next year (see Fig. 8.3). The dry season of 1988 was one of the worst in 12 years with almost no *Coula* nuts produced and barely any other kind of fruits. The chimpanzees scattered in small parties in order to eat leaves and the small fruit production that was available. However, the difference seen between the two wet seasons is more difficult to understand as fruit production was very similar. No obvious reason has been found to explain this difference at the present stage of analysis. Second, seasonal variations in party size were also found in activities: resting and travelling parties are larger in the dry seasons, whereas the opposite is true for the other activities (see Fig. 8.4). As feeding tends to occur for longer periods of time in the wet season, we may speculate that food is less abundant in the rainy season and once the chimpanzees find a patch, they feed for a longer time and in larger parties. But this does not explain the trends observed for the other activities. From this analysis it looks as if general fruit availability especially influences feeding party sizes, but that in all cases other factors have a large influence on party size.

As food availability influences some party sizes, it is important to see if it is particularly food-patch size or tree size that influences party size. I used DBH

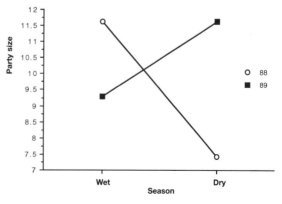

Fig. 8.3. Interactions between year and season on party size in Taï chimpanzees.

Table 8.8. *Mean party size (with standard error) of Taï chimpanzees when feeding in trees of different sizes from August 1987 to February 1989. Criteria for three tree sizes: small tree species have less than 10% of their trunks above 70 cm diameter, medium tree species have 10 to 70% of trunks above 70 cm diameter, and large tree species have more than 70% of the trunks above 70 cm diameter. Parties feeding in larger trees are larger than those feeding in medium and small trees (Student's t test; p < 0.05), but all have the same duration*

Tree size	Number of parties	Party size	Duration (minutes)
Small	176	8.2 ± 0.53	38.4 ± 2.5
Medium	113	10.8 ± 0.66	43.1 ± 3.2
Large	199	11.4 ± 0.50	43.8 ± 1.4

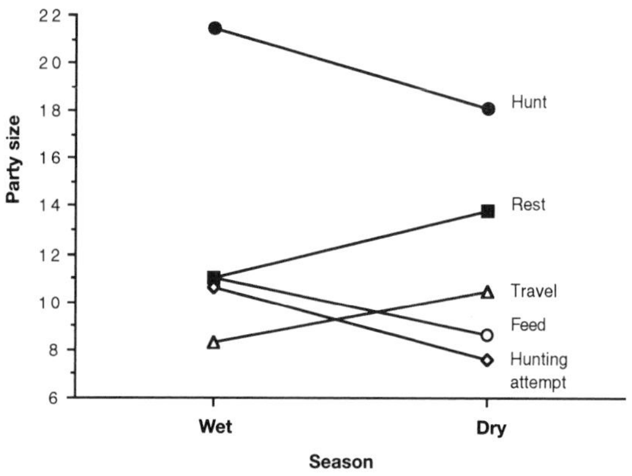

Fig. 8.4. Interactions between activity and season on party size in Taï chimpanzees.

Table 8.9. *Factors influencing feeding party size in Taï chimpanzees, controlling for effects of year, season, duration and tree size (DBH) for wet and dry seasons 1987 to 1989 (ANOVA test: Proc GLM; SAS, 1985). Only results with p-values less than 0.05 shown, all other comparisons not statistically significant*

Source	df	F-value	p
Season	1	12.11	0.0006
DBH	22	2.24	0.002
Year × season	1	14.95	0.0001
Season × DBH	10	3.58	0.0002
Season × DBH × activity	21	2.05	0.005

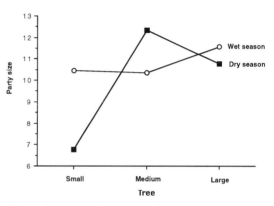

Fig. 8.5. Interactions between tree size and season on party size in Taï chimpanzees.

measurements for each tree species in which I saw the chimpanzees eat (trunk diameters are from van Rompaey, 1993). If larger trees are available in the wet season and chimpanzees feed in larger parties in larger trees, then the previous results would be partly explained. Table 8.8 compares party size and duration for the three sizes of trees for the two-year period. Table 8.9 gives the results of the multiple ANOVA test for feeding parties according to the size of the food patch. The strongest effect is the season, as parties are larger for most tree sizes in the wet season than in the dry season. In addition, the larger the tree, the larger the party that feeds in it. However, some interactions have even

stronger effects on feeding party size: small trees have much smaller parties in the dry seasons than in the wet seasons, while medium trees have larger parties in the dry season than in the wet season, and large trees have about the same party size (see Fig. 8.5). This is affected by the year as well, so that in the first year (1987–88) all parties were smaller in the dry season. In the second year, the dry season had larger parties for medium and large trees and smaller parties only for the small trees compared with the wet season (see Fig. 8.6). This clearly illustrates that tree size plays only a partial role in determining feeding party size. The dry season of 1989 had a much higher fruit availability than did 1988's dry season, and this seems to have strongly affected feeding party size, especially for larger trees. This might be due to parties for all activities being larger in the 1989 dry season than the year before (see Fig. 8.3), and so parties approaching a tree were larger this year. Thus, party size is influenced more by general fruit availability than by tree size in the dry seasons. For the wet seasons, party sizes are similar for different tree sizes and the difference observed between the two years cannot be explained by food availability that remained very similar over those years (see Fig. 8.6). This suggests that another factor influenced party size in this season.

Sex has been proposed to influence strongly the grouping patterns of chimpanzees (Goodall, 1986). Figure 8.7 shows the number of estrous females observed in the community for all months over the year. The average number of estrous females is five per month from September to April, but decreases from May to August. However, for the period analyzed, the wet seasons had more estrous females than the dry seasons, but no seasonal differences in party size could directly be explained by this trend (see Fig. 8.3). If we combined the difference in food availability between the two dry seasons with the difference in estrous females, we could explain what we observe: wet seasons have more estrous females and tend to have larger parties, but a higher fruit availability in a dry season tends to compensate for the low number of estrous females. In the dry season of 1989 party sizes were the same as in the wet season of 1988, despite a strong difference in the number of estrous females available (see Figs. 8.3 and 8.7). In support of this conclusion, party size during May–June 1988 averaged only 2.8 individuals (Doran, 1989), when both general food availability and the number of estrous females were very low (see Fig. 8.7). Therefore, the number of

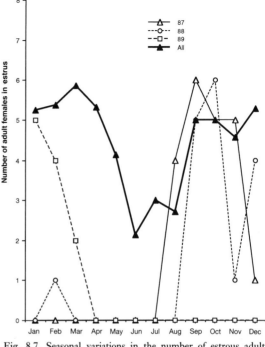

Fig. 8.7. Seasonal variations in the number of estrous adult females in Taï chimpanzees.

estrous females and general food availability are two factors that explain much of the variation in party size observed in Taï chimpanzees. Other factors are still to be found, as party size is significantly different between the two wet seasons and this is not explained by the number of estrous females available.

Taï chimpanzees have a hunting season, the wet season from August to October, in which they usually hunt daily (Boesch, 1994; Fig. 8.8). Successful hunts result in prolonged meat-eating episodes that last much longer than any other activity in the chimpanzees' social life and bring together especially large groups of individuals (see Table 8.6). In meat-eating parties, individuals not only eat meat, they also engage in many social interactions as access to meat is very variable (Boesch, 1994). Hunting may help explain the large party sizes during the wet seasons: the hunting rate in wet seasons is 0.68 hunts per day, whereas for dry seasons it is 0.21. This may also explain the difference between the two wet seasons. In the first season the hunting rate was 0.75 per day, whereas in the second season it was only 0.63 hunts per day. Qualitative

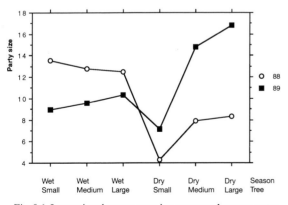

Fig. 8.6. Interactions between tree size, season and year on party size in Taï chimpanzees.

Fig. 8.8. Adult male (right) shares meat of red colobus monkey with adult female chimpanzee at Taï. (Photo by C. Boesch)

Fig. 8.9. Adult male chimpanzees in grooming party at Taï. (Photo by C. Boesch)

observations also support this explanation. At the beginning of the hunting season (August), male hunters first gather to hunt and it is only in the following weeks that other group members join them and party size progressively increases. As a result, hunting rate increases some weeks before party size increases.

In conclusion, general fruit availability, sexual opportunities and hunting rate explain most of the variations observed in party size in Taï chimpanzees. In addition, the activity of the chimpanzees and the size of trees affect the party size within a given season. With such data, it is always difficult to rank the importance of each of these three factors, but it appears that hunting rate and sexual opportunities compensate for a lower fruit availability in the wet season, whereas high fruit availability can compensate for a lower hunting rate and limited sexual opportunities in the dry season. Factors proposed to play a role in social grouping in other populations such as food patch size (measured by the tree size), or proportion of THV (which represents only 3% of the feeding activity in Taï) have a very limited role in explaining party size in Taï. The unifactorial approach is of a limited value in Taï chimpanzees as both ecological parameters (fruit availability and hunting rate) and social par-

ameters (activities and sexual opportunities) interact to explain the social grouping patterns (Fig. 8.9).

Analyses of grouping patterns in other chimpanzee populations do not always consider the impact of such factors, so that it is too early to attempt such a comparison. However, important differences in party size have been reported between different seasons in Kibale and Gombe, suggesting that general food availability is an important factor. Sexual opportunity has been proposed to explain yearly and seasonal variations in party size in Gombe as well (Goodall, 1986; Stanford et al., 1994). In a recent analysis of party size in Gombe chimpanzees, it was suggested that high fruit availability, high hunting rate and high number of estrous females coincide in the fall to produce larger parties (Stanford et al., 1994). In Bossou, seasonal variation in food availability as well as an increased number of estrous females leads to larger party size (Sakura, 1994). Thus, preliminary information tends to indicate that the three factors affecting party size in Taï may also be important in explaining variations in party size in other chimpanzee populations.

The fission–fusion structure in chimpanzees is influenced by the community size in such a way that small

Fig. 8.10. Mixed party of chimpanzees at Taï. (Photo by C. Boesch)

communities tend to lose much of the fluidity typical of large communities. This seems to apply both to chimpanzees and bonobos. Mixed parties are the most common ones, but the tendency for males to associate with females seems to vary, with Taï males the most inclined to do so and Gombe males the least so (Fig. 8.10). This supports the idea that the grouping pattern of chimpanzees differs in the sexual bonding tendency: Taï chimpanzees are characterized by strong bisexual bonding, Gombe chimpanzees are male-oriented in their bonding and Mahale chimpanzees show an intermediate tendency. Thus the Taï system is closer to the one described for bonobos.

ACKNOWLEDGMENTS

This paper was originally prepared for the Wenner-Gren conference, 'The Great Apes Revisited,' held November 12–19, 1994 in Cabo San Lucas, Baja, Mexico. I thank the 'Ministère de la Recherche Scientifique', and the 'Ministère de l'Agriculture et des Ressources Animales' of Côte d'Ivoire, for permitting this study, and the Swiss National Foundation and the Schultz Foundation in Zürich for financing it. In Côte d'Ivoire, this project is integrated in the UNESCO project Taï-MAB under the supervision of Dr Henri Dosso. I am most grateful to A. Aeschlimann, F. Bourlière and H. Kummer for their constant encouragement, P. and B. Lehmann of the CSRS in Abidjan and T. Tiépkan of the station IET in Taï for their logistic support. I thank H. Boesch-Achermann and the editors of this book for commenting upon and correcting the manuscript.

REFERENCES

Boesch, C. 1991. The effect of leopard predation on grouping patterns in forest chimpanzees. *Behaviour*, **117**, 220–42.

Boesch, C. 1994. Cooperative hunting in wild chimpanzees. *Animal Behaviour*, **48**, 653–67.

Caraco, T. & Wolf, L. L. 1975. Ecological determinants of group sizes of foraging lions. *American Naturalist*, **109**, 343–52.

Chapman, C., White, F. & Wrangham, R. W. 1993. Defining subgroup size in fission–fusion societies. *Folia Primatologica*, **61**, 31–4.

Chapman, C., White, F. & Wrangham, R. W. 1994. Party size in chimpanzees and bonobos: a reevaluation of theory based on two similarly forested sites. In: *Chimpanzee Cultures*, ed. R. W. Wrangham, W. C. McGrew, F. B. M. de Waal & P. Heltne, pp. 41–57. Cambridge, MA: Harvard University Press.

Chapman, C., Wrangham, R. W. & Chapman, L. 1995. Ecological constraints on group size: an analysis of spider monkey and chimpanzee subgroups. *Behavioral Ecology and Sociobiology*, **36**, 59–70.

Cheney, D. L. & Wrangham, R. W. 1987. Predation. In: *Primate Societies*, ed. B. B. Smuts, D. L. Cheney, R. M. Seyfarth, R. W. Wrangham & T. T. Struhsaker, pp. 227–39. Chicago: University of Chicago Press.

Clark, A. & Wrangham, R. W. 1994. Chimpanzee arrival pant hoots: do they signify food or status? *International Journal of Primatology*, **15**, 185–206.

Doran, D. 1989. Chimpanzee and pygmy chimpanzee positional behavior: The influence of environment, body size, morphology and ontogeny on locomotion and posture. Ph.D. dissertation, State University of New York, Stony Brook.

Dunbar, R. 1988. *Primate Social Systems*. London: Croom Helm.

Emlen, S. T. 1991. Evolution of cooperative breeding in birds and mammals. In: *Behavioural Ecology: An Evolutionary Approach*, ed. J. R. Krebs & N. B. Davies, pp. 301–37. Oxford: Blackwell Scientific Publications.

Furuichi, T. 1989. Social interactions and the life history of female *Pan paniscus* in Wamba, Zaïre. *International Journal of Primatology*, **10**, 173–97.

Goodall, J. 1968. The behaviour of free-living chimpanzees in the Gombe Stream Reserve. *Animal Behaviour Monographs*, **1**, 161–311.

Goodall, J. 1986. *The Chimpanzees of Gombe*. Cambridge, MA: Harvard University Press.

Halperin, S. D. 1979. Temporary association patterns in free ranging chimpanzees: an assessment of individual grouping preferences. In: *The Great Apes*, ed. D. A. Hamburg & E. R. McCown, pp. 491–9. Menlo Park, CA: Benjamin/Cummings.

Isbell, L. & Young, T. 1993. Human presence reduces predation in a free-ranging vervet monkey population in Kenya. *Animal Behaviour*, **45**, 1233–5.

Kano, T. 1992. *The Last Ape: Pygmy Chimpanzee Behavior and Ecology*. Stanford, CA: Stanford University Press.

Kawanaka, K. 1990. Alpha males' interactions and social skills. In: *The Chimpanzees of the Mahale Mountains*, ed. T. Nishida, pp. 149–70. Tokyo: University of Tokyo Press.

Krebs, J. R. & Davies, N. B. 1993. *An Introduction to Behavioural Ecology*, 3rd edn. Oxford: Blackwell Scientific Publications.

Kummer, H. 1971. *Primate Societies: Group Techniques of Ecological Adaptation*. Chicago: Aldine.

Kummer, H., Banaja, A., Abo-Khatwa, A. & Ghandour, A. 1985. Differences in social behaviour between Ethiopian and Arabian hamadryas baboons. *Folia Primatologica*, **45**, 1–8.

Kuroda, S. 1979. Grouping of the pygmy chimpanzees. *Primates*, **20**, 161–83.

Malenky, R. & Stiles, E. 1991. Distribution of terrestrial herbaceous vegetation and its consumption by *Pan paniscus* in the Lomako Forest, Zaïre. *American Journal of Primatology*, **23**, 153–69.

Nishida, T. 1968. The social group of wild chimpanzees in the Mahali Mountains. *Primates*, **9**, 167–224.

Packer, C., Scheel, D. & Pusey, A. E. 1990. Why lions form groups: food is not enough. *American Naturalist*, **136**, 1–19.

Pulliam, H. R. & Caraco, T. 1984. Living in groups: is there an optimal group size? In: *Behavioural Ecology: An Evolutionary Approach*, ed. J. R. Krebs & N. B. Davis, pp. 122–47. Oxford: Blackwell Scientific Publications.

van Rompaey, R. 1993. Forest gradients in West Africa: a spatial gradient analysis. Ph.D. thesis, University of Wageningen.

SAS. 1985. *SAS User's Guide: Basics*, Version 5 Edition. Cary, NC: SAS Institute Inc.

Sakura, O. 1994. Factors affecting party size and composition of chimpanzees (*Pan troglodytes verus*) at Bossou, Guinea. *International Journal of Primatology*, **15**, 167–83.

van Schaik, C.P. 1983. Why are diurnal primates living in groups? *Behaviour*, **87**, 120–44.

Stanford, C., Wallis, J., Mpongo, E. & Goodall, J. 1994. Hunting decisions in wild chimpanzees. *Behaviour*, **131**, 1–18.

Sugiyama, Y. 1988. Grooming interactions among wild chimpanzees at Bossou, Guinea, with special reference to social structure. *International Journal of Primatology*, **9**, 393–407.

Sugiyama, Y. & Koman, J. 1979. Social structure and dynamics of wild chimpanzees at Bossou, Guinea. *Primates*, **20**, 323–39.

Symington, M. M. 1990. Fission–fusion social organization

in *Ateles* and *Pan*. *International Journal of Primatology*, 11, 47–61.

Takahata, Y. 1990. Adult males' social relations with adult females. In: *The Chimpanzees of the Mahale Mountains*, ed. T. Nishida, pp. 133–48. Tokyo: University of Tokyo Press.

Terborgh, J. & Janson, C. H. 1986. The socioecology of primate groups. *Annual Review of Ecology and Systematics*, 17, 111–35.

Tsukahara, T. 1993. Lions eat chimpanzees: the first evidence of predation by lions on wild chimpanzees. *American Journal of Primatology*, 29, 1–11.

Tutin, C. E. G., McGrew, W. C. & Baldwin, P. J. 1983. Social organization of savanna-dwelling chimpanzees, *Pan troglodytes verus*, at Mt Assirik, Senegal. *Primates*, 24, 154–73.

White, F. 1988. Party composition and dynamics in *Pan paniscus*. *International Journal of Primatology*, 9, 179–93.

White, F. 1989. Social organization of pygmy chimpanzees. In: *Understanding Chimpanzees*, ed. P. G. Heltne & L. A. Marquardt, pp. 194–207. Cambridge, MA: Harvard University Press.

White, F. & Wrangham, R. W. 1988. Feeding competition and patch size in the chimpanzee species *Pan paniscus* and *Pan troglodytes*. *Behaviour*, 105, 148–64.

Wrangham, R. W. 1980. An ecological model of female-bonded primates. *Behaviour*, 75, 262–300.

Wrangham, R. W. 1986. Ecology and social relationships in two species of chimpanzee. In: *Ecological Aspects of Social Evolution: Birds and Mammals*, ed. D. I. Rubenstein & R. W. Wrangham, pp. 352–78. Princeton, NJ: Princeton University Press.

Wrangham, R. W., Clark, A. & Isabirye-Basuta, G. 1992. Female social relationships and social organization of Kibale forest chimpanzees. In: *Topics in Primatology, Vol. 1, Human Origins*, ed. T. Nishida, W. C. McGrew, P. Marler, M. Pickford and F. B. M. de Waal, pp. 81–98. Tokyo: University of Tokyo Press.

9 • Coalition strategies among adult male chimpanzees of the Mahale Mountains, Tanzania

TOSHISADA NISHIDA AND KAZUHIKO HOSAKA

INTRODUCTION

Twenty years ago, at the time of the first Wenner-Gren Symposium on the great apes held in 1974, chimpanzee sociology was still in its infancy. People believed that unlike other non-human primates, great apes were peaceful creatures (e.g. Montagu, 1968; Goodall, 1973), although between-group relationships were known to be antagonistic, at least among the chimpanzees of Mahale (Nishida & Kawanaka, 1972). As examples of agonistic aid, high-ranking males reportedly ran to help their 'friends' against the latters' rivals (Goodall, 1971) and some old males were known to take refuge with particular dominant males against the other males' displays (Simpson, 1973). However, complex coalition tactics were not known. The subsequent 20 years have seen a major breakthrough in chimpanzee sociology and politics.

Coalition is defined as two or more individuals joining forces against one or more conspecific rivals. Now, there are many examples of coalitions in mammals and birds (Connor et al., 1992; Harcourt, 1992). However, coalitions among chimpanzees, adult males in particular, are exceptional for their frequency, complexity and flexibility. Some coalitions are so persistent that one may well call them alliances (e.g. Riss & Goodall, 1977; Goodall, 1986). On the other hand, adult males sometimes show extreme opportunism, changing sides from moment to moment (de Waal, 1982; Nishida, 1983; Uehara et al., 1994). Some coalitions are direct, while others are indirect (de Waal, 1982). In chimpanzees, coalitions are cemented by various social resources such as close association, grooming, food sharing and long-distance calling, in addition to aid in combat (de Waal, 1982; Nishida et al., 1992; Mitani & Nishida, 1993). Moreover, even 'macro-coalition' has been recorded in between-group contexts (Goodall, 1986). Thus, coalitionary strategies are among the most striking examples of the close affinity of chimpanzees with humans (de Waal, 1991; Boehm, 1992). Social tactics are also thought to be a major driving force in the evolution of intelligence (see Byrne, 1995 for review).

Under captive conditions, de Waal (1982) elucidated the characteristics of chimpanzee coalitional tactics. Under natural conditions, continuous and detailed observations of these processes are difficult. However, studies in captive conditions should be supplemented by those under natural conditions, since only in nature can the adaptive significance and distribution of agonistic support be fully understood. Moreover, under natural conditions, social tactics may show a greater variety than in captivity, since living space is not confined and the number of individuals per group is usually larger.

Because of more difficult observational conditions, reports of chimpanzee political behavior in natural habitats are still scant, and detailed, systematic descriptions are needed. This paper aims to fill this gap in knowledge, using new data from a large unit group that includes nine adult males. Agonistic strategies are summarized and compared with information obtained in previous captive (de Waal, 1982) or field studies (Bygott, 1974, 1979; Riss & Goodall, 1977; Nishida, 1983; Goodall, 1986, 1992; Nishida et al., 1992). This paper describes only male–male coalitions. Coalitions among adult females, between sexes, and various age and sex classes will be published later.

METHODS

Study subjects

The chimpanzees of the Mahale Mountains National Park, Tanzania, have been studied since 1965 (for demographic data, see Nishida, 1990). The study group, M Group, consisted of 82 individuals including 9 adult, 8 adolescent, 5 juvenile and 9 infant males together with 29 adult, 9 adolescent, 6 juvenile and 7 infant females during the study period of 1992 (Fig. 9.1).

Sampling methods

The field research on which this paper is based was conducted by the senior author during a 3-month period from mid-August to early November, 1992. Mitani (Mitani & Brandt, 1994) studied vocal behavior of male chimpanzees from April to August, 1992 and Hayaki (1995) made a comparative behavioral study of two adult male chimpanzees, Nsaba and Aji, from July to September, 1992. Related observations were made by the authors from mid-August, 1991 to March, 1992.

All nine adult male chimpanzees of M Group were chosen as targets. One male at a time was followed for as long as possible from morning to evening in a single day. After each male was followed once, the male with the least observation time was selected as a target in the next round of observations, although this was not always possible because of the temporary disappearance of target males. Any social interactions involving a target, as well as the identity of all the adult males and females within 10 m of him, were recorded.

An individual who was within 10 m of the target was regarded as being *in proximity*. A *grooming session* was defined as having ended when grooming between two adult males ceased for more than ten minutes. A *grooming bout* was defined as having ended when either or both males ceased to groom or changed their roles.

Each target was observed for 42–48 hours during the study period, and all focal observation hours totaled 390 hours (see Table 9.1).

'Other-focal' sampling data in this chapter means social behavior of an adult male when the target was another male.

Fig. 9.1. Adult female chimpanzee with infant grooms an adult male at Mahale. (Photo by C. Tutin)

Table 9.1. *Age, dominance rank and observation hours of the target males*

Name	Initial	Age (yr)	Age rank	Dominance rank	Focal hours
Ntologi	NT	>35	2.5	1	48
Nsaba	NS	19	6.5	2	42
Kalunde	DE	29	4	3	42
Jilba	JI	17	8	5	42
Aji	AJ	19	6.5	5	44
Toshibo	TB	16	9	5	44
Bakali	BA	>36	1	7	42
Bembe	BE	20	5	8	43
Musa	MU	>35	2.5	9	43
Total					390

Terminology related to agonistic behavior

Behavior seen in relation to agonistic confrontations is defined below. These terms mostly correspond to those described by van Hooff (1974), Bygott (1974), de Waal & van Hooff (1981), Plooij (1984) and Goodall (1986).

Agonistic confrontations. One or both parties showed at least one 'strictly agonistic element' (de Waal & van Hooff, 1981) such as 'charge', 'bite', 'shrill-bark,' 'flight,' 'grimace' or 'scream'. Agonistic interactions among the same participants were regarded as belonging to the same 'episode' if the interval without agonism did not exceed two minutes (de Waal & van Hooff, 1981).

Side-directed behavior. In addition to agonistic behavior, opponents frequently showed gestures or affinitive behavior towards a third party during, immediately before, or, just after a confrontation, which has been called 'side-directed' behavior (de Waal & van Hooff, 1981; Hemelrijk *et al.*, 1991).

Detour. A male chimpanzee alters his course to avoid another male resting ahead on his path.

Mount. Individual hugs another ventro-dorsally from the rear, usually with teeth-baring and screaming. The mounted individual may show the same expression.

Embrace. Two individuals embrace each other face-to-face in sitting or standing postures ('Embrace-full'), or an individual embraces another around the shoulder ('Embrace-half').

Open-mouth-kiss. Two individuals put their mouths close to each other while panting with mouths wide open.

Mouth. Individual puts his open mouth to the chest, shoulder or other part of another's body.

Hold-out-hand. Individual extends his hand to another party, without touching.

Touch. Individual extends his hand to touch the body of another.

Finger into mouth. Individual puts his finger (usually index) into the mouth of another.

Approach. Individual approaches another male and sits close to him.

Classification of agonistic confrontations

During the study period, 212 agonistic confrontations involving adult males were seen. Of these, 105 were simple dyadic interactions, 102 were polyadic (involving more than 2 individuals), and 5 were semi-polyadic (accompanied by side-directed behavior but no intervention of a third party).

Interactions were divided into 12 different categories (see Table 9.2).

1. One-to-one initiation
   1a. Simple, dyadic interaction.
   1b. One-to-one interaction accompanied by 'side-directed behavior' to a third party by an actor, without any intervention by a third party.
   1c. One-to-one interaction accompanied by side-directed behavior to a third party by a reactor, without intervention by a third party.
   1d. Redirection. A reactor aggresses toward an uninvolved third party.
   1e. Snowballing. One-to-one interactions

Table 9.2. *Categories and frequency of agonistic episodes*

Type of interaction	Frequency	Adult males involved	Other age/sex classes involved
One-to-one initiation			
1a Simple, dyadic	105	–	–
1b SD by A with no IV by T	3	(3)	(0)
1c SD by R with no IV by T	2	(2)	(0)
1d R aggress to T	1	(0)	(1)
1e Snowballing	1	(1)	(0)
One-to-many initiation			
2a Non-directed CD	33	(22)	(11)
2b Separating IV	21	(19)	(2)
2c Neutral IV	3	(2)	(1)
2d Non-directed CD followed by RA	3	(3)	(0)
Many-to-one initiation			
3 Communal attack	7	(6)	(1)
One-to-one initiation joined by T			
4a A-alliance	11	(7)	(4)
4b R-alliance	22	(9)	(13)
Total	212	(74)	(33)

SD, side-directed behavior; A, actor; R, reactor; T, third party; IV, intervention; CD, charging display; RA, reactor-alliance (see text for details).

develop into complicated chain reactions, including redirection, side-directed behavior, and various interventions.

2. One-to-many initiation or 'multiple initiation' (de Waal & van Hooff, 1981)

   2a. Non-directed charging display: an adult male charges at more than one male each of whom is resting but not in close vicinity, some or all of whom respond to the display (e.g. retreat).

   2b. 'Separating intervention' (de Waal, 1982): an adult male charges at two or more males grooming each other or sitting in close proximity (or 'contact-sit').

   2c. Neutral intervention: an adult male intervenes in a dispute, apparently as an arbitrator (cf. 'Impartial intervention,' de Waal & van Hooff, 1981).

   2d. Non-directed charging display followed by reactor-alliance.

3. Many-to-one initiation

Simultaneous, communal attack or threat: two or more males simultaneously attack or threaten one adult male.

4. One-to-one initiation joined by another

   4a. 'Actor-alliance' (de Waal et al., 1976): an aggressing male is joined by another male against the victim.

   4b. 'Reactor-alliance' (de Waal et al., 1976): an aggressed-against male is supported by another male against the aggressor.

Interactions 1b, 1c, 2b, 3, 4a and 4b above were regarded as coalition-related in this study. All the polyadic interactions were reduced to dyadic interactions in which one of the dyad was assigned the role of 'actor' (initiating agonism) and the other that of 'reactor'. Thus, 102 polyadic interactions produced 195 dyadic interactions. Thus, 105 + 195 interactions were used in constructing the agonism-submission matrix (see Table 9.4 below).

Historical background on male hierarchy

To understand the situation in 1992, knowledge of previous years is necessary. Ntologi, the current alpha male, was ostracized by a coalition of two adult males, Kalunde and Nsaba, in May, 1991 (Hamai, 1994). As a result, Kalunde became the alpha, Shike the beta, and Nsaba the gamma male. Although Nsaba occasionally pant-grunted to Kalunde, he sometimes aggressively competed with Kalunde over meat and estrous females. Shike, on the other hand, was very fearful of Kalunde, but confident against Nsaba. Even if Nsaba displayed against Kalunde, Shike sometimes took Nsaba's display as a challenge to himself and chased Nsaba. Thus, Shike's presence guaranteed Kalunde's alpha status. The relationship between Kalunde and Shike was similar to that described for Yeroen and Nikkie, which de Waal (1982) called 'indirect coalition'.

Then, Shike became seriously ill and fell to the bottom of the male hierarchy. At the same time, Nsaba rose in status until by January 1992, it was impossible to determine who was dominant, Kalunde or Nsaba. Shike disappeared and presumably died. Then, Ntologi, who had been roaming alone, returned to M Group. Kalunde and Nsaba now joined forces against Ntologi. However, this time, their coalition was not as successful as it had been in 1991, because Kalunde's superiority over Nsaba had declined. Kalunde appeared to try to take advantage of Ntologi's presence to dominate Nsaba. Thus, Kalunde began to join forces with Ntologi against Nsaba. A few days later, however, he again joined forces with Nsaba against Ntologi. Eventually by March 1992, Ntologi had become alpha, Nsaba second-ranking and Kalunde third-ranking (Hosaka, 1992, unpublished data).

RESULTS

Dominance ranks in 1992

Dominance ranks (Table 9.1) were determined on the basis of the direction of pant-grunt calls (Table 9.3) and the initiation of dyadic agonistic confrontations and detours (Table 9.4).

Only the highest two and lowest three ranks were unequivocal. Relationships among the four middle-ranking males, middle-aged Kalunde and three young adult males Aji, Jilba and Toshibo, were ambiguous. They rarely pant-grunted to each other, except that Aji frequently did so to Kalunde. Although Kalunde was clearly dominant to Aji and Jilba, he was charged at more frequently by Toshibo than he charged at Toshibo. Aji was subordinate to Kalunde, while he was clearly dominant to Toshibo and almost as strong as Jilba. Jilba was

dominant only to Toshibo. Since only Kalunde was clearly dominant to two of the middle-ranking males, he was regarded as third-ranking. Hayaki (1995) reached a similar conclusion on dominance ranks from his research.

The dominance relationships among the four middle-ranking males, therefore, were unstable and non-linear. This non-linearity results from the fact that young adult males do not always rise in rank step by step, and their progress is dynamic. 'Ambitious' young adult males try to defeat past-prime high-ranking males rather than to defeat males immediately above them in rank. When there were many adult males in a group, previous authors have not always arranged them in linear order (Simpson, 1973; Bygott, 1979; Hasegawa & Hiraiwa-Hasegawa, 1983; Hayaki et al., 1989; Nishida et al., 1992; Hayaki, 1995).

Dominance rank was correlated with the total number of pant-grunts received ($N = 9$, $r_s = 0.96$, $p < 0.01$) and with the total number of agonistic initiations ($r_s = 0.96$, $p < 0.01$).

Male rivalries in 1992

Male rivalry was determined on the basis of the frequency of confrontations recorded between two adult males (Table 9.4).

Confrontations occurred 46 times between Ntologi and Nsaba, 35 times between Ntologi and Kalunde, 20 times between Kalunde and Toshibo, and 19 times between Nsaba and Kalunde. All confrontations within other dyads numbered fewer than 16 and were nil for two pairs (Jilba–Musa and Bakali–Bembe).

Thus, antagonism centered around the top three males, and competition was for alpha status and coalition partners. Kalunde, the former alpha male, was involved in three of the four major dyadic rivalries, and was a major party around whom political instability centered in M Group during the study period.

'Side-directed behavior' and related reassurance behavior

Side-directed behavior
Side-directed behavior was seen 21 times (Table 9.5), when the performer appeared to solicit aid from a third

Table 9.3. *Direction of pant-grunt calls, combining focal and other-focal data*

Pant-grunts	Pant-grunted to									
	NT	NS	DE	JI	AJ	TB	BA	BE	MU	Total
NT	–	0	0	0	0	0	0	0	0	0
NS	16	–	0	0	0	0	0	0	0	16
DE	17	1	–	0	0	1	0	0	0	19
JI	14	2	0	–	0	0	0	0	0	16
AJ	42	2	6	0	–	0	0	0	0	50
TB	42	8	0	1	0	–	1	0	1	53
BA	8	0	1	1	0	2	–	0	0	12
BE	15	17	19	4	5	6	2	–	0	68
MU	23	2	2	0	3	3	1	0	–	34
Total	177	32	28	6	8	12	4	0	1	268

Table 9.4. *Initiations of dyadic agonistic confrontations combining focal and other-focal data*

Actor	Reactor									
	NT	NS	DE	JI	AJ	TB	BA	BE	MU	Total
NT	–	45	31+2	9+1	10	13	14	1+1	5	128+4
NS	1	–	7	9	6	5	3	6	3	40
DE	4	12	–	11	3	9	4	4	3	50
JI	0	2	0	–	3	2	2	4	0	13
AJ	0	0	0	4	–	8	3	1	6	22
TB	0	2	11+1	1	1	–	6	6+1	2	29+2
BA	2	1	0	1	0	1	–	0	4	9
BE	1	0	0	0	0	0	0	–	5	6
MU	1	0	0	0	1	0	0	1	–	3
Total	9	62	49+3	35+1	24	38	32	23+2	28	300+6

Note: Detour shown separately by +. If A detours to avoid B, A is the reactor in the table.

party. On nine (43%) occasions, the third party aggressed against or threatened the opponent of the performer of side-directed behavior. For example, when Kalunde encountered Jilba, he mounted Nsaba before he charged against Jilba while throwing a stick. Subsequently, Nsaba charged at Jilba while stamping and drove him up a tree.

De Waal & van Hooff (1981) proposed that side-directed behavior functions as a recruitment of support. If so, Ntologi solicited aid only from Kalunde and only when confronting Nsaba. Nsaba did so only from Kalunde and only when confronting Ntologi. Kalunde's solicitation policy, if any, was inconsistent. He not only solicited aid from Ntologi against Nsaba, but also sol-

icited aid from Nsaba against Ntologi. Apparently, he played each of his superiors off against the other.

In relation to the hypothesis that the major function of side-directed behavior is recruitment of support, it should be noted, however, that the 'recruited' party usually did not respond. Actually, allies intervened between antagonists more often without having been solicited (cf. Hemelrijk *et al.*, 1991). An adult male who confronts his rival may embrace a third party, and then aggress against his rival again. In this example, embracing could be interpreted as 'taking courage' as well as seeking to enlist support.

Moreover, affinitive behavior, such as mounting, embracing, approaching and holding-out-hand, that

Table 9.5. *Side-directed behavior*

A Solicits aid	B Aid solicited	C Contra	Type of behavior by A	B's response
NT	DE	NS	Mount	Neglect
NT	DE	NS	Mount	Threaten C
NT	DE	NS	Mount	Neglect
NT	DE	JL	Embrace	Neglect
NT	DE	BA	Mount	Neglect
NS	DE	NT	Run to B	Neglect
DE	NT	TB	Approach B, Scream	Follow A
DE	NT	NS	Mount	Charge C
DE	NT	NS	Embrace half	Charge C
DE	NS	NT	Run to B, Scream	Charge C
DE	NS	NT	Mount	Neglect
DE	NS	JI	Mount	Threaten C
DE	AJ	TB	Reach hand	Neglect
DE	AJ	NT	Reach hand, Scream	Neglect
DE	JI	TB	Approach B, Scream	Neglect
DE	MU	NT	Run to B, Scream	Charge C
AJ	NS	DE	Scream	Kick C
BA	NT	NS	Scream	Threaten C
BA	NT	AJ	Run to B, Scream	Follow A
BA	DE	TB	Scream	Display C
MU	NT	BA	Touch B, Scream	Neglect

constitutes side-directed behavior was directed not only from one of the combatants to a third party, but also more often the other way round.

On 15 (71%) occasions, the side-directed behavior was directed at a major ally of the antagonist, i.e. an individual likely to interfere against the performer. Possibly, one function of side-directed behavior is to keep a third party on neutral ground during combat: a third party was never seen to attack the performer even if he did not cooperatively attack the opponent of the performer.

Reassurance behavior by a third party to one of the combatants – non-agonistic support, give permission or reassurance

When two males confront each other, a third party may quickly approach and mount one of the fighting males. Thus, the third party appears to express his support for the one he mounts. For example, on seeing Nsaba chasing Kalunde, Ntologi approached and mounted Kalunde. Then, Kalunde turned and chased Nsaba. Thus, Ntologi apparently took sides with Kalunde, and as a result Kalunde could win over Nsaba when his defeat was otherwise likely.

Immediately before a male mates with an estrous female, a higher-ranking male may mount him. This may be called giving permission or reassurance.

Contexts of reassurance behavior

In order to clarify the meaning of gestures shown as side-directed behavior, seven patterns of reassurance were analyzed in terms of the context in which they occurred. Contexts were classified into seven categories on the basis of the relationships between an actor of reassurance A, a reactor B and the third party. Rivalry corresponds to the four most confrontational male pairs mentioned above (Ntologi–Nsaba, Ntologi–Kalunde, Nsaba–Kalunde and Kalunde–Toshibo).

Confrontation with rival. When A was involved in conflict with his rival, and shows reassurance behavior to B. Possible function was to gain reassurance.

Rival arrival. When the rival of A or of both A and B appeared on the scene, or his arrival was judged imminent by the observer from pant-grunts, pant-hoots or other calls in the distance. In

Table 9.6. *Reassurance behavior between two males dependent on context*

	CF	RV	CL	RS	RU	GR	TR	Total
Mount	17	19	1	7	9	0	8	61
Embrace full	2	4	4	1	2	3	1	17
Embrace half	1	1	1	0	1	0	1	5
HH, TC, FI	2	0	1	3	1	0	2	9
Open-mouth-kiss	1	1	1	1	2	9	1	16
Mouth	3	0	1	0	1	0	0	5
Approach	2	0	0	0	0	0	0	2
Total	13	25	9	12	16	12	13	115

Note: HH, hold-out-hand; TC, touch; FI, put finger into mouth; CF, confrontation with rival; RV, arrival of rival; CL, previous conflict; RS, resources; RU, reunion; GR, groom; TR, travel (see text).

response, A showed reassurance behavior to B. Possible function was to enlist support or gain reassurance.

Previous conflict. When A and B aggressed earlier against each other. Possible function was reconciliation.

Attraction to resources. When highly competed-for resources such as estrous females and meat are present close by to A and B. Possible function was to reduce competition.

Reunion. When A and B met after a separation. Possible function was to gain reassurance.

Groom. When A and B engaged in prolonged grooming. Function was ambiguous, but perhaps served to prolong friendship established temporarily by grooming or contact.

Travel. When behavior occurs during travel. Function was ambiguous.

The first three and the last four contexts are not mutually exclusive. If two contexts are applicable to one episode, the context listed earlier above was chosen (Table 9.6). Reassurance behavior shown in the contexts of 'confrontation with rival' and 'rival arrival' may correspond to side-directed behavior.

From this crude analysis, it is apparent that only mounting is clearly related to contexts of competition with a rival, although its function appears to be multifold.

Embrace occurred in various contexts, but most often in those of previous conflict and rival arrival. This suggests that embrace serves to establish reconciliation.

Open-mouth-kissing occurred in various contexts, but most often in the context of prolonged grooming. Perhaps open-mouth-kissing serves to maintain temporarily-established friendships.

Overall frequency of affinitive behavior

Here we analyze, regardless of context, the frequencies of behavior such as mount, embrace and kiss among adult males.

Mounting. Fifty-seven (93%) of 61 mounts involved one or two of the three top-ranking males (see Table 9.7). Thirty-two (52%) involved only two of the top three as the combinations of alpha–gamma or beta–gamma. This suggests that the gamma male Kalunde is a key figure in the process of the current social dyanamics. It is noteworthy that Ntologi and Nsaba, the two major rivals, never mounted each other.

Embracing. Embracing (Table 9.8a) also was most often seen between Ntologi and Kalunde or between Nsaba and Kalunde: Nine (53%) of 17 full embraces occurred in these combinations. Like mounting, embracing never occurred between Ntologi and Nsaba.

Open-mouth-kissing. Unlike embracing, open-mouth-kissing (Table 9.8b) involved Ntologi on 12 of 15 occasions. Open-mouth-kissing occurred between Ntologi and Kalunde seven times. Unlike mounting and embracing, Nsaba and Kalunde never kissed each other.

Table 9.7. *Distribution of mounting*

Mounts	Mounted									Total
	NT	NS	DE	JI	AJ	TB	BA	BE	MU	
NT	–	0	8	3	0	0	0	0	1	12
NS	0	–	5	1	1	0	0	3	0	10
DE	8	11	–	2	0	2	6	0	2	31
JI	1	0	0	–	0	0	0	0	0	1
AJ	0	0	0	0	–	0	0	0	0	0
TB	1	0	1	0	1	–	0	0	0	3
BA	1	0	0	2	0	1	–	0	0	4
BE	0	0	0	0	0	0	0	–	0	0
MU	0	0	0	0	0	0	0	0	–	0
Total	11	11	14	8	2	3	6	3	3	61

Table 9.8a. *Distribution of full embracing*

	NT	NS	DE	JI	AJ	TB	BA	BE	MU
NT	–	0	4	0	0	0	0	0	1
NS		–	5	0	1	0	0	1	0
DE			–	0	1	3	1	0	0
JI				–	0	0	0	0	0
AJ					–	0	0	0	0
TB						–	0	0	0
BA							–	0	0
BE								–	0
MU									–

Table 9.8b. *Distribution of open-mouth kissing*

	NT	NS	DE	JI	AJ	TB	BA	BE	MU
NT	–	0	7	0	0	2	1	0	2
NS		–	0	0	0	0	0	0	0
DE			–	0	0	1	0	0	1
JI				–	0	0	0	0	0
AJ					–	0	0	0	0
TB						–	1	0	0
BA							–	0	0
BE								–	0
MU									–

Overt coalition strategies

Patterns of intervention

When two adult males confronted each other, a third party often intervened between them. Some adult males systematically supported one member of a dyad over the other. Table 9.9 shows 23 interventions observed during the study period.

Ntologi's interventions always involved Kalunde. On four of five occasions he supported Kalunde, in particular, against Nsaba. Once, Ntologi aggressed against Nsaba and Kalunde, supporting Bakali. Kalunde supported Ntologi twice on occasions when the latter aggressed against Nsaba. Thus, Ntologi and Kalunde generally supported each other.

Similarly, Nsaba's intervention almost always involved Kalunde. Nsaba supported Kalunde on three of four occasions, one of which was against Ntologi. On the fourth occasion, Nsaba supported Aji against Kalunde.

The rivalry between Kalunde and Toshibo was another major factor bringing instability to M Group. Interestingly, both Ntologi and Nsaba supported Kalunde against Toshibo. This is understandable because Kalunde was the most important ally for these highest-ranking males.

Interventions may be classified into 'actor vs reactor alliances' or 'winner vs loser support' (de Waal *et al.*, 1976). Among the adult males of this study, actors were as likely to be supported as reactors. It was difficult to determine whether interventions were winner or loser support, since rank was ambiguous among the middle-ranking males. However, if males who have begun to be challenged recently are defined as less dominant, loser support (13 dyads) occurred more frequently than winner support (6 dyads). Age seemed to be an important factor: adult males were more likely to support the older of two combatants. They took sides with the older

Table 9.9 *Intervention in agonistic confrontations*

Intervener	Pro	Contra	AA/RA	WS/LS	OS/YS
NT	DE	NS	AA	LS	OS
NT	DE	NS	RA	LS	OS
NT	DE	NS	AA	LS	OS
NT	DE	TB	RA	LS	OS
NT	BA	NS, DE	AA	LS	OS
NS	AJ	DE	RA	LS	YS
NS	DE	NT	AA	LS	YS
NS	DE	JI	AA	WS	OS
NS	DE	TB	RA	LS	OS
DE	NT	NS	AA	WS	OS
DE	NT	NS	AA	WS	OS
DE	AJ	JI	RA	LS	OS
DE	AJ	JI	RA	LS	OS
DE	NS	JI	RA	WS	OS
DE	BA	TB	RA	LS	OS
DE	BA	NT	RA	LS	OS
JI	DE	TB	RA	LS	OS
JI	TB	BA	AA	WS	YS
JI	DE	NS	AA	LS	OS
TB	DE	JI	AA	WS	OS
TB	DE	NS	AA	LS	OS
BA	DE	JI	AA	WS	OS
MU	DE	NT	RA	LS	YS

Note: Pro, support for; Contra, support against; AA, actor-alliance; RA, reactor-alliance; WS, winner support; LS, loser support; OS, supports older combatant; YS, supports younger combatant.

combatant in 15 of 19 dyadic combinations (binomial test, $p < 0.01$).

Separating intervention

An adult male sometimes suddenly charged into a group of adult males who were peacefully grooming or resting in close proximity. As a result of the charge, the other males scattered in all directions. This behavior, called 'separating intervention' by de Waal (1982), was seen 25 times during this study.

For one adult male to displace two or more adult males, he must be able to dominate all others in combination. Thus, separating intervention is likely to be performed only by high-ranking males. Twenty (80%) of 25 separating interventions observed were performed by the alpha male, Ntologi (see Table 9.10).

Table 9.10. *Separating interventions*

Intervener	Group	No. of cases
NT	NS, DE	11
NT	NS, DE, BA	3
NT	NS, BA	1
NT	NS, BA, AJ, MA	1
NT	NS, JI	1
NT	DE, TB	1
NT	DE, TB, BA	1
NT	AJ, JI	1
NS	AJ, TB	1
NS	DE, AJ, JL, MA, JN	1
DE	NS, BA	1
AJ	JI, TB	1
TB	BA, BE	1
Total		25

Note: MA, adolescent male; JN, adult female.

Table 9.11. *Simultaneous, cooperative attack or threat*

Allies		Common opponent	N
DE + NS	vs	JI	1
DE + NT	vs	NS	1
DE + TB	vs	NS	1
DE + BA	vs	TB	1
DE + MU	vs	NT	1

Ntologi's policy was clear; in 14 of his 20 interventions, the target group included both Nsaba and Kalunde. Nsaba was in the threatened group on 17 occasions and Kalunde on 18 occasions. Thus, Ntologi's major targets of intervention being Nsaba and Kalunde, he appeared to aim at preventing close association between them. This is understandable, given that he had been exiled by the coalition between these males in 1991 (Hamai, 1994).

An adult male (in particular, Kalunde) who had been grooming with Nsaba sometimes left him and rushed to pant-grunt to Ntologi when he noticed the approach of the latter in the distance. On such occasions, Ntologi either severely attacked or mounted and sometimes embraced Nsaba's grooming partner.

Simultaneous, cooperative aggression
On five occasions, two adult males simultaneously charged at a single adult male (see Table 9.11). Each

Table 9.12a. *Percent time spent with other males*

Focal males	NT	NS	DE	JI	AJ	TB	BA	BE	MU
NT	–	14	15	7	9	4	10	4	16
NS	7	–	20	9	14	6	9	6	5
DE	30	25	–	17	10	13	5	2	7
JI	13	6	10	–	6	5	17	4	5
AJ	18	16	4	4	–	8	13	5	16
TB	9	10	13	10	4	–	6	7	4
BA	29	13	9	12	6	10	–	3	6
BE	5	7	7	7	6	13	6	–	1
MU	25	7	10	7	11	12	13	14	–

Table 9.12b. *Percent time spent in two-male party*

Focal males	NT	NS	DE	JI	AJ	TB	BA	BE	MU
NT	–	3.1	6.8	1.9	5.5	3.3	3.1	0.2	6.1
NS	1.0	–	8.7	2.1	10.2	4.8	3.3	6.0	2.4
DE	13.1	11.8	–	2.1	3.5	2.1	0.2	0.1	1.5
JI	1.4	2.8	3.8	–	0.7	2.5	3.8	0.2	1.2
AJ	2.8	7.7	0.4	1.7	–	6.3	2.3	4.2	3.2
TB	1.3	0.6	1.1	2.4	0.5	–	0.3	6.5	0.6
BA	12.2	0.6	2.1	2.9	0.8	2.7	–	1.6	0.2
BE	1.0	0.9	0.5	0.7	1.4	7.3	2.8	–	0.3
MU	8.4	3.1	2.1	1.4	4.5	2.9	1.1	6.2	–

time Kalunde was one of the two males. Three times, cooperative attacks occurred after Kalunde mounted or embraced his allies. Kalunde's coalition tactics seemed opportunistic, since to him Ntologi, Nsaba, and Toshibo were allies on some occasions and rivals on others.

Association among males

Table 9.12a shows the percentage of total time spent by a focal male with other adult males. Two dyads of coalition partners, Ntologi–Kalunde and Nsaba–Kalunde, showed conspicuously high rates as did dyads of the oldest males, Ntologi–Bakali and Ntologi–Musa. However, two low-ranking old males, Bakali and Musa, associated little.

In order to know the association preferences among adult males, percent time spent in a 2-male party (with only one other adult male) may be more relevant (Table 9.12b). Even here, the two dyads of Ntologi–Kalunde and Nsaba–Kalunde had conspicuously high rates, along with Ntologi–Musa and Ntologi–Bakali. The major

rivals, Ntologi and Nsaba rarely formed a 2-male party. Thus, close association is also among the coalition strategies used by adult males. However, new patterns of association such as Aji and Nsaba, and Toshibo and Bembe also emerged. Nsaba's closest associate was Aji during Nsaba's focal period, and similarly Aji's closest associate was Nsaba during Aji's focal period. Similarly, Toshibo and Bembe were each other's closest associate. Toshibo and Bembe were also seen to share a large portion of meat with each other (see Table 9.15 below). Since these two dyads rarely formed coalitions (see Tables 9.5, 9.9, 9.10, 9.11), they cannot be called coalition partners, but may be called friends.

One can speculate about the meaning of such friendships. Many adult males appear to be keen to raise their dominance status. However, there seem to be two types of males, those who are preoccupied with the promotion of their rank and those who appear less interested (Goodall, 1971; Bygott, 1974). Nsaba and Toshibo are among the former type of males, and Aji and Bembe among the latter. Such a combination of males may be

Table 9.13a. *Frequency of grooming (bouts)*

Focal males	NT	NS	DE	JI	AJ	TB	BA	BE	MU	Total
NT	–	0	53*	8	49	24	27	0	16*	177
		0	37	2	50	34	16	0	0	139
NS	0	–	49	36	61*	0*	34	2	9	191
	0		64	18	24	13	24	6	2	151
DE	89	88	–	25	23	49	20	0	11	305*
	59	64		12	24	55	15	0	8	237
JI	5	9	8	–	2	35*	20	0	0	79
	7	22	9		9	14	7	0	0	68
AJ	36	89*	2	26	–	40	21	5	18	237
	25	143	0	22		48	9	4	17	268
TB	5	5	12	23	10	–	12	28	21*	116
	5	8	9	25	7		13	22	5	94
BA	93	20*	14	15	17	60	–	21	0	240*
	83	51	24	6	18	72		25	0	519
BE	0	2	30*	0	2	53	9	–	0	96
	0	0	10	0	1	80	1		0	92
MU	37	3*	3	0*	20*	15*	1	2*	–	81*
	56	29	11	8	42	35	4	13		198

Note: Number of grooming bouts when column males were targets. In each column, figures of upper row are numbers of bouts when focal males groomed and lower row are bouts when they were groomed. Asterisks (*) show asymmetrical grooming relationships (binomial, $p < 0.05$, 2-tailed.)

compatible. The friendship between Ntologi and old declining males Musa and Bakali may also be explained in a similar way.

Grooming among males

Table 9.13a shows that among the three oldest males, Bakali and Musa were groomed significantly more often than they groomed, while the alpha male, Ntologi, was more often a groomer. Such exceptional status for an old alpha male was also reported for the chimpanzees of Gombe (Simpson, 1973).

The two top-ranking males, Ntologi and Nsaba, hardly ever groomed each other. Although Nsaba occasionally followed and even sought contact with Ntologi, Ntologi ignored him and did not allow him to groom. Ntologi's refusal to reconcile appeared to be one of his measures to intimidate Nsaba (cf. de Waal, 1982). However, Kalunde maintained close grooming relationships with both Ntologi and Nsaba. This configuration

of grooming relations, which accords well with the results of Hayaki (1995), supports the idea that Ntologi and Nsaba were major rivals who competed for a coalition with Kalunde. However, this is not the whole story.

For Ntologi, Kalunde was an indispensable ally. Ntologi groomed Kalunde not only frequently but also with unusual intensity. The body parts male chimpanzees groomed were recorded, with the face normally being groomed rarely. However, Ntologi groomed Kalunde's face in 23% of their bouts (Table 9.13b). Moreover, he often used lips and teeth in addition to hands when grooming the face of Kalunde. He not only removed dirt from his face, but also pushed his index finger into Kalunde's nostrils and ate the mucus. (Perhaps as a result, Kalunde's flu was transmitted to Ntologi!) Kalunde also frequently groomed Ntologi and Nsaba on their faces using lips and teeth when doing so. However, Nsaba's attitude was different. Although he spent many hours grooming Kalunde, he rarely groomed his face.

Table 9.13b. *Rate (%) of face grooming among adult males*

Grooms	Groomed									Average
	NT	NS	DE	JI	AJ	TB	BA	BE	MU	
NT	–	X	23	10	8	0	1	X	12	11
NS	X	–	11	2	6	0	0	X	3	3
DE	13	8	–	3	10	0	0	0	0	7
JI	0	8	0	–	4	10	4	X	13	6
AJ	2	5	0	6	–	2	0	0	0	2
TB	0	4	7	3	9	–	6	5	4	5
BA	6	3	3	0	0	3	–	5	4	4
BE	X	0	5	X	0	0	0	–	0	1
MU	7	13	0	X	0	0	0	0	–	4
Average	7	6	12	4	6	3	2	4	6	6

Note: X, no grooming seen (see Table 9.13a).

Table 9.14a. *Initiation of grooming session among adult males: approach*

Approaches	Approached									Total
	NT	NS	DE	JI	AJ	TB	BA	BE	MU	
NT	–	0	9	4	1	1	11	0	13	39
NS	0	–	8	2	8	3	4	1	8	34
DE	9	6	–	3	1	2	8	0	4	33
JI	2	3	1	–	1	1	5	0	2	15
AJ	6	1	3	2	–	2	10	1	4	29
TB	6	3	6	4	1	–	7	7	4	38
BA	13	5	1	4	0	3	–	2	5	33
BE	0	3	3	0	2	3	2	–	2	15
MU	6	1	1	0	2	0	1	2	–	13
Total	42	22	32	19	16	15	48	13	42	249

In the initiation of grooming, the general rule is for younger (Takahata, 1990) or lower-ranking (Simpson, 1973) males to initiate grooming with older, or higher-ranking ones. Initiation of grooming was investigated using two measures: 'approach' and 'first grooming' (Simpson, 1973; Takahata, 1990). In terms of approach to a grooming partner (Table 9.14a), 'approach rate' (approach/approached) was inversely correlated with age ($N = 9$, $r_s = 0.71$, $p < 0.05$). For his age, however, Kalunde's approach to his partner occurred more often than expected. Ntologi also had a higher approach rate (39/42) than his agemates, Musa (13/42) and Bakali (33/48). This may show their motivation to engage in maintaining politically valuable partnerships. 'Approach rate' was not correlated with dominance rank ($r_s = 0.36$).

The 'first-grooming rate' (groom/groomed) was correlated neither with age ($r_s = -0.51$) nor with dominance rank ($r_s = 0.36$) (see Table 9.14b). However, the two rivals, Ntologi and Nsaba, first groomed other males more often than they were groomed, while the opportunist Kalunde was first groomed more often than he initiated grooming.

Thus, although both Kalunde and Ntologi spent many hours grooming each other, their grooming was not symmetrical. Ntologi approached Kalunde as often as Kalunde did so to Ntologi. However, Ntologi first groomed Kalunde far more often than vice versa (see Table 9.14b). Kalunde often approached Ntologi to solicit grooming from him, while Ntologi rarely did so. Perhaps because Ntologi and Nsaba were absolutely incompatible, Kalunde took advantage of the rivalry between these two higher-ranking males. So, Kalunde

Table 9.14b. *Initiation of grooming session among adult males: first grooming*

Grooms	Groomed									Total
	NT	NS	DE	JI	AJ	TB	BA	BE	MU	
NT	–	0	24	9	6	1	11	0	23	74
NS	0	–	10	6	11	5	7	1	10	50
DE	11	10	–	5	3	2	8	0	6	45
JI	2	3	3	–	1	1	5	0	2	33
AJ	7	1	3	5	–	1	8	0	8	17
TB	6	6	14	8	2	–	5	6	9	56
BA	17	5	1	4	1	4	–	2	6	40
BE	0	4	4	0	2	2	2	–	4	18
MU	6	1	1	0	3	0	1	2	–	14
Total	49	30	60	37	29	16	47	11	68	347

Table 9.15. *Adult males as the final possessors of carcasses by hunting, snatching or scavenging*

	Control carcass observed	Method of acquisition			Sharing seen	Shared with
		Snatch	Hunt or scavenge	Method unknown		
NT	13	7	2	4	11	see Table 9.16
NS	3	2	0	1	0	
DE	4	1	3	0	0	
JI	1	0	1	0	0	
AJ	0	0	0	0	0	
TB	4	0	3	1	2	BE, AB[a]
BA	4	1	1	2	0	
BE	3	0	3	0	2	TB, BO[b], TY[c]
MU	2	1	1	0	0	
Total	34	12	14	8	15	

Note: [a], TB's sister; [b], BE's mother; [c], BE's sister.

not only maintained an important political relationship, but also obtained extra benefits.

The two dyads of 'friends' (Nsaba and Aji, and Toshibo and Bembe) frequently engaged in grooming (see Table 9.13a).

Meat sharing

The alpha male Ntologi possessed the carcasses of mammals (usually colobus monkeys) on 13 occasions during the study period (Table 9.15; Fig. 9.2). There were seven chimpanzees whom Ntologi was seen to share meat with more than twice: the two oldest adult males, Bakali and Musa, and five adult females (see Table 9.16).

Nsaba possessed meat only three times: twice he disappeared from the scene, probably monopolizing the meat. Once he was followed by Bakali and two adult females, but Nsaba was not seen to share meat. Kalunde possessed mammalian carcasses on four occasions. Each time, he disappeared from the scene.

This result shows that the alpha male, Ntologi, was able to control the distribution of meat on most occasions.

Mating behavior

The alpha male Ntologi mated with 'stage-3' females (i.e. likely to ovulate; Hasegawa & Hiraiwa-Hasegawa, 1983) more often (20 times) than any other male. The highest reproductive advantage conferred to an alpha male was consistent with the results of previous studies (Tutin, 1979; Hasegawa & Hiraiwa-Hasegawa, 1983; Nishida, 1983).

Fig. 9.2. Alpha male (right) and beta male chimpanzees with
meat of red colobus monkey at Mahale. (Photo by T. Nishida)

Table 9.16. *Chimpanzees with whom the alpha male Ntologi shared or redistributed meat*

Name	Sex	Age*	Reproductive state	No of episodes (large portion shared)	No. of episodes (small portion shared)
ZP	F	Prime	Cycling	10	0
WO	F	Old	Lactating	10	0
BA	M	Old	–	8	1
MU	M	Old	–	8	0
XT	F	Young	Cycling	4	0
DO	F	Old	Lactating	3	0
JN	F	Young	Lactating	1	2
SL	F	Old	Lactating	2	0
CA	F	Old	Cycling	2	0
SV	F	Adol.	–	2	0
GK	F	Old	Cycling	2	0
WL	F	Old	Cycling	1	0
PU	F	Young	Lactating	1	0
WA	F	Old	Cycling	1	0
IK	F	Prime	Cycling	1	0
PI	F	Prime	Lactating	1	0
AJ	M	Young	–	0	1
BT	F	Young	Lactating	0	1
BO	F	Old	Cycling	0	1
GW	F	Old	Cycling	0	1

Note: *Old, old adult ($\geqslant 35$ years); Prime, prime adult ($> 20-< 35$); Young, young adult ($> 15-\leqslant 20$ for males and $> 13-\leqslant 20$ for females); Adol., adolescent female ($> 8-< 13$).

Table 9.17a. *Frequency of mating by stages of female estrous cycle*

	ST1	ST2	ST3	Total	Average stage
NT	4	7	20	31	2.52
NS	4	6	9	19	2.26
DE	3	2	9	14	2.43
JI	6	2	7	15	2.07
AJ	3	3	2	8	1.88
TB	3	7	2	12	1.92
BA	4	0	2	6	1.67
BE	3	4	3	10	2.00
MU	2	2	4	8	2.25

Note: Day of ovulation is defined as Day 0, and Day −1 as one day before Day 0. Thus, ST3 refers to Day −1 to −3, ST2 Day −4 to −7 and ST1 Day −8 and before it (Hasegawa & Hiraiwa-Hasegawa, 1983). Average stage for NT is calculated as $(4 \times 1 + 7 \times 2 + 3 \times 20)/31 = 2.52$.

It appears that Ntologi's presence influenced the mating behaviors of other males: Kalunde mated with stage-3-females as often as Nsaba (9 times each). Two other partners of Ntologi, Musa and Bakali, at the bottom of the male hierarchy, mated with stage-3-females as often as other middle and low-ranking males (see Table 9.17a).

Ntologi's influence is more apparent when the number of other males present is compared during mating with estrous females. There were more adult males present in the proximity of a mating male when the male was one of Ntologi's three partners (i.e. DE, BA or MU) than when he was a non-partner (see Table 9.17b). This suggests that these adult males were able to mate despite the presence of many males because they were less often interrupted in

Table 9.17b. *Number of other adult males present in proximity of a mating male*

Target male	Focal	Focal + other focal
NT	1.10	1.81
NS	0.45	0.74
DE	0.88	1.21
JI	0.27	0.78
AJ	0.62	0.59
TB	0.33	0.58
BA	1.25	1.17
BE	0	1.00
MU	1.20	1.50

mating than were other non-partner males owing to their more frequent association with Ntologi.

DISCUSSION

Coalition strategy in a large versus small group

In such a large group as M Group with as many as nine adult males, the most important political relationships were triangular, involving only the alpha, beta and gamma males. The basic political relationships were similar to those decribed for groups including only three fully adult males (de Waal, 1982; Nishida, 1983), as all major confrontations occurred among the three adult males. Riss and Goodall (1977) reported a case of a power takeover in a large group at Gombe, in which two brothers allied against a single male.

Other complications were unique to large groups. For example, when young, rank-oriented adult males such as Toshibo threatened the position of Kalunde, both Ntologi and Nsaba supported Kalunde against Toshibo. This is probably because for these two rivals (Ntologi and Nsaba), Kalunde was more important as an ally, and failure to support Kalunde could result in the collapse of a current or future coalition with him. Thus, whereas a large number of adult males do provide greater potential for support, they are not the crucial factor.

The presence of old males also added complications. The oldest male, Bakali, once was a beta male and had been the rival of Ntologi from at least 1981 to 1985 (Kawanaka, 1989; Takahata, 1990). However, after declining in rank around 1987, he became one of Ntologi's supporters (Nishida *et al.*, 1992). Even when Ntologi was temporarily isolated from M Group after a *coup d'etat* by Kalunde and Nsaba in 1991, he was occasionally seen roaming only with Bakali (Nishida, 1994). This reflects the strength of Bakali's alliance with Ntologi that lasted for about five years. Musa was not closely associated with Ntologi from 1981 to 1986 (Kawanaka, 1990; Takahata, 1990), but close association between them occurred after Musa declined in rank around 1989 (Mitani & Nishida, 1993). How important these alliances are is unclear, as old males rarely or never get involved in physical confrontations between the alpha male and other males. Usually their participation is limited to barking. However, the presence of these old associates appears to give encouragement and reassurance to the alpha male (see also Kawanaka, 1990). The oldest males tend to support an alpha male, manifesting a winner-support strategy, since this is not only less risky for declining males but also guarantees their status in the male cluster. Perhaps this is the reason why an alpha male and the oldest males are likely to keep tight-knit bonds. During this study period, however, these males exerted scarcely any political influence, because they were too old and feeble by 1992. Both males died in 1993.

Female intervention

There were as many as 29 adult females in M Group, but they were virtually absent from confrontations among adult males (Fig. 9.3). However, there have been cases in which females joined in aggressive confrontations between adult males of M group during times not covered in this study.

In 1991, Kalunde, then alpha, chased Ntologi, who was roaming alone as an exiled ex-alpha male. At that time some adult females, together with some adult males, cooperated with Kalunde against Ntologi (Nishida, 1994). In the same year Jilba was subjected to a gang attack lead by Kalunde, when two adult females cooperated with Kalunde against Jilba. In 1992, just before this study, research assistants observed a similar gang attack on Jilba, in which a few adult females again joined forces with Kalunde (Nishida et al., 1995).

These observations and unpublished data suggest that females are winner-supporters, except when the victims are close relatives, such as their own immature offspring. They do not intervene between competing adult males, unless the winning side is easily predicted as was apparent in the gang attacks. This is understandable, since females might risk both their own and their infants' safety if they intervene in the fighting of the bigger and better-armed adult males.

Thus, females' extreme reluctance to participate in male confrontations appears to be the rule rather than an exception in the natural environment (Riss & Goodall, 1977; Nishida, 1983; Hosaka, 1992; Hamai, 1994).

Grooming strategy

Grooming was seen most often between allies and friends, and so it is regarded as one of the important coalition strategies. Moreover, it appears that grooming of the face has

Fig. 9.3. Adult female chimpanzee in grooming group seeks reassurance from alpha male at Mahale. (Photo by C. Tutin)

a special meaning among male chimpanzees. When a chimpanzee grooms the face of the other, he must gaze at the companion in a face-to-face posture. So, any sign of infidelity on the part of the groomee might be betrayed. The fact that panting is often emitted in face-grooming suggests that tension is high between grooming partners.

Meat-sharing as a coalition strategy?

If meat sharing can be used as a coalition tactic by an alpha male (Nishida et al., 1992), Kalunde should have been granted more meat after he dropped in rank from the second- to third-ranking male. However, he was never given meat during this study period. This suggests that meat may be shared only with partners in an enduring coalition, not with an untrustworthy individual such as Kalunde, who assumes a strategy of 'alliance fickleness'. However, the authors observed Ntologi sharing meat twice with Kalunde, once before and once after this study, so it is likely that the pattern of meat redistribution by Ntologi is dynamic.

Comparison with bonobos

Since there is usually only one silverback male in a gorilla group and even if there are two, they are likely to be an old father and his mature son (Watts, Chapter 2; Tutin, Chapter 5) and orangutans are solitary (van Schaik & van Hooff, Chapter 1), no comparable data on their male support strategies are to be expected (however, see Yamagiwa, 1987, for loser-support strategies in an all-male group consisting of mature and immature male gorillas). On the other hand, male bonobos live in a unit group or community similar to that of chimpanzees (Kuroda, 1979; Kano, 1992), but their relationships within the group appear surprisingly different. There has been scarcely any evidence of male coalitions. Kano (1992) described two examples of a strong association between adult males (one consisting of three members and another of five), but he had no evidence that these groups joined forces against other males.

Ihobe (1992) observed instances of two or more males attacking a single male at the provisioning site at Wamba. However, this happened only four times during 242 hours of observation. Each time, a third party joined forces with an aggressor, thus forming an actor-alliance or winner-support. This contrasts sharply with 29 observations of coalition observed without provisioning during

390 hours in the present study, most of which consisted of loser-support. Moreover, adult female bonobos often intervene in disputes between their adult sons and other adult males; as a result males gain dominance with the aid of influential mothers (Furuichi, 1989; Ihobe, 1992; Kano, 1992). Such maternal intervention has never been reported for wild chimpanzees.

Although more study on bonobos is needed, it is likely that aggression is milder and reconciliation quicker among bonobos than chimpanzees (Ihobe, 1992; de Waal, 1989). It is also likely that bonobo males have evolved a different way of resolving male–male competition than chimpanzees. The ultimate cause for this difference may be that access to estrous females is easier because of their longer receptivity and the stronger gregariousness of bonobo females (Ihobe, 1992; Kano, 1992, Chapter 10).

SUMMARY

1. Three major rivalries were found among the three highest-ranking males, alpha (Ntologi), beta (Nsaba) and gamma (Kalunde) during this study. The most important discord was between alpha and beta, although aggression was directed almost one-sidedly from alpha to beta.

2. The alpha male's political strategies were similar to those described for captive chimpanzees. He frequently associated, groomed and supported his coalition partner, Kalunde. He often took the initiative when initiating interaction with Kalunde. In addition to charging at Nsaba when the latter was alone, he prevented the formation of coalitions between Nsaba and other adult males, Kalunde in particular, by charging into the grooming party including his rival. The alpha also refused to interact with Nsaba who sought contact with him. The neglect or refusal of reconciliation was an important tactic displayed by the alpha to intimidate his rival, as suggested by de Waal (1982).

3. The beta male, Nsaba, was cowed by the alpha and almost gave up challenging him except when he was instigated to do so by Kalunde. However, he sought to associate and groom with his coalition partner Kalunde and his friend Aji. Nsaba also spent a lot of time alone in order to avoid aggression from the alpha.

4. The gamma male, Kalunde, displayed an opportunistic coalition tactic formerly called 'alliance fickleness,' in which the third party changes sides from moment to moment. He frequently supported, associated and groomed with the alpha on one occasion, but did the same thing with the beta on another. He even instigated Nsaba to attack Ntologi. However, his tactic did not work well, unlike the cases for Kamemanfu (Nishida, 1983) and Yeroen (de Waal, 1982). This time, discord between the two highest ranking males did not give strong 'political' power to the gamma male. This is because the strength of the two superiors was not evenly matched.

5. Since gamma was the common coalition partner of alpha and beta, he was supported by either or both of them when he had a dispute with other adult males.

6. Even if there are as many as nine adult males in a group, the basic political relationship was triangular as was previously reported for small groups including only three to four adult males. The key coalition was composed of only two adult males. Mass coalition against a single strong adult male is unlikely to occur in wild chimpanzees unless the target is a socially isolated individual or a member of a different unit group.

7. Females did not intervene in, and had virtually no influence on, male–male confrontations. This was consistent with previous observations in the natural habitat, but different from those reported for captive chimpanzees and bonobos in the wild.

ACKNOWLEDGMENTS

This paper was originally prepared for the Wenner-Gren Conference, 'The Great Apes Revisited,' held November 12–19, 1994 in Cabo San Lucas, Baja, Mexico. I thank Sydel Silverman for her warm invitation. I thank the Tanzania Commission for Science and Technology, Serengeti Wildlife Research Institute, Tanzania National Parks and Mahale Mountains Wildlife Research Centre for permission to study at Mahale; C. Mlay, H. Y. Kayumbo, E. Massawe and A. H. Seki for their assistance in obtaining research permission; H. Hayaki, J. C. Mitani, and K. Norikoshi for cooperation in the field; M. B. Kasagula, R. Kitopeni and H. Rashidi for assistance in collecting data; and W. C. McGrew and L. Marchant, F. B. M. de Waal and M. A. Huffman for useful comments. The study was supported by a grant under the Monbusho International Scientific Research Program (No. 03041046 to TN).

REFERENCES

Boehm, C. 1992. Segmentary 'warfare' and the management of conflict: comparison of East African chimpanzees and patrilineal-patrilocal humans. In: *Coalitions and Alliances in Humans and Other Animals*, ed. A. H. Harcourt & F. B. M. de Waal, pp. 137–73. Oxford: Oxford University Press.

Bygott, J. D. 1974. Agonistic behaviour and dominance in wild chimpanzees. Ph.D. thesis, University of Cambridge.

Bygott, J. D. 1979. Agonistic behavior, dominance, and social structure in wild chimpanzees of the Gombe National Park. In: *The Great Apes*, ed. D. A. Hamburg & E. R. McCown, pp. 405–27. Menlo Park, CA: Benjamin/Cummings.

Byrne, R. 1995. *The Thinking Ape: Evolutionary Origins of Intelligence*. Cambridge: Cambridge University Press.

Connor, R. C., Smolker, R. A. & Richards, A. F. 1992. Dolphin alliances and coalitions. In: *Coalitions and Alliances in Humans and Other Animals*, ed. A. H. Harcourt & F. B. M. de Waal, pp. 415–43. Oxford: Oxford University Press.

Furuichi, T. 1989. Social interactions and the life history of female *Pan paniscus* in Wamba, Zaïre. *International Journal of Primatology*, **10**, 173–97.

Goodall, J. 1971. *In the Shadow of Man*. London: Collins.

Goodall, J. 1973. The behavior of chimpanzees in their natural habitat. *American Journal of Psychiatry*, **130**, 1–12.

Goodall, J. 1986. *The Chimpanzees of Gombe: Patterns of Behavior*. Cambridge, MA: Harvard University Press.

Goodall, J. 1992. Unusual violence in the overthrow of an alpha male chimpanzee at Gombe. In: *Topics in Primatology, Vol. 1, Human Origins*, ed. T. Nishida, W. C. McGrew, P. Marler, M. Pickford & F. B. M. de Waal, pp. 131–42. Tokyo: University of Tokyo Press.

Hamai, M. 1994. 'Attractive' females for male chimpanzees? *Primate Research*, **10**, 33–40. (in Japanese)

Harcourt, A. H. 1992. Coalitions and alliances: are primates more complex than non-primates? In: *Coalitions and Alliances in Humans and Other Animals*, ed. A. H. Harcourt & F. B. M. de Waal, pp. 444–71. Oxford: Oxford University Press.

Hasegawa, T. & Hiraiwa-Hasegawa, M. 1983. Opportunistic and restrictive matings among wild chimpanzees in the Mahale Mountains. *Journal of Ethology*, 1, 75–85.

Hayaki, H. 1995. Social relations of the second ranking male chimpanzee at M Group of the Mahale Mountains National Park, Tanzania. *Primate Research*, 10, 289–305. (in Japanese)

Hayaki, H., Huffman, M. A. & Nishida, T. 1989. Dominance among male chimpanzees in the Mahale Mountains National Park, Tanzania: a preliminary study. *Primates*, 30, 187–97.

Hemelrijk, C. K., Klomberg, T. J. M., Nooitgedagt, J. H. & van Hooff, J. A. R. A. M. 1991. Side-directed behaviour and recruitment of support in captive chimpanzees. *Behaviour*, 118, 89–102.

van Hooff, J. A. R. A. M. 1974. A structural analysis of the social behaviour of a semi-captive group of chimpanzees. In: *Social Communication and Movement*, ed. M. von Cranach & I. Vine, pp. 75–162. London: Academic Press.

Hosaka, K. 1992. A new case of the power takeover in the M Group chimpanzees of the Mahale Mountains. Paper presented at the 11th Congress of Japan Ethological Society, Tsukuba.

Ihobe, H. 1992. Male–male relationships among wild bonobos (*Pan paniscus*) at Wamba, Republic of Zaïre. *Primates*, 33, 163–79.

Kano, T. 1992. *The Last Ape: Pygmy Chimpanzee Behavior and Ecology*. Stanford, CA: Stanford University Press.

Kawanaka, K. 1989. Age differences in social interactions of young males in a chimpanzee unit-group of the Mahale Mountains National Park, Tanzania. *Primates*, 30, 285–305.

Kawanaka, K. 1990. Alpha males' interactions and social skills. In: *The Chimpanzees of the Mahale Mountains*, ed. T. Nishida, pp. 171–87. Tokyo: University of Tokyo Press.

Kuroda, S. 1979. Grouping of the pygmy chimpanzees. *Primates*, 20, 161–83.

Mitani, J. & Brandt, K. L. 1994. Social factors influence the acoustic variability in the long-distance calls of male chimpanzees. *Ethology*, 96, 233–52.

Mitani, J. & Nishida, T. 1993. Contexts and social correlates of long-distance calling by male chimpanzees. *Animal Behaviour*, 45, 735–46.

Montagu, M. F. A. (ed.) 1968. *Man and Aggression*. Oxford: Oxford University Press.

Nishida, T. 1983. Alpha status and agonistic alliance in wild chimpanzees (*Pan troglodytes schweinfurthii*). *Primates*, 24, 318–36.

Nishida, T. (ed.) 1990. *The Chimpanzees of the Mahale Mountains: Sexual and Life History Strategies*. Tokyo: University of Tokyo Press.

Nishida, T. 1994. Review of recent findings on Mahale chimpanzees: implications and future research directions. In: *Chimpanzee Cultures*, ed. R. W. Wrangham, W. C. McGrew, F. B. M. de Waal & P. Heltne, pp. 373–96. Cambridge, MA: Harvard University Press.

Nishida, T., Hasegawa, T., Hayaki, H., Takahata, Y. & Uehara, S. 1992. Meat-sharing as a coalition strategy by an alpha male chimpanzee? In: *Topics in Primatology, Vol. 1, Human Origins*, ed. T. Nishida, W. C. McGrew, P. Marler, M. Pickford & F. B. M. de Waal, pp. 159–74. Tokyo: University of Tokyo Press.

Nishida, T., Hosaka, K., Nakamura, M. & Hamai, M. 1995. A within-group gang attack on a young adult male chimpanzee: ostracism of an ill-mannered member? *Primates*, 36, 207–11.

Nishida, T. & Kawanaka, K. 1972. Inter-unit-group relationships among wild chimpanzees of the Mahali Mountains. *Kyoto University African Studies*, 7, 131–69.

Plooij, F. X. 1984. *The Behavioral Development of Free-Living Chimpanzee Babies and Infants*. Norwood, NJ: Ablex.

Riss, D. & Goodall, J. 1977. The recent rise to the alpha rank in a population of free-living chimpanzees. *Folia Primatologica*, 27, 134–51.

Simpson, M. J. A. 1973. The social grooming of male chimpanzees. In: *Comparative Ecology and Behaviour of Primates*, ed. R. P. Michael & J. H. Crook, pp. 411–505. London: Academic Press.

Takahata, Y. 1990. Social relationships among adult males. In: *The Chimpanzees of the Mahale Mountains. Sexual and Life History Strategies*, ed. T. Nishida, pp. 133–48. Tokyo: University of Tokyo Press.

Tutin, C. E. G. 1979. Mating patterns and reproductive strategies in a community of wild chimpanzees (*Pan troglodytes schweinfurthii*). *Behavioral Ecology and Sociobiology*, 6, 29–38.

Uehara, S., Hiraiwa-Hasegawa, M., Hosaka, K. & Hamai,

M. 1994. The fate of defeated alpha male chimpanzees in relation to their social networks. *Primates*, **35**, 49–55.

Yamagiwa, J. 1987. Male life history and the social structure of wild mountain gorillas (*Gorilla gorilla beringei*). In: *Evolution and Coadaptation in Biotic Communities*, ed. S. Kawano, J. H. Connell & T. Hidaka, pp. 31–51. Tokyo: University of Tokyo Press.

de Waal, F. B. M. 1982. *Chimpanzee Politics*. London: Jonathan Cape.

de Waal, F. B. M. 1989. *Peacemaking among Primates*. Cambridge, MA: Harvard University Press.

de Waal, F. B. M. 1991. The chimpanzee's sense of social regularity and its relation to the human sense of justice. *American Behavioral Scientist*, **34**, 335–49.

de Waal, F. B. M. & van Hooff, J. A. R. A. M. 1981. Side-directed communication and agonistic interactions in chimpanzees. *Behaviour*, **77**, 164–98.

de Waal, F. B. M., van Hooff, J. A. R. A. M. & Netto, W. J. 1976. An ethological analysis of types of agonistic interaction in a captive group of Java-monkeys (*Macaca fascicularis*). *Primates*, **17**, 257–90.

10 • Male rank order and copulation rate in a unit-group of bonobos at Wamba, Zaïre

TAKAYOSHI KANO

INTRODUCTION

The view that male–male sexual competition is lower in bonobos than in chimpanzees has recently been proposed (e.g. Ihobe, 1992; Kano, 1992). This view is based on the estimation that more estrous females are available to a male bonobo than to a male chimpanzee. (1) Maximal or semi-maximal perineal tumescence in adult females lasts for more than 20 days for bonobos at Wamba (Kano, 1992), whereas average durations of maximal tumescence for adult female chimpanzees are 9.6 and 12.5 days at Gombe and Mahale, respectively (Nishida & Hiraiwa-Hasegawa, 1987; Takahata et al., Chapter 11). There is no marked difference in cycle length, and in both species most copulations occur when females show sexual swelling (Takahata et al., Chapter 11). Hence, the sexually receptive proportion of each cycle is greater in bonobo females than in their chimpanzee counterparts. (2) Young nulliparous immigrant females of bonobos show continual estrus and the highest copulation rates (Kano, 1989). Although chimpanzee adolescent females also show semi-continuity or irregularity in genital swelling (Tutin & McGrew, 1973), they mate much less frequently with adult males (Goodall, 1968). This implies that male bonobos are attracted to adolescent swellings, while male chimpanzees are not. (3) Females of both species experience estrous cycles after conception, but sexual receptivity in pregnant bonobos appears to last longer. One nulliparous female continued to copulate until only 19 days before her first live birth (personal observation). (4) There is no significant difference in interbirth intervals between the two species. There is, however, marked difference in the time it takes for females to resume swelling cycles after parturition:

within 1 year for bonobos (personal observation) vs. after 3–5.5 years for chimpanzees (Tutin, 1980; Tutin & McGinnis, 1981; Goodall, 1983; Hiraiwa-Hasegawa et al., 1984; Nishida et al., 1990).

All of the above demonstrate that the female bonobo, compared with her chimpanzee counterpart, exhibits more frequent and prolonged estrus. According to Furuichi's (1992) calculation, the former experiences eight times as many estrous periods in her lifetime than does the latter.

It should be noted that the number of adult females per adult male within a unit group is lower in bonobos (1.0–1.1 for Wamba E group between 1977 and 1984; Kano, 1992) than in common chimpanzees (1.7 for Mahale K group; Nishida, 1979; 2.2 for Gombe Kasakela group Goodall, 1983; 3.9 for Mahale M group; Takahata et al., Chapter 11). The adult sex ratio skewed to females will mitigate male sexual competition in chimpanzees. However, if Furuichi's estimation is valid, many more female-estrous-days are available to an adult male bonobo than to a male chimpanzee, which may result in less sexual competition among male bonobos.

Some direct observations of sexual interaction also support this hypothesis: at Gombe, chimpanzees have three mating types, i.e. possessiveness, consortship and opportunistic mating (Tutin, 1979). Monopolization of estrous females by the alpha male was also observed at Mahale (Hasegawa & Hiraiwa-Hasegawa, 1983; Takasaki, 1985). The development of these mating patterns may be linked to male–male sexual competition. On the other hand, neither possessive behavior nor consortship has been observed so far in bonobos at Wamba, in which all matings occurred opportunistically (Kano, 1992).

Male–male antagonism between groups of chimpanzees

is also thought to be related to male sexual competition, since females of the defeated group may be incorporated into the victorious one (Goodall *et al.*, 1979; Nishida *et al.*, 1985; Uehara *et al.*, 1994). Such antagonism may result in serious injury or even death. In contrast, no fatal aggression between groups has been recorded for bonobos at Wamba; their intergroup encounters varied from group fights to peaceful intermingling (Idani, 1990, 1991; Kano, 1991). Even intergroup, heterosexual copulations were common between sexually mature (adolescent or adult) individuals when groups interacted. I recorded 31 such copulations during 85 hours of contact time between E1 group and other groups (E2, P or B group) between September and December 1988. Sixteen (62 %) of the 26 copulations in which both partners were recognized were intergroup copulations. None of them was harassed or attacked by any member of either group. The above observations were not affected by provisioning, since the bonobos almost completely ignored the artificial food (sugarcane) for that period, depending on a super-abundant natural food, i.e. fruit of *Landolphia owariensis*. Both the high proportion of intergroup copulations and the indifference of other members to them may reflect the low level of sexual competition among male bonobos (Fig. 10.1).

Nevertheless, it is unlikely that male–male competition is absent or that it plays no role in heterosexual mating interactions in bonobos. In many primates that form multi-male groups, higher-ranking males have the advantage of obtaining priority of access to estrous females and, hence, a higher chance of mating than do lower-ranking ones (e.g. Hill, 1987; Paul *et al.*, 1993), although such mating success is not always associated with reproductive success (e.g. Curie-Cohen *et al.*, 1983; Shively & Smith, 1985; Inoue *et al.*, 1991). The purpose of the present study was to explore the existence of sexual competition in bonobo males by seeing how a male's rank affects his mating rate.

STUDY AREA AND SUBJECTS

The present study is based on data collected on the E1 group (or community) of bonobos at Wamba for a total of 341 days during five periods between 1984 and 1991 (Table 10.1). E1 is the smallest of six unit-groups whose entire or partial home range is in the Wamba forest. E1

Table 10.1. *Periods of study of bonobos at Wamba*

	Dates	Days
1	84/11/05 – 85/02/11	99
2	86/11/26 – 87/02/15	82
3A	88/09/10 – 88/10/08	29
3B	88/10/16 – 88/12/02	48
4	90/02/28 – 90/03/19	20
5	90/12/27 – 91/02/27	63
Total		341

Fig. 10.1. Adult male bonobo with fully descended testes. (Photo by F. White)

consisted of 27–33 members; 9–10 adolescent or adult males, 9–11 adolescent or adult females, and 8–13 juveniles or infants (Table 10.2). E1 and E2 were daughter groups resulting from the division of E group between 1991 and 1993. E1's home range was estimated at about 30 km^2 in area, more than 80% of which overlapped with the ranges of P, E2 and B groups. Their range comprised dry primary forest, swamp forest and various stages of secondary vegetation, all of which, except for cultivated fields, were used by the bonobos.

OBSERVATIONAL METHODS

Data were collected under three observational conditions.

Table 10.2. *Composition of E1 group of bonobos at Wamba*

Year	Adult		Adolescent			Juvenile/Infant		
	AM	AF	ADM	YNF	ADF	JIM	JIF	Total
1984	7	7	2	3		4	4	27
1985	7	8	3	2		3	6	29
1986	7	9	3	1	1	5	6	32
1987	7	9	3	1	1	5	7	33
1988	7	10	3			6	7	33
1989	8	10	2		1	6	5	32
1990	7	10	2			5	6	30
1991	7	9	2			6	6	30

Note: Adult (14 years old or more): AM, adult male; AF, adult female. Adolescent (8–13 years old): ADM, adolescent male; YNF, immigrant adolescent (young nulliparous) female; ADF, natal adolescent female. Juvenile (4–7 years old) or Infant (0–3 years): JIM, juvenile/infant male; JIF, juvenile/infant female.

Observations during fixed provisioning

Most observations of E1 group were made at two artificial feeding sites (F1 and F2), which were forest-clearings of 30 m from east to west and 60 m from north to south where artificial foods (sugarcane and occasionally pineapples) were provided by the researchers. The observation points were in the shelters built on the tops of termite mounds at the southern edge of each feeding site. All clearings and adjacent forest edges were easily visible from the observation points. Artificial foods were set out only when bonobos were absent from the sites, in order to avoid direct interaction with humans.

Observations under mobile provisioning

Bonobos were also provisioned in the early morning at or near their sleeping site of the previous night, on occasions when they were traveling far from the artificial feeding sites. This type of artificial provisioning is referred to as 'mobile provisioning' (Nishida, 1979). Its main purpose was to facilitate counting and identifying party membership by bringing the bonobos together on the ground. Otherwise, it was hard to count all members scattered in the dense foliage over a wide range of tall trees. In mobile provisioning, two to six bundles of 1 m-long sugarcane, each of which contained about five pieces, was given in several installments, and each time party members were counted and social interactions were recorded.

Observations under natural conditions

Dependence on artificial foods varied greatly according to the amount and quality of natural foods available. For example, during the peak seasons of *Landolphia owariensis* or *Dialium* spp., bonobos ignored or paid little attention to artificial foods. In such seasons, provisioning was discontinued and bonobos were observed only by following them in the forest. They were also followed in the forest after mobile provisioning.

A total of 851.5 hours of observation was made on E1 group during the combined study periods, with 478.6, 57.3, and 315.6 hours under fixed, mobile provisioning and natural conditions, respectively.

Under the mobile provisioning condition, bonobos repeatedly gathered and jostled each other at a small food spot, and so were put in more competitive situations than at artificial feeding sites where larger amounts of foods were given over a wider area. Under natural conditions, foods were distributed at lower density over the widest areas and, hence, tension among party members appeared to be lower than under provisioned conditions.

Ad libitum or focal animal sampling (Altmann, 1974) was adopted for recording social interactions. The latter was used mainly at the artificial feeding sites. Visibility was best there, since bonobos could be seen from above (at 3 m and 5 m above the ground at F1 and F2, respectively). Moreover, the trained trackers helped by adding information on any new social developments, namely, social activities, and joining and parting of members. Thus, few conspicuous patterns of activity, e.g. aggression, copulation and GG rubbing, probably went unnoticed, whatever sampling method was used. All the data collected in this way were used in this study. In natural conditions, visibility was worse. Both in the trees and on the ground, dense vegetation was a serious obstacle for observation; on most encounters only a few individuals were in view at one time, and long continuous following of individuals was difficult, except when they rested.

Both the frequency of some social interactions and the accuracy of sampling may differ according to the observational conditions. In the present study, therefore, data collected under different conditions were analyzed separately.

TERMS

Sexual interactions in bonobos involve all age-sex classes (Kitamura, 1989; Kano, 1992). Among these, only mature copulations, namely heterosexual mountings with intromission accompanied by pelvic thrusts in which both partners were sexually mature (adolescent or adult), are considered in the present study. Each mount was regarded as a separate copulation, since it was difficult to detect ejaculation (Fig. 10.2).

Rank order among males was judged from aggressive (bite, hit, slap, charge, branch-dragging charge, chase, pseudo-charge, threaten, etc.) or submissive (flee, avoid, scream, crouch, etc.) interactions. Pant-grunting, a conspicuous submissive behavior in chimpanzees, is not seen in bonobos.

Ranking is related to priority of access to desirable resources in competition. Therefore, whether or not an individual can gain access to the resource depends on the spatial distribution of both the resource and the competitors. Even the lowest-ranking male can acquire a desirable resource when other males are not present, so any individual can regulate his rank by placing himself in the proper spot within the group's space. From this point of view, I propose three types of male ranking.

Rank within group

This is position in the ranking order for all sexually mature males of the unit group (or community), that is, the general notion of rank.

Rank within party

In fission–fusion groupings, a bonobo male can manipulate his rank or status by choosing the party he joins, that is, the partners with whom he associates. There are wide discrepancies in the definition of 'party' among researchers or study sites of *Pan* spp. Foraging groups of bonobos at Wamba often extend over wide areas, depending on food distribution and availability. For instance, while foraging in Marantaceae-rich forests with scattered fruit-bearing trees of *Pancovia laurentii*, some members dispersed on the ground to feed on Marantaceae shoots, and others dispersed over several fruit trees, with large inter-individual distances: sometimes the distances between the opposite ends of a foraging group were over 200–300 meters. In this study, all members of such a foraging group were attributed to the same single party, insofar as they were traveling as a whole in the same direction, keeping vocal contact. According to this definition, core parties of E1 group were stable and large in size with occasional split-offs or joinings of small numbers. This definition is similar to that used at Mahale, but very different from those adopted at Lomako (White, 1988, 1989), Taï (Boesch, Chapter 8) and Kibale (Wrangham, personal communication), all of which seem to be only a portion of the party as defined in the present study. Reported differences in grouping patterns between Lomako and Wamba appear to arise mostly from differences in definition of party rather than differences in habituation methods (non-provisioning vs. provisioning). Parties defined at sites other than Wamba and Mahale are somewhat similar to the 'subparty' defined below.

Rank within subparty

Direct physical contact between individuals is available only to members of subparties sharing the same food patch, resting site, etc. Therefore, male ranking within subparty should be considered separately, though it is not easy to determine the size of a subparty. In the present study, all bonobos attending the artificial feeding place at the same time were tentatively considered to be members of a single subparty for the observation under fixed provisioning, and those visible in each tree or on the forest floor were classed as members of an independent subparty. This distinction was based on the observation that agonistic interaction could occur between two males wherever

Fig. 10.2. Bonobos copulate ventro-ventrally at Wamba. (Photo by T. Kano)

they were positioned within the cleared strip for pro-visioning, and that much less agonistic interaction occurred when two males were in different trees or when one was in a tree and the other was on the ground.

Alpha status was stable throughout the study periods and was maintained by a young adult male, Ten. He was top ranked for all three types of rankings, while the ranks of the others varied both within a party and within a subparty. I propose the notion of 'average' rank, which is defined by the following equation:

$$\text{Average rank} = \Sigma \, (X_i Y_i) / Y_i$$

where X_i is the rank of a certain male, and Y_i is the total observation time when the rank of the male was i.

Even the second- or lower-ranking males within the group could be top ranking at least temporarily within a party or subparty. In this paper, they will be termed 'temporary first ranking males' to discriminate them from the alpha Ten.

RESULTS

Three types of male rank

The average ranks within the group, party and subparty, based on the data collected on 54 observation days at the artificial feeding sites between November 5, 1984 and February 11, 1985, are shown in Figure 10.3.

A total of 376 dominant-subordinate episodes were recorded between seven adult and three adolescent males over this period. No dominant-subordinate interaction was observed in eight of the 45 possible dyads, but rank within the group for each male was determined on the supposition that E1 males had a linear rank order. Thus, all adult males were dominant to all adolescent males. There was one tie between the sixth- and seventh-ranking males, Hata and Ika, both of whom were assumed to have the 6.5 ranking. A reversal occurred between the second and third males: the former second

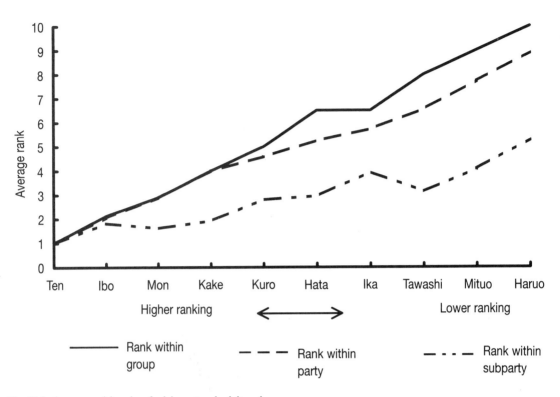

Fig. 10.3. Average social ranks of adolescent and adult male bonobos in three contexts: group, party and subparty, at Wamba.

male, Ibo, was outranked by his younger brother Mon in January 23, 1985. According to the definition of the average rank, Ibo and Mon were assumed to be the 2.1 and 2.9 ranking, respectively (see Figure 10.3).

The differences between rank within-group and rank within-party were slight. This is compatible with the observation that bonobos of Wamba tend to maintain a large core party with rare fission and fusion (e.g. Kano, 1982). In contrast, rank within subparties showed a marked difference from the other two. That is, bonobo males increased their ranks by using the feeding site separately from other males.

Rank within group and copulation rate

A total of 130 copulations by ten sexually mature males (7 adult and 3 adolescent males) was recorded in 209.3 hours of observation at the artificial feeding sites between November 5, 1984 and January 23, 1985 when Ibo was

outranked by Mon (range, 3–33 per adult male, 1–18 per adolescent male; Table 10.3).

Four high-ranking adult males copulated more frequently than three lower-ranking adult males. The adolescent males showed fewer copulations than the three highest-ranking adults, but showed higher frequency than the three lowest-ranking adult males. Overall, there was no significant correlation between rank and copulatory frequency for these 10 males ($N = 10$, $r_s = 0.571$, $p > 0.05$, 2-tailed).

As mentioned above, almost all copulations that occurred within the artificial feeding sites during the study period were recorded due to high observability. Though all party members were not always present simultaneously at the feeding sites, the others rested or fed in nearby bushes or trees, seldom interacting with one another except when they engaged in activities typical of resting time, namely social allogrooming and play. Therefore, the number of unnoticed copulations that

Table 10.3. *Copulations by EI males at artificial feeding sites from November 5, 1984, to January 23, 1985*

Individual name	Estimated age (years)	Average rank within group	Copulation frequency (no.)	Copulation rate (no./h)	Number of other males in the same subparty
Ten	14	1	19	0.44	3.31
Ibo	22	2	33	0.41	3.06
Mon	17	2	20	0.23	2.95
Kake	34	4	15	0.22	3.06
Kuro	29	5	5	0.22	2.8
Hata	30	6.5	3	0.06	2.00
Ika	26	6.5	4	0.32	2.00
Tawashi	10	8	18	0.21	3.00
Mituso	9	9	6	0.09	1.16
Haruo	7	10	7	0.16	3.57
Total (average)			130	(0.23)	(2.69)

occurred outside the feeding places were likely fewer than the number of recorded ones. This implies that in fixed provisioning during the study period, male rank was not correlated with mating success.

However, the number of copulations per observational hour per male (= copulation rate) was more highly correlated with male rank ($N = 10$, $r_s = 0.80$, $p < 0.01$, 2-tailed). This indicates that higher-ranking males had greater chances to mate than lower-ranking ones during the same period in the feeding sites.

A total of 239 copulations was recorded at the feeding sites during five study periods between 1984 and 1991. They were arranged according to male rank order, regardless of individual rank changes. Frequency of copulations was significantly correlated with male rank for these pooled data ($N = 10$, $r_s = 0.84$, $p < 0.01$).

Copulation in relation to rank within subparties

Totals of 239, 39 and 37 copulations in which male partners were recognized were recorded under fixed provisioning, mobile provisioning and natural conditions, respectively, between 1984 and 1991. The alpha male, Ten, was responsible for about 20% of the matings recorded under both fixed and mobile provisioning conditions, and for about 5% of those recorded under natural conditions (see Fig. 10.4(*a*), (*b*), (*c*)). Temporary first-ranking males were responsible for only a small percentage of the recorded copulations, if rank within party

is taken into consideration (see Fig. 10.4(*a*), (*b*), (*c*)). However, if rank within subparties is considered, they were responsible for 32%, 54% and 41% of the total copulations recorded under fixed or mobile provisioning and under natural conditions, respectively (see Fig. 10.4(*a*), (*b*), (*c*)). That is, 54%, 77% and 46% of the copulations recorded in fixed or mobile provisioning and under natural conditions involved the alpha male or temporary first-ranking males. If the 'temporary' second-ranking males (in terms of rank within subparty) are included, most of the copulations recorded in all three observational conditions were performed by the two top-ranking males. The average number of males other than the mating male in the same subparty was 2.7 ($N = 10$, SD = 0.70, Table 10.3), and was not correlated with rank within group ($r_s = 0.29$, $p > 0.05$, 2-tailed).

The data collected between November 1984 and January 1985 show that adult males did most of their copulating when their rank within subparties was first or second, while two of the three adolescent males (Tawashi and Haruo) performed most of their copulations when they were third or lower ranking (see Fig. 10.5). Some adolescent males may have been tolerated in their mating activities by higher-ranking adult males.

Interference in copulations was rare in bonobos: only 5.2% (27) of the 515 mature copulations recorded at F1 between October 1975 and February 1979 were aggressively disrupted by other males (Kano, 1992). However, rare aggressive interference by males does not always indicate a low level of sexual competition among males.

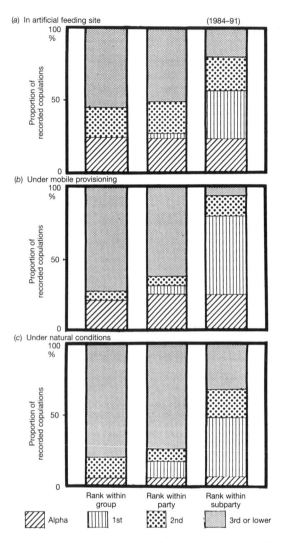

(a) In artificial feeding site   (1984–91)

(b) Under mobile provisioning

(c) Under natural conditions

Rank within group | Rank within party | Rank within subparty

Alpha | 1st | 2nd | 3rd or lower

Fig. 10.4. Male bonobo social ranking and mating success in three contexts: (a) artificial feeding site; (b) mobile provisioning; (c) natural conditions, at Wamba.

Males may refrain from initiating mating when this may provoke another's aggression or harassment. This may be inferred from the above result that only the most dominant males copulated in a group.

DISCUSSION

The present study shows that the dominance rank of male bonobos affects their chances of mating, suggesting that male–male sexual competition is more intense than has previously been thought (Ihobe, 1992; Kano, 1992). Copulation frequency is weakly, and copulation rate is more clearly, positively correlated with male ranking. The effect of dominance is conspicuous in subparties, where only the top-ranking male has clear priority in mating. Any male in a unit-group has a chance to mate with females but temporarily raising one's rank within a subparty makes it much easier to do so.

It is possible that other factors are related to a male's mating opportunity. Three adult males, Ten, Ibo and Mon, whose mothers were alive in the E1 group, were more dominant and exhibited higher mating frequencies and rates than three of the other adult males, Ika, Kuro, and Hata, whose possible mothers were not seen in the group (see Table 10.3). The adolescent males were subordinate to the latter three, but showed higher mating frequencies than they did.

Some late adolescent or young adult males who maintained close associations with their mothers were seen to challenge and to succeed in outranking the motherless prime males (Kano, 1992). Ihobe (1992) also suggests that a mother plays an important role in the rise of her son's rank.

Adult female bonobos closely associate with one another (Kuroda, 1979; Kano, 1982; Kitamura, 1983; White, 1988, 1989; Thompson-Handler, 1990). A son's close association with his mother will result in closer contact to females in his mother's cluster. Hence, a mother may advantage her sons in two ways: (1) in raising his dominance rank; and (2) in facilitating his access to mating partners. Both likely increase the son's chances of mating.

Male–male relationship is another possible factor. Though such clear-cut aggressive alliances as reported for chimpanzees at Gombe (Riss & Goodall, 1977) have not been seen in bonobos at Wamba, some dyads of males showed little aggression to each other. They associated more often with each other in the same party (and probably also in the same subparty) than they did with other males. Mutual tolerance among them may have some effect on their mating opportunities, but this is yet unstudied.

Female choice may also affect a male's mating chances. In bonobos at Wamba, adult females are almost co-dominant to adult males. High-ranking females may

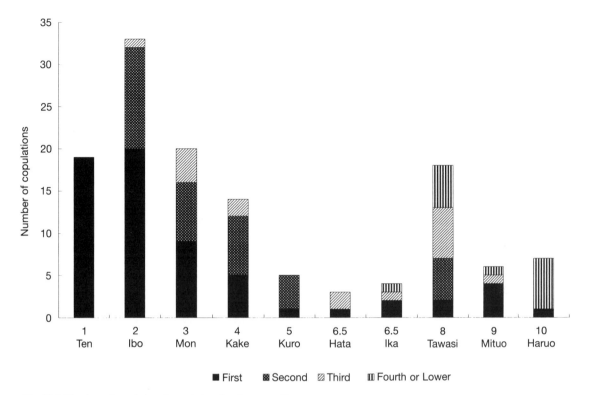

Fig. 10.5. Number of copulations by male bonobos in terms of their rank within subparties at Wamba. Males arranged in order of descending overall social rank.

choose mating partners without objection from any other males. This also needs future analysis of data.

Monopolization through possessiveness of estrous females by high-ranking male chimpanzees reflects severe sexual competition among males. Lower-ranking males adopt counter-strategies. (1) They lure females away and travel with them apart from other community members for periods of days or even weeks (Tutin, 1979). Such consortship is identical to 'taking the first position in rank within party.' (2) They are allowed by higher-ranking males to mate with estrous females in exchange for aggressive assistance in the higher rankers' struggles for dominance (Nishida, 1983).

At Gombe, most births resulted from consortships. Tutin & McGinnis (1981) and Takahata *et al.*, (Chapter 11) found that male rank in Mahale's M group was negatively correlated with male copulation rate. These data indicate that the counter-strategies of lower-ranking males are effective in increasing mating opportunities.

Why have low-ranking male bonobos not developed effective strategies such as consortship or manipulation of the ranks of dominant males? It may be difficult for high-ranking male bonobos to monopolize all estrous females, because of the female's prolonged and frequent estrus. A core party and its space may be too large for higher-ranking males to exert their influence over all the other members. Lower-ranking males can, therefore, have access to sexual partners somewhere in the space of a party. Thus, the disadvantage from subordinance may not have been serious enough for low-ranking males to be forced to exploit counter-strategies such as chimpanzees do.

In this study, analyses were concentrated only on the mating success of male bonobos at Wamba, with no mention of reproductive success. The timing of mating by each male during the female estrous cycle is used indirectly in some other primates to evaluate the latter. In bonobos, many female estrous cycles seem to lack ovulation, which is inferred from the observation that not only young nulliparous females who are apparently in adolescent sterility, but also parous females, exhibit infertile estrous cycles for a few years before conception.

It is difficult not only to determine which cycle has ovulation but also to know the timing of ovulation because of prolonged periods of estrus in bonobo females in the wild. Furuichi (1987) reported on female cycles at Wamba, but a more detailed study will be necessary to understand the sexuality of female bonobos and hence, to estimate the reproductive success of male bonobos. Male reproductive success may be evaluated more directly in the near future by using DNA analysis of pedigree. How a male's mating effort is related to his reproductive success will then be understood.

ACKNOWLEDGMENTS

This study was financially supported by Monbusho Grant-in-Aid for Overseas Scientific Research 1984–90. This paper was originally prepared for the Wenner-Gren conference, 'The Great Apes Revisited,' held November 12–19, 1994 in Cabo San Lucas, Baja, Mexico. I thank the Centre de Recherche en Science Naturelles (CRSN), Zaïre, and especially Delegue General, Dr Zana Ndontoni, for permission to conduct research. I am very grateful to my trackers, Messieurs Nkoy Batolumbo, Bafutsa Bokonda, Ikenge Lokake, Likombe Batuafe, and Batsindelia Lunga, for their assistance in the field, and Drs G. Idani, H. Ihobe, T. Furuichi, and my other field colleagues for cooperation and discussion.

REFERENCES

Altmann, J. 1974. Observational study of behavior: sampling methods. *Behaviour*, **49**, 227–65.

Curie-Cohen, M., Yoshihara, D., Luttrell, L., Benforado, K., MacCluer, J. W. & Stone, W. 1983. The effect of dominance on mating behavior and paternity in a captive troop of rhesus monkeys (*Macaca mulatta*). *American Journal of Primatology*, **5**, 127–38.

Furuichi, T. 1987. Sexual swelling, receptivity and grouping of wild pygmy chimpanzee females at Wamba, Zaïre. *Primates*, **20**, 309–18.

Furuichi, T. 1992. The prolonged estrus of females and factors influencing mating in a wild group of bonobos (*Pan paniscus*) in Wamba, Zaïre. In: *Topics in Primatology, Vol. 2, Behavior, Ecology, and Conservation*, ed. N. Itoigawa, Y. Sugiyama, G. P. Sacket & R. K. R. Thompson, pp. 179–90. Tokyo: University of Tokyo Press.

Goodall, J. 1968. The behaviour of free-living chimpanzees in the Gombe Stream Reserve. *Animal Behaviour Monographs*, **1**, 161–311.

Goodall, J. 1983. Population dynamics during a 15 year period in one community of free-living chimpanzees in the Gombe National Park, Tanzania. *Zeitschrift für Tierpsychologie*, **61**, 1–60.

Goodall, J., Bandora, A., Bergman, E., Busse, C., Matama, H., Mpongo, E., Pierce, A. & Riss, D. 1979. Intercommunity interactions in the chimpanzee population of the Gombe National Park. In: *The Great Apes*, ed. D. A. Hamburg & E. R. McCown, pp. 13–53. Menlo Park, CA: Benjamin/Cummings.

Hasegawa, T. & Hiraiwa-Hasegawa, M. 1983. Opportunistic and restrictive matings among wild chimpanzees in the Mahale Mountains, Tanzania. *Journal of Ethology*, **1**, 75–85.

Hill, D. 1987. Social relationships between adult male and female rhesus macaques: 1. Sexual consortships. *Primates*, **28**, 439–56.

Hiraiwa-Hasegawa, M., Hasegawa, T. & Nishida, T. 1984. Demographic study of a large-sized unit-group of chimpanzees in the Mahale Mountains, Tanzania: a preliminary report. *Primates*, **25**, 401–3.

Idani, G. 1990. Relations between unit-groups of bonobos at Wamba, Zaïre: encounters and temporary fusions. *African Study Monographs*, **11**, 153–6.

Idani, G. 1991. Cases of inter-unit group encounters in pygmy chimpanzees at Wamba, Zaïre. In: *Primatology Today: Proceedings of the XIIIth Congress of the International Primatological Society*, ed. A. Ehara, T. Kimura, O. Takenaka & M. Iwamoto, pp. 235–8. Amsterdam: Elsevier.

Ihobe, H. 1992. Male–male relationships among wild bonobos (*Pan paniscus*) at Wamba, Republic of Zaïre. *Primates*, **33**, 163–79.

Inoue, M., Mitsunaga, F., Ohsawa, H., Takenaka, A., Sugiyama, Y., Gaspard, S. A. & Takenaka, O. 1991. Male mating behaviour and paternity discrimination by DNA fingerprinting in a Japanese macaque group. *Folia Primatologica*, **56**, 202–10.

Kano, T. 1982. Social group of pygmy chimpanzees (*Pan paniscus*) of Wamba. *Primates*, **23**, 171–88.

Kano, T. 1989. The sexual behavior of pygmy chimpanzees. In: *Understanding Chimpanzees*, ed. P. G. Heltne & L. A. Marquardt, pp. 176–83. Cambridge, MA: Harvard University Press.

Kano, T. 1991. Intergroup relationships of bonobos (*pygmy chimpanzees*) at Wamba, Zaïre. *Iden*, **45**, 50–4. (in Japanese)

Kano, T. 1992. *The Last Ape: Ecology and Behavior of Pygmy Chimpanzees*. (translated by E. O. Vineberg). Stanford, CA: Stanford University Press.

Kitamura, K. 1983. Pygmy chimpanzee association patterns in ranging. *Primates*, **24**, 1–12.

Kitamura, K. 1989. Genito-genital contacts in the pygmy chimpanzee (*Pan paniscus*). *African Study Monographs*, **10**, 49–67.

Kuroda, S. 1979. Grouping of the pygmy chimpanzees. *Primates*, **20**, 161–83.

Nishida, T. 1979. The social structure of chimpanzees of the Mahali Mountains. In: *The Great Apes*, ed. D. A. Hamburg & E. R. McCown, pp. 73–121. Menlo Park, CA: Benjamin/Cummings.

Nishida, T. 1983. Alpha status and agonistic alliance in wild chimpanzees (*Pan troglodytes schweinfurthii*). *Primates*, **24**, 318–36.

Nishida, T. & Hiraiwa-Hasegawa, M. 1987. Chimpanzees and bonobos: cooperative relationships among males. In: *Primate Societies*, ed. B. B. Smuts, D. L. Cheney, R. M. Seyfarth, R. W. Wrangham & T. T. Struhsaker, pp. 165–77. Chicago: University of Chicago Press.

Nishida, T., Hasegawa, M., Hasegawa, T. & Takahata, Y. 1985. Group extinction and female transfer in wild chimpanzees in the Mahale National Park, Tanzania. *Zeitschrift für Tierpsychologie*, **67**, 284–301.

Nishida, T., Takasaki, H. & Takahata, Y. 1990. Demography and reproductive profiles. In: *The Chimpanzees of the Mahale Mountains*, ed. T. Nishida, pp. 63–97. Tokyo: Tokyo University Press.

Paul, A., Kuester, J., Timme, A. & Arnemann, J. 1993. The association between rank, mating effort, and reproductive success in male Barbary macaques (*Macaca sylvanus*). *Primates*, **34**, 491–502.

Riss, D. C. & Goodall, J. 1977. The recent rise to the alpha rank in a population of free-living chimpanzees. *Folia Primatologica*, **27**, 134–51.

Shively, C. & Smith, D. G. 1985. Social status and reproductive success of male *Macaca fascicularis*. *American Journal of Primatology*, **9**, 129–35.

Takasaki, H. 1985. Female life history and mating patterns among the M group chimpanzees of the Mahale National Park, Tanzania. *Primates*, **26**, 121–9.

Thompson-Handler, N. 1990. The pygmy chimpanzee: Sociosexual behavior, reproductive biology and life history patterns. Ph.D. dissertation, Yale University.

Tutin, C. E. G. 1979. Mating patterns and reproductive strategies in a community of wild chimpanzees (*Pan troglodytes schweinfurthii*). *Behavioral Ecology and Sociobiology*, **6**, 29–38.

Tutin, C. E. G. 1980. Reproductive behaviour of wild chimpanzees in the Gombe National Park, Tanzania. *Journal of Reproduction and Fertility, Supplement*, **28**, 43–57.

Tutin, C. E. G. & McGinnis, P. R. 1981. Chimpanzee reproduction in the wild. In: *Reproductive Biology of the Great Apes, Comparative and Biomedical Perspectives*, ed. C. E. Graham, pp. 239–64. New York: Academic Press.

Tutin, C. E. G. & McGrew, W. C. 1973. Sexual behavior of group living adolescent chimpanzees. *American Journal of Physical Anthropology*, **38**, 195–200.

Uehara, S., Nishida, T., Takasaki, H., Kitopeni, R., Kasagura, M. B., Norikosi, K., Tsukahara, T., Nyundo, R. & Hamai, M. 1994. A lone male chimpanzee in the wild: the survivor of a disintegrated unitgroup. *Primates*, **35**, 275–82.

White, F. J. 1988. Party composition and dynamics in *Pan paniscus*. *International Journal of Primatology*, **9**, 179–93.

White, F. J. 1989. Social organization of pygmy chimpanzees. In: *Understanding Chimpanzees*, ed. P. G. Heltne & L. A. Marquardt, pp. 194–207. Cambridge, MA: Harvard University Press.

11 • Comparing copulations of chimpanzees and bonobos: do females exhibit proceptivity or receptivity?

YUKIO TAKAHATA, HIROSHI IHOBE AND GEN'ICHI IDANI

INTRODUCTION

Chimpanzees (*Pan troglodytes*) and bonobos (*P. paniscus*) have a similar social structure (patrilineal multi-male and multi-female social unit; Itani, 1985). However, bonobos exhibit quite different copulatory activities from other non-human primates, including chimpanzees: (1) bonobos copulate more often than other anthropoid apes (Kano, 1989); (2) bonobos copulate in varied postures (Savage-Rumbaugh & Wilkerson, 1978); (3) copulation-like behavior often occurs in other social contexts (Kano, 1980; de Waal, 1989); (4) female bonobos exhibit lengthy estrus (Dahl, 1986); (5) suckling mothers resume sexual cycling within a year after giving birth (Kano, 1992). Some authors (e.g. Blount, 1990) have discussed these features in connection with the evolution of the sexuality of human females.

However, how different is the sexual behavior of bonobos from that of chimpanzees in the wild? Here, we compare the copulatory behavior of these two species, based on previous reports and on our field studies.

BACKGROUND

In the Mahale Mountains National Park, Tanzania, the chimpanzees of M group have been observed since 1965. The group contained about 100 individuals in 1981 (see table 3 in Nishida *et al.*, 1990). The adult sex ratio (10 : 39) was significantly skewed toward females compared with the expected ratio of 1 : 1 (Chi-squared = 17.2, df = 1, $p < 0.001$; for discussion of this tendency, see Hiraiwa-Hasegawa *et al.*, 1984; Nishida *et al.*, 1990).

At Wamba in the Luo Forest Reserve, Zaïre, the bonobos of E group have been observed since 1976, and they split into E1 and E2 groups between 1981 and 1983. Another group, P group, also has been observed since 1976. E1 group contained about 30 bonobos and E2 group about 40 bonobos (see fig. 1 in Idani, 1990; table 1 in Ihobe, 1992). In 1986, the adult sex ratios were 7 : 9 (E1 group) and 8 : 11 (E2 group), and did not differ significantly from the expected ratio of 1 : 1 (Chi-squared = 0.25 and 0.47, df = 1, $p > 0.50$ and $p > 0.30$, respectively). In 1988, P group contained about 45 bonobos (see fig. 3 in Idani, 1990). The adult sex ratio was 9 : 13, and did not differ significantly from the expected ratio of 1 : 1 (Chi-squared = 0.73, df = 1, $p > 0.30$).

RESULTS

Genital swelling cycle and estrus of chimpanzees

Adult female chimpanzees exhibit a regular swelling cycle. At Mahale, the mean cycle length was 31.5 days, while maximal perineal swelling lasted for 12.5 days, occupying 40% of the swelling cycle (see Table 11.1; Hasegawa & Hiraiwa-Hasegawa, 1983). At Gombe, mean cycle length was 36 days and maximal swelling averaged 9.6 days, occupying 27% of the typical cycle's length (Tutin & McGinnis, 1981; Goodall, 1986).

Compared with the adult females, the swelling cycles of adolescent females tended to be irregular. Their swellings were smaller than those of adult females, as reported in the laboratory (Nadler *et al.*, 1986), and they rarely mated with mature males (Hasegawa & Hiraiwa-Hasegawa, 1990).

Female chimpanzees tended to copulate during the period of maximal swelling. Of the 147 copulations by

Table 11.1. *Swelling cycle, estrus, and interbirth intervals of chimpanzees and bonobos*

	Mahale[a]	Gombe[b]	Wamba[c]	Wamba[d]
Mean length of swelling cycle (days)	31.5	36	42	32.8
	(44 cycles)	(46)	(6)	(8)
Mean length of maximal swelling (estrous days)	12.5	9.6	> 20	12.9
	(27 estrous periods)	(37)		(9)
Proportion of cycle in maximal swelling (%)	40	27	47	39
Mean length of interbirth interval (year)	6.0	5.5	> 5.0	4.5
	(19 intervals)	(21)		(10)
Mean length of postpartum amenorrhea (month)	53.2	43	< 12	< 12
	(18 intervals)	(10)		

Source: [a] Hasegawa & Hiraiwa-Hasegawa, 1983; Nishida *et al.*, 1990; [b] Tutin & McGinnis, 1981; Goodall, 1986; [c] Furuichi, 1987; Kano, 1992; [d] present study.

Table 11.2. *Number of females on last day of maximal swelling (probable day of ovulation) per observation day; recorded in M group from August to December 1981*

	No. of the females on last day of maximal swelling						
	0	1	2	3	m	V_o	V_o/m
No. of observation days	71	45	9	1	0.524	0.443	0.845

sexually mature males of M group recorded in 1981, 94% were with females during their period of maximal swelling. In Gombe, Goodall (1986) also reported that 96% of the copulations by adult males were with females during or around their period of maximal swelling.

Do female chimpanzees synchronize their estrous cycles or the timing of ovulation? Table 11.2 shows the number of the females of M group who were on the approximate day of ovulation (last day of maximal swelling) per observation day. The ratio of the unbiased estimate of variance (V_0) to the mean number (m) of such females was 0.845, which did not differ significantly from the expectation based on the Poisson distribution ($F = 1.183$, $p > 0.2$). Thus, they seem not to have synchronized the timing of ovulation.

Genital swelling cycle and estrus of bonobos

The swelling cycle length of female bonobos has been reported to be longer than that of chimpanzees (Dahl, 1986). In Wamba, Furuichi (1987) reported that sexual swelling fluctuated periodically with the cycle length of 42 days ($N = 6$, SD = 4.1), which was longer than that

reported for chimpanzees. Although our data were also too few to reassess this issue, the mean length of the cycles recorded in E1 and P groups was 32.8 days ($N = 8$, SD = 12.1 days), which was equal to that of adult female chimpanzees.

Many reports have pointed out that female bonobos also exhibit prolonged estrus. Dahl (1986) described the exhibition of marked swelling for about 75% of the intermenstrual interval in captive bonobos. Kano (1992) also pointed out that maximal swelling usually lasted for 20 days or more.

Our data do not support the conclusion that female bonobos exhibit prolonged estrus. The mean duration of nine maximal swellings recorded for E1- and P-group females was 12.9 days (SD = 9.3 days), which occupied 39% of the whole cycle length (see Table 11.1). However, the greater part of the swelling cycles was occupied by intermediate swelling, which differs from the lower proportions for chimpanzees.

It also has been stressed that female bonobos copulate throughout the swelling cycle (Savage-Rumbaugh & Wilkerson, 1978). On the other hand, Furuichi (1987) reported that 82% of copulations occurred during the period of maximal swelling. Our data roughly support his conclusion. Of the copulations by adult males of E1 and E2 groups, 77% and 67% were with females during their stage of maximal swelling.

As Kano (1992) has discussed, suckling females also may show maximal swelling. Of the four females of E1 group who gave birth in 1986, three females exhibited maximal swelling from November 1986 to February 1987, and one of them often copulated with adult males.

Interbirth intervals of chimpanzees

At Mahale, the mean interval between births was 72 months (Nishida *et al.*, 1990), which consists on average of postpartum amenorrhea (53.2 months), the period from the resumption of swelling to conception (11.2 months) and pregnancy (7.6 months). On the other hand, this value is significantly longer than the 5.1 years reported for the chimpanzees of Bossou, Guinea (Mann-Whitney test, $U = 45$, $p < 0.001$; calculated from Sugiyama, 1994).

While suckling their infants, Mahale females exhibit no swelling, with the exception of one mother who resumed swelling 3–7 months after giving birth (Takasaki *et al.*, 1986). However, if the infant died, most females immediately resumed swelling cycles.

After resuming cycling, females typically became pregnant during the sixth swelling ($N = 12$, $\bar{x} = 5.9$, $SD = 3.5$). During pregnancy, most females displayed maximal swelling an average of 2.6 times ($N = 19$, $SD = 0.9$). Similar post-conception swelling occurred at Gombe (Wallis & Goodall, 1993).

At Mahale, some females exhibited periodic swelling but never gave birth (Nishida *et al.*, 1990). Of these sterile females, one female sometimes did not exhibit swelling for several months. In such cases, she may have aborted her fetus. Furthermore, several old-aged females scarcely copulated, despite having periodic genital swellings. In particular, two females who last gave birth in the 1970s were believed to have been in the post-reproductive stage of their lifespan.

Interbirth intervals of bonobos

Kano (1992) wrote that the interbirth interval of bonobos ranged from four to six years, not differing from that of chimpanzees. For the ten intervals recorded in E1 group from 1976 to 1991, the mean length was 4.5 years (SD = 0.7 years). This value is significantly shorter than the six years reported for Mahale (Mann-Whitney test, $U = 14$, $p < 0.0001$), but it does not differ from the 5.1 years reported for Bossou ($U = 73.5$, $p = 0.386$; calculated from Sugiyama, 1994).

As stated above, female bonobos resume swelling cycles within one year after parturition. Thus, Kano (1992) speculated that they continued to exhibit non-ovulatory estrus for 3–4 years. Female bonobos also exhibited post-conception estrus, but we have no data about its duration. In one case, No (adult female of E1 group) was seen to copulate on December 30, 1984, and to give birth on January 18, 1985.

Mating patterns and initiation of copulation by chimpanzees of Mahale and Gombe

Both chimpanzees and bonobos show a basically promiscuous mating pattern. However, the following data indicate that the chimpanzees of Gombe, those of Mahale and the bonobos of Wamba exhibit rather different mating patterns from one another.

First, there are notable differences in the mating patterns of the Gombe and Mahale chimpanzees. At Gombe, Tutin (1979) identified three mating patterns (opportunistic mating, possessiveness and consortship), and she described that the chimpanzees often form consortships, in which a male and female form a temporary but exclusive pair bond and avoid others. Only a few copulations but about 33–50% of conceptions occurred in such consortships (Goodall, 1986). At Mahale, more than 80% of conceptions occur in opportunistic matings and consortships are absent (Hasegawa & Hiraiwa-Hasegawa, 1983, 1990; Takasaki, 1985).

Second, at Gombe, adult males typically take the initiative in copulations (Goodall, 1986), while at Mahale adult females often take the initiative. For example, Nishida (1980) reported that, of the 147 copulations in which the initiator could be determined, 77 were initiated by males and 70 were by females. Table 11.3 shows that estrous females of M group unilaterally approach the adult males, who are widely dispersed within the home range, copulate with them and then leave them (see Hasegawa, 1992). Males rarely prevent females from leaving. Possessiveness, in

Table 11.3. *Social interactions in 51 copulations by nine adult and one late adolescent male chimpanzees of M group, Mahale, in 1981*

	Male : Female		Chi-squared	(df = 1)
Approach	10	28	38.8	$p < 0.001$
Approach to within 3 m of other	12	31	40.9	$p < 0.001$
Solicit copulation	22	26	16.0	$p < 0.001$
Leave	12	26	30.3	$p < 0.001$

Note: Expected values calculated from mean rate of number of males to that of females in maximal swelling per observation day.

Table 11.4. *Copulatory sequences of bonobos in E1 and P groups, Wamba*

Type	1986–87	1988–89		Total
	E1 group	P group	E1 group	
Male approaches and displays sexually to female, then mates with her.	49	34	27	110
Male displays sexually to female. She approaches or follows him, then mates with him.	23	31	19	73
Female approaches and presents to male, then mates with him.	6	13	4	23
Total	78	78	50	206

which a male forms a short-term relationship with an estrous female and tries to prevent lower-ranking males from copulating with her, was seen mostly in the alpha male (Hasegawa & Hiraiwa-Hasegawa, 1983; Takasaki, 1985).

Thus, the female chimpanzees of Mahale may have developed proceptive behavior as defined by Beach (1976).

Mating patterns and initiative of copulation by bonobos of Wamba

At Wamba, all observed copulations are opportunistic (Kano, 1989), and they tend to be initiated by males rather than by females. Of 101 copulations, 62% were initiated by a male's display or approach, 26% were initiated by a female's approach or solicitation, and in 12% it was impossible to judge which sex initiated (Kano, 1989). Furuichi (1992) also described males as being the initiators in 36 of 37 copulations in E1 group. Our data agree with their reports. Of the 206 copulations, 89% were preceded by male approach or display

(see Table 11.4). Thus, the female bonobos of Wamba may have developed receptivity as defined by Beach (1976).

Females display much pseudo-copulatory behavior, especially genito-genital (GG) rubbing, in which the females rapidly rub their perinea together (Kuroda, 1980). At times of group excitement, GG-rubbing frequently occurs, often followed by the participants feeding in close proximity (Kano, 1987).

Is GG-rubbing a sexual behavior? This depends on the definition of sexual behavior. Kano (1989) speculated that GG-rubbing may be reassurance between females, and he pointed out that GG-rubbing was frequent in the feeding periods when bonobos were forced to lessen interindividual distances, rather than in resting periods. Female bonobos do not form exclusive homosexual female–female pairs, as reported for the estrous females of Japanese macaques (*Macaca fuscata*; e.g. Wolfe, 1984), and the females who are not maximally swollen often participate in GG-rubbing (see Table 11.5). Thus, GG-rubbing may function to decrease social tension between females, as Kano (1989) supposed.

Table 11.5 *Swelling states and GG-rubbing: number of days (pooled) when females of E1 and P groups were in each swelling state; in parentheses, number and proportion of days in which females displayed GG-rubbing*

	Swelling state		
	No swelling	Intermediate swelling	Maximal swelling
Adult females	334 (88; 26%)	191 (63; 33%)	159 (53; 33%)
Adolescent females	13 (7; 54%)	34 (10; 29%)	26 (4; 15%)

Selectivity in copulation by chimpanzees of Mahale and bonobos of Wamba

In order to compare promiscuity in chimpanzees and bonobos, we calculated Shannon–Weaver index of diversity, H' ($= -\Sigma(n_i/N)\log_2(n_i/N)$; N is the total number of observed copulations, and n_i is the total number of copulations observed for each dyad; Shannon & Weaver, 1949). A high value of H' means that copulations are more dispersed among many dyads, i.e. they exhibit more promiscuous mating.

Our data indicated that H' was 5.85 for the M group chimpanzees in 1981, and was 3.44 and 3.52 for the bonobos of E1 and E2 groups from November 1986 to February 1987. Similarly, H' was 4.53 for the E1 group bonobos from November 1984 to February 1985. These results suggest that the bonobos of Wamba do not copulate more promiscuously than do the chimpanzees of Mahale.

On the other hand, Kano (1992) and Furuichi (1992) considered that the prolonged estrus of female bonobos may decrease male competition within the unit group. Does this mean that male bonobos are sexually less competitive than male chimpanzees?

First, there is no clear correlation between copulation rate and male rank order in chimpanzees. In Gombe, male rank order showed no consistent correlation with copulation rates (Tutin, 1979; Fig. 11.1). In 1981 at Mahale, the copulation rates of the ten adult or late adolescent male chimpanzees of M group were negatively correlated with their rank order ($r_s = -0.67$, $p < 0.05$).

Of course, this does not mean that male chimpanzees do not compete with one another for estrous females. In particular, the alpha-male sometimes monopolized the females who were near ovulation and prevented the females from leaving him. In spite of such direct or indirect competition, however, overall male rank order was not reflected in copulation rates.

Fig. 11.1. Adolescent male chimpanzee copulates with adult female at Gombe. (Photo by C. Tutin)

In Wamba, significant correlations were found between male rank order and the number of copulations in some cases ($r_s = 0.92$, $p < 0.01$, in E group from 1978 to 1979, calculated from Kano, 1992; $r_s = 0.79$, $p < 0.05$ in E1 group in 1987, calculated from Furuichi, 1992). In the present data, however, there were no statistically significant correlations ($r_s = 0.29$ and 0.33, $p > 0.05$). Thus, in Wamba, there is no consistent relationship between male rank order and mating success, and it is unclear if male bonobos are less sexually competitive than male chimpanzees.

Do bonobos copulate more often than chimpanzees?

Kano (1992) concluded that bonobos copulate more frequently than do chimpanzees. Here, we compare the

Table 11.6. *Copulation rates of adult and adolescent male bonobos and chimpanzees*

| | No. of males | Copulation rates (no./hour) | | Source |
		Mean	Range	
Adult male chimpanzees				
M group (Mahale)	9	0.20	0.10–0.51	Present study
M group (Mahale)	2	0.29	0.22–0.31	Huffman, unpublished
M group (Mahale)	–	0.22	–	Hasegawa, 1991
Gombe	–	–	0.26–0.72	Goodall, 1986
Adult male bonobos				
E1 group (Wamba)	7	0.11	0.0–0.20	Present study
E2 group (Wamba)	8	0.10	0.0–0.27	Present study
E group (Wamba)	14	0.21	0.0–0.42	Kano, 1992
Adolescent male chimpanzees				
M group (Mahale)	1	0.70	–	Present study
M group (Mahale)	6	0.53	–	Hayaki, 1985
Adolescent male bonobos				
E1 group (Wamba)	3	0.05	0.0–0.065	Present study
E2 group (Wamba)	2	0.08	0.04–0.10	Present study
E group (Wamba)	5	0.19	0.0–0.26	Kano, 1992

copulation rates of the chimpanzees of Mahale and Gombe with those of the bonobos of Wamba.

First, the copulation rates of adult male bonobos of Wamba were equal to or lower than those of adult male chimpanzees of Mahale and Gombe (see Table 11.6). Adult male chimpanzees copulated 0.20–0.28 times/hour at Mahale, and copulated with ejaculation 0.26–0.72 times/hour at Gombe. At Wamba, adult male bonobos copulated 0.10–0.21 times/hour. Note that these findings do not result from differences among the groups in the number of estrous females.

Second, adolescent male chimpanzees copulate more frequently than do adolescent male bonobos (see Table 11.6). At Mahale, adolescent males copulated 0.53–0.70 times/hour, and at Gombe, Tutin (1979) reported that the highest copulation rates were shown by juvenile and adolescent males. By contrast, adolescent male bonobos at Wamba copulated only 0.05–0.19 times/hour.

How often do adult female chimpanzees and bonobos copulate during estrus? At Mahale, three adult female chimpanzees who were in maximal swelling copulated 0.79 times/hour with sexually mature males (see Table 11.7; Hasegawa, 1991). Since the period of maximal swelling occupied 40% of the swelling cycle (see Table 11.1),

females may have copulated 0.31 times/hour throughout the swelling cycle. If a female chimpanzee swells maximally 8.6 times throughout an interbirth interval as stated above, they may have copulated 0.04 times/hour $(= 0.31 \times [8.6 \text{ cycles} \div 69.5 \text{ cycles}])$ throughout the interbirth interval. Table 11.7 shows that the adult female bonobos of E1 group copulated 0.53 times/hour with sexually mature males throughout the swelling cycle (or throughout the interbirth interval). Excluding one female who copulated very frequently (2.4 times/hour), the remainder copulated 0.28 times/hour. On the other hand, Kano (1992) reported that adult female bonobos of E group copulated 0.14 times/hour (see Table 11.7).

In conclusion, the copulation rates of adult female bonobos are approximately equal to those of adult female chimpanzees throughout the swelling cycle, but these should be much higher than those of adult female chimpanzees throughout the interbirth interval.

The copulation rates of adolescent female chimpanzees and bonobos may differ notably. Adolescent female chimpanzees rarely mated with adult males. Of the five adolescent females of M group who showed constant swelling cycles in 1981, only one copulated once with a sexually mature male.

Table 11.7. *Copulation rates of adult or adolescent female bonobos and chimpanzees*

	No. of females	Copulation rates (no./hr)	
		Mean	Range
Female Chimpanzees[a]			
At maximal genital swelling	3	0.79	0.48–1.02
Over genital swelling cycle	–	0.31	–
Over interbirth interval	–	0.04	–
Bonobos El group[b]			
With newborn or 1-year-old infant	2	0.22	0.20–0.25
With 2- or 3-year-old infant	3	0.98	0.19–2.36
With 4- or 6-year-old infant	2	0.43	0.33–0.67
All adult females	7	0.53	0.16–2.36
Bonobos E group[c]			
With newborn or 1-year-old infant	15	0.09	0.0–0.22
With 2- or 3-year-old infant	5	0.24	0.0–0.56
With 4- or 6-year-old infant	5	0.08	0.0–0.29
Without infant	3	0.12	0.05–0.39
All adult females	28	0.14	0.0–0.56
Adolescent bonobos El group[d]	3	0.60	0.35–1.13
Adolescent bonobos E group[e]	13	0.50	0.0–1.33

Note: [a]adult female chimpanzees of Mahale (calculated from Hasegawa, 1991); [b]adult female bonobos of El group of Wamba (present study); [c]adult female bonobos of E group of Wamba (Kano, 1992); [d]adolescent female bonobos of El group (present study); [e]adolescent female bonobos of E group (Kano, 1992).

By contrast, adolescent female bonobos exhibited high copulatory rates and these were much higher than those of adult females (see Table 11.7). Since adult male bonobos actively solicited young females (Idani, 1991), adolescent females appear to have 'attractivity' for the adult males (cf. Beach, 1976).

DISCUSSION

Estrous female chimpanzees of Mahale: proceptive and promiscuous mating

It has been pointed out that bonobos' copulations differ from that of chimpanzees. In particular: (1) female bonobos copulate throughout the major part of their swelling cycle (Savage-Rumbaugh & Wilkerson, 1978; Thompson-Handler *et al.*, 1984); (2) bonobos copulate more promiscuously and more often than chimpanzees (Kano, 1992); and (3) male bonobos are sexually less competitive than male chimpanzees (Kano, 1989). How-

ever, the present data do not always support these conclusions.

On the other hand, there are great differences in the copulatory activities of female bonobos of Wamba and female chimpanzees of Mahale. First, female bonobos resume swelling cycles only a year after parturition. Since the female bonobos become pregnant 4–5 years after their last parturition, they continue to exhibit non-ovulatory estrus for 3–4 years (Kano, 1992). By contrast, adult female chimpanzees resume cycling about 4.4 years after parturition in Mahale (Nishida *et al.*, 1990). Thus, throughout interbirth intervals, the copulation rate of adult female bonobos is likely to be much higher than that of adult female chimpanzees.

Second, the present data suggest that the female chimpanzees of Mahale exhibit proceptive copulatory behavior, while the female bonobos of Wamba exhibit rather receptive behavior. Since the chimpanzees of a unit group or community tend to be dispersed within their home range (Wrangham, 1979), estrous females may have to search more actively for mating partners.

Is such proceptive behavior by female chimpanzees female choice? There are no clear data to suggest that females exclusively choose particular males who may have good genes, including the alpha male (Hasegawa, 1992; de Waal, 1982; Goodall, 1986; also see Small, 1989 for a review), and this departs from the expectations of the 'best male hypothesis' (Clutton-Brock & Harvey, 1976, cited in Wallis, 1992).

Our data suggest that estrous females of M group copulate promiscuously, as expected by the 'many males hypothesis' proposed by Hrdy (1981) and by Harcourt (1981). The 'possessive' behavior performed by the alpha-male suggests that female choice does not always focus on him.

Why do the female chimpanzees copulate so much? Hasegawa (1992) speculated that promiscuous mating may promote sperm competition and so females are likely to be impregnated by the good genes of males who win out in sperm competition (cf. Small, 1988). Paternity exclusion analysis may supply answers to this question.

Receptivity of adult female bonobos in Wamba

Some female primates have flexible patterns of sexual receptivity and 'concealed ovulation' (Hrdy & Whitten, 1987; but see Steklis & Whiteman, 1989). Based on the timing of ovulation and the behavioral and physical signs of estrus, female primates can be divided into several types, as follows: (1) 'honest' females who show conspicuous physical and behavioral cues, i.e. estrus around ovulation, and do not show estrus during pregnancy (e.g. *Papio* spp.; Hrdy & Whitten, 1987); (2) females who show conspicuous estrus around ovulation, but also show estrus during pregnancy and thus may 'deceive' the males (e.g. *Pan troglodytes*; Wallis & Lemmon, 1986); and (3) females who show rather unreliable cues of estrus, sometimes apart from ovulation and even during pregnancy, so that males may be 'deceived' by their apparent behavioral estrus (e.g. *Cercopithecus aethiops*; Andelman, 1987).

Female bonobos may have developed a fourth type of estrus pattern: females that show rather receptive estrus throughout almost all the birth interval, which may not be linked to ovulation (Furuichi, 1987, 1992; Kano, 1992). We have no data to gauge their attractivity, but, since copulations tend to be initiated by male bonobos,

these females' lengthy estrus may also mean ongoing attractivity. Thus, female bonobos may have developed 'continual receptivity and attractivity'.

Why have female bonobos developed such an estrous pattern? Kano (1992) and Furuichi (1987, 1992) pointed out that prolonged estrus may: (1) decrease male–male competition within the unit group; (2) decrease the possibility of infanticide; and (3) form or maintain male–female social relationships. Reviewing the sexual activity of bonobos, Blount (1990) also stressed that their sexual behavior may function in proximate terms as a tension-reduction mechanism. Unfortunately, at present, there are few data to test these hypotheses.

On the other hand, another possiblility is that bonobos' estrus functions to decrease social tension that originates from female–female competition. In contrast to female chimpanzees, female bonobos aggregate in the large-sized main party. In terms of kin selection theory, unrelated female bonobos may be potential competitors with one another (White & Lanjouw, 1992). In fact, grooming among female bonobos is less frequent than among males and between males and females, and agonistic interactions between females are much more severe than those within other dyads (Furuichi, 1989).

In particular, female bonobos may have developed GG-rubbing to reduce such social tension (Kano, 1980; Furuichi, 1989; White and Lanjouw, 1992). Although anestrous females may be involved in GG-rubbing (see Table 11.5), most participants are estrous females (Kuroda, 1980; Kano, 1992). Thus, their prolonged estrus may result in decreasing social tension among female bonobos. Furuichi (1992) wrote that some bonobo females showed maximal swelling in the early stages of postpartum amenorrhea but that they copulated rarely. They may show maximal swelling in order to exchange GG-rubbing with other females, rather than to copulate with males.

In conclusion, there are many unresolved questions as to the copulatory behavior and reproductive biology of chimpanzees and bonobos. Further studies are needed in order to compare the sexual behavior of these two *Pan* species and other anthropoid apes.

ACKNOWLEDGMENTS

This paper was originally prepared for the Wenner-Gren conference 'The Great Apes Revisited,' held November 12–17, 1994 in Cabo San Lucas, Baja, Mexico. We thank

W. C. McGrew and T. Nishida, for giving us this opportunity and for comments on the manuscript; D. S. Sprague for valuable comments; T. Kano, M. A. Huffman, T. Hasegawa, M. Hiraiwa-Hasegawa, H. Hayaki, H. Takasaki, K. Kawanaka, S. Uehara and T. Furuichi for kind permission to use their unpublished data; staff of the Mahale Mountains Wildlife Research Centre and to the people of Wamba for help in the field; Tanzania and Zaïre Governments for permission to carry out studies. This study was supported by the Grand-in Aid to Research of the Japan Ministry of Education (Nos.56041019 and 57043014 to T. Nishida and No.61041071 to T. Kano) and by the Japan International Cooperation Agency.

REFERENCES

Andelman, S. J. 1987. Evolution of concealed ovulation in vervet monkeys (*Cercopithecus aethiops*). *American Naturalist*, **129**, 785–99.

Beach, F. A. 1976. Sexual attractivity, proceptivity, and receptivity in female mammals. *Hormones and Behavior*, **7**, 105–38.

Blount, B. G. 1990. Issues in bonobo (*Pan paniscus*) sexual behavior. *American Anthropologist*, **92**, 702–14.

Clutton-Brock, T. H. & Harvey, P. H. 1976. Evolutionary rules and primate societies. In: *Growing Points in Ethology*, ed. P. P. G. Bateson & R. A. Hinde, pp. 195–237. Cambridge: Cambridge University Press.

Dahl, J. F. 1986. Cyclic perineal swelling during the intermenstrual intervals of captive female pygmy chimpanzees (*Pan paniscus*). *Journal of Human Evolution*, **15**, 369–85.

Furuichi, T. 1987. Sexual swelling, receptivity, and grouping of wild pygmy chimpanzee females at Wamba, Zaïre. *Primates*, **28**, 309–18.

Furuichi, T. 1989. Social interactions and the life history of female *Pan paniscus* in Wamba, Zaïre. *International Journal of Primatology*, **10**, 173–97.

Furuichi, T., 1992. The prolonged estrus of females and factors influencing mating in a wild group of bonobos (*Pan paniscus*) in Wamba, Zaïre. In: *Topics in Primatology, Vol. 2, Behavior, Ecology, and Conservation*, ed. N. Itoigawa, Y. Sugiyama, G. P. Sackett & R. K. R. Thompson, pp. 179–90. Tokyo: University of Tokyo Press.

Goodall, J. 1986. *The Chimpanzees of Gombe*. Cambridge, MA: Harvard University Press.

Harcourt, A. H. 1981. Inter-male competition and the reproductive behavior of the great apes. In:

Reproductive Biology of the Great Apes, ed. C. E. Graham, pp. 301–18. New York: Academic Press.

Hasegawa, T. 1991. Rankon-shakai no nazo [Sexual behavior of wild chimpanzees]. In: *Saru No Bunka-shi [The Cultural History of Primates]*, ed. T. Nishida, K. Izawa & T. Kano, pp.371–88. Tokyo: Heibonsha. (in Japanese)

Hasegawa, T. 1992. Mesu nitotteno rankon [The evolution of female promiscuity: chimpanzees and Japanese macaques]. In: *Doubutsu-shakai Niokeru Kyodo To Kogeki [Aggression and Cooperation among Animal Societies]*, ed. Y. Ito, pp. 223–50. Tokyo: Tokaidaigaku-Shuppankai. (in Japanese)

Hasegawa, T. & Hiraiwa-Hasegawa, M. 1983. Opportunistic and restrictive matings among wild chimpanzees in the Mahale Mountains, Tanzania. *Journal of Ethology*, **1**, 75–85.

Hasegawa, T. & Hiraiwa-Hasegawa, M. 1990. Sperm competition and mating behavior. In: *The Chimpanzees of the Mahale Mountains*, ed. T. Nishida, pp. 115–32. Tokyo: University of Tokyo Press.

Hayaki, H. 1985. Copulation of adolescent male chimpanzees, with special reference to the influence of adult males, in the Mahale National Park, Tanzania. *Folia Primatologica*, **44**, 148–60.

Hiraiwa-Hasegawa, M., Hasegawa, T. & Nishida, T. 1984. Demographic study of a large-sized unit group of chimpanzees in the Mahale Mountains, Tanzania: A preliminary report. *Primates*, **25**, 401–13.

Hrdy, S. B. 1981. *The Woman That Never Evolved*. Cambridge, MA: Harvard University Press.

Hrdy, S. B. & Whitten, P. L. 1987. Patterning of sexual activity. In: *Primate Societies*, ed. B. B. Smuts, D. L. Cheney, R. M. Seyfarth, R. W. Wrangham & T. T. Struhsaker, pp. 370–84. Chicago: University of Chicago Press.

Idani, G. 1990. Relations between unit-groups of bonobos at Wamba, Zaïre: Encounters and temporary fusions. *African Study Monographs*, **11**, 153–86.

Idani, G. 1991. Social relationships between immigrant and resident bonobo (*Pan paniscus*) females at Wamba. *Folia Primatologica*, **57**, 83–95.

Ihobe, H. 1992. Male–male relationships among wild bonobos (*Pan paniscus*) at Wamba, Republic of Zaïre. *Primates*, **33**, 163–79.

Itani, J. 1985. The evolution of primate social structures. *Man*, **20**, 593–611.

Kano, T. 1980. Social behavior of wild pygmy chimpanzees (*Pan paniscus*) of Wamba: a preliminary report. *Journal of Human Evolution*, **9**, 243–60.

Kano, T. 1987. Social organization of the pygmy chimpanzee and the common chimpanzee: Similarities and differences. In: *Evolution and Coadaptation in Biotic Communities*, ed. S. Kawano, J. H. Connell & T. Hidaka, pp. 53–64. Tokyo: University of Tokyo Press.

Kano, T. 1989. The sexual behavior of pygmy chimpanzees. In: *Understanding Chimpanzees*, ed. P. G. Heltne & L. A. Marquardt, pp. 176–83. Cambridge, MA: Harvard University Press.

Kano, T. 1992. *The Last Ape: Pygmy Chimpanzee Behavior and Ecology*. Stanford, CA: Stanford University Press.

Kuroda, S. 1980. Social behavior of the pygmy chimpanzees. *Primates*, **21**, 181–97.

Nadler, R. D., Herndon, J. G. & Wallis, J. 1986. Adult sexual behavior: hormones and reproduction. In: *Comparative Primate Biology, Vol. 2A, Behavior, Conservation, and Ecology*, ed. G. Mitchell & J. Erwin, pp. 363–407. New York: Alan R. Liss.

Nishida, T. 1980. The leaf-clipping display: a newly-discovered expressive gesture in wild chimpanzees. *Journal of Human Evolution*, **9**, 117–28.

Nishida, T., Takasaki, H. & Takahata, Y. 1990. Demography and reproductive profiles. In: *The Chimpanzees of the Mahale Mountains*, ed. T. Nishida, pp. 63–97. Tokyo: University of Tokyo Press.

Savage-Rumbaugh, E. S. & Wilkerson, B. J. 1978. Socio-sexual behavior in *Pan paniscus* and *Pan troglodytes*: a comparative study. *Journal of Human Evolution*, **24**, 327–44.

Shannon, C. E. & Weaver, W. 1949. *The Mathematical Theory of Communication*. Chicago: University of Illinois Press.

Small, M. F. 1988. Female primate sexual behavior and conception: are there really sperm to spare? *Current Anthropology*, **29**, 81–100.

Small, M. F. 1989. Female choice in nonhuman primates. *Yearbook of Physical Anthropology*, **32**, 103–27.

Steklis, H. D. & Whiteman, C. H. 1989. Loss of estrus in human evolution: too many answers, too few questions. *Ethology and Sociobiology*, **10**, 417–34.

Sugiyama, Y. 1994. Age-specific birth rate and lifetime reproductive success of chimpanzees at Bossou, Guinea. *American Journal of Primatology*, **32**, 311–8.

Takasaki, H. 1985. Female life history and mating patterns among the M group chimpanzees of the Mahale National Park, Tanzania. *Primates*, **26**, 121–9.

Takasaki, H., Hiraiwa-Hasegawa, M., Takahata, Y., Byrne, R. W. & Kano, T. 1986. A case of unusually early postpartum resumption of estrous cycling in a young female chimpanzee in the wild. *Primates*, **27**, 517–9.

Thompson-Handler, N., Malenky, R. K. & Badrian, N. 1984. Sexual behavior of *Pan paniscus* under natural conditions in the Lomako Forest, Equateur, Zaïre. In: *The Pygmy Chimpanzee: Evolutionary Biology and Behavior*, ed. R. L. Susman, pp. 347–68. New York: Plenum Press.

Tutin, C. E. G. 1979. Mating patterns and reproductive strategies in a community of wild chimpanzees (*Pan troglodytes schweinfurthii*). *Behavioral Ecology and Sociobiology*, **6**, 29–38.

Tutin, C. E. G. & McGinnis, P. R. 1981. Sexuality of the chimpanzee in the wild. In: *Reproductive Biology of the Great Apes: Comparative and Biomedical Perspectives*, ed. C. E. Graham, pp. 239–64. New York: Academic Press.

de Waal, F. B. M. 1982. *Chimpanzee Politics*. London: Jonathan Cape.

de Waal, F. B. M. 1989. Behavioral contrasts between bonobos and chimpanzees. In: *Understanding Chimpanzees*, ed. P. G. Heltne & L. A. Marquardt, pp. 154–75. Cambridge, MA: Harvard University Press.

Wallis, J. 1992. Chimpanzee genital swelling and its role in the pattern of sociosexual behavior. *American Journal of Primatology*, **28**, 101–13.

Wallis, J. & Goodall, J. 1993. Anogenital swelling in pregnant chimpanzees of Gombe National Park. *American Journal of Primatology*, **31**, 89–98.

Wallis, J. & Lemmon, W. B. 1986. Social behavior and genital swelling in pregnant chimpanzees (*Pan troglodytes*). *American Journal of Primatology*, **10**, 171–83.

White, F. J. & Lanjouw, A. 1992. Feeding competition in Lomako bonobos: variation in social cohesion. In: *Topics in Primatology, Vol. 1: Human Origins*, ed. T. Nishida, W. C. McGrew, P. Marler, M. Pickford & F. B. M. de Waal, pp. 67–79. Tokyo: University of Tokyo Press.

Wolfe, L. D. 1984. Japanese macaque female sexual behavior: a comparison of Arashiyama East and West. In: *Female Primates: Studies by Women Primatologists*, ed. M. F. Small, pp. 141–157. New York: Alan R. Liss.

Wrangham, R. W. 1979. Sex differences in chimpanzee dispersion. In: *The Great Apes*, ed. D. A. Hamburg & E. R. McCown, pp. 481–9. Menlo Park, CA: Benjamin/Cummings.

- PART IV

Minds

12 • Conflict as negotiation

FRANS B. M. DE WAAL

INTRODUCTION

Many animals, both primates and non-primates, are characterized by permanent associations between individuals with partly conflicting interests. While the existence of these associations must mean that the parties gain from them, this does not imply that competition has disappeared. Often, animals continue to compete over resources, such as food and mates, despite association and cooperation.

For such an arrangement to work, aggressive conflict needs to be controlled and regulated so that it does not undermine relationships. Paralleling interspecific symbiosis and mutualism, individual members of group-living species appear to enter 'social contracts' that optimally represent their respective interests. No doubt, some of these compromises have been worked out at the genetic level, in the course of evolution, but negotiations about the terms of a relationship may also take place as a social process. The present paper seeks to lay out a heuristic framework from which to conduct research in this area.

In order to do so, I will first critically review the current state of two approaches to social conflict: one by psychologists and social scientists, the other by evolutionary biologists. Traditionally, these frameworks have neglected the contractual perspective: not because the idea of compromise and negotiation is alien to these fields of science, but because of continued adherence to a dichotomy between aggressive competition and sociality.

Psychological approaches

When psychologists coined the term *prosocial behavior* for friendly and cooperative acts, aggression was by implication relegated to the category of antisocial behavior (Krebs & Miller, 1985). Aggression is generally presented as a socially negative tendency that poses serious problems to society, and hence needs to be contained or eliminated.

Even though aggressive behavior is undeniably social (i.e. aggression is generally defined as harmful behavior directed at members of the actor's own species), psychologists have historically ignored how it arises within and affects relationships. Psychologists implicitly adhere to what may be called an *Individual Model* of aggression, i.e. aggressive behavior is viewed as an expression of the individual due to both internal and external factors, such as personality, experience and frustration. Social relationships do not figure prominently in this model. Indeed, research subjects, both human and animal, often have neither a history nor future of interaction together. Human aggression is typically studied in experiments in which strangers are instructed how to interact (e.g. punish one another), or in which subjects are exposed to visual materials, such as violent movies (reviewed by Berkowitz, 1993). Similarly, animals are typically introduced to strangers, sometimes belonging to a different species (e.g. a mouse is placed in a rat's home cage), or electrically shocked so as to study pain-induced aggression (reviewed by Johnson, 1972).

Essentially, then, aggressive behavior is often investigated in a social vacuum. This neglect of social context is surprising given that most human conflict concerns familiar individuals. This not only applies to minor forms of aggression, but also to violent crimes, such as rape and murder, which are mostly committed by family members or acquaintances of victims (Braught *et al.*, 1980).

However, not all psychologists and social scientists try to divorce aggressive conflict from other aspects of social life, or view it as necessarily destructive and anti-social (Lyons, 1993). Recently, there are signs of a change in perspective. Gottman's (1994) extensive longitudinal data on human marital relationships, for example, have begun to challenge the conventional notion that conflict is incompatible with a lasting marriage. More important than *whether* people fight, is *how* they do it. The investigator distinguishes several types of marriages, one of which is characterized by frequent upheavals. He notes that these so-called 'volatile' relationships are by no means less stable than others:

> It turns out that these couples' volcanic arguments are just a small part of an otherwise warm and loving marriage. The passion and relish with which they fight seems to fuel their positive interactions even more. Not only do they express more anger but they laugh and are more affectionate than the average validating couple. These couples certainly do not find making up hard to do – they are masters at it. As intense as their battles may be, their good times are that much better.
>
> *Gottman, 1994, p. 41*

Evolutionary approaches

A false dichotomy between aggressive conflict and sociality has also dominated evolutionary approaches. Initially, Struggle-for-Life language was directly transferred to the social domain resulting in an over-emphasis on clashing individual interests. The possibility of shared interests was so far from the minds of early evolutionary biologists that when it came to accounting for the rarity of lethal violence, rather than assuming a need for cooperation, explanations focused exclusively on the physical risks of combat (Maynard Smith & Price, 1973).

Popp & DeVore's (1979) contribution to the original *Great Apes* symposium is a typical example of this line of argument. The authors wondered why dominants fail to kill subordinates, and suggested as a possible drawback that subordinates might fight for their lives. They added: 'By contrast, the only benefit for the potential assassin would be the elimination of just one of many competitors' (p. 329). In this rather Hobbesian view, then, each individual is surrounded not by friends and family, but by competitors. If true, there would be absolutely no need for conflict resolution: victory and defeat would be the only options worth considering.

Van Rhijn & Vodegel (1980) were the first to demonstrate that these early models rested on overly simplistic assumptions. They argued that individual recognition is a common capacity in the animal kingdom, and that animals remember against which individuals they have won or lost confrontations, resulting in systematic avoidance of stronger individuals. The widespread existence of stable territories and dominance hierarchies attests to the validity of this model.

The models become even more complex if the value of cooperation is factored in, that is, the dependency of social animals on group-life in general (van Schaik, 1983), and on the assistance of certain individuals, in particular (de Waal, 1989a). As Kummer (1978) pointed out, long-term social relationships are investments worth maintaining and defending. Primates often engage in alliances, in which two or more parties join each other against a third to reach goals that cannot be reached through individual action. Furthermore, chimpanzees (*Pan troglodytes*) form macro-alliances (involving all adult males of a community) against neighboring communities. See Harcourt & de Waal (1992) for an overview of alliances in primates and other animals, and de Waal (1982), Nishida (1983), Goodall (1986), Boehm (1992), and Nishida & Hosaka (Chapter 9) for examples of both intra and intercommunity alliances in chimpanzees.

The presence of alliances has profound implications for competition, not only in the triadic sense (i.e. both contestants need to take into account the presence or absence of potential allies), but also within the dyad itself. The two contestants may have an alliance themselves. If so, they need to take into account not only the value of the resource over which they are competing, and the risk of bodily harm in case of a fight, but also the value of their relationship. Specifically, each party needs to weigh the benefits of access to the resource against the possibility that use of force may reduce the other's tendency (or ability, in case of physical harm) to provide assistance in the future. Thus, the basic dilemma facing competitors is that they sometimes cannot win a fight without losing a friend and supporter.

THE RELATIONAL MODEL

As exemplified by Lorenz's (1967) *On Aggression*, ethologists used to assume that aggressive behavior is produced by an innate drive, hard or impossible to control, that seeks an outlet that in humans ranges from sports to gang violence. This view is now considered outdated. Attention has shifted from aggression as the expression of an internal state, to aggression as the product of conflicting individual interests. Increasingly, aggressive behavior is regarded as a product of social decision-making: it is one of several ways in which conflict can be resolved. This framework will be referred to here as the *Relational Model* (see Fig. 12.1). It differs from the Individual Model in that it focuses on the way aggressive behavior functions within social relationships.

One of the discoveries underlying this model is that primates engage in non-aggressive reunions between former opponents not long after an aggressive incident. For example, two male chimpanzees engage in charging displays in which they throw rocks and sticks at each other. It ends with screaming and chasing. Ten minutes after the pursuit, one male approaches his rival with an outstretched hand, panting in a friendly manner. This overture results in an embrace and a peace sealed with a kiss.

As it turns out, such post-conflict reunions are not limited to the great apes; they take place in other primates and perhaps in non-primates as well. Systematic evidence now exists for chimpanzees, bonobos, gorillas (Watts, Chapter 2), and a variety of Old World primates (reviewed by de Waal, 1993a; Fig. 12.2). Depending on the species, reunions include mouth-to-mouth contact, embracing, sexual contact, grooming, holding hands, clasping the hips of the other and so on. Opponents thus engage in rather intense forms of physical contact following conflict. Comparison of post-conflict episodes with behavior during control observations (i.e. periods not preceded by aggression) demonstrate major differences. These results contrast sharply with earlier views of aggression as an antisocial force. In the short run, dispersal is indeed the most conspicuous effect of aggression. In the long run, however, there is a significant tendency towards reunion following aggressive conflict. On average, aggression reduces rather than increases interindividual distances. This has been demonstrated in many species in captivity, and field studies have thus far supported this conclusion.

On the assumption that these post-conflict reunions serve to protect relationships against the undermining effects of aggression, they have been labeled

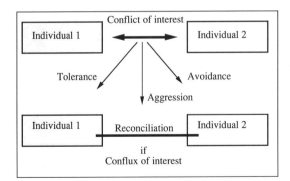

Fig. 12.1. In the relational model, aggressive behavior is one way of settling conflicts of interest. Other ways are tolerance (e.g. sharing of resources) or avoidance of confrontation. If aggression occurs, the nature of the relationship determines whether or not repair attempts will be made. If strong mutual interest in maintenance of the relationship exists, then reconciliation is most likely. Parties negotiate the terms of their relationship by going through cycles of conflict and reconciliation.

Fig. 12.2. Juvenile female bonobo. (Photo by F. de Waal)

reconciliations (de Waal & van Roosmalen, 1979). Even though primates vary dramatically in conciliatory tendency and behavior, all species seem to follow one rule: *reconciliation aims at restoring the closest and most valuable relationships* (de Waal & Yoshihara, 1983; reviewed by Kappeler & van Schaik, 1992). Thus, in species in which kin relationships are highly cooperative, related individuals reconcile more often than unrelated ones (e.g. Aureli *et al.*, 1989, 1992; Demaria & Thierry, 1992). And in chimpanzees, in which males form alliances that serve both intragroup and intergroup competition, conflicts among males are more often reconciled than those among females (de Waal, 1986a; Goodall, 1986; but see Baker & Smuts, 1994).

Experimental support for the Valuable Relationship Hypothesis has been produced by Cords & Thurnheer (1993), who trained macaques to cooperate during feeding. The investigators measured a dramatic increase in conciliatory tendency between partners that needed each other for food acquisition.

The Relational Model changes the kind of questions we ask about aggression and violence. For example, the question why aggression sometimes escalates to violence, or becomes so frequent as to damage relationships beyond repair, achieves special significance within this framework as it automatically translates into questions about the value individuals attach to their relationships and the social skills required to settle disputes in an alternative manner. The days of a deterministic view of aggression, isolated from the social context, are clearly behind us (de Waal, 1989b; Silverberg & Gray, 1992; Mason & Mendoza, 1993).

NEGOTIATION ROOM

It is easy to see how mild aggression can function as a tool for negotiation and behavioral modification of the partner. In combination with powerful peacemaking mechanisms, aggressive behavior makes it possible to arrive at social arrangements that might be impossible to achieve if all parties were simply to adhere to immutable objectives. Aggressive behavior is one of the means to test and to provoke the other, and to make each party's interests clear. To do so forcefully yet non-destructively, and to adjust one's objectives when the original objec-

tives cannot be realized, is the key to a mutually satisfactory coexistence.

Even though the present volume focuses on apes, it must be pointed out that the phenomenon of negotiated relationships is by no means restricted to this taxonomic group; it is probably widespread among long-lived social animals. Apes are of particular interest, however, because of the degree of symmetry in their social relationships. This was noted early by Maslow (1940), who contrasted rhesus macaques (*Macaca mulatta*) and chimpanzees. Whereas a dominant rhesus monkey terrorizes its subordinates, a dominant chimpanzee often acts as friend and protector: chimpanzees typically share food and reassure one another at times of tension. It is not unusual for a frightened subordinate to actually rush into the arms of a dominant.

Let us postulate that a large proportion of aggressive conflict in primate groups expresses individual interests and serves to modify the behavior of those with discordant interests. If the target of aggression is absolutely powerless, we expect this to result in great asymmetry in the degree to which each individual's interests is represented in the relationship. We may then speak of a 'despotic' relationship, that is, one in which the dominant exploits the subordinate with impunity.

Social symmetry is increased by (a) tolerance by dominants, and (b) alliances among subordinates. The first condition may apply to a great many species, including some of the closest relatives of the quintessential primate despot, the rhesus macaque. For example, stumptail macaques (*Macaca arctoides*) show greater tolerance around resources, reconcile more often after fights, and groom each other more than rhesus: they appear to give social closeness priority over enforcement of status, whereas rhesus seem to have reversed priorities (de Waal & Luttrell, 1989). Despite the contrast between these two species, however, their alliance structure is similar. Alliances in a stumptail group are as downward directed in the hierarchy as those in a rhesus group: third parties overwhelmingly intervene on behalf of winners rather than losers (see Fig. 12.3).

Chimpanzees, in contrast, show both sources of social symmetry mentioned above: a tolerant attitude as well as a considerable amount of upward directed alliances (de Waal & Luttrell, 1988; Fig. 12.3). Compared with the

Table 12.1: *Dominance style refers to the way that dominants treat subordinates and vice versa, particularly in competitive contexts*

Dominance style	Characterization	Typical species
Despotic	Dominants exploit subordinates with impunity and punish them severely for transgressions (such as when a subordinate takes precedence in a competitive situation). Reconciliation after fights and sharing of resources are relatively uncommon.	Rhesus macaque
Tolerant	Dominants and subordinates have close relationships with low levels of violence, frequent reconciliation and a high level of tolerance around contested resources. Room exists for subordinates to protest against aggression, yet they rarely band together against dominants.	Stumptail macaque
Egalitarian	Many characteristics shared with tolerant species, but also subordinates are capable of mounting effective alliances to curb the power of dominants.	Chimpanzee

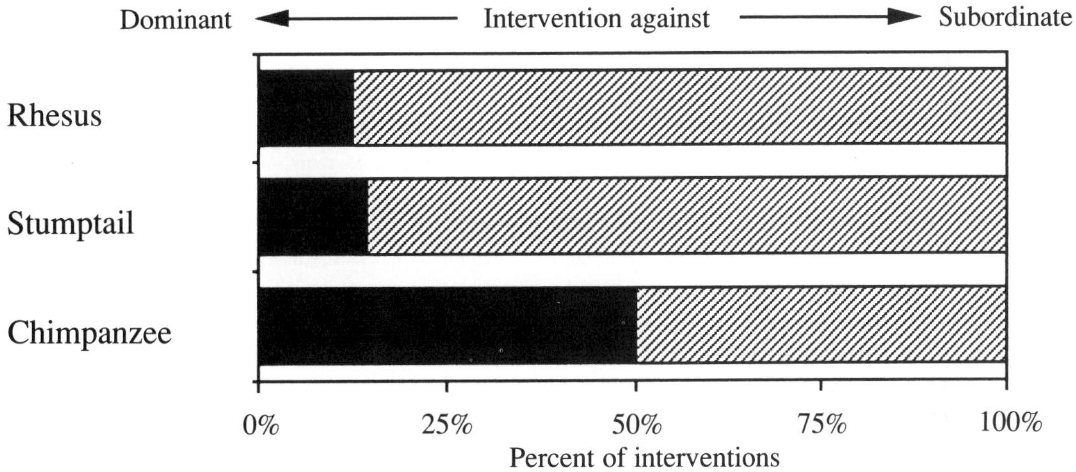

Fig. 12.3. If interventions in agonistic conflicts are random, half will be against dominant targets and half against subordinate targets. Based on 2053 interventions in adult rhesus monkeys, 535 in stumptail macaques and 1504 in adult chimpanzees, this graph shows that chimpanzee interventions are independent of dominance rank, whereas macaques mostly direct interventions down the hierarchy (from de Waal & Luttrell, 1988).

rhesus macaque's 'despotism,' and the stumptail macaque's 'tolerance,' the chimpanzee's dominance style leans towards egalitarianism (see Table 12.1). Chimpanzees are by no means absolutely egalitarian (a social hierarchy is clearly in evidence, particularly among adult males, e.g. Bygott, 1979; Noë *et al.*, 1980; de Waal, 1986a), yet subordinates do have ways of curbing the power of dominants. They are capable of making their voice heard, if necessary, through collective force (de Waal, 1982).

The capacity to mitigate the hierarchy through alliances is relevant to human social evolution. It is assumed that hunter–gatherer societies marked by weakly developed hierarchies (e.g. Woodburn, 1982) achieve this special arrangement through active *leveling mechanisms*. Leaders who become too proud or bossy, or fail to redistribute food equally, lose the respect and support of their 'inferiors'. In extreme cases, they pay with their lives (Boehm, 1993; Erdal & Whiten, 1994; see also Knauft, 1991). Human egalitarian societies go much further in this regard than chimpanzees, yet the underlying alliance mechanisms may have much in common.

Because of the hierarchical plasticity of apes, the study of negotiation is particularly relevant for an understanding of their social relationships. Negotiation is about influence and leverage; the more two-sided these are, the more room there is for mutually satisfactory solutions to conflicts of interest. 'Dominance style' thus sets the outer boundaries of the negotiation process. Humans attach great value to balanced outcomes, classifying them as equitable or just.

I leave aside here the question why some species have evolved a steep hierarchy and others a more level one, or why some human societies have developed an egalitarian ethos, and others have not. For plausible socio-ecological explanations of this variability see Vehrencamp (1983) and van Schaik (1989).

AREAS OF NEGOTIATION

Terms such as 'negotiation' and 'bargaining' characterize processes of communication and interaction in which individuals A and B go through repeated exchanges until they reach some sort of outcome or equilibrium that benefits both (albeit not necessarily to an equal degree). The animals are mutually responsive to each other's actions while following their own private agendas, which may overlap little or much, depending on their relationship and the issue at hand. Explicit references in the primate literature to negotiation and bargaining can be found in Maxim (1982), de Waal (1982), Hinde (1985), Dunbar (1988), Smuts & Watanabe (1990), Chadwick-Jones (1991), Noë et al. (1991) and Colmenares (1991).

Below I review a number of different relationships that combine status conflict or contest over resources with a motivation to get along and to cooperate. The tension between competition and cooperation works itself out in a negotiation process characterized by agonistic conflict, temper tantrums, greeting ceremonies with uncertain outcomes and so on. The end result is often a state of mutual acceptance and respected boundaries of behavior.

Weaning compromise

Apart from physical strength and the backing of alliances – both of which may translate into social dominance of one party over another – there exist sources of social leverage that do not depend on force. Shared reproductive interest is the most obvious one.

Trivers' (1974) analysis of parent–offspring conflict as a genetic tug-of-war captures the trade-off between a parent's natural physical dominance and the offspring's leverage. The offspring derives power from its status as recipient of genetic and behavioral investments that the parent cannot afford to lose. This results in profoundly conflicting tendencies in the parent between exerting dominance, on the one hand, and meeting the offspring's demands, on the other. Not that the two parties are aware of the genetic implications of their conflict: at the motivational level it is expressed in a conflict between a desire to dictate the other's behavior and an inhibition to harm the other.

The sort of compromise mother and offspring strike is easy to see in chimpanzees (Clarke, 1977). Some of the older juveniles (4–5 years of age) in our group at the field station of the Yerkes Regional Primate Research Center regularly suck on their mother's lower lip or ear, or nap with the skin right next to her nipple in their mouth (see Fig. 12.4). Such substitute nursing is the result of an extended period of discrepant desires between mother and offspring. Access to milk has become increasingly restricted despite vociferous protest by the youngster. Towards the end of the process, youngsters have to content themselves with something that merely resembles nursing. They bury their face close to their mother's breast, in her armpit for example, only to sneak up to a nipple when they get a chance.

If the mother brings superior strength to the weaning battlefield, the offspring brings a well-developed larynx and blackmailing tactics. The youngster cajoles its mother with signs of distress, such as pouts and whimpers. If everything else fails, the youngster may resort to a temper tantrum at the peak of which he almost chokes in his screams, or vomits at her feet. This is the ultimate threat: a literal waste of maternal investment. Moreover, the commotion may attract unwanted attention to the mother. Sometimes other adults attack a mother of a particularly noisy juvenile. Thus, apart from undermining his mother's determination by raising concern about his welfare, the offspring exerts social pressure.

The mother has weapons, too. Goodall (personal communication) offers a striking example with regards

abrupt drop in maternal attention and opportunities for nursing caused unprecedented tantrums in Faustino. To him, life definitely must have seemed 'unjust'. Fifi did not completely wean him, however, but continued to allow him to sleep in her nest, and even to nurse along with his new sibling.

One of Fifi's answers to Faustino's tantrums was to climb high up a tree and throw him to the ground, while at the last instant holding on to an ankle. The young male would hang upside down for 15 seconds or more, screaming his head off, before his mother would retrieve him. Goodall saw Fifi employ this scare tactic twice in a row, after which Faustino stopped having tantrums that day.

Weaning conflict is the first negotiation in a young mammal's social life. It is conducted with a social partner of paramount importance to survival and illustrates all the crucial elements of the Relational Model: a combination of conflict and conflux of interests, and a cycling through positive and negative interactions resulting in some sort of agreement about the 'proper' time for nursing (Altmann, 1980) and the general terms of the relationship. Mother and offspring share so many interests that the conflict mostly works itself out to their mutual advantage, with increased juvenile independence and the mother's readiness for the next pregnancy.

Testing of dominance

It is striking how many of the studies that discuss negotiation concern baboons (see references at the beginning of this section). For students of ape behavior it is important to look into this literature; similar processes probably take place in their subjects. The most detailed analyses thus far concern greetings among male baboons documented by Smuts & Watanabe (1990) in wild olive baboons (*Papio cynocephalus anubis*), in Kenya, and by Colmenares (1991) in a mixed, hybridized group of hamadryas (*Papio hamadryas*) and cynocephalus baboons at Madrid Zoo, in Spain.

Male baboons typically approach one another with friendly gestures, such as lipsmacking, after which they proceed to hip-grasping and mounting postures. Sometimes one male will fondle the other's genitals. Contact lasts only a few seconds, after which the two males separate again. Both Smuts & Watanabe (1990)

Fig. 12.4. Two examples of weaning compromise between mother and offspring chimpanzees. In both cases, the offspring has developed a habit of sucking on a part of the mother's body other than the nipple because of repeated nursing conflicts with the mother. Both offspring are four years of age: (*a*) a female (*b*) a male. (Photos by F. de Waal)

to the interbirth intervals of Fifi at Gombe National Park. Fifi had had regular 5-year intervals, almost like clockwork. Faustino was the first offspring born after a shorter interval: 4.5 years after the previous birth. This was only a minor deviation, however. She next gave birth in 1992, when Faustino was only 3.5 years old. The and Colmenares (1991) found these encounters to be very tense, occasionally erupting into fights. The apparent reason is that greetings serve to test and confirm who is on top, hence the jockeying for position to decide which male is going to be the mounter (generally the dominant) and which the mounted. Greetings thus seem a way of assessing the state of the relationship: a male who used to elicit presentations in another learns from his partner's refusal that their previous roles are no longer working, and that there may be a challenge in the air. The advantage of this sort of ritualized information-exchange could be that rank issues can be settled without physical confrontation.

Smuts & Watanabe (1990) found the youngest, most aggressively competitive adult males to be most preoccupied with dominance and to complete gestural exchanges during greetings least often. Senior males, on the other hand, who generally have known each other for a long time and who tend to support each other against younger and stronger males (Smuts, 1985; Bercovitch, 1988; Noë & Sluijter, 1990), are less nervous in each other's presence and complete greeting rituals more often. The older males' greetings are conducted rather leisurely, as they have changed from negotiations about dominance to confirmations of partnerships. A few combinations of older males even reach the point at which dominance hardly seems to matter anymore; they have moved from asymmetrical to almost perfectly symmetrical exchanges.

Male baboons, then, seem to test each other continuously, measuring vulnerabilities and strengths through the other's performance during greetings, oftentimes arriving at some form of settlement without the need to actually underline positions by means of force. Study of this process led Colmenares (1991, p. 59) to summarize the main ingredients of negotiation as follows: (a) conflict of interest between two parties; (b) non-aggressive means to assess and influence the partner's motivations and intentions; and (c) ability to adjust initial goals to those of the partner.

Chimpanzees, too, show greeting ceremonies in which there seems great hesitancy about which partner will play which role, particularly between males with unstable dominance relationships. The chimpanzee's most reliable status indicator is pant-grunting, sometimes referred to as a 'submissive greeting,' as the

vocalization is typically uttered when apes meet after a separation (Bygott, 1979; de Waal, 1982; Noë et al., 1980; Hayaki, 1990; Fig. 12.5). Pant-grunts may be exchanged prior to or during reconciliation. Indeed, when a formalized status relationship is lacking (thus no pant-grunting occurs), reconciliations are often tense and problematic, if they occur at all. De Waal (1986a) proposed a mechanism of *conditional reassurance*, that is, that dominant individuals will be prepared to achieve a cooperative and relaxed relationships only with subordinates who formally accept a lower status.

The mechanism of conditional reassurance was investigated during a period of great instability between two male chimpanzees, Yeroen and Luit, at the Arnhem Zoo. As usual during power struggles, the exchange of status signals had completely ceased. Friendly contacts had become rare as well, whereas aggressive confrontations and intimidation displays between the two males reached a peak. The low contact level was due to the future dominant, Luit, walking away each time Yeroen approached.

Luit's attitude changed dramatically the day Yeroen uttered his first submissive pant-grunts: Luit suddenly became receptive to Yeroen's overtures. Aggression diminished and contact between the rivals increased over the next few days (see Fig. 12.6). Significantly, the longest grooming session of the entire 3-month period occurred only a few hours after Yeroen's acknowledgment of Luit's status. This case study confirmed that peaceful coexistence among male chimpanzees depends on a formal clarification of their dominance relationship (de Waal, 1986a).

Whereas the mechanism of conditional reassurance is still under debate (e.g. York & Rowell, 1988; Colmenares, 1991; Kappeler & van Schaik, 1992; Cords & Thurnheer, 1993; de Waal, 1993b), greetings of male baboons and chimpanzees seem to have one point in common: their occurrence and completion depend on an agreed-upon status of the relationship. As such, these encounters constitute an area of negotiation in which animals test one another and collect information about how each one of them assesses the relationship. From this perspective, the most interesting encounters are those that do not unfold in a routine manner. Irregularities and tensions during the greeting process may reflect unresolved issues in the relationship relating to status,

Fig. 12.5. Chimpanzee 'greeting' ceremonies can be tense inter-actions between adult males. These involve submissive behavior, such as pant-grunting and bobbing or bowing, by the subordi-nate (right), and a more erect and intimidating posture by the dominant (left). The dominant may then move an arm over or jump over the subordinate, whose acceptance of these histrionics demonstrates recognition of his inferior position. Afterwards, the two males may engage in contact, such as an embrace or grooming. (Photo by F. de Waal)

level of cooperation and the balance of individual interests.

Accepted dominance

Chimpanzees have clearly moved away from despotism, creating a social arrangement with room for sharing, tol-erance and alliances from below. High-ranking individ-uals do have disproportionate privileges and influence, yet dominance also depends to some degree on accept-ance from below.

An example is the position of the current alpha male of the chimpanzee group at the Yerkes Field Station. After removal of the previous alpha male, two new adult males were introduced. The nine females, who had lived together for many years, expelled both of them (i.e. the males had to be treated for serious injuries). Several months later, two new males were introduced. One received the same reception as the previous two males, but the other, named Jimoh, was permitted to stay. Within minutes of his introduction, two older females contacted and briefly groomed Jimoh, after which one of them fiercely defended him against an attack by the alpha female.

Years later, during a background check on the chim-panzees in our study, it was discovered that Jimoh had been housed with those same two females at another institution before coming to the Yerkes Center. Appar-ently, this contact of 14 years earlier had made all the difference.

Although nearly 30 years old, Jimoh is an unusually small male, considerably smaller than several of the adult females he has come to dominate through sheer energy and persistence. Every day he made charging displays, asserting himself, sometimes supported by his female allies, but increasingly on his own. One by one, the females began to bow and pant-grunt to him, including eventually the alpha female. Jimoh also took up the so-called 'control role': few males are so alert at breaking up the smallest squabbles before they get out of hand (see Boehm, 1994, for a discussion of the control role in chimpanzees).

Only on rare occasions, such as when he fiercely

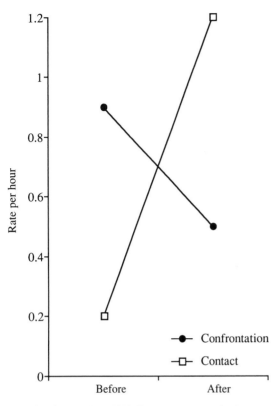

Fig. 12.6. On March 16, 1978, Yeroen was observed to bow and pant-grunt to Luit for the first time after three months of intense rivalry. The graph compares mean hourly rates of aggressive confrontations and affiliative contacts between the males during the seven observation days before and after Yeroen's first submission (data from de Waal, 1986a).

attacks a juvenile (usually a young male trying to copulate with an estrous female), do we see females band together against him. These incidents, in which Jimoh is invariably defeated, confirm that he is dominant only by the grace of the females' acceptance: a power balance reminiscent of the *primus inter pares* arrangement of human egalitarian societies (Boehm, 1993). However atypical this undersized male's relation with the females may be, it illustrates that dominance is not always a matter of one party imposing conditions upon the other: rank relationships resemble a mutual contract.

That such processes may also operate in the field is illustrated by the case of Goblin, in Gombe National Park. Having lost his top position, Goblin tried to stage a come-back but he was defeated by a ferocious mass-attack. Goodall (1992) comments that the unusually hostile reception may have have been caused by Goblin having been a very tempestuous alpha male, who frequently disrupted the group with charging displays: 'Possibly, his return would have roused a less dramatic response had he himself been a more peaceful and calm individual' (Goodall, 1992, p. 139).

Tit-for-tat transactions

In chimpanzees, male access to estrous females is affected by social dominance (Nishida, 1979; Tutin, 1979; de Waal, 1982), but also by alliances in the context of sexual competition. As illustrated by the Nikkie–Yeroen alliance at the Arnhem Zoo, such sex-related alliances are not necessarily identical to those formed during dominance struggles.

Nikkie attained alpha rank with the help of Yeroen, and remained dependent on Yeroen in order to dominate their common rival, Luit. Initially, Yeroen gained considerably from supporting Nikkie. His contribution to the mating activity in the group increased from 9% during Luit's alpha period to 39% under Nikkie. This changed during the second part of Nikkie's rule, however. Yeroen's mating success dropped to 19% while Nikkie's rose to 46%. The reason Yeroen's success decreased was that Nikkie had begun to cooperate with either other male in the sexual context. It was not unusual for Nikkie and Luit to walk shoulder to shoulder in Yeroen's direction to keep him away from a female.

Nikkie's tolerance of Luit increased steadily until one summer day, Luit started to climb towards a sexually receptive female who sat high up in a tree. Yeroen vocalized and looked alternately at Nikkie and Luit. In response, Luit returned to the ground, approached the other two males and joined in their hooting chorus. After a couple of minutes, however, Luit returned to the female. Now Yeroen burst into loud screaming against Luit, while holding out his hand in a request for support from Nikkie. Nikkie, however, walked away from the scene. This led to a highly unusual surprise attack by Yeroen on Nikkie, jumping on him from behind and biting him in the back.

Two days after the above incident, an unobserved fight took place in the night-quarters of the males. Although no winner or loser could be determined on the basis of an injury count, Nikkie behaved as the loser. Even though Luit probably had not been involved in the physical battle (he was the only male without serious injuries), he emerged as the new dominant. While this may be hard to understand in dyadic terms, if one considers the triadic structure, the explanation is simple. Measures of association and cooperation in the group subsequent to the night-quarter incident showed that the Nikkie–Yeroen coalition had collapsed (de Waal, 1986b).

My interpretation of these observations is one of a broken *transaction* of reciprocal favors. The benefits Yeroen derived from backing Nikkie had been reduced to nearly zero, particularly in terms of access to estrous females. Nikkie's attitude towards Yeroen's sexual activities had become indistinguishable from that towards Luit's. Sexual privileges may have constituted Yeroen's primary compensation for supporting Nikkie. One would expect Yeroen to closely monitor the fulfillment of this deal. Incidents such as the one in which Nikkie, in spite of his ally's appeal, failed to interrupt their rival's sexual activities may have constituted the last straw for Yeroen. When Nikkie failed to keep his end of the deal, leaning more and more toward Luit in the sexual context, Yeroen simply terminated the collaboration.

A transaction is defined here, then, as a cooperative relationship that serves the interests of both parties, continuation of which is contingent upon continued payoffs. The violent end to the Nikkie–Yeroen alliance suggests that transactions are regulated not only by rewards but also by punishments. This may be part of an overall tendency towards social reciprocity in the chimpanzee involving agonistic support, food sharing, grooming and sanctions against cheating (de Waal, 1982, 1989c; de Waal & Luttrell, 1988).

CONCLUSIONS

During negotiations, the objectives of one party are being tested against those of another. Mutual adjustment will occur only if there exist shared interests or reasonably symmetrical power relations. These two conditions are not independent: leverage of subordinate or weaker parties increases with the need of dominant or stronger parties for their cooperation. Negotiation takes place whenever it is more costly to disregard than to accommodate the partner's interests.

As a result of this process, each party gives in to some of the demands or objectives of the other so as to make room for a relationship that in the long run benefits both. Failure to accommodate the other's interests may be disadvantageous. The mother who abruptly weans her offspring endangers her own reproductive investment. The dominant who terrifies subordinates may reduce cooperative tendencies upon which his own survival and that of his offspring depend. The subordinate who fails to acknowledge a dominant's rank may face continued challenges and potentially destructive conflict.

So, animals settle for less than their initial objectives in order to secure the cooperation, tolerance, or survival of those on whom they depend. This dynamic process usually leads to an equilibrium that can be characterized as a compromise or transaction. The mother allows her infant to nurse but not whenever or as much as it wants. The dominant asserts himself but only insofar as necessary to maintain his rank. The subordinate defers to a dominant while at the same time setting limits to his behavior. The discrepancy between actual and realized objectives means that inter-individual negotiation probably goes together with intra-individual ambivalences (Mason, 1993). One can almost see individuals being pulled between conflicting inclinations, such as between care and punishment, or between monopolization and sharing of food.

The term 'transaction' implies more than a mere equilibrium or compromise between conflicting tendencies. The term suggests *expectations*, that is, each party has a set of predictions about the other's behavior, deviations from which may prompt adjustment of their own attitude. Thus, Yeroen changed his attitude towards Nikkie when Nikkie became more favorably inclined towards Luit, and a particular male baboon may become reluctant to support another who has begun to put up resistance during greeting ceremonies.

Whenever a violation of expectations leads to protest, scientists should pay special attention. Protest may derive from an image of how the other ought to behave. Anecdotal evidence for 'outraged' reactions in chimpanzees are the sudden attack by Yeroen on Nikkie described in the previous section, a female's protest against a male's lack of reciprocation described by de Waal (1982, p. 207), and Nishida's (1994, pp. 390–1) observations of subordinates who single-handedly put to flight dominants who apparently broke rules of friendship. Such unusual sequences suggest that chimpanzee transactions may be governed by a crude moral sense.

Unfortunately, expectations are hard or impossible to measure. What we can measure, though, is the stages through which negotiations unfold. We can focus on the behavioral tools, such as greeting ceremonies, protest reactions, reconciliations, temper tantrums and so on. The study of conflict as negotiation starts with the careful documentation of how and when these tools are being used. This is a formidable task indeed as it requires analysis of behavioral exchanges in their entirety over long intervals.

But then, this may be exactly how the minds of social animals keep track of behavior: not in terms of mere frequencies and durations, but in terms of how the behavior functions within the relationship, and how it fits or violates expectations about one another.

ACKNOWLEDGMENTS

This paper was originally prepared for the Wenner-Gren conference 'The Great Apes Revisited,' held November 12–19, 1994 in Cabo San Lucas, Baja, Mexico. The author would like to thank the editors, Filippo Aureli and Josep Call for helpful comments on the manuscript. Writing was supported by the Base Grant of the National Institutes of Health (RR00165) to the Yerkes Regional Primate Research Center.

REFERENCES

Altmann, J. 1980. *Baboon Mothers and Infants.* Cambridge, MA: Harvard University Press.

Aureli, F., van Schaik, C. P. & van Hooff, J. A. R. A. M. 1989. Functional aspects of reconciliation among captive long-tailed macaques (*Macaca fascicularis*). *American Journal of Primatology,* **19**, 39–51.

Aureli, F., Das, M. & Veenema, H. C. 1992. Interspecific differences in the effect of kinship on reconciliation frequency in macaques. *Abstracts of the 30th Annual Meeting of the Animal Behavior Society,* p. 5.

Baker, K. & Smuts, B. B. 1994. Social relationships of female chimpanzees: diversity between captive social groups. In: *Chimpanzee Cultures*, ed. R. W. Wrangham, W. C. McGrew, F. B. M. de Waal & P. Heltne, pp. 227–42. Cambridge, MA: Harvard University Press.

Bercovitch, F. 1988. Coalitions, cooperation, and reproductive tactics among male baboons. *Animal Behaviour,* **36**, 1198–209.

Berkowitz, L. 1993. *Aggression: Its Causes, Consequences, and Control.* New York: McGraw-Hill.

Boehm, C. 1992. Segmentary warfare and the management of conflict: comparison of East African chimpanzees and patrilineal–patrilocal humans. In: *Coalitions and Alliances in Humans and Other Animals*, ed. A. H. Harcourt & F. B. M. de Waal, pp. 137–73. Oxford: Oxford University Press.

Boehm, C. 1993. Egalitarian behavior and reverse dominance hierarchy. *Current Anthropology,* **34**, 227–54.

Boehm, C. 1994. Pacifying interventions at Arnhem Zoo and Gombe. In: *Chimpanzee Cultures*, ed. R. W. Wrangham, W. C. McGrew, F. B. M. de Waal & P. Heltne, pp. 211–26. Cambridge, MA: Harvard University Press.

Braught, G. N., Loya, F. & Jameson, K. J. 1980. Victims of violent death: a critical review. *Psychological Bulletin,* **87**, 309–33.

Bygott, J. D. 1979. Agonistic behavior, dominance, and social structure in wild chimpanzees of the Gombe National Park. In: *The Great Apes*, ed. D. A. Hamburg & E. R. McCown, pp. 405–28. Menlo Park, CA: Benjamin/Cummings.

Chadwick-Jones, J. K. 1991. The social contingency model and olive baboons. *International Journal of Primatology,* **12**, 145–61.

Clarke, C. B. 1977. A preliminary report on weaning among chimpanzees of the Gombe National Park, Tanzania. In: *Primate Bio-Social Development*, ed. S. Chevalier-Skolnikoff & F. E. Poirier, pp. 235–60. New York: Garland.

Colmenares, F. 1991. Greeting behaviour between male baboons: oestrous females, rivalry, and negotiation. *Animal Behaviour,* **41**, 49–60.

Cords, M. & Thurnheer, S. 1993. Reconciling with valuable partners by long-tailed macaques. *Ethology,* **93**, 315–25.

Demaria, C. & Thierry, B. 1992. The ability to reconcile in Tonkean and rhesus macaques. *Abstracts of the 30th Annual Meeting of the Animal Behavior Society,* p. 101.

Dunbar, R. I. M. 1988. *Primate Social Systems*. London: Croom Helm.

Erdal, D. & Whiten, A. 1994. On human egalitarianism: An evolutionary product of Machiavellian status escalation? *Current Anthropology*, **35**, 175–8.

Goodall, J. 1986. *The Chimpanzees of Gombe: Patterns of Behavior*. Cambridge, MA: Harvard University Press.

Goodall, J. 1992. Unusual violence in the overthrow of an alpha male chimpanzee at Gombe. In: *Topics in Primatology, Vol. 1, Human Origins*, ed. T. Nishida, W. C. McGrew, P. Marler, M. Pickford & F. B. M. de Waal, pp. 131–42. Tokyo: University of Tokyo Press.

Gottman, J. 1994. *Why Marriages Succeed or Fail*. New York: Simon & Schuster.

Harcourt, A. H. & de Waal, F. B. M. (eds.) 1992. *Coalitions and Alliances in Humans and Other Animals*. Oxford: Oxford University Press.

Hayaki, H. 1990. Social context of pant-grunting in young chimpanzees. In: *The Chimpanzees of the Mahale Mountains*, ed. T. Nishida, pp. 189–206. Tokyo: University of Tokyo Press.

Hinde, R. A. 1985. Expression and negotiation. In: *The Development of Expressive Behavior: Biology–Environment Interactions*, ed. G. Zivin, pp. 103–16. Orlando, FL: Academic Press.

Johnson, R. N. 1972. *Aggression in Man and Animals*. Philadelphia, PA: Saunders.

Kappeler, P. M. & van Schaik, C. P. 1992. Methodological and evolutionary aspects of reconciliation among primates. *Ethology*, **92**, 51–69.

Knauft, B. M. 1991. Violence and sociality in human evolution. *Current Anthropology*, **32**, 391–428.

Krebs, D. L. & Miller, D. T. 1985. Altruism and aggression. In: *Handbook of Social Psychology, Vol. II*, 3rd edn., ed. G. Lindzey & E. Aronson. New York: Random House.

Kummer, H. 1978. On the value of social relationships to nonhuman primates: a heuristic scheme. *Social Science Information*, **17**, 687–705.

Lorenz, K. 1967. *On Aggression*. London: Methuen.

Lyons, D. M. 1993. Conflict as a constructive force in social life. In: *Primate Social Conflict*, ed. W. A. Mason & S. P. Mendoza, pp. 387–408. Albany, NY: State University of New York Press.

Maslow, A. H. 1940. Dominance-quality and social behavior in infra-human primates. *Journal of Social Psychology*, **11**, 313–24.

Mason, W. A. 1993. The nature of social conflict: a psycho-ethological perspective. In: *Primate Social Conflict*, ed. W. A. Mason and S. P. Mendoza, pp. 13–47. Albany, NY: State University of New York Press.

Mason, W. A. & Mendoza, S. P. (eds.) 1993. *Primate Social Conflict*. Albany, NY: State University of New York Press.

Maxim, P. E. 1982. Contexts and messages in macaque social communication. *American Journal of Primatology*, **2**, 63–85.

Maynard Smith, J. & Price, G. R. 1973. The logic of animal conflict. *Nature*, **246**, 15–18.

Nishida, T. 1979. The social structure of chimpanzees at the Mahale Mountains. In: *The Great Apes*, ed. D. A. Hamburg & E. R. McCown, pp. 73–121. Menlo Park, CA: Benjamin/Cummings.

Nishida, T. 1983. Alpha status and agonistic alliance in wild chimpanzees. *Primates*, **24**, 16–34.

Nishida, T. 1994. Review of recent findings on Mahale chimpanzees. In: *Chimpanzee Cultures*, ed. R. W. Wrangham, W. C. McGrew, F. B. M. de Waal & P. Heltne, pp. 373–96. Cambridge, MA: Harvard University Press.

Noë, R. & Sluijter, A. A. 1990. Reproductive tactics of male savanna baboons. *Behaviour*, **113**, 117–70.

Noë, R., de Waal, F. B. M. & van Hooff, J. A. R. A. M. 1980. Types of dominance in a chimpanzee colony. *Folia Primatologica*, **34**, 90–110.

Noë, R., van Schaik, C. P. & van Hooff, J. A. R. A. M. 1991. The market effect: an explanation for pay-off asymmetries among collaborating animals. *Ethology*, **87**, 97–118.

Popp, J. L. & I. DeVore, I. 1979. Aggressive competition and social dominance theory: synopsis. In: *The Great Apes*, ed. D. A. Hamburg & E. R. McCown, pp. 317–38. Menlo Park, CA: Benjamin/Cummings.

van Rhijn, J. & Vodegel, R. 1980. Being honest about one's intentions: an evolutionary stable strategy for animal conflicts. *Journal of Theoretical Biology*, **85**, 623–41.

van Schaik, C. P. 1983. Why are diurnal primates living in groups? *Behaviour*, **87**, 120–2.

van Schaik, C. P. 1989. The ecology of social relationships amongst female primates. In: *Comparative Socioecology: The Behavioral Ecology of Humans and Other Mammals*, ed. V. Standen & R. Foley, pp. 195–218. Oxford: Blackwell Scientific Publications.

Silverberg, J. & Gray, J. P. (eds.) 1992. *Aggression and Peacefulness in Humans and Other Primates*. New York: Oxford University Press.

Smuts, B. B. 1985. *Sex and Friendship in Baboons.* New York: Aldine.

Smuts, B. B. & Watanabe, J. M. 1990. Social relationships and ritualized greetings in adult male baboons (*Papio cynocephalus anubis*). *International Journal of Primatology*, **11**, 147–72.

Trivers, R. L. 1974. Parent-offspring conflict. *American Zoologist*, **14**, 249–64.

Tutin, C. E. G. 1979. Mating patterns and reproductive strategies in a community of wild chimpanzees. *Behavioral Ecology and Sociobiology*, **6**, 39–48.

Vehrencamp, S. A. 1983. A model for the evolution of despotic versus egalitarian societies. *Animal Behaviour*, **31**, 667–82.

de Waal, F. B. M. 1982. *Chimpanzee Politics.* London: Jonathan Cape.

de Waal, F. B. M. 1986a. Integration of dominance and social bonding in primates. *Quarterly Review of Biology*, **61**, 459–79.

de Waal, F. B. M. 1986b. The brutal elimination of a rival among captive male chimpanzees. *Ethology and Sociobiology*, **7**, 237–51.

de Waal, F. B. M. 1989a. Dominance 'style' and primate social organization. In: *Comparative Socioecology: The Behavioral Ecology of Humans and Other Mammals*, ed. V. Standen & R. A. Foley, pp. 243–64. Oxford: Blackwell Scientific Publications.

de Waal, F. B. M. 1989b. *Peacemaking among Primates.* Cambridge, MA: Harvard University Press.

de Waal, F. B. M. 1989c. Food sharing and reciprocal obligations among chimpanzees. *Journal of Human Evolution*, **18**, 433–59.

de Waal, F. B. M. 1993a. Reconciliation among primates: a review of empirical evidence and unresolved issues. In: *Primate Social Conflict,* ed. W. A. Mason & S. P. Mendoza, pp. 111–44. Albany, NY: State University of New York Press.

de Waal, F. B. M. 1993b. Co-development of dominance relations and affiliative bonds in rhesus monkeys. In: *Juvenile Primates: Life History, Development, and Behavior*, ed. M. E. Pereira & L. A. Fairbanks, pp. 259–70. New York: Oxford University Press.

de Waal, F. B. M. & Luttrell, L. M. 1988. Mechanisms of social reciprocity in three primate species: symmetrical relationship characteristics or cognition? *Ethology and Sociobiology*, **9**, 101–18.

de Waal, F. B. M. & Luttrell, L. M. 1989. Toward a comparative socioecology of the genus *Macaca*: different dominance styles in rhesus and stump-tailed macaques. *American Journal of Primatology*, **19**, 83–109.

de Waal, F. B. M. & van Roosmalen, A. 1979. Reconciliation and consolation among chimpanzees. *Behavioral Ecology and Sociobiology*, **5**, 55–66.

de Waal, F. B. M., & Yoshihara, D. 1983. Reconciliation and redirected affection in rhesus monkeys. *Behaviour*, **85**, 224–41.

Woodburn, J. 1982. Egalitarian societies. *Man*, **17**, 431–51.

York, A. & Rowell, T. 1988. Reconciliation following aggression in patas monkeys (*Erythrocebus patas*). *Animal Behaviour*, **36**, 502–9.

13 • Language perceived: *Paniscus* branches out

E. SUE SAVAGE-RUMBAUGH, SHELLY L. WILLIAMS, TAKESHI
FURUICHI AND TAKAYOSHI KANO

LANGUAGE AS THE DOMAIN OF *HOMO*

Human beings have traditionally viewed their capacity for language as something special and unique amongst the creatures of the planet. The communication systems of all other animals have been described as 'non-verbal,' a term implying the absence of intentionality, symbolic encoding and internal structure (Hinde, 1972; Feldman & Rime, 1991). Because other animals purportedly were unable to produce acts of communication equivalent to language, linguists have assumed that all forms of communication employed by other species are primitive or biologically prewired. Information not generated by an extant emotional state is assumed to be beyond the communicative capacity of animals. Thus, it is thought that animals cannot, for example, intentionally communicate desires or plans to travel to a certain tree or to look for a certain type of food. In fact, even to speak of animals as having 'plans' is generally discouraged by the scientific community.

These presuppositions have prevented scientists from interpreting the communication systems of other animals, especially primates, in the same way that they interpret their own communications. The difficulties that result from such limitations are many. They can be realized by imagining what would happen if a scientist tried to decode an unknown human language by looking only at the relationship between the words of a speaker and the behavior of a listener. If one observes adults conversing, one often finds that the listener does nothing in response to what the speaker says, except to reply with other words.

If an outside observer does not know any of the words, nor even whether one is hearing individual words as opposed to phonemes, it is impossible to rely upon the correlations between the words of the speaker and the listener to gain an understanding of the meaning of the sounds emanating from the speaker. Yet this is precisely what scientists studying animal communication do. They try to correlate the vocal sounds made by one individual with the ensuing behavior of another individual. Let us suppose, for purposes of conjecture, that one animal was saying to another, 'And where have you been today?' and the other were to reply, 'Oh, down to the river.' Could this ever be decoded by looking at correlations between these two vocalizations or between any pairs of vocalizations that were symbolic as opposed to ritualized utterances?

Increasingly sophisticated grammatical analysis cannot provide greater insight into the basic phenomenon of symbolic communication, because the meaning transferred by the words often does not lie in grammar but in the mutually understood intent of the speaker.

LANGUAGE IN THE EYES OF THE PERCEIVER

The communication systems of non-human primates, and perhaps those of other organisms too, will continue to elude science as long as studies of animal communication look for relationships between sounds and specific objects or ensuing events that follow quickly in time (Menzel,1987). Our species' symbolically based systems of meaning do not depend upon a clear correspondence between symbol and referent. Indeed, it is only since the invention of dictionaries and written linguistic systems that people have commonly assumed that individual words have specific meanings.

[173]

Developmental psychologists, linguists and others studying the onset of human language have no commonly accepted theory on how it is that a child learns which meanings to assign to which words (Pinker, 1984; Bruner, 1990). The puzzle of language acquisition, often termed 'Quine's dilemma' (as he was one of the first philosophers to pose it succinctly), is that there is no way to determine accurately the meaning of new words, if one does not already know the meaning of other words (Quine, 1960). Thus, one can look up the meaning of a word like 'telephone' in a dictionary, if one has not heard it before, but what if one does not know the meanings of any words contained in the definition? How is it that one initially breaks into language?

This primary problem, that of breaking into language, has been used to support the contention that *Homo sapiens* possesses a special portion of cerebral cortex devoted to grammatical analysis. It is said to be this biologically-based ability that enables us to group words into categories and thus to perform grammatical analysis (Chomsky, 1978). From grammatical analysis comes meaning and its partner – rational thought. Few would disagree that *Homo sapiens* possesses something that seems to be absent in apes; after all, our brains are significantly larger and heavier.

Yet, by linking grammar with both neural function and rational thought, linguists have concluded that other animals not only do not possess language, they *cannot* possess it (Chomsky, 1978; Pinker, 1984). Furthermore, they have concluded that if other animals *did* have language, any such language would have to contain a grammar like that of human beings, and would have to be recognizable by us as language. That is to say, non-human language would need to share important characteristics of human languages, such as a clear distinction between nouns and verbs.

Looking for a grammar in the vocalizations of animals is difficult, especially since science has not yet found any 'words' to work with. To take this sort of heuristic approach is to remain forever bound in Quine's dilemma, since it is impossible to look for the grammar without the words, and it is impossible to understand the words without knowing the grammar. Thus, it is a chicken-and-egg problem. Scientists are assumed to be freed from this problem when they study other human languages, because their uniquely human 'grammar analyzing' brain permits them to make the initial leap into another language.

Scientists who are interested in the study of communication by other species must therefore begin with one hand tied behind their backs. They must assume that, unless they are able to prove otherwise, other animals have no words, no grammar and no intentional system of communication. Thus, the first course of business is not to look at animal communication from the top down, by asking what sort of information animals might wish to convey to each other. By contrast, scientists undertake the study of another human language by trying to learn what it is that people are saying to each other or to the researchers.

Had scientists gone about the study of human languages using the same sorts of techniques that have been applied in investigations of animal communication, they most likely would have concluded that humans have no real language. Instead, they would have found that this primate species (*Homo sapiens*) produces a wide variety of emotional sounds which grade into each another and that various individuals tend to broadcast these sounds in order to keep in contact with one another. Indeed, the need to stay in vocal contact seems to be so great amongst *Homo sapiens* that they have invested a great deal of effort in making devices that allow them to make contact sounds to one another across long distances.

ASSUMPTIONS UNDERLYING LANGUAGE AS WE KNOW IT

In order to study the communication systems of non-human primates or other complex organisms in a way that permits us to find a symbolically-based system of communication, should it exist, we will have to approach the problem in the same way we approach the learning of language in our own species. This means we will have to begin with the following assumptions – all of which we apply to human language from the outset:

1. That communication events are purposeful and that the speaker has a specific intent – to convey information to the listener.
2. That similar intentions can be encoded in complex ways that differ from time to time, the

meanings of which depend largely upon the context within which they are embedded.

3. That the effect of the message may not alter the listener's immediate behavior in any specific manner, but that the information received may cause the listener to make different decisions at some future point than would otherwise have been made.

4. That it will be very difficult, in most cases, to determine the specific point in time that a symbolic communication will cause a predictable effect upon the listener.

Our current analytical and quantitative methods of determining whether or not a purported 'fact' about animal communicaton is verifiable do not permit an analysis of the sort described above. Lacking the scientific tools to decode the communicative systems of other species in an objective manner, we have instead drawn the conclusion that other animals lack language. Alternatively, we might have concluded that science currently lacks the tools to discern such a system.

What shall scientists do if they wish to study the communication systems of other animals? How can we avoid either a conclusion that is compromised by the current limitations of our scientific method or one that may get at the communication only to be subject to criticism for being flawed by anthropomorphism?

The first thing scientists must do is to permit themselves to ask, 'If an ape (or other animal) were to have a language, what might it look like?' Having dared to ask such a question, they then must address an even more basic question, 'What might an ape (or other animal) need to do that would require a symbolic system of communication?' This is a difficult question, for it needs an understanding of the kinds of information that symbols alone can carry. Humans are so adept at symbol use that they sometimes use symbols in unnecessary ways, for example, to say 'I am happy with you.' A symbolic system is not needed to express such information.

If some aspect of animal behavior could be discerned that requires a form of symbolic communication for satisfactory execution, we could then begin to look for actions that might have symbolic significance and relate those actions to the behavior of interest. To date,

examples of animal behavior that require a symbolic system are few. Since apes are more closely related to human beings than they are to other animals (Zihlman, Chapter 21), instances of behavior requiring such a symbol system are most likely to be found among apes. For example, Boesch (1991) reported that male chimpanzees (*Pan troglodytes*) in the Taï forest appeared to utilize specific drumming patterns to signal direction of travel and duration of resting stops. However, shortly after these patterns were noted, the males who were using them vanished from the group and the patterns were not continued by the remaining males.

Apes are reported to travel somewhat unpredictably about their environment in search of ripe food (Schaller, 1963; Goodall, 1968; Galdikas, 1995; Kano, 1992). If they find a particularly rich area of fruit, they may stay there several hours; however, they generally move on to another location as soon as they have consumed the available fruit. Little has been said about how they choose their travel paths; indeed, there has been no clear indication that they choose them in anything at all like the sense that *Homo sapiens* chooses to travel to a neighboring village. Apes may be communicating something about the availability of food in different locations with specific calls, after they arrive at a feeding site, but no one has yet reported evidence for anything more than a relationship between the affective state of excitement produced by large quantities of food and the coupled increase in volume and frequency of vocalizations (Hauser *et al.*, 1993; Mitani, Chapter 18). Indeed, many scientists have remarked that there seems to be little about the apes' way of life that would require a linguistic system (Humphrey, 1983; Davidson & Noble, 1989). Apart from the use of tools, nothing has been reported about the natural life of apes that is distinctly different from that of monkeys living in the same areas. Is this because apes are really no different from monkeys in spite of their much larger brains? Or, is it because the kinds of questions that have been posed by field researchers have failed to address the differences that do exist across primates?

LANGUAGE IN THE LABORATORY

Studies of at least one species of ape in captivity, the bonobo (*Pan paniscus*), have led to very different findings

(Savage-Rumbaugh & Lewin, 1994). In captivity, it is now clear that bonobos are capable of complex symbolic communication with the following qualities:

1. Understanding of symbolic communication arises spontaneously in the bonobo, without the need for shaping or planned reinforcement of specific skills.
2. Human speech can be readily understood by apes when exposure occurs at an early age. The correct meanings are assigned to spoken words spontaneously, without shaping or planned reinforcement by human beings.
3. Rapid pairing between a printed symbol and a speech sound occurs in the same manner.
4. Grammaticality of simple spoken sentences is well within the ability of the bonobo.
5. Bonobos can modify their vocal expressions and use vocal sounds in answer to speech directed toward them.
6. Bonobos generate novel communications and have some grammatical structure.
7. Bonobos have a large degree of voluntary control over their vocal output.

The ease with which such skills are acquired in the captive setting suggests that the potential for symbolic communication is present in the bonobo. Given that the demands of resource procurement, group coordination and predator evasion are far greater in the wild than in captivity, it seems likely that a creature having such a highly developed ability for symbolic communication would utilize it in the wild (Fig. 13.1).

Have studies of wild apes failed to find examples of symbolic communication because scientists have been hesitant to describe what they see or to ask questions about how a symbolic communication system might occur? Or have they failed because apes do not have such a system? A survey of the studies of vocal behavior amongst great apes (Mitani, Chapter 18) suggests the former. According to Mitani (p. 252), '... there is also a pressing need for construction of a robust theory that can be used to generate *a priori* predictions about what animals should be talking about in order to survive and reproduce successfully in their natural habitats.'

Fig. 13.1. Three-year-old male bonobo at Lomako. (Photo by G. Hohmann)

QUESTIONS FOR THE FIELD

The first three authors spent six weeks at Wamba in Zaïre, where wild bonobos have been studied consistently by Takayoshi Kano (1992) and his colleagues since 1973. The goal of this initial study was to seek evidence for the possibility of symbolic communication in nature. From the outset, the guiding question of the research was, 'What forms of symbolic information might bonobos need to convey in order to conduct their daily lives as they do?'

The results of this study are based on 120.8 hours of observation of bonobos between July 30, 1994 and September 8, 1994, when bonobos were observed on 26 days for an average of 4.6 hours per day. Another 100 hours were spent in the forest becoming familiar with bonobo trails, nesting sites and food resources. Due to length considerations, the data are presented here in summary form only.

Four groups at Wamba were observed, but most of the data were taken on E2, and most of the rest on E1. Identification of the groups was made by the trackers who have been following these groups for two decades. The bonobos were not always in sight during the whole observation period, but they were always within hearing range, so that even when out of sight it was still possible to discern their location and direction of travel as well as to record vocalization data.

BONOBOS IN THE FIELD – WHY COMMUNICATE?

When following the bonobos in the field, two things are soon apparent to anyone attempting to study either gestural or vocal symbolic communication. First, unless the bonobos stay close together, gestural communication is precluded by the denseness of the vegetation. Second, the bonobos rarely vocalize when they travel on the ground, but they vocalize often while in the trees (Kano, 1992).

In order to understand the implications of the second point, it is necessary to know that bonobos typically travel for much of each day and that most of their travel is on the ground. Also, while traveling, they do not wander about, but go directly from one feeding site to another, eating terrestrial vegetation and digging for insects and tubers as they travel. Observers generally have the impression that bonobos are headed in a particular direction as they travel. This impression seems to arise from the observation that the direction they take generally minimizes the distance between feeding sites.

They tend to travel in single file with the group splitting up into parties. These parties frequently leave the feeding or nesting sites at different times. Resting and play occur on the ground as well as in the trees, especially if the distance between feeding sites is long. When following another party on the ground, bonobos appear attuned to the overall travel movements of the group and know precisely the direction taken by other group members, although these members are out of sight and are silent. Such knowledge is evidenced by the fact that parties always reunite at distant feeding or nesting sites.

The reunions at feeding sites take place before the group again divides to travel to the next feeding site (Fig. 13.2). Such aggregation differs from the independent convergence of chimpanzees at provisioning sites in that it entails the entire group, and is accomplished generally several times each day at different locations in the forest.

Fig. 13.2. Adolescent female bonobo in fruiting tree at Lomako. (Photo by G. Hohmann)

Additionally, when the group splits into parties, the parties often take the same path to the next destination.

How do bonobos know the travel patterns of parties that are out of sight ahead of them? If the bonobos communicate their intended destination while still in the trees, and if they always travel the same routes from point A to point F, then they could infer that other parties were always somewhere on the trail between A and F. However, if there were multiple routes from A to F (as is the case) or if one of the parties changed their mind enroute, advance communication would not suffice. Some additional way of communicating new information to other parties would be needed.

HOW DO HUMANS FOLLOW BONOBOS?

Trackers generally know where the parties have gone, although they quickly lose sight of the apes once terrestrial travel begins, and they always follow the last party to leave an area. (Humans traveling in between parties disturbs the bonobos.) The trackers cannot say where the bonobos were headed on the basis of their vocalizations prior to leaving a site. Trackers are able to follow the apes because of signs that the humans see in the vegetation.

Are the bonobos doing the same thing? Are they tracking each other? This is possible, yet the bonobos never seemed to lose each other, although the trackers often lost the apes. The bonobos followed each other through swampy areas where there were no footprints and in streams where there was no vegetation. They did so quickly and were not seen to bend over and study the ground carefully or to smell it, as is characteristic of chimpanzees. Each party always seemed to know precisely where to go.

The trackers, though skilled in their knowledge of the forest, as well as in the ways of the bonobos, often temporarily lost the group that they were tracking on the ground. Only rarely did bonobos permit the trackers to follow them within a few meters. At any greater distance the dense vegetation made visual tracking impossible, yet the trackers, like the bonobos, generally did not bend over to study the ground for the slightly bent leaf or the faint impression of a foot. Indeed, the vegetation was sufficiently dense and the ground sufficiently spongy as to make such cues exceedingly difficult to discern. Only when the trackers lost the party that they were following did they pause long enough to study the ground carefully and only then did they travel in circles looking for faint signs of vegetation disturbance. Most of the time, they, like the bonobos, seemed to know precisely where everyone was headed and they proceeded speedily on their way, hardly even glancing down, chopping rapidly with their machetes to clear the trail as they went. How did they know where to go through what seemed to be a virtually trackless forest, and how did they do so without even stopping to look for tracks if these were indeed there?

The trackers had two things to assist them. First, their general knowledge of the forest and of the habits of bonobos often helped them make accurate guesses regarding intended destinations. If the bonobos were traveling in a particular direction, the trackers sometimes surmised that they had decided to head toward one or more known feeding sites. The trackers' general knowledge of the tendency of the bonobos to forage regularly in swamps and of the location of these swamps, enabled them accurately to predict when the bonobos were headed toward a swamp.

Second, since the bonobos often walk in single file, they create trails as they travel repeatedly between different feeding sites. Their paths are not well-worn like the human trails, nor is vegetation removed from them as characterizes human trails. Bonobo trails are not easy to discern; unlike human trails they vanish at certain points only to reappear later. Sometimes, these points are in muddy or swampy areas; other times, they are in areas where the composition of the ground cover makes it difficult to discern a faint trail. Unlike human trails, sometimes there were places where the bonobos moved for a short distance through the trees instead of on the ground. All of the trails were faint compared with human trails, and there were many more of them. In addition, the trails frequently criss-crossed one another.

Because the usage of such trails had not been previously reported, trackers were queried about the existence of bonobo trails and whether or not the bonobos often followed such trails. They responded that there were bonobo trails and that the bonobos used them often. Once the existence of bonobo trails was verified, we realized that whenever the trackers were moving rapidly, without appearing to study the vegetation, they were following a bonobo trail. If trackers could easily and quickly follow parties of apes using such trails, then bonobos probably could do so as well.

Still, the trails presented a problem because there were so many and they often intersected each other. What does one do when two trails cross? What does one do when the trail is so faint that it is impossible to distinguish between them and any slight indentation of the ground cover? Even the capacity to predict the likely travel path of the bonobos and to follow bonobo trails can prove insufficient. The trackers sometimes followed the wrong bonobo trail for several hundred meters before realizing their error, and at times they lost the bonobos altogether for several hours and, in some cases, for several days.

One advantage that the bonobos had was that those in the lead appeared to care about whether or not their followers were able to find their way. Bonobos from leading parties sometimes turned back to interact with members of a following party.

SYMBOLS IN THE SUBSTRATE

When trails crossed or became too faint to read, trackers relied on altered vegetation. Occasionally, while walking along a trial, they would point out with the tips of their machetes a group of leaves that had been completely flattened, stamped down with the force of a bonobo foot being scuffed firmly over the plant.

Sometimes there would be a line of three such plants, each only a few centimeters from the other. However, usually there was no such clear indication, but only a very slightly bent leaf that by itself meant nothing. Only if many of these slightly bent leaves were seen did the trackers attempt to follow them. Occasionally there were also bonobo dungs or leafy branches that had been broken off and left on the ground, or even deliberately stuck perpendicularly into the ground. Sometimes partially eaten fruits or shoots were left on the ground and sometimes a plant had been broken for feeding purposes during travel. The trackers used all such cues, in combination with their intuition as to where the bonobos might go, to track them through the dense forest.

In order to be sure that such vegetation disturbance resulted from the activities of the bonobos, and not the trackers, trackers were asked not to cut the vegetation, contrary to their custom, while following bonobos. Without the vegetation slashed by the trackers to distract the eye, it could be seen that flattened leaves and broken branches occurred at fairly regular intervals, often at points where the bonobos appeared to have stopped to dig, rest, feed or change their direction of travel.

The leaves that appeared to have been firmly flattened, indeed trampled, differed from those that were merely bent or otherwise minimally disturbed. The firmly flattened vegetation was unusual; typically wild bonobos move easily through the forest with virtually no vegetation displacement. Were the wild bonobos doing such things for some purpose?

In an effort to answer this question, all cases of apparently intentional flattening of vegetation by the bonobos were recorded. Vegetation was judged to have been intentionally flattened when it was pressed down and broken in a manner clearly beyond that required to traverse the area. This differed from area to area as the dryness of the forest and the stiffness of the vegetation constantly varied. In addition, other information at the site of such occurrences (such as whether a trail was visible, whether it crossed another trail, etc.) was collected. The goal was to determine if bonobo parties could gain information about the travel patterns of other bonobos from these pieces of altered vegetation.

Even if such a pattern of vegetation alteration was found, the pattern would not prove that bonobos were intentionally altering the vegetation in a symbolic way. Similarly, if sticks were found in digging holes, it would not prove that the bonobos used these sticks for digging. However, the added fact that they readily do so in the laboratory setting without instruction suggests that the capacity to conceptualize the need for a stick and the means to obtain and to utilize it are well within the bonobos' cognitive competencies. Similarly, the readiness for language acquisition evidenced by captive bonobos suggests that such a communicatory need exists in the wild and that bonobos are prepared to exercise this capacity.

To determine if symbols are being utilized, it is essential to begin by looking for some sort of telling pattern in the occurrence of events. If a pattern of similar symbolic events exists across many dissimilar instances, can a symbol system be assumed to be operative? This is true of human communication systems as well as those of bonobos. For example, one of the communication systems used by the trackers was vegetation-based. For example, three fruits placed in a line on the ground meant that bonobos nested nearby. This system was very inconspicuous, both by its nature and by the manner in which the trackers left the signals. The meanings of such symbols could not have been deciphered without extensive effort had the trackers not explained it.

A language-based explanation of a symbol defines for the listener the regularities of occurrence that are permitted for that symbol and thereby the events to which it can be linked and which define its meaning. Lacking a bonobo translator, one can only work diligently to try and discover any patterns of regularity for oneself.

With another species, one does not even know whether a system for such communication exists, or if it does, what its medium and units are. Slight changes in the orientation of the branch, for example, could mean many things. If data regarding orientation are not taken, it will then be difficult to find a consistent pattern. Imagine trying to determine the pattern inherent in Morse code without knowing the language in which it was being sent.

RECORDING FLATTENED AND BROKEN VEGETATION (FBV)

All various forms of vegetation that appeared to be intentionally displaced along the trail, as well as the type and location of displacement were noted (Fig. 13.3). These notations were made by drawing the shape of the vegetation relative to the direction of travel. 'Intentionally

Fig. 13.3. Terrestrial plants along a bonobo trail, available for use as FBV. (Note human hand in upper left corner, for scale.) (Photo by S. Savage-Rumbaugh)

displaced' was defined as any vegetation that was completely flattened on the trail, with all the flattened leaves or foliage oriented in a clear direction, or any branches that were broken off and placed on or near the trail. Vegetation that was simply bent over, but not completely broken and pressed down, was not included, since passing through the forest inevitably resulted in some vegetation being slightly displaced (Fig. 13.4).

In order to create intentional damage to the vegetation, it was necessary for bonobos to alter their stride and to scuff their foot firmly in a particular spot, or to use both hands to break a large branch off of a tree, or

to pull a large plant or a sapling deliberately toward and onto the trail, or to thrust perpendicularly, a large, leafy branch (often as much as 1.5 meters long), proximal end first, into the ground.

It should be noted that while any instance of FBV was being observed and recorded, the future direction of travel was not yet known. Therefore, the observer's notes were not biased by the knowledge that any instance of FBV signaled a certain consequence. FBV was recorded when it was noticed, where it occurred. For example, a flattened leaf was seen and recorded, although the trail did not split for another 3 meters.

The following events were also recorded:

1. Vegetation that was simply bent over, but not completely broken and pressed down, was not included, since passing through the forest inevitably resulted in some vegetation disturbance.
2. All instances in which a bonobo trail could be discerned, whether or not we were following that trail or crossing it.
3. All instances in which a bonobo trail could be discerned, whether or not it crossed a human trail.
4. All instances in which the bonobo trail being followed was so faint that it was difficult to follow.

Since vegetation can be disturbed by other species too, it was essential to place some constraints upon the conditions under which data were recorded. These were as follows:

1. Data were gathered only while following an identified and habituated party that was traveling immediately ahead of the observer. This constraint meant that it was necessary to arrive before dawn at the sleeping site. Also, the group had to descend and take the sugarcane that was offered, so that individuals could be seen in order to verify the identity of the group. (There are five habituated groups at Wamba, but these observations concentrated on E2, which had not been provisioned for the previous four years.)
2. A party had to leave the sleeping site and begin terrestrial travel while others remained in their

Fig. 13.4. Adult male bonobo inadvertently displaces vegetation while eating fruit on the ground. (Photo by G. Hohmann)

nests. The bonobos were followed only when the last party left.

3. Data were limited to occasions on which it was possible to follow continually within vocal range of the group as it traveled. After following, it was essential to find and to identify the group at the new site. Sometimes a group was lost around mid-morning and never recovered, or if recovered, not seen close enough to ascertain the group's identity. Data from such occasions were discarded. This constraint was essential for ensuring that the vegetation displacement was not made by any source other than bonobos.

4. All bonobo vocalizations during travel between sites were recorded, since these were also sources of potential communication about the group's movement.

5. Trackers refrained from cutting vegetation, and the observers stayed abreast of the trackers as they traveled to make certain that no vegetation destruction resulted from their activities.

These conditions were met on September 6 and September 8, 1994. The initial contact period of sugarcane consumption was videotaped on both days, permitting confirmation of the group's identity and documentation of the parties' departures. The first and second authors independently recorded data on the direction of group movement and on vocalizations from the initial contact. For the time-coded events reported here, the two observers agreed on 87% of scored events. Vegetation displacement was noted by the second author as well, but was not scored independently, since the second author was inevitably aware of the points at which the first author stopped to draw quickly the leaf and branch patterns.

FINDINGS

The data are presented here in summary form only.

Often the FBV in a given section of trail was limited to 2–3 species of plants, even though many other kinds were found along the trail. However, the character of the forest changed so rapidly that on another section of the trail, the species used for FBV might be entirely different. The number of species that were utilized as FBV was always much lower than the number of available species along any given portion of the trail.

The forms of flattened or broken vegetation encountered varied considerably and included the following:

1. Short plants, sometimes flattened in a consecutive line, that did not coincide with normal stride length.
2. Large branches (2 or more meters) broken off and laid on the trail. These branches were freshly broken, usually with a combination of hands and teeth.
3. Small branches (less than 2 meters, usually about 0.5 meters) laid on or by the trail.
4. Trees bent over toward the trail. These were generally saplings that were 3–4 meters high and flexible enough to be bent down with own arm.
5. Large branches deliberately broken off and stuck into the ground in an upright position. These were generally shoved down about 8 cm, a depth and position that only a deliberate act would achieve.
6. Plants of the same species flattened in the same position on either side of the trail. In these cases, intentional scuffing was required to selectively flatten the same species of plant on either side of the trail. (Bonobos generally walk on the trails, not along the sides of them.)
7. Tall plants bent over on the trail. These appeared to have been grasped and pulled toward the trail as a bonobo walked past them. None of these FBV alterations would have resulted from chance vegetation disturbance while walking, though some of them could have been the result of feeding activities. Many of them were non-food plants, however, and the appearance of the vegetation disturbance suggested that it was done purposefully.

A number of distinct uses of FBV were found with apparent regularity at various locations along the trail. These uses included:

1. Flattened or broken vegetation on the trail was pressed down in the direction of travel, the function of which appeared to be that of confirming the direction of travel – *35 cases.*
2. FBV that was not on the trail but indicated or pointed in the direction of a bonobo trail – *2 cases* (both at nesting sites).
3. FBV appeared to indicate that the trail being followed should be maintained, when another trail joined or crossed it – *20 cases.*
4. FBV appeared to indicate that a new trail should be taken at a point where another trail joined or crossed the trail – *13 cases.*
5. FBV served to replace the trail in wet areas – *7 cases.*
6. FBV appeared to confirm that direction of the trail when a large object such as a termite mound or log was in the middle of the trail – *12 cases.*
7. FBV indicated a digging location – *3 cases.*
8. Footprints replaced the trail – *14 cases.* (Footprints are not normally seen on the trail, as the trail is a narrow area of soil mildly packed down from usage. However, in some areas the ground is sufficiently soft and muddy and footprints left by the bonobos can be followed.)
9. FBV appeared to indicate a resting or stopping area – *4 cases.*
10. FBV was absent when trails crossed or ended – *3 cases.*
11. FBV did not appear to indicate direction of travel – *4 cases.*

The fact that FBV occurred at various points along the trail does not provide enough evidence to conclude that the bonobos were purposefully disturbing the vegetation in order to communicate information. The case would be strengthened if it could be shown that the vegetation alteration exhibited a pattern not readily accounted for simply by a random breaking and dropping of branches and leaves. If vegetation was flattened or broken for its own sake and no other purpose, one would expect to find it intermittently along the trail. However, if it was being used to leave messages, one would be more likely to find FBV at places where changes in group activity occurred, such as changing direction of travel, resting, digging for tubers and insects, etc.

Were there instances in which FBV occurred, but either did not provide any specific information, provided incorrect information, or did not occur when information was needed? The frequency of such occasions is likely to be low if vegetation alteration was being used as a signal of travel intentions.

There were 35 instances in which FBV simply confirmed the direction of travel. That is, had it not appeared, individuals following that trail could probably have continued in the correct direction, as the trail did not appear too faint to follow. There were 61 cases in which FBV was needed because of objects in the trail, the presence of water, a crossing trail, etc. That is, had FBV not occurred specifically at that location, a follower depending upon FBV could have possibly gone in the wrong direction had the FBV that indicated the correct travel path not been present. There were also 11 cases in which FBV indicated something other than the direction of travel (for example, resting, digging, etc.). When these instances are grouped together, it can be seen that FBV served to signal something both specific and necessary on 72 occasions. On 35 other occasions, it served to confirm what was already known, and on only 7 occasions was it absent when needed or present but potentially misinformative. Clearly, FBV was not randomly distributed along the trails.

However, not only was FBV much more likely to occur at trail junctions, digging sites and rest areas, *but it also rarely occurred when the trail was clear and easily followed.* Due to the rapid pace of the follow, we made no accurate record of the distance traveled between each instance of FBV or the distance traveled when no FBV was observed, but over 50 meters were often traversed with no FBV seen.

In addition, if FBV is employed to indicate trail changes, cases in which it is not followed should result in loss of the bonobos. This occurred twice. Though FBV clearly signaled a direction at a trail intersection, twice the trackers failed to notice this and took the other direction. Both times the trackers soon realized that they were no longer following the bonobos and began to travel in large circles until they regained the bonobos' trail.

CONCLUSION

The above accounts are not presented as definitive evidence that bonobos intentionally use vegetation in a symbolic manner. More information is needed before such a conclusion can be firmly established, including additional observation of bonobos producing FBV.

During the data collection period, two instances of bonobos producing FBV were observed; the remaining data were collected after the FBV had already been produced. However, during the entire study period, bonobos were observed to produce FBV on a number of occasions, although data were not gathered during these episodes since at the time the significance of the behavior was not understood. Adult male bonobos were seen to carefully select certain branches, break them off slowly and deliberately at the base, carry them to another location, drop them and continue to travel.

What is clear is that bonobos traverse their environment in a complex manner utilizing a subgrouping pattern not reported for other apes. That is, the group often breaks up into parties when traveling from one feeding or resting site to another, and these parties rejoin at another destination later on in the day. This may occur 2–5 times or more in any given day. It is also evident that broken pieces of vegetation are sometimes placed in a conspicuous manner along their trails. Also, if humans want to follow the bonobos, they can do so successfully by using these silent signals. Moreover, it is not likely that these alterations of vegetation are simply the result of normal travel, as they appear deliberate in both structure and location.

To date, humans have generally considered language to be a vocal system and have concluded that similar skills are absent in other animals because they do not speak. Yet humans do not communicate only through spoken signals. In the forests of Wamba, humans often whistle or leave silent signals in the vegetation, and many (though not all) of the these signals are similar to those recorded for bonobos, such as two or three leaves of a specific plant flattened in a row near the edge of a trail at a turn-off point, a foot dragged briefly in the vegetation, a branch stuck vertically in the ground to indicate where bonobos were last seen, etc. Such signals indicate location as well as intended destination of travel and are often combined, though whether or not they have a grammar remains to be discovered (cf. Boesch, 1991).

Language can be written as well as spoken, because the verbal context alone can provide all of the required interpretive information. In trying to determine how humans evolved from a system of communication that was context-bound to one that was context-free, or at least one that incorporated its own context into the system, various scholars have proposed schemes based on 'associative learning,' gestures, animal sounds, etc. (Davidson, 1991). However, none of these schemes has clearly indicated how language might have arisen; all demand something like a 'leap of faith' at the transition point from signals that work in the here-and-now to symbols that work across time in a context-free sense.

If bonobos are leaving symbolic messages systematically encoded in the vegetation, this would be the potential basis for a communication system that could serve to bridge the cognitive gap between a system that is dependent upon immediate knowledge of a common context by all participants versus one that is context-independent, such as language. Even if the bonobos are not intentionally leaving such signals in the forest, the fact that human beings are doing so suggests that the cognitive prerequisites for language could have been shaped by such communications.

A central ingredient in messages or signals left for others who will come later is that the leaver of the signal must not only be concerned with the here-and-now as experienced, but with the world as it will be experienced by the interpreter of the signal – who will receive it later. Thus the signaler must, in a sense, enter into the mind of the potential receiver at the time of signal creation. The signaler must realize such things as the fact that the point where trails cross is a point at which someone following might be tempted to turn in a different direction. For this sort of realization to occur, a complex 'theory of mind' of other individuals must emerge.

ACKNOWLEDGMENTS

This paper was originally prepared for the Wenner-Gren conference 'The Great Apes Revisited,' held November 12–19, 1994 in Cabo San Lucas, Baja, Mexico. Preparation of this chapter was supported by Grant HD-06016 from the National Institute of Child Health and Human Development, and by additional support from the College of Arts and Science, Georgia State University.

REFERENCES

Boesch, C. 1991. Symbolic communication in wild chimpanzees? *Human Evolution*, **6**, 81–90.

Bruner, J. 1990. *Acts of Meaning*. Cambridge, MA: Harvard University Press.

Chomsky, N. 1978. *Syntactic Structures*. The Hague: Mouton & Co.

Davidson, I. 1991. The archaeology of language origins: a review. *Antiquity*, **48**, 63–5.

Davidson, I. & Noble, W. 1989. The archaeology of perception: traces of depiction and language. *Current Anthropology*, **30**, 125–55.

Feldman, R. S. & Rime, B. 1991. *Fundamentals of Nonverbal Behavior*. Cambridge: Cambridge University Press.

Galdikas, B. M. F. 1995. *Reflections of Eden*. Boston: Houghton Mifflin.

Goodall, J. 1986. The behavior of free-living chimpanzees in the Gombe Stream Reserve. *Animal Behavior Monographs*, **1**, 161–311.

Hauser, M., Texidor, P., Field, L. & Flaherty, R. 1993. Food-elicited calls in chimpanzees: effects of food quantity and divisibility. *Animal Behaviour*, **45**, 817–9.

Hinde, R. (ed.) 1972. *Non-verbal Communication*. Cambridge: Cambridge University Press.

Humphrey, N. 1983. *Consciousness Regained: Chapters in the Development of Mind*. Oxford: Oxford University Press.

Kano, T. 1992. *The Last Ape: Pygmy Chimpanzee Behavior and Ecology*. Stanford, CA: Stanford University Press.

Menzel, E. 1987. Communication and cognitive organization in humans and other animals. In: *Origins and Evolution of Language and Speech*, ed. S. R. Harnad, H. D. Steklis & J. B. Lancaster, pp. 131–42. New York: New York Academy of Sciences.

Pinker, S. 1984. *Language Learnability and Language Development*. Cambridge, MA: Harvard University Press.

Quine, W. V. 1960. *Word and Object*. Cambridge, MA: MIT Press.

Savage-Rumbaugh, E. S. & Lewin, R. 1994. *Kanzi: The Ape at the Brink of the Human Mind*. New York: Wiley & Sons.

Schaller, G. 1963. *The Mountain Gorilla: Ecology and Behavior*. Chicago: University of Chicago Press.

14 • Reciprocation in apes: from complex cognition to self-structuring

CHARLOTTE K. HEMELRIJK

INTRODUCTION

Although the great apes, as our nearest relatives, are a valuable subject for cognitive research, it remains difficult to gain insights into their mental abilities. Some of the problems are reflected in a debate between Kummer *et al.* (1990) and de Waal (1991) over the question of at what stage of research (anecdotal, correlational or experimental) anthropomorphic labels for behavior are allowable, if ever. Whereas de Waal uses these terms deliberately throughout his writings, Kummer and his co-authors argue that such terms should not be used unless more parsimonious alternative hypotheses are excluded by experimental evidence.

Although this discussion lays bare some of the difficulties in cognitive primatological research, it does not attack its central problem: how to go beyond anthropomorphic labeling by generating meaningful alternative hypotheses. The central thesis of the present paper is that adopting a view based on self-organization provides us with a tool to generate such alternatives.

Thinking in terms of self-structuring asks for a different way of thinking about complexity. By convention it is believed that a complex system can be understood by taking it apart and studying the pieces. This analytic procedure, also known as the 'top-down' or reductionist approach, boils down to a static description of the system. Supporters of the self-structuring view, however, advocate an opposite route: by studying interactions on a lower level, the emergence of a macrostructure on a higher level is perceived and therefore better understood (Hogeweg, 1988). In this so-called 'bottom-up' approach, parts are understood from the dynamics of the whole. Thus, there is a shift of focus from objects to

relationships (Cohen & Stewart, 1994), whereby relationships are often considered to be self-reinforcing or autocatalytic.

For instance, social organization in humans and animals is conventionally held to be predetermined by differences in the attributes of the group members. From the perspective of self-organization, however, many forms of social organization are considered to emerge from local interactions between individuals and their environment (Chase, 1993); in turn, the developing social organization feeds back to the individuals and shapes the interactions, thus affecting the group structure.

An illustrative example comes from Seyfarth's (1977) research on grooming patterns of adult female monkeys. In a number of primates it has been found that: (a) high-ranking individuals receive more grooming than others, and (b) most grooming occurs between females of adjacent rank. Instead of focusing on the properties of individual females, Seyfarth studied the group as a whole and made a plausible case that this pattern may be a side-effect of the following principles: (a) a female's attractiveness to others is directly related to her rank, because potentially more benefits can be received from females of higher rank (e.g. receiving support); and (b) access to preferred grooming partners is restricted by competition. Because higher-ranking females win this competition, each female will in the end groom most frequently with partners adjacent in rank and be groomed most often by the female ranking just below her. The interactions between the monkeys are local, in the sense that each female directly competes with another and no one has an overview of the global grooming patterns. Therefore, less complex mental processes are assumed than in the traditional approach in which the usual thing is to suppose that each individual distributes its grooming according to the height and adjacency of the rank of each partner by keeping track of its grooming records.

Later in this paper I will show that it is unnecessary to assume the future return of benefits and that a dominance hierarchy may arise through self-structuring.

The shift from explanations based on individual properties to those stressing local interactions between individuals and their surroundings reflects a paradigmatic change; although the principle of economy of hypotheses is best served by the paradigm of self-structuring, there is a reluctance among ethologists to generate parsimonious hypotheses for animal behavior. This is especially true in studies of monkeys and apes: complex individual abilities are more easily accepted in primates than in any other taxon and are even the main source of interest for many primatologists. Yet, local interactions may have profound impact on the behavior and social structure of primates as will be reviewed below.

In this chapter, I will explore whether or not it is meaningful to apply the notion of self-structuring to another phenomenon: reciprocation in primates. Conventionally, if exchange of social acts is observed among non-related primates, the theory of reciprocal altruism (Trivers, 1971) is invoked, in which, as is typical for a top-down approach, it is (implicitly) assumed that animals are especially equipped to reciprocate beneficial acts. For example, reciprocity of 'altruistic' acts may be 'calculated' because primates perform some kind of recordkeeping.

Yet in our research on exchange of grooming and support in captive chimpanzees (Hemelrijk & Ek, 1991; Hemelrijk *et al.*, 1991), evidence for the theory of reciprocal altruism and for recordkeeping to reciprocate remained inconclusive. Therefore, it is worthwhile to use a computer model to see if patterns of reciprocation may in principle emerge from simple behavioral rules and self-structuring processes alone, that is, in the absence of any specific motivation to reciprocate.

The chapter is divided into two parts. The first part concerns the top-down, static approach. I will review statistical and experimental methods used to study reciprocity in primates and show that they have led to only indecisive results.

In the second part, I will illustrate the self-structuring

view with additional examples of simulation studies of primate social behavior. Next, I will propose a model for the self-organized emergence of reciprocation and other complex social behavior. This model concerns simple artificial entities 'living' in a virtual world. They lack an explicit motivation to reciprocate or interchange and behave through simple rules and local self-reinforcing dominance interactions. From the latter a spatial distribution of individuals will result. In turn, this spatial arrangement will appear to be of central importance in structuring patterns of social interactions. If statistically significant patterns of reciprocity are detected from the behavior of these entities, the implemented rules form a useful, parsimonious alternative hypothesis to the anthropomorphic, intentional ones and deserve further study.

RECIPROCITY IN PRIMATES: THE TOP-DOWN APPROACH

The central problem of research on reciprocity is that neither costs and benefits, nor proximate mechanisms to reciprocate are known. Indications for the correctness of the theory of reciprocal altruism can only be derived from the observed pattern of supposed beneficial acts. However, there are neither hints about the rate at which different kinds of acts should be exchanged, nor expectations about the patterns that should result if individuals differ in their tendency to confer beneficial acts on others (Seyfarth & Cheney, 1988).

In past studies on reciprocal altruism, how such patterns should look was decided intuitively, and implications of the criteria used were not considered. I have reviewed examples of such operationalizations (Hemelrijk, 1990a) and shown that they are based on a so-called 'actor–reactor' model in which each individual has to trace all acts of others. Under this model, however, complete reciprocity can occur in even group sizes only and exchange strategies require complex cognitive capacities.

Fewer problems result if one starts from a simple type of mechanism proposed *a priori*, in which individuals perform some kind of crude recordkeeping (on an ordinal scale) of acts given to and received from every partner, and direct acts relatively more often to those from whom they receive more frequently (Hemelrijk,

1990a). In this 'actor–receiver' model, individuals only direct 'relatively' more acts to those from whom they receive acts more frequently, but do not necessarily return as many acts as they receive. In this way, individual variation in the tendency to direct acts is dealt with. In addition, no assumptions are made about the rate at which different kinds of acts should be exchanged (from now on called 'interchange' in contrast to 'reciprocity' that concerns one type of act only). Other advantages of the 'relative actor–receiver' model are that (a) complete reciprocation can be framed for odd as well as even group sizes, and (b) exchange strategies require fewer complex cognitive capacities than under the actor–reactor model. This is so, since instead of recording all acts among others, only acts in which the individual is involved have to be traced. Statistical testing for this pattern can be done using an especially adapted correlation test, the so-called Kr-correlation (Hemelrijk, 1990a); this method reckons with the dependency in the matrix data and establishes whether a significant positive correlation between rows in an actor versus receiver matrix is found.

Such a statistically significant correlation, however, does not guarantee that a mechanism for reciprocation or interchange exists, since it may still be a side-effect of correlations with a third variable. This may be corrected for by using a partial Kr-correlation test (Hemelrijk, 1990b).

As an illustration of this statistical approach, findings from my study on captive chimpanzees in Arnhem Burgers' Zoo are summarized below. The main conclusion is that patterns of reciprocity and interchange may arise easily as a statistical side-effect.

A case study: reciprocity and interchange in captive chimpanzees

Chimpanzee social behavior has been described as 'political' and 'complex social manoeuvering' by de Waal (1992). Coalitions are considered to be a main tool in these political endeavors. Chimpanzees form coalitions by intervening in fights against others. According to de Waal & Luttrell (1988), chimpanzees reciprocate support by keeping track of the records of these social acts.

Negotiation about transactions concerning reciprocal favors of all kinds are mediated by special communicative gestures, categorized under 'side-directed behavior' (de Waal & van Hooff, 1981), such as 'hold-out-hand' and agitated vocalizations.

Goodall (1986) describes 'sexual transactions'. She remarks that females use their conspicuous sexual swelling in bargaining for favors. Being thus privileged during the estrous period, females are groomed more often by males and are more successful when begging to males for food.

To test these ideas, I have used data on relative reciprocity and interchange in a group of captive chimpanzees living in an open-air enclosure of about 0.7 ha (Adang et al., 1987) in the Arnhem Burgers' Zoo, The Netherlands. The study concerns only mature individuals (i.e. males at least seven years old and females after menarche; see Goodall, 1968). Data were collected over 14 years (1976–89) by various investigators and the group size varied between 20–30 individuals, containing 3–7 mature males and 9–12 mature females. In addition to observational study on the undisturbed group, an experiment using a food-apparatus also was performed in order to measure rates of exchange more precisely.

Observations on the undisturbed group

The female chimpanzees reciprocated grooming (Hemelrijk & Ek, 1991). Perhaps this reflects reciprocal altruism, but it is questionable whether grooming should be considered altruistic at all. Although it may be argued that time spent grooming could otherwise have been spent doing some other activity such as feeding, the 'lost' time need not be costly, because it is not correlated with feeding time; also any time 'lost' to grooming is equally borne by the groomee (Dunbar, 1988). Although the benefit of grooming for the recipient is obvious, grooming may also be beneficial for the groomer by having a placatory effect (Goosen, 1975; Terry, 1970) and by having a physiologically rewarding effect through the brain's opiate system (Keverne et al., 1989). In this way, individuals may reinforce themselves to groom with certain partners and this may lead to the observed reciprocity (Fig. 14.1).

Supporting behavior by males appeared to be reciprocal only in periods without a clear-cut alpha–male (Hemelrijk & Ek, 1991), but not when the position of the alpha–male was strong. There were insufficient

Fig. 14.1. Adult female and adult male chimpanzees groom mutually at Arnhem Burgers' Zoo. (Photo by C. K. Hemelrijk)

indications that reciprocation resulted from complying with requests from others (Hemelrijk *et al.*, 1991). Instead, both males may have joined in one another's fights in order to attack common rivals and this may have been the cause of the reciprocity as found in the data. Since both males may benefit directly from such joint attacks, supporting behavior is selfish and there is no need for the participants to keep records.

Males groomed more often those estrous females with whom they copulated more frequently, but this did not seem to result in long-term bonds. Hence, male grooming of females may simply function directly to suppress the male's aggressive tendencies or the tendency of the female partner to flee. Therefore, no anthropomorphic label such as 'bargaining for sex' is needed to explain these results (Hemelrijk *et al.*, 1992).

Females aimed their side-directed behavior more often at those from whom they received support more frequently (Hemelrijk *et al.*, 1991). This may seem a sign of negotiation, but females did not appear to increase

their chances of receiving support by showing side-directed behavior, compared with those cases in which they did not use recruitment behavior. Instead of being a means to negotiate, side-directed behavior may be directly effective: side-directed behavior was significantly more often directed to higher-ranking partners, and the close proximity of these high-ranking partners may thwart further aggression from the opponent.

Only the fact that females groomed more often those from whom they received support more frequently (Hemelrijk & Ek, 1991) could not be explained as a side-effect of other processes. It was, however, difficult to understand as interchange as well, since partners that were groomed did not support more often those females by whom they were groomed more frequently (this being the reverse correlation between support and being groomed). Thus, support was given independent of being groomed and grooming was not 'paid back'. This does not comply with what one expects in balanced 'barter' between individuals.

Experiments with the food apparatus

A food-dispensing apparatus was used in order to gain insight into rates of exchange (Hemelrijk, unpublished; see Fig. 14.2). Initially, both animals could provide themselves with onions when they cooperated by

Fig. 14.2. Two chimpanzees simultaneously tug at a handle on each side of a dispensing apparatus in order to receive food (for further explanation, see text).

simultaneously tugging a handle at each side of the machine. Individuals who performed at the machine were all youngsters (except for one adult female), and they cooperated more frequently with partners the higher the rank of the associate. This appeared to be a by-product of the monopolization of the device by the higher-ranking group members, since they chased lower-ranking individuals away from it. In later experiments, food appeared only at one side of the vending machine, and non-rewarded pulling of the handle at the other side was considered 'altruistic' behavior. This assistance was not reciprocated, nor was it exchanged for other behavior, such as grooming or play, when analyzed in terms of matrix correlations.

Within-pair reciprocity and interchange

Most of the matrix correlations described above were hard to interpret as proof for either the theory of recipro-cal altruism or the performance of recordkeeping to reciprocate or to interchange. This may be due to two important shortcomings of the 'relative' actor–receiver model. First of all, the tendency to reciprocate and to interchange acts with each partner and after every act received is taken to be fixed. Second, it still assumes a kind of 'recordkeeping' (although of a simpler type than in the actor–reactor model, see above). However, an even simpler mechanism may be considered. Suppose that instead of recordkeeping, subjects simply experience directly an increased tendency to reciprocate shortly after having received an act and that this tendency exting-uishes, for instance, exponentially over time if opportun-ities to reciprocate are lacking. Furthermore, assume that the intensity of this tendency depends on the identity of the partner and the specific act received. Under these conditions and summed over a period of time, this does not have to result in patterns of 'relative' reciprocity or interchange (although it is likely to result in patterns of 'qualitative' reciprocity or interchange; see Hemelrijk 1990a). To test for such 'within-pair' reciprocity and interchange, frequencies of beneficial acts by each indi-vidual to each partner should be compared, before and after a beneficial act has been received from that particu-lar partner. Analyzing 'within-pair' reciprocity and inter-change at a group level means that these comparisons should be done for all combinations of pairs of individ-uals. Unfortunately, so far there is no statistical test available for such an arrangement. As an alternative, one could choose a random set of exclusive pairs of individuals and test for 'within-pair' reciprocity in this subset using the Wilcoxon matched-pairs signed-ranks test.

There are more methodological problems besides statistical testing. Often, the opportunity to return cer-tain kinds of acts is absent, for example support in fights cannot be given if no fights are going on. Therefore, I used an experimental procedure, inspired by that from Seyfarth & Cheney (1984), in order to study whether the actual frequency of support after being groomed was higher than without antecedent grooming (Hemelrijk, 1994). The experiment was carried out on exclusive pairs of individuals from a captive colony of long-tailed macaques, *Macaca fascicularis*. Animals were separated in triads and grooming was stimulated by dropping a sticky mixture on the back of the individual to be groomed. Tiny food parcels were used to induce fights to create an opportunity to support. Although the results confirmed those of Seyfarth & Cheney (1984), some problems remained. Most importantly, the experimental data were incompatible with observations collected on the complete group: individuals in the experimental con-text supported partners that they never supported in the intact group. Obviously, a whole group is more than the sum of its component triads. This raises doubts about what such experiments may tell us about what is going on in a complete group.

SOCIALITY AND EXCHANGE: THE BOTTOM-UP APPROACH

Since the top-down approach did not provide us with clear evidence for the theory of reciprocal altruism or for a motivation to reciprocate (for instance by means of a crude recordkeeping mechanism), it is worthwhile to apply the bottom-up perspective.

Before doing so, I will illustrate with examples how a self-structuring approach produces more parsimonious hypotheses than the conventional sociobiological and anthropomorphic way of thinking. I will first remark on the origins of specific grouping structures and second on patterns of social interactions within groups.

Then I will present the outline of a model in which

reciprocity and interchange may arise as a side-effect from self-reinforcing aggressive interactions, spatial structure and grooming between artificial entities that lack every motivation to reciprocate.

Group formation

Chimpanzee societies are characterized by relatively solitary females, social males and an unpredictable joining and leaving of the parties by its members (Goodall, 1986; Wrangham, 1987). Male grouping is thought to be advantageous for territorial defense (Wrangham, 1987) and dependent on close relatedness among males (Goodall, 1986). As an alternative, te Boekhorst & Hogeweg (1994b) studied whether grouping patterns similar to those of chimpanzees (in particular males) are reproducible among non-territorial, unrelated artificial entities. They set up a virtual community that has some features in common with the chimpanzee community of Gombe, such as the size of the habitat, the community composition (Halperin, 1979) and cycling dynamics of females (Tutin, 1980). Imaginary data about tree size are partly in correspondence with Ghiglieri (1984). The entities behave according to the following rules. (1) Both sexes search for food, although their diet partially differs: females rely more on food sources with a high renewal rate or omnipresence – such as insects and leaves – than males, cf. McGrew (1979) and Goodall (1986). (2) Only males search for females (Trivers, 1972): after spotting a conspecific, a male approaches in order to see whether the other is an estrous female. If so, he stays with her until she is no longer in estrus, if not he goes back to his foraging routine.

Despite this minimalistic approach, it is unpredictable what grouping patterns will result, because of the complex interplay between the habitat's structure and the behavior of its artificial inhabitants. Yet, in correspondence to grouping in real chimpanzees: (a) artificial females were relatively solitary; (b) males aggregated, especially around estrous females (cf. Goodall, 1986); and (c) party size fluctuated in a random-like fashion. In summary, the study of te Boekhorst & Hogeweg (1994b) suggests that patterns of grouping similar to that of chimpanzees result in these artificial creatures as a side-effect of dietary differences between the sexes and mate-

seeking behavior of the males, and without assuming benefits in terms of close relatedness and territorial defense.

Interestingly, tree size appeared to influence party formation: the artificial entities travel together especially when many small-sized fruit trees bear fruit simultaneously. Coordinated traveling is due to the fact that a small tree is depleted by co-feeding entities before they are satiated. Consequently, they synchronously leave a tree, and because foraging entities always go to the next nearest food tree in their direction of movement, a travel band originates. Large trees, however, permit each entity to feed until satiated and hence each one departs separately.

Subsequently, te Boekhorst & Hogeweg (1994a) studied whether travel bands in orangutans may originate in the same way as described above. In this case the authors were not interested in sex differences, but in the fact that orangutans in the Ketambe area (Sumatra, Indonesia) travel together especially in seasons of high fruit availability (Sugardjito et al., 1987). They designed a virtual environment that resembled Ketambe concerning the distribution of food and the population composition of its orangutans and implemented the following rules. Entities search for the huge, but uncommon fig trees in preference to the small but seasonally abundant fruit trees (cf. Sugardjito et al., 1987). If neither is present, then they look for a conspecific, since the presence of another might indicate the presence of a food tree. The artificial creatures were observed to form temporary aggregations in the fig trees. However, in seasons of high fruit availability, they encounter many fruit trees and feed in them with others. Thus, the mechanism of the first model about chimpanzee grouping also operated under orangutan-like conditions. A number of hypotheses derived from the model, among others, that the probability of remaining in a travel band depends on the size of the former tree in which the party members fed together, were confirmed with field data in the same paper. Note that these grouping patterns result from the direct interaction between subjects and environment, without assuming additional benefits for being in a group. This type of modeling is therefore pre-eminently suited to check the necessity of assumptions about various costs and benefits of group life, such as protection against predators and competition for food (van Schaik, 1983; Wrangham, 1987).

Social behavior in groups

A self-structuring approach to explain patterns of social behavior observed at a group level was advocated implicitly by Hinde (1982). Hinde distinguished four different levels of complexity, each with its own emergent properties: individual behavior, interactions, relationships and social structure. Each level is described in terms of the level below it, and levels influence each other mutually. For instance, the nature of the participants' behavior influences their relationships, but these relationships in turn also affect the participants' behavior. A caution that follows from this view is that observed social structure can vary dramatically with circumstances, without any changes in the underlying motivational mechanisms or strategies.

Dominance structure provides an example. Certain types of dominance hierarchy have been considered to be species-specific by primatologists. For instance, Hrdy & Hrdy (1976) contrasted the age-graded pattern found among Hanuman langurs (*Presbytis entellus*) with the genealogical dominance structures of female rhesus macaques (*Macaca mulatta*). The age-graded system depends mainly on female age, with females in their reproductive prime deferred to by both young and old females, whereas rank in the genealogical structure is predominantly determined by kinship. Hrdy & Hrdy (1976) suggested that the contrast might be linked to the differences in the mating system. (In both species, females are philopatric and males migrate, but macaque groups are multimale, whereas langurs are often unimale.) This affects the degree of relatedness among the females and thus the potential benefit of yielding rank to a related female of higher reproductive value.

As an alternative, Datta & Beauchamp (1991) and Datta (1992) have shown by means of a bottom-up simulation that the two types of dominance structure may be an effect of demographic variation through the availability of allies. In their simulation, the behavioral rules were based on realistic assumptions, such as 'older, larger females tend to dominate younger, smaller females, but younger females can dominate older, larger females, pro-

vided that they have powerful allies against the latter . . .' (Datta, 1992, p. 62) An important result of their study is the effect of population growth. In a declining population, lack of allies constrains females in their competition for high rank, and so they have to rely mainly on their own individual fighting abilities, resulting in an age-graded hierarchy. In contrast, in an expanding population, female matrilines are large, and thus the dominance hierarchy comes about through support from kin.

The model of Datta & Beauchamp (1991) is more than just the proximate counterpart of Hrdy & Hrdy's (1976) ultimate explanation, because it makes the suppositions concerning female relatedness and the species-specificity of different dominance structures superfluous. Note that the life history characteristics of langurs compared to macaques may be such that macaque populations are more expansive than those of langurs. If so, the difference in dominance structure may be a species difference after all. However, in this case (and this is crucial) dominance structure is not a result of direct selection favoring a certain type of social behavior, but an effect of life-history traits that are under quite different selection pressures. This shows that bottom-up models do not serve just to provide anti-evolutionary arguments in a simplistic nature–nurture debate.

Rank within a given dominance system is often thought to result from individual variation in the predisposition to become dominant (Slater, 1986). Hogeweg & Hesper (1983), however, have shown that a dominance hierarchy may emerge in a virtual 'world' even among initially completely identical individuals due to chance and self-reinforcing interactions. They represented self-reinforcing dominance interactions by the so-called DODOM interaction. DODOM implies that if two entities meet, they display and observe their mutual DOM, a variable reflecting the tendency to dominate in a hierarchical interaction. Subsequent winning or losing is determined by chance and mutual DOM, as follows:

$$\text{if RAND}(0,1) < \text{DOM1}/(\text{DOM1} + \text{DOM2})$$
$$\text{then} \quad k = 1 \text{ (entity 1 wins)} \qquad (1)$$
$$\text{else} \quad k = 0 \text{ (entity 2 wins)}$$

Thus, a random number between 0 and 1 is drawn from a uniform distribution and is compared to the relative dominance of both entities. If this ratio is smaller than the random number, entity 1 wins, but otherwise entity 2 wins.

Updating of DOM is done so that:

$$\text{DOM1:} = \text{DOM1} + (k\text{-DOM1}/(\text{DOM1} + \text{DOM2})) \quad (2)$$

$$\text{DOM2:} = \text{DOM2-}(k\text{-DOM1}/(\text{DOM1} + \text{DOM2}))$$

Suppose for instance, that DOM1 = 1 and DOM2 = 2, i.e. the ratio is 1/3 and p (ratio > 1/3) = 2/3. This implies a larger probability for subject 2 to win. If subject 2 indeed wins, its DOM value increases from 2 to 2.33, while the value of entity 1 decreases to 0.67. However, if by chance entity 2 loses, its DOM drops to 1.33 and that of entity 1 rises to 1.67. Thus, expected outcomes reinforce the relative DOM only slightly, whereas unexpected outcomes give rise to a relatively large change in DOM, leading to a damped positive feedback.

What are the indications that such a system operates in the real world? Referring to the so-called 'loser effect', Rowell (1974) and Chase *et al.* (1994) suggest that in animals a dominance rank may result from effects of chance and dynamic reinforcements. Losing a fight increases the probability of losing again in an immediate subsequent fight against an otherwise equally matched individual. This 'loser effect' is well documented in a wide variety of taxa. Its dynamics may correspond to the damped positive feedback as in DODOM.

To gain insight in the dynamical relation between dominance rank and social structure, Hogeweg (1988) used another 'world,' this time inhabited by entities named SKINNIES. This simulation does not bear a relation to any particular mammalian species, but should be considered an exercise in getting acquainted with patterns of self-structuring resulting from the interaction between dominance rank and spatial arrangement of creatures that know each other individually and remember each other's DOM from the last fight. SKINNIES are inherently social, thus if there is no other SKINNY nearby, they will move to the nearest others. Upon meeting another SKINNY, each estimates whether it will win, using the DOM-value of the partner when they last met. If they lose this mental 'battle' they flee; if not, they perform DODOM 'in reality' (by matching their own DOM-value to the current one of the partner). This simple interaction structure generated an interesting

social structure: SKINNIES tended to form faithful pairs or interaction groups, which meet once in a while. In small groups (> 2), the dominant SKINNY interacted with several submissives, whereas the submissives only interacted with the 'boss'. Such patterns are reminiscent of the social structure of gorillas (Stewart & Harcourt, 1987) and hamadryas baboons, *Papio hamadryas* (Kummer, 1968). In larger groups, the most dominant individuals appeared to be in the center and subordinates were on the periphery, as is found in a number of primate species (Japanese macaques, *Macaca fuscata*: Itani, 1954; Yamada, 1966; baboons, *Papio* spp. and capuchin monkeys, *Cebus* spp.: Janson, 1990). The important insight obtained from Hogeweg's simulation is that such a spatial group structure does not necessarily have to be a consequence of competition for an advantageous spatial position *per se* (as for instance suggested by Janson, 1990), but may result from any unspecified type of displacement.

Can we think of an explanation for reciprocity and interchange in terms of the types of dynamical relationships as sketched above? Imagine a group of primates with dominants in the center and subordinates at the periphery. The spatial distribution has an array of effects. First, because they dwell mostly in the center of the group, more dominant individuals will encounter more individuals in general and therefore also more that are involved in a fight; this and the 'winner-effect' (that is, the chance to win increases after winning a fight, Chase *et al.*, 1994) leads them to support others in fights more often than do low-ranking ones (as found for chimpanzee males; see Hemelrijk & Ek, 1991). Second, assume the individuals groom others in proportion to their encounter rate; then higher-ranking individuals will be groomed more often than lower-ranking ones. Because in this model those nearest in space generally will be those nearest in rank, a pattern might emerge in which grooming occurs preferentially among individuals of similar dominance rank. Note that this is an alternative to the explanation by Seyfarth (1977) for the association between adjacency in rank and grooming. Whereas Seyfarth assumes that females are competing to groom high ranking partners to receive something valuable in return, I leave out any assumption about future social benefits. Instead, I suggest that any form of unspecified competition leads to hierarchical displacements. From these, a spatial structure results. The spatial arrangement of animals, in turn, determines through proximity the grooming pattern at a group level.

The power of this scenario becomes more clear by realizing that it is likely to result also in the correlation between grooming and received support mentioned before (Hemelrijk & Ek, 1991), because each individual grooms others more often the higher its rank, and higher-ranking individuals intervene more frequently in fights (Hemelrijk, 1990b).

The exact patterns of social interaction produced by this model are unpredictable due to the complex inter-play of self-reinforcing social and spatial effects. I have implemented the above rules in a model (Hemelrijk, in preparation), and I am currently collecting data on the artificial entities. These will be analyzed in the same way as were the data for the captive chimpanzees. In this fashion, I investigate whether patterns of reciprocation and interchange can occur without invoking correspond-ing proximate motivations or specific ultimate cost-benefit considerations to return acts. If this is confirmed, this model will offer an alternative to the assumed overly complex cognitive capacities. Furthermore, it raises doubt about the validity of using the theory of reciprocal altruism to explain social behavior in primates, possibly including humans.

CLOSING REMARKS

To go beyond simplistic sociobiological cost-benefit reasoning and anthropomorphic labeling, parsimonious alternative hypotheses are needed. The self-structuring approach is the suitable tool, since it relates maximum complexity at a group level to maximum individual simplicity. This is done by studying the effects of local animal–environment interactions in a bottom-up approach.

ACKNOWLEDGMENTS

This paper was originally prepared for the Wenner-Gren conference 'The Great Apes Revisited,' held November 12–19, 1994 in Cabo San Lucas, Baja, Mexico. I am grateful to Rolf Pfeifer for supporting my research and to the editors of this book for inviting me to the Wenner-Gren conference. Thanks to René te Boekhorst, Bill McGrew and Toshisada Nishida for correcting former drafts. This research is supported by the Swiss National Science Foundation, grant number 21-34119.92.

REFERENCES

Adang, O. M. J., Wensing, J. A. B. & van Hooff, J. A. R. A. M. 1987. The Arnhem Zoo colony of chimpanzees *Pan troglodytes*: development and management techniques. *International Zoo Yearbook*, **26**, 236–48.

te Boekhorst, I. J. A. & Hogeweg, P. 1994a. Effects of tree size on travelband formation in orangutans: data-analysis suggested by a model. In: *Artificial Life*, ed. R. A. Brooks & P. Maes, pp. 119–29. Cambridge, MA: MIT Press.

te Boekhorst, I. J. A. & Hogeweg, P. 1994b. Self-structuring in artificial 'CHIMPS' offers new hypotheses for male grouping in chimpanzees. *Behaviour*, **130**, 229–52.

Chase, I. D. 1993. Generating societies: collective social patterns in humans and animals. In: *Selforganisation & Life: From Simple Rules to Global Complexity. European Conference on Artificial Life*, pp. 174–91. Bruxelles: Université Libre de Bruxelles.

Chase, I. D., Bartelomeo, C. & Dugatkin, L. A. 1994. Aggressive interactions and inter-contest interval: how long do winners keep winning? *Animal Behaviour*, **48**, 393–400.

Cohen, J. S. & Stewart, I. 1994. *The Collapse of Chaos: Discovering Simplicity in a Complex World*. New York: Viking Penguin.

Datta, S. B. 1992. Effects of availability of allies on female dominance structure. In: *Coalitions and Alliances in Humans and in Other Animals*, ed. A. H. Harcourt & F. B. M. de Waal, pp. 61–82. Oxford: Oxford University Press.

Datta, S. B. & Beauchamp, G. 1991. Effects of group demography on dominance relationships among female primates. I. Mother–daughter and sister–sister relations. *American Naturalist*, **138**, 201–26.

Dunbar, R. I. M. 1988. *Primate Social Systems*. London: Croom Helm.

Ghiglieri, M. P. 1984. *The Chimpanzees of Kibale Forest: A Field Study of Ecology and social Structure*. New York: Columbia University Press.

Goodall, J. 1968. Behaviour of free-living chimpanzees of the Gombe Stream Reserve. *Animal Behaviour Monographs*, **1**, 163–311.

Goodall, J. 1986. *The Chimpanzees of Gombe: Patterns of Behavior*. Cambridge, MA: Harvard University Press.

Goosen, C. 1975. After effects of allogrooming in pairs of adult stump-tailed macaques. A preliminary report. In: *Contemporary Primatology*, ed. S. Kondo, M. Kawai & A. Ehara, pp. 263–8. Basel: Karger.

Halperin, S. D. 1979. Temporary association patterns in free-ranging chimpanzees: an assessment of individual grouping preferences. In: *The Great Apes*, ed. D. A. Hamburg & E. R. McCown, pp. 491–9. Menlo Park, CA: Benjamin/Cummings.

Hemelrijk, C. K. 1990a. Models of and tests for reciprocity, unidirectionality and other social interaction patterns at a group level. *Animal Behaviour*, **39**, 1013–29.

Hemelrijk, C. K. 1990b. A matrix partial correlation test used in investigations of reciprocity and other social interaction patterns at a group level. *Journal of Theoretical Biology*, **143**, 405–20.

Hemelrijk, C. K. 1994. Support for being groomed in long-tailed macaques, *Macaca fascicularis*. *Animal Behaviour*, **48**, 479–81.

Hemelrijk, C. K. & Ek, A. 1991. Reciprocity and interchange of grooming and 'support' in captive chimpanzees. *Animal Behaviour*, **41**, 923–35.

Hemelrijk, C. K., Klomberg, T. J. M. , Nooitgedagt, J. H. & van Hooff, J. A. R. A. M. 1991. Side-directed behaviour and recruitment of support in captive chimpanzees. *Behaviour*, **118**, 89–102.

Hemelrijk, C. K., van Laere, G. J. & van Hooff, J. A. R. A. M. 1992. Sexual exchange relationships in captive chimpanzees? *Behavioral Ecology and Sociobiology*, **30**, 269–75.

Hinde, R. A. 1982. *Ethology*. New York: Oxford University Press.

Hogeweg, P. 1988. MIRROR beyond MIRROR, puddles of LIFE. In: *Artificial Life, SFI Studies in the Sciences of Complexity*, ed., C. Langton, pp. 297–316. Redwood City, CA: Addison-Wesley.

Hogeweg, P. & Hesper, B. 1983. The ontogeny of interaction structure in bumble bee colonies: a MIRROR model. *Behavioral Ecology and Sociobiology*, **12**, 271–83.

Hrdy, S. B. & Hrdy, D. B. 1976. Hierachical relations among female Hanuman langurs (Primates: Colobinae, *Presbytis entellus*). *Science*, **193**, 913–4.

Itani, J. 1954. *The Monkeys of Mt Takasaki*. Tokyo: Kobunsha.

Janson, C. H. 1990. Social correlates of individual spatial choice in foraging groups of brown capuchin monkeys, *Cebus apella*. *Animal Behaviour*, **40**, 910–21.

Keverne, E. B., Martensz, N. D. & Tuite, B. 1989.

Beta-endorphin concentrations in cerebrospinal fluid of monkeys are influenced by grooming relationships. *Psychoneuroendocrinology*, **14**, 155–61.

Kummer, H. 1968. *Social Organization of Hamadryas Baboons*. Chicago: University of Chicago Press.

Kummer, H., Dasser, V. & Hoyningen-Huene, P. 1990. Exploring primate social cognition: some critical remarks. *Behaviour*, **112**, 84–98.

McGrew, W. C. 1979. Evolutionary implications of sex differences in chimpanzee predation and tool use. In: *The Great Apes*, ed. D. A. Hamburg & E. R. McCown, pp. 440–63. Menlo Park, CA: Benjamin/Cummings.

Rowell, T. E. 1974. The concept of social dominance. *Behavioral Biology*, **11**, 131–54.

van Schaik, C. P. 1983. Why are diurnal primates living in groups? *Behaviour*, **87**, 120–44.

Seyfarth, R. M. 1977. A model of social grooming among adult female monkeys. *Journal of Theoretical Biology*, **65**, 671–98.

Seyfarth, R. M. & Cheney, D.L. 1984. Grooming, alliances and reciprocal altruism in vervet monkeys. *Nature*, **308**, 541–3.

Seyfarth, R. M. & Cheney, D. L. 1988. Empirical tests of reciprocity theory: problems in assessment. *Ethology and Sociobiology*, **9**, 181–7.

Slater, P. J. B. 1986. Individual differences and dominance hierarchies. *Animal Behaviour*, **34**, 1264–5.

Stewart, K. J. & Harcourt, A. H. 1987. Gorillas: variation in female relationships. In: *Primate Societies*. ed. B. B. Smuts, D. L. Cheney, R. M. Seyfarth, R. W. Wrangham & T. T. Struhsaker, pp. 155–65. Chicago: University of Chicago Press.

Sugardjito, J., te Boekhorst, I. J. A. & van Hooff, J. A. R. A. M. 1987. Ecological constraints on the grouping of wild orangutans (*Pongo pygmaeus*) in the Gunung Leuser National Park, Sumatra, Indonesia. *International Journal of Primatology*, **8**, 17–41.

Terry, R. L. 1970. Primate grooming as a tension reduction mechanism. *Journal of Psychology*, **76**, 129–36.

Trivers, R. L. 1971. The evolution of reciprocal altruism. *Quarterly Review of Biology*, **46**, 35–57.

Trivers, R. L. 1972. Parental investment and sexual selection. In: *Sexual Selection and the Descent of Man*., ed. B. Campbell, pp. 136–79. Chicago: Aldine.

Tutin, C. E. G. 1980. Reproductive behaviour of wild chimpanzees in the Gombe National Park. *Journal of Reproduction and Fertility*, Supplement, **28**, 43–57.

de Waal, F. B. M. 1991. Complementary methods and

convergent evidence in the study of primate social cognition. *Behaviour*, **118**, 297–320.

de Waal, F. B. M. 1992. Coalitions as part of reciprocal relations in the Arnhem chimpanzee colony. In: *Coalitions and Alliances in Humans and Other Animals*, ed. A. H. Harcourt & F. B. M. de Waal, pp. 233–57. Oxford: Oxford University Press.

de Waal, F. B. M. & van Hooff, J. A. R. A. M. 1981. Side-directed communication and agonistic interactions in chimpanzees. *Behaviour*, **77**, 164–98.

de Waal, F. B. M. & Luttrell, L. M. 1988. Mechanisms of social reciprocity in three primate species: symmetrical relationship characteristics or cognition ? *Ethology and Sociobiology*, **9**, 101–18.

Wrangham, R. W. 1987. Evolution of social structure. In: *Primate Societies*, ed. B. B. Smuts, D. L. Cheney, R. M. Seyfarth, R. W. Wrangham & T. T. Struhsaker, pp. 282–96. Chicago: University Press.

Yamada, M. 1966. Five natural troops of Japanese monkeys of Shodoshima Island (I): distribution and social organization. *Primates*, **7**, 315–62.

15 • Chimpanzee intelligence in nature and in captivity: isomorphism of symbol use and tool use

TETSURO MATSUZAWA

INTRODUCTION

Chimpanzee intelligence can be studied from two very different points of view. One is the experimental analysis of the cognitive skills of captive chimpanzees. The other is the field observation of behavioral patterns in wild chimpanzees. This paper aims to form a new bridge between the two separate fields by presenting a synthesized view of chimpanzee intelligence through research in both the laboratory and the wild.

People have believed that human beings are a unique species (Sebeok & Umiker-Sebeok, 1980; Lieberman, 1991). In comparison with other animals, humans seem to have special features: language, tools, culture and so on. These human traits have been attributed to our species' unique intelligence based on a highly developed neocortex. Are we really so unique? If so, how and why has such a unique creature evolved? What is the ultimate difference between humans and other creatures? To answer these questions, one should look not only at humans but also at the species with whom we last shared a common ancestor in our evolutionary history. Chimpanzees are the most appropriate species, because they are the closest relative of humans.

My approach may be described as Comparative Cognitive Science (CCS). This new discipline aims to understand human cognition from an evolutionary perspective. The research method is characterized by the comparison of the performance of different species based on an unified objective scale. For that purpose, human performance and chimpanzee performance were compared using the same test apparatus and following the same test procedure.

I have studied chimpanzees in the laboratory (Matsuzawa, 1985a, b) and also in the wild (Matsuzawa,

1994). The following sections will focus on the hierarchical nature, and the constraints that can be found both in symbol–referent relationships and in tool–object relationships.

SYMBOL USE FOR THE STUDY OF VISUAL COGNITION

This section describes a series of experiments in captivity and aims to demonstrate the visual world of chimpanzees in comparison with that of humans. The project was named the 'Ai-Project' after the main subject, a female chimpanzee named 'Ai' (which is Japanese for 'love'). The long-term project started in 1978 when Ai was about one and a half years old. We have ten chimpanzees in a group now at the Primate Research Institute of Kyoto University. The first question was, how do chimpanzees see the world? My colleagues and I have studied color perception (Matsuzawa, 1985b), form perception (Matsuzawa, 1990; Tomonaga & Matsuzawa, 1992), complex pattern recognition (Matsuzawa, 1989; Fujita & Matsuzawa, 1990), face recognition (Tomonaga *et al.*, 1993) and the concept of number (Matsuzawa, 1985a; Matsuzawa *et al.*, 1991) in both humans and chimpanzees in the same test situation using a computer-controlled apparatus. The Ai-Project also covered some topics on the language-like skills learned by the chimpanzees (Asano *et al.*, 1982; Tomonaga *et al.*, 1991; Itakura & Matsuzawa, 1993). In parallel to the visual cognition studies, auditory perception of the chimpanzees was also investigated by Kojima (Kojima & Kiritani, 1989; Kojima *et al.*, 1989; Kojima, 1990).

The primary method of the Ai-Project is keeping a single subject in front of a computer-controlled apparatus (see Fig. 15.1). There is essentially no direct inter-

Fig. 15.1. Ai learns to discriminate Kanji characters when a sample character is presented in the left display interfaced with a computer.

action between the experimenter and the subject without the interfacing apparatus. The visual symbols called lexigrams, which the chimpanzees acquired, do not always function like 'words' in natural human language. However, the symbols that the chimpanzees learned were sufficient to let us know how they perceived the outer world. The Ai-Project is characterized by the experimental and objective approach to chimpanzees' cognitive functions, and contrasts with the daily-life context and more-or-less episodic approaches used in the so-called 'ape-language' studies (Gardner & Gardner, 1969; Premack, 1971; Fouts, 1973; Rumbaugh, 1977; Patterson, 1978; Terrace, 1979; Savage-Rumbaugh, 1986; Savage-Rumbaugh & Lewin, 1994).

A series of experiments revealed that the visual world of the chimpanzees was fundamentally similar to that of humans. Chimpanzees are trichromatic, and they learned to use visual symbols to express how they perceive various colors (Matsuzawa, 1985b). Their categories of color classification were similar to and as stable as those of humans. Form perception and visual acuity were also comparable to those of humans (Matsuzawa, 1990). There was no significant difference between species in the fundamental perception of such things as color and form.

A new method called the 'constructive matching-to-sample' procedure made it possible to analyze the detailed process of pattern perception (Matsuzawa, 1989; Fujita & Matsuzawa, 1990). When a complex geometric pattern was presented as a sample, humans and chimpanzees preferred to make the copy of the sample pattern beginning with the outer contour first, although the order of reconstruction was the subject's choice. This may mean that both species perceive patterns as starting from the outer contour.

In sum, the fundamental perceptual abilities do not differ between humans and chimpanzees. However, comparison of higher cognitive functions provided a different view of the visual world in each of the two species. The chimpanzee can easily recognize chimpanzee faces, but has somewhat more difficulty recognizing human faces (Matsuzawa, 1990). The chimpanzee recognized upside-down pictures and forms faster than human subjects (Tomonaga *et al.*, 1993). They showed no monotonic increase in reaction time in the mental rotation task. In comparison with humans, chimpanzees are relatively adept at assuming various orientations in the 3-dimensional space of the forest's canopy. The species-specific adaptation to the environment might explain the constraints found in the visual cognitive tasks in each species.

HIERARCHICAL NATURE OF VISUAL COGNITION

Complex pattern recognition might be a unique feature of human cognitive processes. Human language has a highly developed hierarchical structure of generating and perceiving complex signals, whether it is based on the manual-visual mode or the vocal-auditory mode. The chimpanzee, Ai, learned to use visual symbols to express various aspects such as number, color, object, individuals and so on. For example, she named 11 colors by both lexigrams and Kanji characters (see Fig. 15.2). She learned to discriminate 42 lexigrams, 26 letters of the English alphabet, 41 Kanji characters and 10 Arabic numerals from 0 through 9. At present, the total number of visual symbols that are actually used as 'words' is 84. Ordinary Japanese children learn to read and write about 3000 Kanji letters by the age of 12. Each Kanji has a different meaning just as do lexigrams. Although Ai's vocabulary is small in comparison with humans, it is obvious that the chimpanzee can master the visual symbol system in its rudimentary form.

Ai learned to combine the visual symbols. The chimpanzee was required to name the number, color and object of 125 types of samples (Matsuzawa, 1985a). Fifteen keys on the computer terminal for chimpanzees were operative at one time: 5 numbers (1, 2, 3, 4, 5), 5 colors (red, yellow, green, blue, black) and 5 objects (pencil, paper, brick, spoon, toothbrush). The sequence in which the keys were pressed was left up to the chimpanzee, but she was asked to select the three keys correctly describing each of the three attributes of $5 \times 5 \times 5 = 125$ types of sample items. For example, when 5 red toothbrushes were shown in a display window as a sample item, it was necessary for Ai to press keys of '5', 'red' and 'toothbrush' in any order. Although no particular 'word order' was required, the chimpanzee favored two particular sequences almost exclusively among six possible alternatives: color/object/number and object/color/number. In both sequences, numerical naming was always located in the last position.

Ai also learned to construct the visual symbols (in this case, lexigrams of complex geometric patterns) by choosing the corresponding component elements from the alternatives (Matsuzawa, 1989). Moreover, she learned to construct the visual symbols when the corresponding object was shown as a sample item. For example, Ai chose a rectangle, a circle and a dot to construct a complex symbol 'apple' when an actual apple was presented as a sample. She chose a circle, a lozenge and a wave for 'banana'. These skills demonstrated the

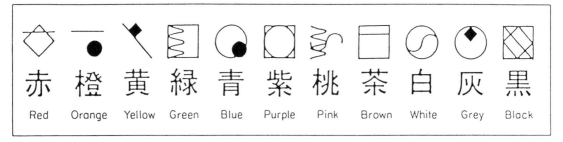

Fig. 15.2. The chimpanzee learned to name 11 colors by both lexigrams (top row) and Kanji characters (middle row). The lexigram system is functionally equivalent to the Kanji system in which each character has a meaning.

chimpanzee's capability of constructing hierarchical patterns from basic elements called 'graphemes'. These might correspond to the distinctive levels of phoneme, word and phrase in human vocal language. Premack's work on 'constructing a face' (Premack, 1975) also demonstrated the chimpanzees' capability of construction from scratch, although his research did not contain verbal materials.

In a recent study (Iversen & Matsuzawa, unpublished data), two chimpanzees were trained to touch a video-monitor screen to draw a letter of the alphabet. Both subjects successfully drew letters on a screen when a sample letter with thin white lines appeared on a screen. One of them, Ai, can draw a parallel line with the same length when a sample line and a starting point were shown on the monitor. These skills are a step for the chimpanzees actually to construct visual symbols.

Language and mathematical systems have a common feature. The hierarchical nature of chimpanzees' cognitive function was further investigated in learning a number system. Ai is the first chimpanzee to learn to describe numbers using Arabic numerals (Matsuzawa, 1985a). Her cognitive skills were tested in a rather small and strictly defined system of numbers, instead of in the more complex system of language-like skills.

Ai learned to label from 1 to 9 with accuracy of more than 90%, and then to touch three numerals from small to large. The relationship of lesser to greater numbers was trained in the two adjacent numbers such as '2 and 3' and '3 and 4.' Ai spontaneously transferred the skill of transitive relationship of adjacent numbers to that of nonadjacent ones such as '2 and 7.' Moreover, the transitivity shown in two numerals was transferred to three and then four numerals without explicit training. This means that the chimpanzee acquired the number system whose structure was based at least on the ordinal scale (Matsuzawa et al., 1991). The numerical competence of chimpanzees was also investigated by Boysen and her colleagues (Boysen & Berntson, 1989; Boysen & Capaldi, 1993).

FIELD EXPERIMENTS ON TOOL USE

In the Ai-project described above, chimpanzee intelligence was explored in an individual who was isolated from a social context and worked with a computer system. Chimpanzees as well as humans are social animals. There should be different aspects of chimpanzee intelligence in the social context in which interaction among individuals is essential. Social intelligence in chimpanzees has been investigated in various ways: leadership (Menzel, 1971, 1973), imitation (Tomasello et al., 1993), mirror self-recognition (Gallup, 1970; Povinelli et al., 1993; Inoue, 1994; Parker et al., 1994), theory of mind or mental state attribution (Premack & Woodruff, 1978; Povinelli et al., 1993, 1994), intentional communication or tactical deception (Woodruff & Premack, 1979; Byrne & Whiten, 1988; Povinelli et al., 1990), etc.

This section aims to clarify how chimpanzee intelligence is utilized in their natural habitat. The behavior of interest is stone tool use by wild chimpanzees at Bossou in Guinea and Mt Nimba in Ivory Coast, West Africa.

During the last 30 years, long-term studies on wild chimpanzees have accumulated data on tool use and shown differences in local populations (Goodall, 1986; Nishida, 1990; McGrew, 1992). Although the list of tool use patterns by wild chimpanzees has become long, most of the patterns consist of a single detached object used as a tool, such as a leaf for drinking water or a twig for fishing termites. Stone tool use may be the most complex tool use by wild chimpanzees because a set of two detached objects is needed.

Chimpanzees at Bossou are known to use a pair of stones as a hammer and anvil to crack open oil-palm nuts, *Elaeis guineensis* (Sugiyama & Koman, 1979). The nut species cracked are different across local populations in West Africa. Chimpanzees at Taï Forest in Ivory Coast crack open five species of nuts instead of oil-palm nuts: *Coula edulis* (Olacaceae), *Panda oleosa* (Pandaceae), *Parinari excelsa* (Rosaceae), *Sacoglottis gabonensis* (Humiriaceae) and *Detarium senegalense* (Caesalpiniaceae) (Boesch & Boesch, 1983; Boesch, 1991b; Boesch et al., 1994). To analyze the tool use in detail, my colleagues and I started a field experiment on stone tool use (Sakura & Matsuzawa, 1991; Matsuzawa, 1994). We took stones and nuts to the top of a hill in the central part of the chimpanzees' home range (see Fig. 15.3).

Field experiments on stone tool use revealed various aspects of chimpanzee intelligence in the wild (see Table 15.1). The chimpanzees showed a consistent hand preference when using a hammer stone: each individual

Table 15.1. *Stone tool use by wild chimpanzees at Bossou, Guinea. Hand holding the hammer stone in nut-cracking was recorded from field experiments beginning in 1988. Data were usually collected from January to March in dry season of each year*

Name	Sex	Age	Mother	Year observed						
				1988	1990	1991	1992	1993	1994	1995
Tua	m	adult	unknown	?	L	L	L	L	L	L
Kai	f	adult	unknown	?	R	R	R	R	R	R
Nina	f	adult	unknown	?	X	X	X	X	X	X
Fana	f	adult	unknown	?	L	L	L	L	L	L
Jire	f	adult	unknown	?	L	L	L	L	L	L
Velu	f	adult	unknown	?	R	R	R	R	R	R
Yo	f	adult	unknown	?	L	L	L	L	L	L
Pama	f	adult	unknown	?	X	X	X	X	X	X
Kie[a]	f	18	Kai	?	R	R	–	–	–	–
Foaf	m	13.5	Fana	?	R	R	R	R	R	R
Puru[a]	m	13.5	Pama	R	R	R	–	–	–	–
Vube	f	12	Velu	?	L	–	–	–	–	–
Ja[a]	f	11	Jire	?	R	R	R	R	–	–
Yunro[a]	f	10	Yo	?	X	X	X	X	–	–
Na	m	9.5	Nina	?	R	R	R	R	R	R
Kakuru[a]	f	8.5	Kie	?	AR	R	–	–	–	–
Vui	m	8.5	Velu	?	X	X	L	L	L	L
Pili	f	8	Pama	–	X	R	R	R	R	R
Jokro[a]	f	6	Jire	–	X	X	X	–	–	–
Yela[a]	f	5.5	Yo	–	X	–	–	–	–	–
Fotayu	f	3.5	Fana	–	–	–	X	X	X	AR
Vuavuaf	f	3.5	Velu	–	–	–	X	X	X	AL
Yoro	m	3.0	Yo	–	–	–	X	X	X	X
Poni	m	2.0	Pama	–	–	–	–	X	X	X
Nto	f	1.5	Nina	–	–	–	–	–	X	X
Juru	f	1	Jire	–	–	–	–	–	X	X

Notes: [a], individuals who disappeared or died by February, 1995; L, always used left hand for hammer; R, always used right hand for hammer; AR, ambidextrous but right hand dominant; AL, ambidextrous but left hand dominant; X, no successful hammer use but ate nuts cracked by others; ?, data unavailable because not seen at cracking site; –, data unavailable because subject disappeared or died or was not yet born, in research period. Age was estimated in February 1995.

Fig. 15.3. Adult female chimpanzee cracks open an oil-palm nut using stone, hammer, and anvil. (*a*) Reaches to pick up nut with right hand while holding hammer in left; (*b*) cracks nut with left hand; (*c*) eats nut with right hand.

always used one hand for a hammer (Fushimi *et al.*, 1991; Sugiyama *et al.*, 1993). Chimpanzees' hand preference in simple reaching and various tasks in captivity (Finch, 1941; Marchant & Steklis, 1986; Bard *et al.*, 1990; Tonooka & Matsuzawa, 1995) and in tool use in the wild (Boesch, 1991a: McGrew & Marchant, 1992) is one of the most important research topics in seeking the origin of human hemispheric lateralization (MacNeilage *et al.*, 1987; Hopkins & Morris, 1993; Ward & Hopkins, 1993).

Furthermore, each chimpanzee had his or her own favorite stone tools and transported them, which may indicate a rudimentary form of possession. Three of the chimpanzees used a third stone as a wedge to keep the surface of the anvil stone flat and stable (see Fig. 15.4). The wedge could be classified as a metatool: a tool that was used to improve the function of another tool (Matsuzawa, 1991). This finding is the most complex form of tool use among wild chimpanzees.

Stone tool use requires a long time for learning, almost ten years to reach the refined level of adult chimpanzees. There are at least three developmental stages needed to acquire the skill. First is the action of manipulating a single object, such as a nut or a stone (see Fig. 15.5). Second is the action of relating two objects; a nut and a stone, or a stone and another stone. Third is coordinating the multiple actions of manipulated objects.

Direct comparison of stone tool use between human children and chimpanzee youngsters shows that both species first successfully use stones as tools at the age of three years. Longitudinal experimental field studies on stone tool use (Matsuzawa, 1994; Inoue *et al.*, unpublished data) showed the ontogenetic development of the skill and suggested that there is a critical period of learning the skill at the age of three to five years.

At the age of 3.5 years, one-third of the infant chimpanzees start to use a pair of stones as a tool. The youngest chimpanzee who used the third stone as a wedge for the anvil was 6.5 years old. Through direct experience and observational learning from other community members, the chimpanzees at Bossou acquired the skills of using stone tools.

The chimpanzees at Bossou use many other tools in addition to those in nut cracking (Sugiyama & Koman, 1979). They use leaves as a sponge or as a container for drinking water, a twig as a wand for dipping for ants, etc. They extract sap from oil-palm trees with a 'pestle'

Fig. 15.4. Chimpanzees at Bossou, Guinea, used a third stone as a wedge to keep the surface of the anvil stone flat and stable. Wedge is a meta-tool, that is, a tool used to improve the function of another tool. Set of stones was left by a 6.5-year-old juvenile male, Na, on January 13, 1992.

Fig. 15.5. Typical behavior shown in the developmental stage before acquiring nut-cracking skills in a young chimpanzee at Bossou. Yo, 2-year-old male, manipulates a kernel taken from his mother who cracked open the oil-palm nut.

and then drink it using a 'fibre sponge' (Sugiyama, 1994; Yamakoshi & Sugiyama, 1995). A close examination shows that many details of tool use in the Bossou group differ from the tool use reported elsewhere (McGrew, 1992). In a sense, the sum total of tool use (or local tool-kit) constitutes a unique cultural tradition in each chimpanzee community.

For example, to see the details of leaf tool use for drinking water, we set up an outdoor laboratory for observing and video-recording the behavioral processes of the tool use (Tonooka et al., 1994). A water container usually used by local people was buried in the ground at the outdoor laboratory. We observed five episodes by three chimpanzees in which leaves were used as a drinking tool. The observation revealed a third method of using leaves: the chimpanzees folded a leaf in the mouth as in paper-folding and put it into the container for capturing water. The chimpanzees preferred to use a particular leaf (*Hybophrynium braunianum*), a wide, soft and hairless leaf, as a tool (4 out of 5 leaves used in the outdoor-laboratory, 17 out of 23 leaves used in a tree hollow under natural conditions, 21 out of 28 episodes in total). The high selectivity of leaves and the newly-found folding method is yet another example of the chimpanzees' ability to find a suitable material and modify it into a useful tool.

CULTURAL DIFFERENCES AND THE MECHANISM OF TRANSMISSION

To explore the cultural differences between communities, we also carried out a field survey of the chimpanzees of Mt Nimba, Ivory Coast. The distance between the two communities of Bossou and Nimba is about 10 km. We found that there are behavioral differences between the two neighboring communities (Matsuzawa & Yamakoshi, 1995).

Chimpanzees at Nimba built nests on the ground in addition to nests in the trees. Of the 464 chimpanzee nests Matsuzawa counted at Nimba in February 1994, 164 (35%) were built on the ground. Ground-nests are an unusual feature of Nimba chimpanzees, as other chimpanzee populations have rarely been seen to use them. Nimba chimpanzees were also found to use stone tools to crack open at least two kinds of nuts: *Coula edulis* and *Carapa procera*.

Chimpanzees at Nimba and the neighboring community of chimpanzees at Bossou may have different traditions in various respects such as in food repertoire, tool use and nest building methods. A field experiment on cultural transmission was carried out to analyze these differences (Matsuzawa, 1994; Yamakoshi & Matsuzawa, unpublished data). We provided a group of Bossou chimpanzees with unfamiliar *Coula* nuts that, unlike oil palm nuts, are encased within a thick fruit. Most of the chimpanzees examined the fruits and tried to bite them, but did not attempt to crack them, despite being skilled oil-palm nut crackers. But one adult female named Yo immediately placed the *Coula* nut on her stone anvil, cracked it and ate it. A group of juveniles gathered around and peered at what she was doing, but they did not try to take the *Coula* nuts. The next day a 6-year-old male named Vui, unrelated to Yo, successfully cracked open a nut without any practice. Four days later a 5-year-old female named Pili followed suit. Both of these juveniles cracked a nut open, sniffed its kernel, chewed it, then spat it out. Although we provided *Coula* nuts, continuously for another two weeks, these two juveniles were the only group members who learned to crack them. In general, adult chimpanzees ignored the new nuts, whereas youngsters watched Yo's new behavior and then tried it themselves.

A further test of *Coula* nut knowledge was conducted a year later. We provided the chimpanzees with wooden balls (3 cm diameter) of the same shape and size as *Coula* nuts (Tonooka et al., unpublished data). Yo simply ignored the wooden balls. Other adult chimpanzees similarly ignored the balls, or picked them up, sniffed them, then dropped them. Three young chimpanzees, Vui, Pili and Na (an 8-year old) tried to crack the wooden balls as soon as they appeared. This supplementary experiment suggests that Yo was not the sort of individual to try to crack any unfamiliar object with stone tools. Yo seemed to 'know' the *Coula* nuts. The youngsters, on the other hand, seemed ready to try to crack any objects resembling edible nuts even if the objects were unfamiliar. Their attempts to crack open wooden balls may reveal an abiding tendency to try to crack open unfamiliar nut-like objects which was facilitated by their observing Yo's cracking new nuts in the last year.

In chimpanzees, adolescent females emigrate from their natal groups. Our interpretation of the above results is as follows: the *Coula*-cracking female, Yo, was born in another community such as Mt Nimba, 10 km away from Bossou, where a tradition of cracking *Coula* nuts already existed. She grew up and learned to crack *Coula* nuts there, before her migration to Bossou. Once at Bossou she had no further opportunity to crack *Coula* nuts, because *Coula* trees are absent there. Our experimental manipulation reintroduced her to *Coula* nuts; as a result, she functioned as an innovator by introducing a new kind of nut cracking to the Bossou community. Hannah & McGrew (1987) also described a similar process in the nut-cracking by captured chimpanzees who were reintroduced as a group onto an island in Liberia.

New behavior was transmitted from an immigrant female to other members of her adopted community. Moreover, the field experiments show the transmission of knowledge from one generation to another. Although all the Bossou chimpanzees had the opportunity to access *Coula* nuts, only younger chimpanzees learned to crack them and they learned to crack them by observing the informant, an adult female.

There can be a dynamic cultural interchange between neighboring chimpanzee communities having different traditions, as our *Coula* nut experiments suggest. Regional differences may be maintained or changed across generations through learning processes during social interactions among the members in each community. Experiments using 'taste-aversion learning' in Japanese monkeys (*Macaca fuscata*) have experimentally shown the social transmission of food habits from mothers to their offspring in other primates (Matsuzawa *et al.*, 1983; Hikami *et al.*, 1990).

TREE-STRUCTURE ANALYSIS OF HIERARCHICAL COGNITIVE PROCESSES

Through my work in the laboratory and in the wild, I have tried to analyze experimentally the hierarchical nature of cognitive processes in chimpanzees. The hierarchical nature was clearly shown by the developmental processes of stone tool use when we examined the way chimpanzees manipulate objects. Both in the symbol use skills in the laboratory and in the tool use skills in the wild, the chimpanzees must learn how to manipulate detached objects in an appropriate temporal order and

spatial arrangement. Examination of these skills allowed me to discern the hierarchical structure of actions relating multiple objects and symbols.

Similar to Chomsky's method of describing the syntactic structure of human language by a generative grammar (Chomsky, 1978), I devised a notation system using a tree-structure for describing cognitive processes like symbol use and tool use by chimpanzees. The system is a kind of action grammar to find structural rules in human actions (Greenfield, 1991). The tree-structure analysis is a way of describing the cognitive processes involved in a series of actions or in a behavioral pattern dealing with multiple objects.

To explain the basic idea of the tree-structure analysis, the actions involved in tool use were analyzed by focusing on the objects (see Fig. 15.6). The tree-structure analysis proceeds in the following three steps. First, describe a behavioral pattern in a sentence, in either an active or a passive voice. For example, the observed behavior in an episode can be described such as 'A chimpanzee uses a hammer stone to hit a nut on an anvil stone.' Second, identify the targets of the sentence. The targets are the detached objects in tool use. You 'neglect' the actor and the particular action in the sentence. For example, there are three objects in the episode: 'hammer stone,' 'nut' and 'anvil stone.'

Third, connect the targets one-by-one to make a cluster. To make the connections, you should look at the temporal order of relating the targets. The relationships among the targets can be represented by the connecting lines. The connection is called a 'node' in the tree-structure. There is a node between a nut and an anvil stone. The cluster of the nut on an anvil stone is connected with a hammer stone at the higher level of another node. The tree-structure analysis recognizes two nodes in this episode of nut cracking.

By following the above three steps in the tree-structure analysis, the complexity of the actions can be represented by the depth of the nodes in the tree-structure. In the notation system of the tree-structure analysis, hierarchical level is shown by the depth of nodes, in other words, the number of nesting clusters in the tree-structure.

Examples of Level 0 structure of the actions in object manipulation or tool use are as follows. 'A chimpanzee picked up a nut.' 'A chimpanzee touched or hit a stone.'

Examples of observed behavior

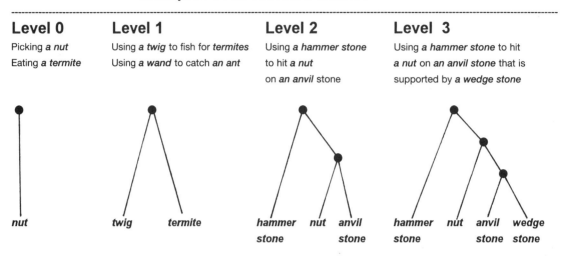

Level 0

Picking *a nut*

Eating *a termite*

Level 1

Using *a twig* to fish for *termites*

Using *a wand* to catch *an ant*

Level 2

Using *a hammer stone*
to hit *a nut*
on *an anvil* stone

Level 3

Using *a hammer stone* to hit
a nut on *an anvil stone* that is
supported by *a wedge stone*

nut

twig   termite

hammer   nut   anvil
stone   stone

hammer   nut   anvil   wedge
stone   stone   stone

Fig. 15.6. In tree-structure analysis, the actions involved in tool use were analyzed by focusing on the objects (see text for details). In the notation system of the tree-structure analysis, hierarchical level is shown by the depth of nodes, that is, the number of nested clusters in the tree-structure. A node is represented by a solid circle. Each level in the structure for object manipulation or tool use is shown.

'A chimpanzee picked up and ate a termite.' 'A chimpanzee clipped a leaf.' All the actions described in the above sentences involve only one target object. When you ignore the actor and the particular action, there should be no relationship because a relationship requires two or more objects.

Level 1 structure is as follows. 'A chimpanzee uses a twig to fish for termites.' You may omit the subject from the sentence because the actor should be neglected in the tree-structure analysis: 'Using a wand to catch a safari ant,' 'Using a leaf to drink water,' 'Throwing a stone at another chimpanzee' and so on.

Level 2 structure is as follows. 'A chimpanzee uses a hammer stone to hit a nut on an anvil stone.' Note that the choice of the active or passive voice in a sentence or the choice of word order has no effect on the depth of nodes connecting the target objects.

Level 3 structure is as follows. 'A chimpanzee uses a hammer stone to hit a nut on an anvil stone which is supported by a wedge stone.' From the viewpoint of tree-structure analysis, this is the most advanced stage of tool use ever found in wild chimpanzees, a meta-tool used in nut cracking.

The tree-structure analysis tells us that most of the tool use reported in wild chimpanzees belongs to the Level 0 structure, like 'leaf grooming' and 'leaf clipping,' and the Level 1 structure like 'termite fishing,' 'ant dipping,' 'leaf sponging' and so on. Level 0 and Level 1 structures cover a wide range of object manipulation by chimpanzees in the wild.

Tool use is a kind of object manipulation in a broader sense. Level 1 structure covers another form of object manipulation called 'orienting manipulation,' such as an infant chimpanzee pushing a stone against another stone. This is not tool use because it has no explicit function. However, this is a real example of the behavior shown by the less-than-3-year-old chimpanzees in the stage before acquiring the nut cracking skills. According to the definition in the tree-structure analysis, this behavior can be classified as a Level 1 structure because one object was related to the other one. Although there should be no difference in the depth of nodes (that is, in the level of cognitive process relating objects), the orienting manipulation at Level 1 of stone-stone play is considered a precursor of tool use at Level 2 of nut cracking with two stones. The tree-structure analysis can identify the level of complexity in both tool use and orienting manipulation on the basis of a unified objective scale.

APPLICATION OF THE TREE-STRUCTURE ANALYSIS

The tree-structure of symbol use

The tree-structure analysis should be applied to symbol use as well as tool use. The first step of symbol use is the establishment of the referent–symbol relationship, just as in the object-tool relationship in tool use. Symbolic matching, of a real apple to the lexigram apple, is equivalent to a tool use of retrieving an apple with a stick from the viewpoint of the tree-structure analysis. The structure can be described as Level 1 because the two targets, a referent and a visual symbol, were connected in a behavioral pattern.

Level 1 structure covers a wide range of symbol use in the language-like skills shown by the apes in captivity. The previous studies on signing apes have reported that the mean length of utterance (MLU) seldom exceeds two. It means that not more than two signs were used in each episode of gestural communication (Patterson, 1978; Terrace, 1979).

The most complex symbol use shown by Ai was as follows. The chimpanzee can match 5 red toothbrushes to the 3 lexigrams, 'red,' 'toothbrush' and 'five' in particular sequences. If the tree-structure analysis is applied, the depth of nodes of color/object/number naming is the same as that of the above-mentioned meta-tool use in the nut cracking: cracking a nut with the three stones, 'hammer,' 'anvil' and 'wedge'.

Embedding and self-embedding structures

When you carefully look at the tree-structure, you will also find the rules in constructing the tree-structure. As an organism cannot behave randomly, there should be some rules determining the hierarchical structure connecting the component actions involved in a behavioral pattern. There should also be some constraints on the rules. An action grammar is defined as a set of rules for generating a variety of behavioral patterns in chimpan-zees. Among the rules, the self-embedding rule seems to take on an important and critical role in the hierarchical structure of cognitive processes.

In the case of human symbol use and tool use, we can easily identify the set of actions that can be expanded by a single rule, such as an embedding rule. For example, there could be a sentence describing a behavioral pattern like, 'I found a piece of cheese that was eaten by the mouse that was chased by the cat that was bitten by the dog.' There is a recurring structure of embedded phrases in the sentence. In such sentences, one part of the sentence refers back to another part. A sequential behavior following an embedding rule can have an infinite number of nodes in the tree-structure analysis. The complexity in cognitive processes of decoding such embedded references can be shown by the depth of nodes in the self-embedding structure.

Humans not only have a language but also a meta-language that is used to describe the language. They can use a tool, a meta-tool and even a meta-meta-tool, and so on. The self-embedding structure, a kind of recurring structure, is following a self-embedding rule in which the same word is used in different levels. To recognize the difference of levels, the prefix 'meta-' is added.

The number of levels of self-embedding structure may be a useful indicator of cognitive capabilities. In the case of chimpanzees, they can use a tool, they might use a meta-tool, a tool for another tool. However, our field observation on chimpanzees' nut cracking with stones suggests that there should be a limit on the depth of the self-embedding structure of tool use by chimpanzees.

To tackle the more general idea, let us think about the meta-level of 'relationship'. Chimpanzees recognize the relationship of two things AA' (at Level 1 structure) like 'the relationship of a tool and an object' and 'the relationship of a symbol and a referent.' However, at the next level of the self-embedding structure, they cannot easily recognize 'the relationship of the relationships' (at Level 2 structure). Chimpanzees can match A with A', and B with B' at Level 1 structure. However, at Level 2 structure, they have to match AA' with BB' rather than CD, and to match CD with EF rather than BB. Past studies have shown that some chimpanzees can master the 'conceptual matching-to-sample' task based on the conceptual equivalence rather than physical resemblance

(Premack, 1983). At present, there is no good evidence that chimpanzees recognize 'the relationship of the relationships of the relationships' (at Level 3 structure). There should be a limit or a constraint on the self-embedding structure of chimpanzees' cognitive skills.

Tree-structure analysis of social intelligence

The tree-structure analysis can be extended for describing the cognitive processes involved in social intelligence in chimpanzees. For example, their social intelligence is well documented by de Waal (1982) in his descriptions of alliance formation with coalition partners at Arnhem. The targets of the tree-structure analysis are individuals involved in a social interaction instead of objects in tool use.

Words like friend, opponent, mother, offspring, brother and so on, represent a particular individual and also the implicit relationship to the other individual. A 'friend,' for instance, means the individual who is a friend of the other individual. In that sense, the cognitive process of 'friend' implies the recognition of a social relationship between the two individuals.

The connections among the individuals can be analyzed from the viewpoint of the tree-structure analysis. The relationship between individuals in a social context can be treated as equivalent to the relationship between objects and tools in tool use, and also the relationship between referent and visual symbols in symbol use. Again, you can determine the limit of the depth of nodes of the tree-structure of individuals making an alliance formation or whatever relationships in a social context.

Perspective

Future studies will focus on finding the rules constituting the tree-structure of cognitive processes, especially from the developmental perspective. Formation of a tree-structure of cognitive processes in young individuals may be compared with the corresponding deterioration of the tree-structure in aged or handicapped individuals. It will also be important to know the limit of nodes in the tree-structure of cognitive processes from the phylogenetic perspective. For that purpose, it is necessary to compare various kinds of primate species, in addition to humans and chimpanzees, based on a unified objective scale. From a viewpoint of grammatical rules of actions, the tree-structure analysis can provide a systematic description of various aspects of cognitive processes across species and can illuminate the hierarchical nature of intelligence.

ACKNOWLEDGMENTS

This paper was originally prepared for the Wenner-Gren conference 'The Great Ape Revisited,' held November 12–19, 1994 in Cabo San Lucas, Baja California Sur, Mexico. I thank S. Silverman, the president, and other staff of the Foundation. This study was financed by grants from the Ministry of Education, Science, and Culture, Japan (Nos. 63510057, 63626504, 03610048, 04551002, 05044006, 06260222, 04651017 to Matsuzawa, and No. 01041058 to Sugiyama for field research). I thank the following people for their assistance in the laboratory work: S. Nagumo, K. Murofushi, T. Asano, S. Kojima, K. Kubota, K. Fujita, M. Tomonaga, T. Kojima, S. Yoshikubo, S. Itakura, K. Hikami, T. Fushimi, M. Tanaka, I. Iversen, J. Inagaki, K. Kumazaki and N. Maeda. Thanks are also due to the following people for their assistance in the field work in Guinea and Ivory Coast: Y. Sugiyama, G. Yamakoshi, R. Tonooka, N. Inoue, O. Sakura, T. Fushimi, G. Goumi, T. Kamara, J. Koman, C. Habout and B. Ouayogode. Thanks are also due to D. Premack and J. Goodall for inspiring me with perspectives of the study. I am grateful to W. McGrew, L. Marchant and T. Nishida for their constant encouraging support for preparing the manuscript.

REFERENCES

Asano, T., Kojima, T., Matsuzawa, T., Kubota, K. & Murofushi, K. 1982. Object and color naming in chimpanzees (*Pan troglodytes*). *Proceedings of the Japan Academy*, **58(B)**, 118–22.

Bard, K., Hopkins, W. & Fort, C. 1990. Lateral bias in infant chimpanzees (*Pan troglodytes*). *Journal of Comparative Psychology*, **104**, 309–21.

Boesch, C. 1991a. Handedness in wild chimpanzees. *International Journal of Primatology*, **12**, 541–58.

Boesch, C. 1991b. Teaching among wild chimpanzees. *Animal Behaviour*, **41**, 530–2.

Boesch, C. & Boesch, H. 1983. Optimization of

nut-cracking with natural hammers by wild chimpanzees. *Behaviour*, 83, 265–86.

Boesch, C., Marchesi, P., Marchesi, N., Fruth, B. & Joulian, F. 1994. Is nut cracking in wild chimpanzees a cultural behaviour? *Journal of Human Evolution*, 26, 325–38.

Boysen, S. & Berntson, G. 1989. Numerical competence in a chimpanzee. *Journal of Comparative Psychology*, 103, 23–31.

Boysen, S. & Capaldi, E. 1993. *The Emergence of Numerical Competence: Animal and Human Models.* Hillsdale, NJ: Lawrence Erlbaum.

Byrne, R. & Whiten, A. 1988. *Machiavellian Intelligence: Social Expertise and the Evolution of Intellect in Monkeys, Apes, and Humans.* Oxford: Oxford University Press.

Chomsky, N. 1978. *Syntactic Structures.* The Hague: Mouton & Co.

Finch, G. 1941. Chimpanzee handedness. *Science*, 94, 117–18.

Fouts, R. 1973. Acquisition and testing of gestural signs in four young chimpanzees. *Science*, 180, 978–80.

Fujita, K. & Matsuzawa, T. 1990. Delayed figure reconstruction by a chimpanzee and humans. *Journal of Comparative Psychology*, 104, 345–51.

Fushimi, T., Sakura, O., Matsuzawa, T., Ohno, H. & Sugiyama, Y. 1991. Nut-cracking behavior of wild chimpanzees (*Pan troglodytes*) in Bossou, Guinea (West Africa). In: *Primatology Today*, ed. A. Ehara, T. Kimura, O. Takenaka & M. Iwamoto, pp. 695–6. Amsterdam: Elsevier.

Gallup, G. 1970. Chimpanzee: Self-recognition. *Science*, 167, 86–7.

Gardner, R. & Gardner, B. 1969. Teaching sign language to a chimpanzee. *Science*, 165, 664–72.

Goodall, J. 1986. *The Chimpanzees of Gombe: Patterns of Behavior.* Cambridge, MA: Harvard University Press.

Greenfield, P. 1991. Language, tools and brain: The ontogeny and phylogeny of hierarchically organized sequential behavior. *Behavioral and Brain Sciences*, 14, 531–51.

Hannah, A. C. & McGrew, W. C. 1987. Chimpanzees using stones to crack open oil-palm nuts in Liberia. *Primates*, 28, 31–46.

Hikami, K., Hasegawa, Y. & Matsuzawa, T. 1990. Social transmission of food preferences in Japanese monkeys (*Macaca fuscata*) after mere exposure or aversion learning. *Journal of Comparative Psychology*, 104, 233–7.

Hopkins, W. & Morris, R. 1993. Handedness in great apes: A review of findings. *International Journal of Primatology*, 14, 1–26.

Inoue, N. 1994. Mirror self-recognition among infant chimpanzees: application of longitudinal and cross-sectional methods. *Japanese Journal of Developmental Psychology*, 5, 51–60. (in Japanese with English summary)

Itakura, S. & Matsuzawa, T. 1993. Acquisition of personal pronouns by a chimpanzee. In: *Language and Communication: Comparative Perspectives*, ed. H. Roitblat, L. Herman & P. Nachtigall, pp. 347–63. Hillsdale, NJ: Lawrence Erlbaum.

Kojima, S. 1990. Comparison of auditory functions in the chimpanzee and human. *Folia Primatologica*, 55, 62–72.

Kojima, S. & Kiritani, S. 1989. Vocal-auditory functions in the chimpanzee: vowel perception. *International Journal of Primatology*, 10, 199–213.

Kojima, S., Tatsumi, I., Kiritani, S. & Hirose, H. 1989. Vocal-auditory functions of the chimpanzee: consonant perception. *Human Evolution*, 4, 403–16.

Lieberman, P. 1991. *Uniquely Human: The Evolution of Speech, Thought, and Selfless Behavior.* Cambridge, MA: Harvard University Press.

MacNeilage, P., Studdert-Kennedy, M. & Lindblom, B. 1987. Primate handedness reconsidered. *Behavioral and Brain Sciences*, 10, 247–303.

Marchant, L. & Steklis, H. 1986. Hand preference in a captive island group of chimpanzees (*Pan troglodytes*). *American Journal of Primatology*, 10, 301–13.

Matsuzawa, T. 1985a. Use of numbers by a chimpanzee. *Nature*, 315, 57–9.

Matsuzawa, T. 1985b. Color naming and classification in a chimpanzee (*Pan troglodytes*). *Journal of Human Evolution*, 14, 283–91.

Matsuzawa, T. 1989. Spontaneous pattern construction in a chimpanzee. In: *Understanding Chimpanzees*, ed. P. Heltne & L. Marquardt, pp. 252–65. Cambridge, MA: Harvard University Press.

Matsuzawa, T. 1990. Form perception and visual acuity in a chimpanzee. *Folia Primatologica*, 55, 24–32.

Matsuzawa, T. 1991. Nesting cups and metatools in chimpanzees. *Behavioral and Brain Sciences*, 14, 570–1.

Matsuzawa, T. 1994. Field experiments on use of stone tools by chimpanzees in the wild. In: *Chimpanzee Cultures*, ed. R. W. Wrangham, W. C. McGrew, F. B. M. de Waal & P. G. Heltne, pp. 351–70. Cambridge, MA: Harvard University Press.

Matsuzawa, T. & Yamakoshi, G. 1995. Comparison of

chimpanzee material culture between Bossou and Nimba, West Africa. In: *Reaching into Thought*, ed. A. Russon, K. Bard & S. Parker, Cambridge: Cambridge University Press.

Matsuzawa, T., Hasegawa, Y., Gotoh, S. & Wada, K. 1983. One-trial long-lasting food aversion learning in wild Japanese monkeys (*Macaca fuscata*). *Behavioral and Neural Biology*, **39**, 155–9.

Matsuzawa, T., Itakura, S. & Tomonaga, M. 1991. Use of numbers by a chimpanzee: A further study. In: *Primatology Today*, ed. A. Ehara, T. Kimura, O. Takenaka & M. Iwamoto, pp. 317–20. Amsterdam: Elsevier.

McGrew, W. 1992. *Chimpanzee Material Culture: Implications for Human Evolution*. Cambridge: Cambridge University Press.

McGrew, W. & Marchant, L. 1992. Chimpanzees, tools, and termites: hand preference or handedness? *Current Anthropology*, **32**, 114–9.

Menzel, E. 1971. Communication about the environment in a group of young chimpanzees. *Folia Primatologica*, **15**, 220–32.

Menzel, E. 1973. Chimpanzee spatial memory organization. *Science*, **182**, 943–5.

Nishida, T. (ed.) 1990. *The Chimpanzees of the Mahale Mountains*. Tokyo: University of Tokyo Press.

Parker, S., Mitchell, R. & Boccia, M. (eds.) 1994. *Self-awareness in Animals and Humans: Developmental Perspectives*. Cambridge: Cambridge University Press.

Patterson, F. 1978. The gestures of a gorilla: language acquisition in another pongid. *Brain and Language*, **5**, 72–97.

Povinelli, D., Nelson, K. & Boysen, S. 1990. Inferences about guessing and knowing by chimpanzees (*Pan troglodytes*). *Journal of Comparative Psychology*, **104**, 203–10.

Povinelli, D., Rulf, A., Landau, K. & Bierschwale, D. 1993. Self-recognition in chimpanzees (*Pan troglodytes*): distribution, ontogeny, and patterns of behavior. *Journal of Comparative Psychology*, **107**, 347–72.

Povinelli, D., Rulf, A. & Bierschwale, D. 1994. Absence of knowledge attribution and self-recognition in young chimpanzees (*Pan troglodytes*). *Journal of Comparative Psychology*, **108**, 74–80.

Premack, D. 1971. Language in chimpanzee? *Science*, **172**, 808–22.

Premack, D. 1975. Putting a face together. *Science*, **188**, 228–36.

Premack, D. 1983. The codes of man and beasts. *Behavioral and Brain Sciences*, **6**, 125–67.

Premack, D. & Woodruff, G. 1978. Does the chimpanzee have a theory of mind? *Behavioral and Brain Sciences*, **1**, 515–26.

Rumbaugh, D. M. (ed.) 1977. *Language Learning by a Chimpanzee*. New York: Academic Press.

Sakura, O. & Matsuzawa, T. 1991. Flexibility of wild chimpanzee nut-cracking behavior using stone hammers and anvils: an experimental analysis. *Ethology*, **87**, 237–48.

Savage-Rumbaugh, E. S. 1986. *Ape Language: From Conditioned Response to Symbol*. New York: Columbia University Press.

Savage-Rumbaugh, E. S. & Lewin, R. 1994. *Kanzi*. New York: John Wiley & Sons.

Sebeok, T. & Umiker-Sebeok, J. (eds.) 1980. *Speaking of Apes: A Critical Anthology of Two-way Communication with Man*. New York: Plenum Press.

Sugiyama, Y. 1994. Tool use by wild chimpanzees. *Nature*, **367**, 327.

Sugiyama, Y. & Koman, J. 1979. Tool-using and making behavior in wild chimpanzees at Bossou, Guinea. *Primates*, **20**, 513–24.

Sugiyama, Y., Fushimi, T., Sakura, O. & Matsuzawa, T. 1993. Hand preference and tool use in wild chimpanzees. *Primates*, **34**, 151–9.

Terrace, H. 1979. *Nim: A Chimpanzee who Learned Sign Language*. New York: Alfred A. Knopf.

Tomasello, M., Savage-Rumbaugh, S. & Kruger, A. 1993. Imitative learning of actions on objects by chimpanzees, enculturated chimpanzees, and human children. *Child Development*, **64**, 1688–705.

Tomonaga, M. & Matsuzawa, T. 1992. Perception of complex geometric figures in chimpanzees and humans: analyses of visual similarity on the basis of choice reaction time. *Journal of Comparative Psychology*, **106**, 43–52.

Tomonaga, M., Matsuzawa, T., Fujita, K. & Yamamoto, J. 1991. Emergence of symmetry in a visual conditional discrimination by chimpanzees (*Pan troglodytes*). *Psychological Reports*, **68**, 51–60.

Tomonaga, M., Itakura, S. & Matsuzawa, T. 1993. Superiority of conspecific faces and reduced inversion effect in face perception by a chimpanzee. *Folia Primatologica*, **61**, 110–4.

Tonooka, R. & Matsuzawa, T. 1995. Hand preferences of captive chimpanzees in simple reaching for food: analysis with manipulative patterns and development. *International Journal of Primatology*, **16**, 17–35.

Tonooka, R., Inoue, N. & Matsuzawa, T. 1994. Leaf-folding behavior for drinking water by wild chimpanzees at Bossou, Guinea: a field experiment and leaf selectivity. *Primate Research*, **10**, 307–13. (in Japanese with English summary)

de Waal, F. 1982. *Chimpanzee Politics: Power and Sex Among Apes*. New York: Harper & Row.

Ward, J. & Hopkins, W. (eds.) 1993. *Primate Laterality: Current Behavioral Evidence of Primate Asymmetries*. New York: Springer-Verlag.

Woodruff, G. & Premack, D. 1979. Intentional communication in the chimpanzee: the development of deception. *Cognition*, **7**, 333–62.

Yamakoshi, G. & Sugiyama, Y. 1995. Pestle-pounding behavior of wild chimpanzees at Bossou, Guinea: A newly observed tool-using behavior. *Primates*, **36**, 489–500.

- PART V

Apes compared

16 • Comparative positional behavior of the African apes

DIANE M. DORAN

INTRODUCTION

Studies of positional behavior, that is the locomotor and postural behavior of an animal, are generally undertaken in order to understand how variations in postcranial musculoskeletal anatomy are expressed in behavior. For students of human evolution, the positional behavior of the African apes is especially interesting because the defining adaptation in the transition from apes to hominids is a locomotor feature, the advent of bipedalism (see Susman, 1984 for further discussion). Thus, studies of African ape positional behavior are frequently an attempt to elucidate issues that have direct implication for studies of human evolution (e.g. Doran, 1992a, 1993a; Hunt, 1994).

Before the late 1960s, most locomotor information was gathered anecdotally by observers primarily interested in other aspects of primate behavior (e.g. Schaller, 1963; Reynolds & Reynolds, 1965; Goodall, 1968). These studies upset the then-currently held ideas that apes were morphologically adapted as brachiators, and thus, that early hominids evolved from brachiators (see Fleagle et al., 1981; Hunt, 1991 for review). Experimental studies of non-human primate locomotion led Fleagle et al. (1981) to hypothesize that a human ancestor primarily adapted for climbing would show a forelimb morphology similar to that previously associated with brachiation, and a hindlimb morphology that is morphologically and functionally preadapted for bipedalism. They concluded that a climbing theory of human ancestry was more consistent with both experimental and field data. However, at that time, there were few quantitative field data to evaluate their hypothesis.

In a separate issue, the discovery and subsequent study of early hominid fossils suggested a remarkable sex

difference in the degree of arboreality and locomotor behavior of *Australopithecus afarensis*. This finding prompted questions of whether intraspecific sex differences in body size and morphology are associated with sex differences in positional behavior and habitat use in extant species, and whether the proposed sex differences in the positional behavior of *A. afarensis* fell within the normal range of sex differences in the extant African apes. Again, there were no quantitative data available to address this question.

These two issues prompted a series of field studies, and since that time, long-term quantitative studies of positional behavior have been conducted on all of the African apes, including bonobos, *Pan paniscus* (Susman et al., 1980; Susman, 1984; Doran, 1992a), two subspecies of chimpanzees, *Pan troglodytes schweinfurthii* and *P. t. verus* (Hunt, 1992; Doran, 1993a; Doran & Hunt, 1994), and both mountain and western lowland gorillas, *Gorilla gorilla beringei* and *G. g. gorilla* (Tuttle & Watts, 1985; Remis, 1994).

There are now adequate data available to describe the locomotor and postural features that unite nearly all of the African apes, as well as to consider how positional behavior varies within and across most sites and species. Thus, the importance of climbing in apes and sex differences in positional behavior can now be more closely examined.

Several factors influence positional behavior including morphology, body size and ecological factors including habitat, seasonality and foraging strategies (see Doran, 1989, for review).

African apes are so similar in their postcranial anatomy (long forelimbs relative to hindlimbs, mobile shoulders, broad thorax with dorsally placed scapula, reduced number of lumbar vertebrae, and the absence of

Table 16.1. *Summary of species and subspecies studied*

	Pan paniscus ♀	Pan paniscus ♂	P.t. schweinfurthii (Mahale) ♀	P.t. schweinfurthii (Mahale) ♂	P.t. schweinfurthii (Gombe) ♀	P.t. schweinfurthii (Gombe) ♂	P.t. verus ♀	P.t. verus ♂	G.g. gorilla ♀	G.g. gorilla ♂	G.g. beringei ♀	G.g. beringei ♂
Species	Pan paniscus		Pan troglodytes						Gorilla gorilla			
Subspecies			P.t. schweinfurthii				P.t. verus		G.g. gorilla		G.g. beringei	
[a]Body size	33.2	45	33.2			43	?	?	71.5	169.5	97.7	159.2
[b-f]Site	Lomako		Mahale		Gombe		Taï		Bai Hokou		Karisoke	
Habitat	primary lowland rainforest		closed forest woodland		thicket woodland semideciduous forest		lowland evergreen rainforest		semideciduous rainforest		montane rainforest open canopy	
[b,g]Degree of arboreality	?	?	48%	33%	68%	37%	65%	49%	?	?	7%	2%
Sex difference in arboreality	probably		yes		yes		yes		probably		yes	
[b,c,e,g,i]Height used most often when above ground (meters)	?	?	5–20	<5	<5	<5	>20	>20	>20	>20	<7	<7
Sampling method and sample size	Locomotor bouts; 1615	993	2 minute instantaneous; 4175	6124	1234	1543	1 minute instantaneous; 6543	5975	1 minute instantaneous; 529	405	1 minute instantaneous; 7210	4470
[h,j-m]Day range	Wamba 2.0 km (not influenced by seasonality)		?	3.0 km	3.0 km	4.9 km	≈1.9 km	≈3.0 km	2.3 km		0.6 km	
[d,j-m]Home range	58 km² (with overlap)		19.4 km²		6.8 km² ♀	10.3 km² ♂ (year range)	27 km²		22.9 km² (with overlap) minimum estimte		9.4 km² (with overlap)	

?, Data not available.

Information is based on the following: [a] Junger & Susman, 1984; [b] Doran & Hunt 1994; [c] Collins & McGrew, 1988; [d] Boesch & Boesch, 1983; [e] Remis, 1994; [f] Watts, 1985; [g] Doran (in prep); [h] Doran 1989; [i] Tuttle & Watts, 1985; [j] Kano, 1992; [k] Goodall, 1986; [l] Watts, 1991; [m] Hasegawa, 1990.

a tail) that some authors consider them as size variants of a single type (Fleagle, 1988). Thus, one would expect to find broad similarities in African ape positional behavior.

The African apes are anatomically distinguished from each other primarily by differences in body size and in the degree of sexual dimorphism in body size (see Table 16.1). Gorillas are roughly three times larger than chimpanzees and bonobos (but see Jungers & Susman, 1984 for discussion of subspecific differences in chimpanzee body size and Zihlman, Chapter 21). The degree of sexual dimorphism in body size is much greater in gorillas than in chimpanzees. Chimpanzee (*Pan troglodytes schweinfurthii*) and bonobo males are 1.3 times larger than females on average, whereas western lowland male gorillas are nearly 2.5 times larger than females. Western lowland gorillas are also more sexually dimorphic in body size than are mountain gorillas, which are, on average, only 1.6 times larger than females. The degree of dimorphism in western lowland gorillas is so great that there is a much greater body size difference between male and female western lowland gorillas than between female gorillas and either male or female chimpanzees (see Table 16.1). Given that (a) sexual dimorphism in body size may be correlated with differences in locomotor and postural behavior (Doran, 1993a), and (b) that the African apes are morphologically similar, it is feasible to suggest that positional behavior differences may be greater between male and female western lowland gorillas than between female western lowland gorillas and either male or female chimpanzees.

Several authors have suggested that differences in primate body size are correlated with differences in locomotor behavior (Napier, 1967; Cartmill, 1974; Cartmill & Milton, 1977; Fleagle & Mittermeier, 1980; Fleagle, 1985). Generally, larger animals are more terrestrial, and when in an arboreal setting, they may use larger substrates or engage in different locomotor and postural activities than smaller animals (see Doran, 1993a for review). Thus, one would expect to find differences in the degree of arboreality both within and between species of African apes, with chimpanzees more arboreal than gorillas, females more arboreal than males in each species, and a striking sex difference in male and female western lowland gorilla arboreality.

Of course body size influences more than positional behavior. The influence of body size on diet is well documented (Bell, 1971; Jarman, 1974). Large animals have a lower basal metabolic rate, and although they require more total energy, they require less energy per kilogram of body mass than smaller animals (Kleiber, 1975). Large animals have longer guts, longer gut retention time and a higher coefficient of gut differentiation than smaller animals and thus can exploit lower quality diets than smaller animals (Chivers & Hladik, 1984; Milton, 1984). In primates an increase in body size is correlated with increased folivory, which is in turn correlated with decreased overall activity levels and reduced day ranges (see Oates, 1987 for review). Thus, differences in the African apes' body size and foraging strategies should also influence their positional behavior.

There are considerable differences in habitat and foraging strategy of the African apes. The greatest difference is between mountain gorillas and all other African apes. Mountain gorillas are folivorous and live in montane forests where fruits (and trees) are scarce. Although chimpanzees have a wider geographic distribution and are found in a wider variety of habitat types than are the other African apes, they are consistently primarily frugivorous across sites. Bonobos, restricted in their distribution to the lowland rainforest of Zaïre, are also primarily frugivorous, although they routinely incorporate more terrestrial herbaceous vegetation (THV) in their diet than do chimpanzees (Malenky & Wrangham, 1994; Wrangham et al., Chapter 4). Although the diet of the western lowland rain forest gorilla is less well-documented, there is strong indirect evidence to suggest that fruit is an important component of its diet (Williamson et al., 1990; Tutin et al., 1991; Tutin & Fernandez, 1993; Remis, 1994). Clearly these differences in habitat type and diet should also influence positional behavior.

STUDIES OF AFRICAN APE POSITIONAL BEHAVIOR

The results discussed in this paper are taken primarily from the work of three researchers: Kevin Hunt on chimpanzees in Tanzania (Mahale and Gombe); Melissa Remis on lowland gorillas in C.A.R. (Bai Hokou); and the author, bonobos in Zaïre (Lomako), chimpanzees in

Table 16.2. *Locomotor categories*

I.   Quadrupedalism (Quad): mode of locomotion employing all four limbs in a definable gait on horizontal or diagonal substrates and includes: knuckle-walking, tripedalism, palmigrade quadrupedalism, crutch walk and running.

II. Quadrumanous Climbing and Scrambling[a] (QC/Scram): mode of locomotion that uses hands and feet in vertical climbing or in varying combinations of unpatterned diverse gaits. All activities occur above substrate. Activities include: vertical climbing, fire slide, scrambling, bridging, tree-swaying and pull-up.

III. Suspensory Behavior (Susp): trunk is vertical and suspended below a substrate. Weight is borne by forelimbs. Activities include: arm-swinging or brachiating, dropping, and riding or crashing foliage to ground.

IV. Bipedalism (Biped): locomotion with weight borne on hindlimbs with trunk vertical. Includes bipedalism and aided bipedalism.

V.   Leaping and Diving (Le): includes leaping, diving and hopping.

VI. Miscellaneous (Misc): sommersaulting.

[a] Remis' locomotor categories do not include a separate leaping category, rather her category 'acrobatic' includes leaping plus bridging, tree-swaying and fire slide.
Source: modified after Susman 1984; Doran 1993b.

the Ivory Coast (Taï Forest), mountain gorillas in Rwanda (Karisoke) and western lowland gorillas (briefly) in Congo (Ndoki). The species and subspecies studied and differences in the sites where they were studied are summarized in Table 16.1. Results from these studies are comparable because methods used by the three researchers were similar and discussed prior to data collection. Postural and locomotor behaviors included in each category of behavior are listed in Table 16.2. The major difficulty in comparing results from these studies stems from the differing levels of habituation at the different sites. Chimpanzees and mountain gorillas were well habituated and could be followed throughout the day. Bonobos and western lowland gorillas were not habituated and generally could not be followed during terrestrial travel, resulting in observations of these species that are biased towards arboreal sightings. Thus, no measures of the degree of arboreality or overall postural and locomotor behavior can be made for bonobos and western lowland gorillas.

OVERVIEW OF AFRICAN APE POSITIONAL BEHAVIOR

Similarities in African ape overall positional behavior

The African apes are remarkably similar in their overall locomotor and positional behaviors (see Tables 16.3 and 16.4). They travel terrestrially between feeding and resting sites by quadrupedal knuckle-walking and generally more than 85% of their locomotor activity is quadrupedal (see Table 16.3). The next most frequently used locomotor behavior for all African apes, except mountain gorillas, is climbing, whereas suspensory locomotor behavior accounts for less than 1.5% of all locomotor activity. Thus, climbing is an important and relatively frequent locomotor activity common to all African apes.

The postural activity of the African apes is also very similar; sit and lie account for more than 90% of postural activity (see Table 16.4). Although chimpanzees use suspensory postures more frequently than mountain gorillas, suspensory activity accounts for very little of the overall postural activity of either chimpanzees or mountain gorillas (with the possible exception of female chimpanzees at Mahale).

Differences in African ape positional behavior

Degree of arboreality

The major differences in the positional behavior of the African apes are related directly to how arboreal each species or subspecies is. At each site studied, there is a sex difference in arboreality; females consistently spend more time than males above ground (see Table 16.1). Although chimpanzees differ in their degree of arboreality at different sites, all chimpanzees studied spend considerably more time in the trees than the mountain gorillas (see Table 16.1). This is not surprising given that mountain gorillas are terrestrial folivores, and that there are few trees in the habitat in which they were studied.

Although quantitative data on the degree of arboreality (the actual amount of time spent above

Table 16.3 *Summary of combined terrestrial and arboreal (overall) locomotion*

Species	Pan paniscus		Pan troglodytes						Gorilla gorilla			
Subspecies			P.t. schweinfurthii				P.t. verus		G.g. gorilla		G.g. beringei	
Site	Lomako		Mahale[a]		Gombe[a]		Tai[a]		Bai Hokou		Karisoke[b]	
	♀	♂	♀	♂	♀	♂	♀	♂	♀	♂	♀	♂
Quad	?	?	91.3	93.6	89.5	96.5	85.6	86.6	?	?	95.6	97.3
QC/Scram	?	?	7.7	5.1	8.9	3.5	10.9	11.1	?	?	2.7	0.7
Susp	?	?	0.9	0.8	0.5	–	1.4	1.1	?	?	0.2	–
Biped	?	?	0.2	0.3	1.1	–	1.2	1.2	?	?	1.4	1.7
Le	?	?	–	0.2	–		0.6	–	?	?	–	–
Misc									?	?	0.1	0.2
Number sampled	?	?	560	922	190	257	685	732	?	?	1101	747

Data from instantaneous sampling; ? indicates data not available.
Information is based on: [a] Doran & Hunt, 1994; [b] Doran (in prep.).

Table 16.4. *Summary of combined terrestrial and arboreal (overall) posture*

Species	Pan paniscus		Pan troglodytes						Gorilla gorilla			
Subspecies			P.t. schweinfurthii				P.t. verus		G.g. gorilla		G.g. beringei	
Site	Lomako		Mahale[a]		Gombe		Tai[b]		Bai Hokou		Karisoke[c]	
	♀	♂	♀	♂	♀	♂	♀	♂	♀	♂	♀	♂
Sit	?	?	72.6	79.6	?	?	80.6	70.8	?	?	76.4	70.4
Stand	?	?	2.1	1.2	?	?	5.6	6.1	?	?	5.3	7.7
Lie	?	?	18.6	16.6	?	?	11.6	22.1	?	?	18.2	21.9
Arm-hang	?	?	6.0	2.4	?	?	2.1	1.0	?	?	0.1	0.02
Number sampled	?	?	3506	4929	?	?	4367	4293	?	?	6607	4067

Data from instantaneous sampling; ? indicates data not available.
Information is based on: [a] Hunt, 1989; [b] Doran 1989; [c] Doran (in prep.).

ground versus on the ground) of bonobos and western lowland gorillas are not available, there are adequate qualitative data to rank confidently all the African apes on their degree of arboreality.

Bonobos are probably the most arboreal of the African apes (Fig. 16.1). Although they, like chimpanzees, travel primarily between feeding and resting sites by terrestrial knuckle-walking, they are capable of and may engage in some arboreal travel, unlike adult chimpanzees (Doran, 1993b). This behavioral difference is

also coupled with morphological differences (hands with more curved phalanges and, in males, a narrower scapula), which are consistent with the bonobos' increased arboreality compared with chimpanzees (Doran, 1993b). It is likely that there is a sex difference in bonobo arboreality, with males engaging in more terrestrial travel than females (Doran, 1993b; Fruth, personal communication).

Female chimpanzees are the next most arboreal of the African apes, followed by male chimpanzees (see Table

Fig. 16.1. Adult bonobo moves quadrupedally and arboreally on a bough at Lomako. (Photo by F. White)

16.1). There is considerable inter-site variation in how frequently chimpanzees are above ground and at what height. The Taï chimpanzees, lowland rainforest dwellers, are the most arboreal of chimpanzees studied to date. Taï males and females use the trees frequently and, when above ground, are most often at heights of greater than 20 meters (see Table 16.1). This contrasts with frequent use of much smaller trees at Gombe and Mahale. Sex difference in height use at Taï is not a result of a sex difference in feeding (although there are sex differences in diet), but rather females spend more time resting arboreally than males. This may be related to the females' smaller body size and smaller party sizes (often with infants) and resulting increased risk of predation. Clearly the degree of arboreality is related to many factors, including habitat type, diet, body size and predator pressure. An interesting question to consider is to what extent does large body size constrain arboreal behavior in gorillas? Gorillas frequently have been classified as the 'essentially terrestrial' African ape (in comparison with chimpanzees), on the basis of mountain gorilla studies (e.g. Reynolds, 1979). However, western lowland gorillas are more arboreal than mountain gorillas. They are routinely observed in trees at heights of up to 35 meters (e.g. Tutin & Fernandez, 1985; Kuroda, 1992; Remis, 1994). Although there are few quantitative data to

address this issue (it is still unclear how much of their day is spent in trees), qualitative accounts hint at interesting differences between mountain and western lowland gorillas. Remis (1994) has noted that even in the dry season, when fruits were largely unavailable, all gorillas continued to eat arboreal foods.

Given the large degree of sexual dimorphism in western lowland gorillas' body size, and the trend of greater male terrestriality seen in all African apes studied to date, it is likely that female western lowland gorillas are more arboreal than their male counterparts. Results from a preliminary study of western lowland gorillas, in which height was recorded at first sighting, suggest that this is the case (Doran, unpublished report). Arboreal sightings accounted for 43% of all female sightings ($N = 30$) and 19% of male sightings ($N = 16$). During sightings where both males and females were present (there was never more than one adult male present), the group was seen most frequently on the ground (54% of 13 sightings), although during 31% of these mixed-sex sightings females were arboreal (at heights greater than 20 m), whereas the silverback was on the ground ($N = 13$; no sighting included more than one male). Remis (1994) has noted a similar trend. This difference raises interesting questions about potential sex differences in gorilla foraging strategy. Given the western lowland gorilla females' greater arboreality and considerably smaller body size in comparison with males, are female gorillas more frugivorous than males? There is some evidence to suggest that males incorporate larger quantities of fibrous material in their diet than do females (Remis, 1994), although complete verification awaits further study.

Arboreal locomotor behavior

The African apes differ, not only in how frequently they are above ground, but also in the locomotor activities they perform when they are above ground (see Table 16.5). Chimpanzees show a remarkable similarity between the sexes and across sites in their activities. Mahale males and females, Gombe males and females and Taï females do not differ in the frequency of arboreal locomotor activities performed. However, Taï males differ significantly from the others in that they use less quadrupedalism, and more tree-swaying, aided bipedalism, bridging and climbing (Doran & Hunt, 1994). It is puzzling why Taï males moved differently in the trees

Table 16.5 *Arboreal locomotion*

Species	*Pan paniscus*		*Pan troglodytes*						*Gorilla gorilla*			
Subspecies			*P.t. schweinfurthii*				*P.t. verus*		*G.g. gorilla*		*G.g. beringei*	
Site	Lomako		Mahale[a]		Gombe[a]		Taï[a]		Bai Hokou[b]		Karisoke[c]	
	♀	♂	♀	♂	♀	♂	♀	♂	♀	♂ (in group)	♀	♂
Quad	44.4	26.1	31.2	31.2	31.1	31.2	30.3	11.7	16	21	47.9	68.2
QC/Scram	42.8	57.9	58.8	58.8	58.8	58.8	59.8	76.7	71.1	68.4	45.0	21.9
Susp	7.8	10.0	6.8	6.8	6.8	6.8	7.4	5.8	4.4	7.9	5.2	5.6
Biped	1.9	1.1	2.6	2.6	2.6	2.6	0.8	5.8	9	2.6	1.4	4.2
Le	1.3	4.8	0.5	0.5	0.5	0.5	1.6	–	–	–	–	–
Misc											0.5	–
Number sampled	468	993	192	192	192	192	122	103	45	38	118	35

Data from instantaneous sampling, except Lomako where locomotor bout sampling used.
Information is based on: [a] Doran & Hunt, 1994; [b] Remis, 1994; [c] Doran (in prep.).

compared with all other chimpanzees sampled. One possible explanation is that Taï males may be much heavier than the other chimpanzees studied, since the behavioral differences seen are those that one would predict if males were significantly larger than females. There are currently no body weight data from Taï available to address this issue.

Regardless of the explanation as to why Taï male chimpanzees differ so from the other chimpanzees, it is clear that there is a range of locomotor behavior within chimpanzees, largely based on their degree of arboreality. It would be interesting to test whether body size is the primary reason for the difference, by determining whether *Pan troglodytes troglodytes*, the heaviest of the three chimpanzees subspecies (Jungers & Susman, 1984), is more like Taï males rather than Taï females and Mahale and Gombe males and females in its locomotor behavior.

Both male and female bonobos differ significantly from chimpanzees in their arboreal locomotor behavior (see Table 16.5). Male bonobos are more suspensory and engage in more leaping than chimpanzees. Female bonobos are more quadrupedal and leap more, and climb and scramble less than chimpanzees.

Bonobos, at least in the Lomako Forest, actually travel occasionally between adjacent trees and along 'arboreal highways,' a behavior that is not seen in chim-

panzees. Of all arboreal travel locomotor bouts (as opposed to arboreal feeding locomotor bouts) sampled in the Lomako Forest, 27–30% were when exiting one tree and entering the next, and 24% for males and 44% for females were for travel along arboreal highways (Doran, 1989). Bonobos travel between trees by bridging, leaping and suspensory behavior, which accounts for their more frequent use of these activities compared with chimpanzees. In addition, female bonobos travel arboreally primarily by palmigrade quadrupedalism, which accounts for their high incidence of quadrupedalism during arboreal locomotion.

The most striking intersite difference of bonobo positional behavior is that at the Wamba (Zaïre) study site, bonobos are reported to travel exclusively on the ground (Kuroda, personal communication; personal observations). This difference is probably related to differences in habitat and in the degree of habituation of the study subjects. Although the Lomako and Wamba study sites are only separated by about 300 km, there are major habitat differences at the two sites. The Lomako site is situated at a distance of 35 km from the nearest road or village and consists largely of undisturbed primary forest. The Wamba study site comprises about 100 km², which surrounds and includes the five hamlets of the village of Wamba (Kano & Mulavwa, 1984). There is more secondary forest at Wamba than at Lomako and the area has been influenced greatly by human inhabitants, as is evidenced by the relative paucity of wildlife in comparison with Lomako. Thus, more primary forest with resultant different forest structure (more continuous arboreal routes) and possibly higher density of non-human predators (due to the undisturbed nature of forest with less human intervention) in the Lomako Forest may have some influence on the greater incidence of arboreal travel at Lomako Forest compared with Wamba.

It is interesting to note that Kuroda (personal communication) reported that prior to habituation at Wamba, bonobos fled from human observers by travelling in the trees. This suggests that the distances (up to 1200 m) travelled arboreally by Lomako bonobos may be an artifact of the lack of habituation. Although this is possible, it is notable that arboreal travel, when it occurred at Lomako, frequently occurred at a leisurely pace, after subjects fed and rested in one tree and then continued on to another tree. Usually, if unhabituated individuals flee human observers, they do not stop to feed. In one instance during the Lomako study, arboreal travel with interspersed feeding continued for 12 hours, during which the bonobos travelled arboreally 1200 meters without descending to the ground.

Thus, differences in forest structure and degree of habituation make it difficult to interpret the differences in ranging at the two sites. Clearly, these inter-site differences cannot be resolved without further study. However, it clearly points out a sharp contrast in the behavior of bonobos and chimpanzees. Whether habituated or not, no equivalent behavior of arboreal travel for forest-dwelling chimpanzees has been reported.

There is a sharp distinction between the arboreal locomotor behavior of mountain and western lowland gorillas. Mountain gorillas spend very little of their time (2% males, 7% females) above ground, and when above ground they are found at heights of less than 7 meters. The majority of their above-ground locomotion consists of pulling up onto large horizontal or slightly sloped substrates (fallen trunks) and then walking quadrupedally, although females engage in more pulling up and less quadrupedalism than males.

Western lowland gorillas' arboreal locomotion primarily occurs when climbing up into or down from food trees and between feeding sites within trees, with only occasional travel between adjacent trees before descending (Remis, 1994), a pattern which is similar to that described for chimpanzees in lowland forest. Lowland gorillas spend most of their (arboreal) locomotor time climbing, scrambling and walking quadrupedally (Remis, 1994). No statistical sex difference in behavior was noted, probably because of small sample sizes, since Remis (1994) noted trends of males engaging in more suspensory activity and less quadrupedalism and bipedalism than females. Remis (1994) also suggested that male arboreal locomotor behavior may be more constrained by body size than that of females.

Ranging patterns

Day range is influenced by several factors, including metabolic rate, group size (and biomass), and density and seasonal variation in food (reviewed in Oates, 1987). Average chimpanzee day ranges vary considerably both seasonally and between sites, and males, on average, have longer day ranges than females (see Doran, 1989, for review). Bonobo day ranges (mean of 2.0 km; Kano, 1992) appear to fall within the chimpanzee range although at the lower end of the range, which may be related to provisioning at Wamba. One would predict smaller sex differences in bonobos' (as compared with chimpanzees') day ranges because bonobo foraging parties include both males and females more frequently than is the case for chimpanzees.

There is also variation in reported day ranges for western lowland gorillas. Remis (1994) estimates western lowland gorillas' day range in C.A.R. to average 2.3 km/day, which is similar to that of chimpanzees and bonobos, and considerably greater than that of mountain gorillas (mean of 0.57 km, Watts, 1991). Tutin's (Chapter 5) measured day range in Lopé is 1.1 km/day on average, with groups travelling further when fruit was abundant than when fruit was scarce.

DISCUSSION

How important is climbing in apes?

The similarities in African apes' postcranial anatomy, and particularly those of the shoulder and forelimb, have historically been cited as morphological adaptations to suspensory locomotion, i.e. brachiation (reviewed in

Fleagle *et al.*, 1991). However, qualitative field studies did not provide support for frequent brachiation. Fleagle *et al.* (1981) reviewed the experimental electromyographic (EMG) data on shoulder muscular activity, and made a very convincing argument that climbing was the single activity that most strongly influenced the evolution of the anatomy of the shoulder and forelimb (and hindlimb) of apes.

In a study of chimpanzee and baboon positional behavior, Hunt (1991) revived and adapted the brachiationist theory, and proposed that rather than a suspensory *locomotor* activity (brachiation), it was a suspensory *postural* activity (arm-hanging) that was the primary selective force that shaped chimpanzee morphology. He based this conclusion on his finding that arm-hanging was a more frequent activity than vertical climbing in chimpanzees at Mahale and Gombe, and that it was the most distinctive shared postural mode of apes, constituting 5% or more of postural behavior for all apes (except mountain gorillas).

However, there are two major factors which suggest that this is not a viable hypothesis. First, EMG studies by Tuttle *et al.* (1979) indicate that the shoulder and forelimb muscles of great apes are largely inactive during passive suspension (and most active during hoisting), whereas studies summarized in Fleagle *et al.* (1981) demonstrate these muscles are active with high levels of muscle recruitment during climbing. Thus, it is difficult to argue that the large size and specialized morphology of ape shoulder muscles is a response to arm-hanging rather than climbing.

Second, Hunt's (1991) reported frequencies of arm-hanging (5.3–12.1% of all postural behavior) are not species-typical. Chimpanzees in the Taï forest climbed more than they engaged in arm-hanging; arm-hanging accounted for only 0.9–1.8% of total postural activity (Doran, 1989). Furthermore, no arm-hanging was reported in a study of chimpanzee behavior by Sabater Pi (1979, cited in Hunt, 1991). Results of all African ape behavior summarized in this paper show relatively infrequent suspensory postural activity (less than 2.5% of all postural activity, except for Mahale females). Climbing is a relatively frequent and important activity (for climbing into and out of food sources) in all of the African apes that consume fruit. Hunt (1994) stated that chimpanzees rarely ascend relatively large trunks. This

is not the case in the Taï forest, where chimpanzees frequently climb vertical trunks (at heights greater than 20 m) to ascend into fruit trees. The differences found in these two studies are most likely habitat-related, since Mahale and Gombe chimpanzees are found in more open habitats with much smaller trees than in the Taï forest. When Mahale and Gombe chimpanzees were above ground, they were found at much lower heights (<5 m vs. >20 m) than were Taï chimpanzees. This difference probably also explains the more frequent use of arm-hanging at Mahale, where it appears that chimpanzees use a combination of arm-hang and aided bipedalism, to feed in the terminal branches of small trees, a behavioral pattern that is not seen in the Taï Forest chimpanzees.

This broader survey of African ape positional behavior lends further support to the hypothesis that climbing, rather than brachiation (or arm-hanging) is the activity for which African ape shoulder and forelimb muscles evolved and are being maintained.

To what degree do African apes show sex differences in positional behavior?

All positional behavior studies of the African apes indicate clear sex differences in behavior, although this is much less distinctive in the overwhelmingly terrestrial (and cohesively grouped) mountain gorillas.

Female chimpanzees, western lowland gorillas (at least during the dry season) and probably female bonobos are more arboreal than their male counterparts. Reasons for this greater arboreality vary. Female chimpanzees in the Taï Forest did not spend more time feeding arboreally than male chimpanzees, but did spend more time resting arboreally than males, perhaps as a result of their smaller party sizes and increased risk of predation (Doran, 1993a). Female chimpanzees at Mahale and Gombe also spent more time resting arboreally than males, but also spent more time feeding arboreally as well, a trait which Hunt suggests is related to the females' lower social rank. It is likely that bonobo females may travel arboreally more frequently than males (Doran, 1993b). Thus, there is a clear trend of increased female arboreality among the African apes.

In addition to sex differences in arboreality, there are also sex differences in the frequency of arboreal activities performed and types of substrates used. During arboreal feeding at Taï, males and females use similar substrates (presumably related to the location of food in the trees) and perform different activities on them, whereas when resting arboreally, males and females choose different substrates in predictable fashion based on body size (Doran, 1993a). These results differ from those of Hunt, who found that males and females used different substrates when feeding (but did not indicate whether there were differences during resting).

Evolutionary implications of sex differences in African ape positional behavior

African apes are frequently used to construct models of early hominid behavior (e.g. Kinzey, 1987). Thus, it is interesting to consider the hypothesis that the two sexes of one of the earliest hominid ancestors, *Australopithecus afarensis*, 'partitioned their time between the trees and ground differently,' and that the heavier males spent more time on the ground than the females (Susman *et al.*, 1984), given the findings presented here of African ape positional behavior.

Morphological evidence (i.e. limb proportions, relative toe length, relative curvature of manual phalanges, etc.) has led to the suggestion that *Australopithecus afarensis* engaged in arboreal activity (Stern & Susman, 1983). Based on an apparently remarkable degree of sexual dimorphism in body size, one would predict sex differences in the positional behavior of *A. afarensis*.

Chimpanzees, one of the closest, if not the closest living relative of hominids, have less morphological and body size sexual dimorphism in comparison with *Australopithecus afarensis*, and yet are sexually distinct in their degree of arboreality, distance travelled per day, arboreal locomotor activity and substrate choice. Female chimpanzees travel less distance each day and, at least in the Taï Forest where leopard predation on chimpanzees has been documented (Boesch, 1991), spend more resting time above ground than males. Results summarized here also indicate sex differences in the arboreality of western lowland gorillas, bonobos, and to a lesser extent, mountain gorillas. Therefore, based on these findings, it seems likely that there were sex differences in the arboreal locomotor behavior of *A. afarensis*.

Although it has been shown before that primate species of differing body size differ in locomotor and postural activities (Fleagle & Mittermeier, 1980; Crompton, 1984), these studies clearly demonstrate that body size differences within a species can also result in positional behavior differences. These findings may influence how we interpret sex differences in body size of extinct species.

ACKNOWLEDGMENTS

This paper was originally prepared for the Wenner-Gren Conference 'The Great Apes Revisited,' held November 12–19, 1994 in Cabo San Lucas, Baja Mexico. I thank Bill McGrew and Toshisada Nishida for their invitation to take part in the conference. The field studies were made possible by the governments of the Republics of Congo, Ivory Coast, Rwanda and Zaïre. I gratefully acknowledge the Dian Fossey Gorilla Fund, Christophe and Hedwige Boesch, Randy Susman, Richard Malenky and Nancy Thompson Handler. Generous financial support was provided for this work by the L. S. B. Leakey Foundation, the Wenner-Gren Foundation for Anthropological Research and the National Geographic Society.

REFERENCES

Bell, R. H. 1971. A grazing ecosystem in the Serengeti. *Scientific American*, **225(1)**, 153–70.

Boesch, C. 1991. The effects of leopard predation on grouping patterns in forest chimpanzees. *Behaviour*, **117**, 220–42.

Boesch, C. & Boesch, H. 1983. Optimisation of nut-cracking with natural hammers by wild chimpanzees. *Behaviour*, **83**, 265–86.

Cartmill, M. 1974. Pads and claws in arboreal locomotion. In: *Primate Locomotion*, ed. F. A. Jenkins, pp. 45–84. New York: Academic Press.

Cartmill, M. & Milton, K. 1977. The lorisiform wrist joint and the evolution of brachiating adaptations in Hominoidea. *American Journal of Physical Anthropology*, **47**, 249–72.

Chivers, D. J. & Hladik, C. M. 1984. Diet and gut morphology in primates. In: *Food Acquisition and Processing in Primates*, ed. D. J. Chivers, B. A. Wood & A. Bilsborough, pp. 213–30. New York: Plenum Press.

Collins, D. A. & McGrew W. C. 1988. Habitats of three groups of chimpanzees (*Pan troglodytes*) in western Tanzania compared. *Journal of Human Evolution*, **17**, 553–74.

Crompton, R. H. 1984. Foraging, habitat structure, and locomotion in two species of Galago. In: *Adaptations for Foraging in Nonhuman Primates*, ed. P. S. Rodman & J. G. Cant, pp. 73–111. New York: Columbia University Press.

Doran, D. M. 1989. *Chimpanzee and Pygmy Chimpanzee Positional Behavior*. Ann Arbor, MI: University Microfilms International.

Doran, D. M. 1992a. The ontogeny of chimpanzees and pygmy chimpanzee locomotor behavior: a case study of paedomorphism and its behavioral correlates. *Journal of Human Evolution*, **23**, 139–57.

Doran, D. M. 1992b. A comparison of instantaneous and locomotor bout sampling methods: a case study of adult male chimpanzee locomotor behavior and substrate use. *American Journal of Physical Anthropology*, **89**, 85–99.

Doran, D. M. 1993a. Sex differences in adult chimpanzee positional behavior: the influence of body size on locomotion and posture. *American Journal of Physical Anthropology*, **91**, 99–115.

Doran, D. M. 1993b. The comparative locomotor behavior of chimpanzees and bonobos: the influence of morphology on locomotion. *American Journal of Physical Anthropology*, **91**, 83–98.

Doran, D. M. & Hunt, K. D. 1994. The comparative locomotor behavior of chimpanzees and bonobos: species and habitat differences. In: *Chimpanzee Cultures*, ed. R. W. Wrangham, W. C. McGrew, F. B. M. de Waal & P. G. Heltne, pp. 93–108. Cambridge, MA: Harvard University Press.

Fleagle, J. G. 1985. Size and adaptation in primates. In: *Size and Scaling in Primate Biology*, ed. W. L. Jungers. pp. 1–19. New York: Plenum Press.

Fleagle, J. G. 1988. *Primate Adaptation and Evolution*. New York: Academic Press.

Fleagle, J. G. & Mittermeier, R. A. 1980. Locomotor behavior, body size and comparative ecology of seven Surinam monkeys. *American Journal of Physical Anthropology*, **52**, 301–14.

Fleagle, J. G., Stern, J. T., Jungers, W. J., Susman, R. L., Vangor, A. K. & Wells, J. P. 1981. Climbing: a biomechanical link with brachiation and with bipedalism. *Symposia of the Zoological Society of London*, **48**, 359–75.

Goodall, J. 1968. The behaviour of free-living chimpanzees in the Gombe Stream Reserve. *Animal Behaviour Monographs*, **1**, 161–311.

Goodall, J. 1986. *The Chimpanzees of Gombe: Patterns of Behaviour*, Cambridge, MA: Harvard University Press.

Hasegawa, T. 1990. Sex differences in ranging patterns. In: *The Chimpanzees of the Mahale Mountains*, ed. T. Nishida, pp. 99–114. Tokyo: University of Tokyo Press.

Hunt, K. D. 1989. *Positional Behavior in Pan troglodytes at the Mahale Mountains and the Gombe Stream National Parks, Tanzania*. Ann Arbor, MI: University Microfilms International.

Hunt, K. D. 1991. Positional behavior in the Hominoidea. *International Journal of Primatology*, **12**, 95–118.

Hunt, K. D. 1992. Positional behavior of *Pan troglodytes* in the Mahale Mountains and Gombe Stream National Parks, Tanzania. *American Journal of Physical Anthropology*, **87**, 83–105.

Hunt, K. D. 1994. The evolution of human bipedality: ecology and functional morphology. *Journal of Human Evolution*, **26**, 183–202.

Jarman, P. J. 1974. The social organization of antelope in relation to their ecology. *Behaviour*, **48**, 215–67.

Jungers, W. L. & Susman, R. L. 1984. Body size and skeletal allometry in African apes. In: *The Pygmy Chimpanzee*, ed. R. L. Susman. pp. 131–77. New York: Plenum Press.

Kano, T. 1992. *The Last Ape*. Stanford, CA: Stanford University Press.

Kano, T. & Mulavwa, M. 1984. Feeding ecology of the pygmy chimpanzee (*Pan paniscus*) of Wamba. In: *The Pygmy Chimpanzee*, ed. R. L. Susman, pp. 233–74, New York: Plenum Press.

Kinzey, W. G. (ed.) 1987. *The Evolution of Human Behavior: Primate Models*. Albany, NY: State University of New York Press.

Kleiber, M. 1975. *The Fire of Life*. New York: Kriger.

Kuroda, S. 1992. Ecological interspecies relationships between gorillas and chimpanzees in the Ndoki-Nouabale Reserve, Northern Congo. In: *Topics in Primatology*, *Vol. 2*, ed. N. Itoigawa, Y. Sugiyama, G. Sackett & R. K. R. Thompson, pp. 385–94, Tokyo: University of Tokyo Press.

Malenky, R. K. & Wrangham, R. W. 1994. A quantitative comparison of terrestrial herbaceous food consumption by *Pan paniscus* in the Lomako Forest, Zaïre and *Pan troglodytes* in the Kibale Forest, Uganda. *American Journal of Primatology*, **32**, 1–12.

Milton, K. 1984. The role of food-processing factors in primate food choice. In: *Adaptations for Foraging in Nonhuman Primates*, ed. P. S. Rodman & J. G. Cant, pp. 249–79, New York: Columbia University Press.

Napier, J. R. 1967. Evolutionary aspects of primate

locomotion. *American Journal of Physical Anthropology*, **27**, 333–42.

Oates, J. F. 1987. Food distribution and foraging behavior. In: *Primate Societies*, ed. B. B. Smuts, D. C. Cheney, R. M. Seyfarth, R. W. Wrangham & T. T. Struhsaker, pp. 197–210. Chicago: University of Chicago Press.

Remis, M. J. 1994. *Feeding Ecology and Positional Behavior of Western Lowland Gorillas* (Gorilla gorilla gorilla) *in the Central African Republic*. Ann Arbor, MI: University Microfilms International.

Reynolds, V. 1979. Some behavioral comparisons between the chimpanzee and the mountain gorilla in the wild. In: *Primate Ecology: Problem-Oriented Field Studies*, ed. R. W. Sussman, pp. 323–39, New York: John Wiley & Sons.

Reynolds, V. & Reynolds, F. 1965. Chimpanzees of the Budongo Forest. In: *Primate Behavior*, ed. I. DeVore, pp. 368–424, New York: Holt, Rinehart and Winston.

Schaller, G. 1963. *The Mountain Gorilla*. Chicago: University of Chicago Press.

Stern, J. T. & Susman, R. L. 1983. The locomotor anatomy of *Australopithecus afarensis*. *American Journal of Physical Anthropology*, **60**, 279–317.

Susman, R. L. 1984. The locomotor behavior of *Pan paniscus* in the Lomako Forest. In: *The Pygmy Chimpanzee*, ed. R. L. Susman, pp. 369–98, New York: Plenum Press.

Susman, R. L., Badrian N. L. & Badrian A. J. 1980. Locomotor behavior of *Pan paniscus* in Zaïre. *American Journal of Physical Anthropology*, **53**, 69–80.

Susman, R. L., Stern, J. T. & Jungers, W. L. 1984. Arboreality and bipedality in the Hadar hominids. *Folia Primatologica*, **43**, 113–56.

Tutin, C. E. G. & Fernandez, M. 1985. Foods consumed by sympatric populations of *Gorilla g. gorilla* and *Pan t. troglodytes* in Gabon: some preliminary data. *International Journal of Primatology*, **6**, 27–43.

Tutin, C. E. G. & Fernandez, M. 1993. Composition of the diet of chimpanzees and comparisons with that of sympatric lowland gorillas in the Lopé Reserve, Gabon. *International Journal of Primatology*, **30**, 195–211.

Tutin, C. E. G., Fernandez, M., Rogers, M. E., Williamson, E. A. & McGrew, W. C. 1991. Foraging profiles of sympatric lowland gorillas and chimpanzees in the Lopé Reserve, Gabon. *Philosophical Transactions of the Royal Society, Series B*, 179–86.

Tuttle, R. H. & Watts, D. P. 1985. The positional behavior and adaptive complexes of *Pan gorilla*. In: *Primate Morphophysiology, Locomotor Analyses and Human Bipedalism*, ed. S. Kondo, pp. 261–88, Tokyo: University of Tokyo Press.

Tuttle, R. H., Basmajian J. V. & Ishida, H. I. 1979. Activities of pongid thigh muscles during bipedal behavior. *American Journal of Physical Anthropology*, **50**, 123–36.

Watts, D. P. 1985. Relations between group size and composition and feeding competition in mountain gorilla groups. *Animal Behaviour*, **33**, 72–85.

Watts, D. P. 1991. Strategies of habitat use by mountain gorillas. *Folia Primatologica*, **56**, 1–16.

Williamson, E. A., Tutin, C. E., Rogers, M. E. & Fernandez, M. 1990. Composition of the diet of the lowland gorillas at Lopé in Gabon. *American Journal of Primatology*, **21**, 265–77.

17 • Nest building behavior in the great apes: the great leap forward?

BARBARA FRUTH AND GOTTFRIED HOHMANN

INTRODUCTION

Over decades apes have served as either referential or conceptual models in attempts to reconstruct the path of human evolution (Ghiglieri, 1987; Wrangham, 1987). In the search for behavioral traits shared by all members of the great apes, few have turned out to be conservative, that is, common features seen in all extant hominoids, and by inference present in our common ancestor. Of these shared traits, skilled object manipulation has been of great interest in comparative analyses as a basic criterion for hominization. Tool use and tool production, however, vary tremendously not only among the four species but also within a single species. Thus the trait in common is not tool use itself, but the general ability for environmental problem solving (McGrew, 1992). Nest building is part of this ability. It is probably the most pervasive form of material skill in apes. Whether or not this trait should be considered as tool use is much disputed (Goodall, 1968; Alcock, 1972; Beck, 1980; Galdikas, 1982).

Nest building is called 'bed building' by some investigators (Itani, 1979; Hiraiwa-Hasegawa, 1986). It is a daily habit of weaned great apes to build a place in which to rest. The technique employed depends on the site and on the available materials. Orangutans, chimpanzees and bonobos start their arboreal constructions by preparing a foundation of solid sidebranches or forks, bending, breaking and inter-weaving sidebranches crosswise. They complete the structure by bending most of the smaller twigs in a circular pattern over the rim. Detached twigs are added for lining (Davenport, 1967; Goodall, 1968; Horn, 1980). Gorillas, who more often establish terrestrial nests, build similarly. They gather mostly herbaceous materials by pulling, bending and breaking stems to arrange them around and under their bodies. They concentrate on the rim rather than on a foundation (Schaller, 1963). Our knowledge of ape nest building comes from numerous descriptions from early expeditions and dates to the past century (Du Chaillu, 1861). Some of these early descriptions were a misinterpretation of function and use (Garner, 1896), but most studies reported the morphology and use of these constructions for orangutans, gorillas and chimpanzees in accurate descriptions (Hornaday, 1879; Savage & Wyman, 1843). Yerkes & Yerkes (1929, p. 220) summarized the observations on ape nests done by the various explorers of the past century as follows: 'No phase of the mode of life [of the chimpanzee] and no behavior pattern has attracted more attention or produced more useful literature than that of nest construction.'

Beginning with ape-focused field research, nests continued to be an indicator of the subjects' presence, but behavioral descriptions lagged behind. Since apes mostly sleep at night, nest building has been eclipsed by the day-to-day observations on social behavior. With some exceptions (Schaller, 1963; Goodall, 1968; MacKinnon, 1974), nest construction at night and nest leaving in the morning became more the curfew times of behavioral observation than a topic of research itself. Thus, even today, our knowledge of nest building behavior and its social implications is limited. Instead, census methods have been developed that use nests as a way to estimate population density (Ghiglieri, 1984; Tutin & Fernandez, 1984). Height and age of nests, and sometimes species of trees used for nest construction, have often been noted. These have been compared with availability and thus have led to a more comprehensive knowledge of the environmental factors that influence the choice of raw material, nest sites, and locations. One aspect rarely discussed concerns the evolutionary implications of nest building behavior. Yerkes & Yerkes (1929, p. 564) showed great foresight when they noted that:

> The observed range of variation is from almost complete lack of tendency and capacity for construction of nest or bed to their definite presence in the anthropoid apes and their relatively high development in man. This comparison is peculiarly significant because nesting behaviour illustrates the appearance and phylogenetic development of constructivity, and, coincidentally, the transition from complete dependence on self-adjustment to increasing dependence on manipulation or modification of environment as a method of behavioral adaptation.

This paper tries to revive the thinking of more than 65 years ago by considering nest building as one of the crucial steps in hominoid evolution. The chapter's title indicates our attempt to pick up the thread and to pay the attention to this behavioral pattern that it deserves.

We formulate and present a provocative hypothesis: nest building is not only properly placed within the realm of tool use, but it is also the original tool that led to the mental and physical ability to use the tools we see today.

Given space limitation here, a comprehensive review of studies on nest building in apes will be published elsewhere (Fruth, in preparation). This review includes data provided by previous studies as well as unpublished data gathered by a questionnaire distributed in December 1993. Here we present: (a) a summary of the review in detailed tables; (b) a comparison of the results that will attempt to filter the species-specific features and discuss them according to their adaptive or cultural implications; (c) an evolutionary scenario, in which we speculate on how nest building could have been developed and established; and (d) some ideas on the implications nest building may have had for the evolution of apes' mental capacities.

METHODS

The data for interspecific comparison come from previous publications (Tables 17.2–5). To update these publications, responses to the questionnaire (Q) are summarized in Tables 17.6–8. Unless noted, data on bonobos are from our 29-month study at Lomako (1990–94). Interspecific comparison is based on up to 20 variables that include morphological, ecological, metrical and behavioral features. To help assess the data, length of study and type of investigation are indicated. Special terms are defined as follows:

Custom.   Special habit within one population or study group, e.g. rain cover.

Nest group.   All nests built in close proximity in one night.

Nest site.   Place in the home range where the nest group is built. This includes type of habitat, topography and characteristic features, e.g. proximity to water or food resources.

Nest location.   Position of the nest in terms of vertical and horizontal distribution, e.g. ground vs. tree, top of the crown vs. bottom of the crown, etc.

Reuse.   Abandoned nests used again by the same or another individual on consecutive nights or even later. This term excludes a nest made by an individual, left briefly for defecation and then reoccupied.

RESULTS AND SUMMARY

Table 17.1 gives a detailed description of the abbreviations used in Tables 17.2–17.8. Tables 17.2–5 review published information on great ape nests. Tables 17.6–8 report the results from the questionnaire.

COMPARISON

Table 17.9 summarizes the main results for the four species. Nest building behavior in the four species of great apes appears to be rather uniform throughout all levels of comparison.

Morphology

All four species build both day and night nests. Orangutan and gorilla males may rarely spend the night without a nest. In orangutans, chimpanzees and bonobos, nest sites are not distributed randomly throughout the habitat but are chosen selectively. In gorillas, distribution of nest sites appears to be arbitrary. The degree of selectivity of nest sites seems to depend on the variation in

Table 17.1. *Abbreviations used in (a) Table 17.2–17.5 and (b) Table 17.6–17.8*

(a)

Type of habitat	F, forest; R, rain; P, primary; S, secondary; E, evergreen; M, montane; L, lowland; G, grassland; W, woodland; SV, savanna; G, gallery; PL, plantation (sometimes combined)
Duration	Duration of the study in months
Direct/indirect observation	+, yes; −, no
Nests	Sample sizes of day nests/night nests; ?, information not quantified; −, no data
Height	Height of nests in meters; numbers indicate the range; bracketed numbers the height of most nests
Choice of material	Opportunistic or selective
No. of species used	?, information not quantified; −, no data
Feeding trees	Feeding trees used for day nests (DN) or night nests (NN) or both
Type of construction	Number indicates the number of construction types distinguished; ?, information not quantified; −, no data
Ontogeny	Information available: +, yes; −, no
Time of construction	In minutes for one nest
Reuse	Number indicates how often reuse has been observed; percentage (%) restricts reuse to either day or night nests (DN/NN); yes, reuse happens but not quantified; −, no data
Activities in nest	See Table 17.9
Nest groups	Sample size of groups in study: +, data available; ?, information not quantified; −, no data
Group size	Numbers indicate range; bracketed numbers the median; +, data not specified
Nearest neighbor	Nearest distances between nests in meters

(b)

Type	Day nests (DN) or night nests (NN)
Integrated trees	Maximal number of integrated trees per nest; where no trees are used, nests are made of herbaceous vegetation
Percent. int. nests	Percentage of nests that are integrated
Nest group size vs. day-travel party	Size of night nest group (NNGR) versus size of day-travel party (DTP); possible categories: smaller, same, larger

habitat structure: the more variation, the more evident selectivity towards specific site features. Data from Lopé show that western lowland gorillas choose secondary forest more often than expected, while chimpanzees choose primary forest (Tutin *et al.*, 1995). Here, selectivity may be a result of niche differentiation between the sympatric species. Gorillas also differ from the other apes in the location of nests. They mostly sleep on the ground, while the three other species sleep in trees. Analyzing the height of tree nests, gorillas are again the only species to construct nests, on average, below 10 m, while the three other apes usually build between 10 and 20 m. No quantitative data of availability and choice of material were at our disposal for orangutans or for most gorilla and chimpanzee subspecies. By pooling descriptive material with the few data available, however, we conclude that in general, orangutans, chimpanzees and bonobos are highly selective toward specific tree species for nest construction, while gorillas seem to be opportun-

istic. Feeding trees or plants are used for both day and night nests in all species. The only difference was whether or not feeding trees bearing ripe fruit were used for nest construction. While this behavior is reported for day nests of all species, chimpanzees and bonobos seem to be reluctant to construct night nests within those trees. However, orangutans do so. In all species, proximity to feeding trees is reported to be close. Concerning the type of construction, differences emerge at the species level: with few exceptions, orangutans build nests in single trees, while chimpanzees have been reported to use two trees in 5–10% of all cases. With the exception of Lilungu, bonobos integrate trees in over 30% of all cases. Sometimes they use up to seven trees for a single nest, something never reported for the three other species. For herbaceous ground nests, gorillas integrate stems of different plants in over 50% of all cases. If nests are constructed within trees however, the rate is below 5% and the maximum of integrated trees is three. We

Table 17.2. *Published reports of orangutans*

	Borneo (*Pongo pygmaeus pygmaeus*)						Sumarate (*Pongo pygmaeus abelii*)		
Reference	Schaller, 1961	Yoshiba, 1964	Davenport, 1967	Rodman, 1979	Galdikas, 1975, 1982	MacKinnon, 1974	Mackinnon, 1974	Rijksen, 1978	Sugardjito, 1983
Study site	Sarawak	NE Sabah	Sepilok, Sabah	E. Kalimantan	Tanjung Puting-NP, Kalimantan	Ulu Segama, Sabah	Ranun River, N. Sumatra	Ketambe, N. Sumatra	Ketambe, N. Sumatra
Country	Indonesia	Malaysia	Malaysia	Indonesia	Indonesia	Malaysia	Indonesia	Indonesia	Indonesia
Type of habitat	PF	PF	PF	PF	PF	PF	PF	PF	PF
Years	1960–61	1963	1964	1970–71	1971–80	1968–70	1971	1971–74	1980–82
Duration	3	2	7	15	108	18	8	38	6
Study's goal	Survey	Survey	Behavior	Ecology, behavior	Behavior	Ecology, behavior	Ecology, behavior	Behavior	Ecology, behavior
Direct observation	+	–	+	+	+	+	+	+	+
Indirect observation	+	+	+	–	–	+	+	+	+
Nest morphology									
Nests	1/228	–/614	?/28	–/?	?/?	?/510	?/?	?/?	?/172
Height	4–40 (15–25)	4–34 (15–30)	–	–	–	4–30 (15–30)	–	13–15	≤30
Choice of material	–	–	–	–	–	selective	–	–	–
No. of species used	–	–	–	–	–	–	–	–	–
Feeding trees	DN/NN	DN/NN	DN/NN	DN/NN	DN/NN	DN/NN	?/?	DN/–	DN/NN
Type of construction	–	?	?	–	–	?	?	?	–
Nest building									
Info. on ontogeny	–	–	–	–	+	+	–	–	–
Time of construction	–	–	6.5 min	–	<5 min	2–3 min	–	–	–
Reuse	–	–	50% of DN	–	R	8	–	yes	–
Activities within nest	R/I	–	R/E	–		R/P/G	R/P/G	R/P/E	–
Nest groups									
Nest groups	–	?	–	?	?	?	–	?	–
Group size	–	3	–	1–5	≤5	1–3	–	1–2	–
Nearest neighbor	–	–	–	–	–	–	–	–	–

Table 17.3. *Published reports of gorillas*

	Mountain (Gg beringei)			Eastern lowland (Gg graueri)			Western lowland (Gg gorilla)			
Reference	Donisthorpe, 1958	Kawai & Mizuhara, 1959	Schaller, 1963	Casimir, 1979	Yamagiwa, 1983	Mwanza et al., 1992	Groves & Sabater Pi, 1985	Remis, 1993	Williamson, 1988	Tutin et al., 1995
Study site	Kisoro Virungas	Kisoro Virungas	7 diff. areas	Kahuzi Biega	Kahuzi Biega	Itebero		Bai Hokou, Dzanga-S.	Lopé	Lopé
Country	Uganda	Uganda	Uganda, Zaïre	Zaïre	Zaïre	Zaïre	Equ. Guinea	C.A.R.	Gabun	Gabun
Type of habitat	MRF	MRF	MRF	MPF	MPF	PF/SF	PF	PF/SF	PF	PF
Year	1956	1959	1959–61	1979	1978–79	1987–90	1963–69	1990–92	1984–85	1983–94
Duration	8	3	18	15	7	10	18	27	16	48
Study's goal	Pilot study	Ecology behavior	Survey, behavior	Ecology, behavior	Survey, ecology	Survey, ecology	Behavior	Ecology, behavior	Ecology	Survey, ecology
Nest morphology										
Direct observation	+	+	+	+	+	–	+	–	+	–
Indirect observation	+	+	+	+	+	+	+	+	+	+
Nests	?/225	?/365	22/3012	–/964	–/171	–/375	?/448	–/1231	24/748	–/2435
Height	–	0–16	0–15	0–15	–	0–30	0–15	0–30	0–16	0–35
Choice of material	Opportunistic	Opportunistic	Opportunistic	Opportunistic	–	Opportunistic	Selective	Selective	Opportunistic	Selective
No of species used	?	?	?	50	–	–	22	–	38	98
Feeding trees	DN/NN	DN/NN	DN/NN	DN/NN	?/?	?/?	DN/NN	?/?	?/?	DN/NN
Type of construction	3	–	5	2	–	–	4	3	8	7
Nest building										
Ontogeny	–	–	+	–	–	–	–	–	–	
Time of construction	–	–	5 min	–	–	–	–	–	–	
Reuse	0	1	–	–	–	–	12	–	0	
Activities in nest	–	R	R/E	–	–	–	R/E	–	–	
Nest groups										
Nest groups	–	36	400	63	58	83	49	163	113	373
Group size	3–12	4–15	(16, 9)	(15)	(14, 3)	–	2–16	(8)	1–19	(7)
Nearest neighbor	–	0–50	0–38.5	0–40	–	–	1.5–15	–	–	–

Table 17.4. *Published reports of chimpanzees*

	Eastern (*Pan troglodytes schweinfurthii*)	Central (*Pan troglodytes troglodytes*)		Western (*Pan troglodytes verus*)		
Reference	Goodall, 1968	Baldwin *et al.* 1981	Wrogemann 1992	Baldwin *et al.* 1981	Anderson *et al.*, 1983	Marchesi *et al.*, 1995
Study site	Gombe		Lopé	Mt Assirik	Sapo	Nationwide
Country	Tanzania	Equ. Guinea	Gabon	Senegal	Liberia	Ivory Coast
Type of habitat	GWF	PF/SF	PF	SV/GF	PF	PF/SF/PL
Year	1960–67	1963–69	1988–89	1976–79	1982	1989–90
Duration	?	?	16	40	2	15
Study's goal	Behavior	Behavior, ecology	Behavior, ecology	Ecology	Survey	Survey
Direct observation	+	+	+	−	−	−
Indirect observation	−	+	+	+	+	+
Nest morphology						
Nests	?/384	−/195	−/523	−/252	−/67	?/611
Height	0–25	2–40	2–45	2–44	6–20	−
Choice of material	opportunistic	−	selective	−	−	−
No. of species used	−	−	45	−	−	−
Feeding trees	DN/−	?/?	DN/NN	?/?	?/?	?/?
Type of construction	?	?	?	?	−	−
Nest building						
Info. on ontogeny	+	−	−	−	−	−
Time of construction	1–5 min	−	−	−	−	−
Reuse	20×	−	−	−	−	−
Activities within nest	R	−	−	−	−	−
Nest groups						
Nest groups	?	66	68	83	?	182
Group size	1–17 (2–6)	1–12 (2)	1–26 (1)	1–18 (4)	1–10 (1)	1–22 (2–5)
Nearest neighbor	−	4	6	4	4	−

suggest that the higher frequency of integrated nests in bonobos is a response to higher sociality expressed also by larger parties at night (Fruth & Hohmann, 1994). If individuals seek to increase proximity of night nests, then the combination of adjacent trees may compensate for the lack of suitable nest sites. Again, with the exception of the gorilla, customs are reported from all four species. However, variation within a species exceeds variation across species and thus should be considered on the cultural level rather than on that of species-specificity.

Behavior

Ontogeny is similar in all species. Day nest construction by immatures reaches a peak between the first and second year after birth in orangutans and gorillas respectively, and in the third year in chimpanzees. Infants of orangutans and gorillas construct their own night nests slightly earlier (between 3 and 4 years of age) than do chimpanzees and bonobos. Time of construction for a nest varies between 1 and 7 minutes. Time of utilization is similar for all species and is influenced in the same way by season, weather and light conditions. All apes usually rest from dusk to dawn. If light is reduced by rain or clouds, they may leave their nests later in the morning and start to build new ones earlier in the evening. Reuse is reported for all populations but with different frequencies. Orangutans seem to reuse nests most often, followed by chimpanzees in dry habitats, then gorillas and bonobos. Reuse of nests by chimpanzees is mostly a question of habitat and availability of suitable nest material; the drier the habitat the more often reuse

Table 17.5. *Published reports of bonobos*

	Pan paniscus					
Reference	Horn, 1980	Badrian & Badrian, 1977	Fruth & Hohmann, 1993, in prep.	Kano, 1983	Kano, 1992	Sabater Pi & Veá, 1990
Study site	Lake Tumba	Lomako	Lomako	Yalosidi	Wamba	Lilungu
Country	Zaïre	Zaïre	Zaïre	Zaïre	Zaïre	Zaïre
Type of habitat	PF	PF	PF	G/PF	PF	PF
Years	1972–74	1974–75	1990–94	1973–75	1974–86	1988–89
Duration	24	11	31	3.5	*c.* 70	13
Study's goal	Behavior, ecology	Behavior, ecology	Behavior, ecology	Ecology	Behavior, ecology	Ecology
Direct observation	+	+	+	+	+	−
Indirect observation	+	+	+	+	+	+
Nest morphology						
Nests	?/107	−/174	164/1156	19/2380	637/3357	−/611
Height	0–25	5–34	3–50	0–50	0–50	2–36
Choice of material	Opportunistic	−	selective	selective	selective	−
No of species used	−	26	24	103	108	?
Feeding trees	?/?	DN/NN	DN/−	?/?	DN/NN	?/?
Type of construction	−	−	5	5	5	−
Nest building						
Info. on ontogeny	−	−	+	−	−	−
Time of construction	−	−	1–7 min	−	0–7 min	−
Reuse	−	−	0.2% of NN	−	−	yes
Activities within nest	−	−	R/P/G/S/E/T	−	R/G/S/P	−
Nest groups						
Nest groups	−	−	266	−	−	−
Group size	−	−	1–25 (17.5)	−	−	−
Nearest neighbor	−	−	0–86 (9.8)	−	−	−

occurs. (In drier habitats trees are scarcer, semideciduous and show a much lower rate of regeneration than in wetter habitats; nest raw materials are therefore limited.) Although a variety of associated activities occur in the context of nest building as shown in Tables 17.2–17.8, all species use nests for similar purposes. Differences on the species level may refer to the frequency of different activities (Fig. 17.1). Sex differences are reported for all species, mostly involving frequency of nest building. In chimpanzees and bonobos, females construct day nests more often than males. Orangutan and gorilla males build night nests less often. Height of females' nests is on average higher in orangutans, gorillas and bonobos. While orangutan and bonobo females start nest construction at dusk, gorilla females wait until the silverback has chosen the place and has built the first nest. Detailed investigations on further sex differences are available only for bonobos. Thus, interspecific comparison awaits further investigations. Feces are dropped outside the nest by orangutans, chimpanzees and bonobos. Gorillas often defecate in their nests, but there appear to be population differences. Whether or not nests are fouled has been discussed with reference to the fruit content in diet and has its roots at the environmental rather than the species-specific level.

Nest groups

Only recently have primatologists started to consider nest groups, instead of single nests, and to be aware that a nest is not constructed independently from others (Schaller, 1963; Goodall, 1968; Fruth & Hohmann, 1994). The data for nest groups seem to reflect the differences in social organization and are thus a good mirror

Fig. 17.1. Adolescent female (right) begs for chewed fruit from an adult female bonobo at Lomako. (Photo by G. Hohmann)

of species-specificity. While the semisocial orangutans hardly ever show accumulations of nests, chimpanzees show a wide range, but favor resting in small parties. Gorillas usually stay together as whole groups, while bonobos often congregate to form large overnight parties. Nearest neighbor analyses suggest that interindividual proximity during the day is reflected in night nest association. The more familiar the individuals, the closer the night nests. However, with the exception of gorillas, patterns within the group have not yet been analyzed. In gorillas, nest position is related to age and sex classes. Data on bonobos suggest, however, that not only social organization is reflected (e.g. females built higher than males) but also social structure (e.g. low-ranking individuals nested on the periphery).

Population-specific differences

All species show interpopulational differences in nest building behavior. Orangutans use fruiting trees for night nesting more often on Borneo than on Sumatra; western lowland gorillas build more often in trees than do mountain gorillas; chimpanzees at Gombe and Guinea

Bissau build nests in oil palms (*Elaeis guineensis*) but others do not; bonobos at Lilungu do not integrate more than two trees, while others do. Intraspecific differences may indicate the range of ecological adaptation, but considering the variety of ape habitat and differences in social organization and diet, nest building appears to be surprisingly uniform on the interspecific level. Therefore, it is likely that nest building has developed under evolutionary constraints and is thus a long-standing trait for all the great apes.

EVOLUTION OF NEST BUILDING BEHAVIOR: A SCENARIO

Nest building with its basic features is shared by the four species of great apes. Because of this activity's persistence and uniformity among these closely related species, it is improbable that nest building developed independently four times during evolution. When was it invented and why?

Searching for a common ancestor brings us to the great ape stock in the early Miocene (Martin, 1990). At that time eight to ten species of comparatively small apes

Table 17.6. *Gorilla questionnaire*

Questionnaire	Gorilla gorilla beringei	Gorilla gorilla graueri		Gorilla gorilla gorilla		
Reference	Fletcher, unpubl.	Yamagiwa, publ. partly	Williamson, unpubl.	Tutin, publ. partly	McFarland, unpubl.	Williamson, unpubl.
Study site	Virunga NP	Kahuzi Biega NP	Kahuzi Biega outside NP	Lope NP	Cross River State	Dja Reserve
Country	Rwanda, Zaïre	Zaïre	Zaïre	Gabon	Nigeria	Cameroon
Type of habitat	Hagenia/MF.	MF/PLF	SF/PLF	PF	??	LRF
Year	1989–92	1978–94	1994	1983–94	1993	1994
Direct observation	+	+	−	−	−	−
Indirect observation	+	+	+	+	+	+
Nest morphology						
Type	DN/−	DN/NN	DN/NN	−/NN	−/NN	DN/NN
Location in forest	Undergrowth	Undergrowth	Lower canopy	middle canopy	Lower canopy	Undergrowth
Position in crown	−	Bottom	Middle	−	−	−
Height	<10	<10	<10	<10	<10	<10
Choice of species	−	Opportunistic	−	Opportunistic	−	−
Feeding plants/trees	?/?	DN/NN	?/?	DN/NN	?/?	?/?
Integrated trees	no	3	4 shrubs	3	2	no
Percent. int. nests	>50%	<5%	10–30%	<5%	>50%	−
Nest building						
Info. on ontogeny	+	+	−	−	−	−
Time of construction	−	1–2 min	−	−	−	−
Reuse	Yes	0	−	Yes	−	−
Activities within nest	R/P/G/E	R/P/E	−	−	−	−
Nest groups						
Nest groups	?	+	+	+	+	+
Group size	−	5–20	5–10	+	2–5	5–10
Nest group size versus day-travel party	Same	Same	Same	Same	Same	Same

inhabited large forested parts of Africa and Arabia. The Miocene shows us the greatest species diversity as well as inferred association patterns of apes. Up to five species inhabited early Miocene forests occupying different strata of the canopy (Andrews, 1987). During the middle Miocene 11 to 13 species were known; they covered a larger geographical area than that of the early Miocene, and some had increased in body size. Their size ranged from that of small Old World monkeys to female gorillas. These apes radiated in size as they adapted to the different strata of the forest. Miocene apes were predominantly frugivorous and differences in body size reduced competition between closely related species (MacKinnon, 1977). Some species became more terrestrial, a pattern that opened up a new ecological niche. Larger body size also may have been favored as a result of selec-

tion pressure by terrestrial predators. Thus, some apes also succeeded in interspecific competition with other ground-dwelling mammals.

Monkeys were uncommon during the early Miocene, but expanded by the middle Miocene. From three unspecialized species of cercopithecoids during the early Miocene, monkeys had increased to 14 species by the late Miocene. While the colobines specialized in the abundant food source of leaves, cercopithecines adapted to the frugivorous niche. Both colobines and cercopithecines became better adapted to this diet than did the hominoids by evolving a greater tolerance to plant secondary compounds. The cercopithecoids' capacity to feed on fruits with high tannin or alkaloid content allowed them to consume fruit before it ripened. This adaptation helped during periods of food shortage

Table 17.7. *Chimpanzee questionnaires*

Questionnaire	Pan troglodytes schweinfurthii				Pan troglodytes troglodytes		Pan troglodytes verus					
Reference	Nishida, unpubl.	Moore, unpubl.	Wrangham, publ. partly	Yamagiwa, publ. partly	Wrogemann, unpubl.	Williamson, unpubl.	Sugiyama, publ. partly	McGrew, publ. partly	Boesch, unpbl.	Moore, publ. partly (south-west)	Alp, unpubl.	Pavy, unpubl.
Study site	Mahale	Ugalla	Kanyawara	Kahuzi-Biega	Lopé	Dja Reserve	Bossou	Mt Assirik	Taï	Mali	Tenkere	Bafing area
Country	Tanzania	Tanzania	Uganda	Zaïre	Gabon	Cameroon	Guinea	Senegal	Ivory Coast	Mali	Sierra Leone	Mali
Type of habitat	GWF	W	MRF	MF/PLF	PLRF	LF	SF	SV	PF	G/W	SV/W/F	W/F
Year	1966–94	1985–94	1987–94	1987–94	1988–89	1994	1975–94	1976–79	1979–94	1984	1989–94	1992
Direct observation	+	–	+	–	–	–	+	+	+	–	+	+
Indirect observation	+	+	+	+	+	+	+	+	–	+	+	+
Nest morphology												
Type	DN/NN	DN/NN	DN/NN	DN/NN	DN/NN	DN/NN	DN/NN	DN/NN	DN/–	DN/NN	–/NN	DN/NN
Location in forest	Middle canopy	Middle canopy and monolayer	–	Lower and middle canopy	Middle canopy	Lower canopy	Middle canopy	Upper canopy	–	Canopy	Middle canopy	Middle canopy
Position in crown	Bottom	No preference	–	Middle and bottom	Middle	Bottom	Top and bottom	Top	–	–	Top	Middle
Height	10–20	10–20	–	10–20	10–20	<10	10–20	10–20	–	<10	10–20	10–20
Choice of species	Selective	Selective	–	Opportunistic	Selective	Selective	Selective	Selective	–	–	–	Selective
Feeding trees	DN/NN	DN/NN	?/?	DN/NN	DN/NN	?/?	DN/NN	DN/NN	?/?	?/?	DN/NN	DN/NN
Integrated trees	2	2	–	2	2	3	3	4	–	?	2	–
Percent. int. nests	<5%	?	–	5–10%	5–10%	30–50%	5–10%	5–10%	–	<5%	<5%	–
Nest building												
Ontogeny	+	–	–	–	–	–	+	–	–	–	–	–
Time of construction	–	–	–	>2 min	–	–	3–5 min	4 min	–	–	3–4 min	–
Reuse	Yes	Yes	0	Yes	Yes	–	Yes	Yes	Yes	–	Yes	Yes
Activities in nest	R/P/G/S/E	–	R/P/G/E	R	R/P	–	R/P/G	–	–	–	R/G/E	–
Nest groups												
Nest groups	?	+	+	+	+	+	+	+	–	+	+	–
Group size	?	2–5	2–10	1–5	1	1	>10	2–5	–	2–5	5–10	–
Nest group size vs. day-travel party	Smaller	?	Smaller	Smaller	?	?	Same	Same	–	?	?	–

Table 17.8. *Bonobo questionnaires*

Questionnaire	*Pan paniscus*		
Reference	Kano, publ	Sabater Pi, publ.	Fruth, publ. partly
Study site	Wamba	Lilingu	Lomako
Country	Zaïre	Zaïre	Zaïre
Type of habitat	EPF	SF	EPF
Year	1973–94	1988–90	1990–94
Direct observation	+	+	+
Indirect observation	+	+	+
Nest morphology			
Type	DN/NN	DN/NN	DN/NN
Location in forest	Middle canopy	Middle canopy	Middle canopy
Position in crown	Middle	Top	Middle
Height	10–20	10–20	10–20
Choice of species	Selective	Selective	Selective
Feeding trees	DN/ –	?/?	DN/ –
Integrated trees	>5	2	>5
Percent. int. nests	30–50%	<5%	30–50%
Nest building			
Info. on ontogeny	–	–	+
Time of construction	–	–	1–7 min
Reuse	Yes	0	Yes
Activities within nest	R/P/G/S/E	R/P	R/P/G/S/E/T
Nest groups			
Nest groups	+	–	+
Group size	>10	–	1–25
Nest group size vs. day-travel party	Same	–	Larger

(Wrangham, 1980), but also put them in competition with large frugivorous mammals like apes. The increase in monkey species' diversity became directly proportional to the decrease in ape diversity (Andrews, 1987), which may indicate the importance of the adaptation to tolerate plant secondary compounds. Extrapolating the extinction rate during the Miocene, apes 'should' have become extinct 3.2 million years ago (Andrews, 1987, fig. 1). However, some of them have survived. How did they manage to compete successfully against the better adapted monkeys?

One answer to this selection pressure could be the flexibility of ape social organization (Di Fiore & Rendall, 1994). Another answer could be adaptation to new ecological niches. Gorillas may have specialized in superabundant food and chimpanzees may have started nut cracking. A third answer, however, could be the ability

to monopolize food resources – not solely by body size, group size and physical power – but by the construction of nests.

Nest construction was not 'invented' all at once in ape evolution. It probably occurred as a byproduct of feeding habits.

Miocene apes came at least occasionally to the ground. With the exception of the orangutan, extant great apes spend at least half of their waking lives on the ground. They feed on fruit that ripens on the ground (e.g. *Parinari, Autranella, Irvingia*; Fig. 17.2). In a party of several individuals they search for ripe fruit, carry as much as they can by hand or mouth, and sit elsewhere, placing all of the fruit in a little heap in front of them. Seated at that one spot, they consume their concentration of 'prey' (see also Hunt, 1994). Perhaps this was the '*proto-nest?*'

Other fruit species ripen in trees. Most fruits and sometimes the most tasty ones are found on the peripheral branches. These distal twigs are often inaccessible for other fruit-eaters. Apes bend and break these branches towards their body, sitting on the more solid, proximal parts toward the trunk (Wrangham, 1975; Rjiksen, 1978; personal observations). Thus their hands are free to reach and to grasp what is wanted. Finally they end up sitting on the end of a limb enhanced by broken branches. This feeding spot may result in a solid platform, a '*feeding nest.*'

Both the 'proto-nest' on the ground and the 'feeding nest' in the tree allow food consumption in a secure and comfortable place and may have been the predecessors of sleeping nests. All modern great apes build such feeding nests. Some apes even remain in feeding trees overnight. In orangutans, nests in feeding trees have been reused. The largest numbers of nests observed in one tree are for day nests within feeding trees. One could argue that the apes who survived were those large enough to bend and break branches to open a niche inaccessible to many others. Additionally, they were able to monopolize and defend food resources against competitors, if necessary, even overnight, by their presence at the resource. Observations of bonobos suggest that the nests built in the feeding context sometimes marked a property for the nest builder and left the 'owner' undisturbed by the approach of other group members (Fruth & Hohmann, 1993). From this scenario onwards

Table 17.9. *Comparison of nest building in the great apes*

Comparison	Orangutan	Gorilla	Chimpanzee	Bonobo
Nest sites				
Distribution	Discontinuous	Arbitrary	Discontinuous	Discontinuous
Choice	Selective	Opportunistic	Selective	Selective
Nests				
Type	DN/NN	DN/NN	DN/NN	DN/NN
Frequency of nest construction	Daily, adult males sometimes not	Daily, sometimes not	Daily, all weaned individuals	Daily, all weaned individuals
Location	In trees only	Predominantly on the ground, in trees also	NN: in trees only DN: sometimes on the ground	NN/DN: predominantly in trees
Height	10–20 m	< 10 cm	10–20 m	10–20 m
Choice of material	Selective	Opportunistic (selective)	Selective	Selective
Feeding plants or fruiting trees used for	DN/NN	DN/NN	DN/–	DN/–
Type of construction	Mostly one tree, foundation, rim, lining	Mostly mixed plants, if tree then single, (foundation) rim	Mostly one tree, foundation, rim, lining	Often several trees, foundation, rim, lining
Custom	Roof, raincover	Soil and turf	Oil palm	(Rain) cover
Behavior/Ontogeny				
1st attempts	12 months	8 months	8 months	No analyzed data
DN peak	2 years	1½ years	3 years	
Own NN	3–4 years	3–4 years	4–5 years	
Time to construction	2–3 min	1–5 min	1–5 min	1–7 min
Period of use	Whole night	Whole night	Whole night	Whole night
Reuse	Often	Rarely	Rarely–often	Rarely
Activities	R/P/I/G/O	R/E/P/G	R/E/P/S/G/B/I	R/E/P/S/G/I/T
Sex differences	F: higher F: more often F: before M	F: more often above ground F: after silverback M	F: more DN than M	F: higher, longer, earlier, more elaborate than M
Nest group				
Entity	Party (subgroup)	Group	Party	Party
Nests per group Range (M)	1–5 (1)	1–2 (4–7)	1–26 (2–5)	1–25 (5–13)
Nearest neighbor Range (M)	Same tree	1–10 m	4–6 m	4.5 m
Population				
Differences	Sociality Group size	Feces Selectivity	Reuse Group size	Integration Ground nests

Abbreviations used: DN, day nest; NN, night nest; F, female; M, male; (M), median; m, meter; min, minutes; R, rest; P, play; G, groom; S, sex; I, ill or wounded; T, taboo; B, birth; E, eat; O, other.

we should consider the creation of these restricted zones ('taboo-nests') not as a highly derived evolutionary trait, but as the original use of day nests. Defense of food patches shifted from the level of interspecific or intergroup competition, to the level of interindividual competition.

Most modern great apes construct night nests outside fruit-bearing trees. Rjiksen (1978, p. 151) suggested that this may be to avoid the risk of agonistic encounters with other species, which seems to contradict our hypothesis. However, Rjiksen paid particular attention to man as the threatening species, and he did not discuss nocturnal fruit-eating competitors such as fruit bats (Pteropodidae). Small nocturnal mammals may be successful with or without apes occupying the tree. There-

Fig. 17.2. Adolescent male bonobo eats *Irvingia* fruit on the ground at Lomako. (Photo by G. Hohmann)

fore apes may prefer to sleep apart, not to avoid conflicts but to avoid a restless night in the midst of active frugivores. However, apes may stay close enough to reoccupy the tree early in the morning and consequently prevent other competitors such as monkeys from entering the tree. Proximity to feeding trees may thus be a result of the calculation of costs (sleeplessness, loss of food) and benefits (sleep, gain of food). It may change with the quality and interspecific popularity of the resource.

THE GREAT LEAP FORWARD: SPECULATIONS ON THE IMPLICATIONS OF NEST BUILDING FOR HOMINID EVOLUTION

Group living has been discussed either as an adaptation to cooperative defense of food trees (Wrangham, 1980, 1983) or as an adaptation to predator pressure (van Schaik, 1983). At least in *Pan paniscus*, an additional factor may have influenced group size and cohesion. Bonobos who forage in different parts of the home range often gather at night nest sites. The next morning they depart in a specific direction, often in parties of a differ-

ent composition from that of the day before. We suggest that aggregation at night may facilitate transfer of information on the quality of food patches visited during the day (Hohmann & Fruth, 1994). Accordingly, we consider nest groups as potential 'information-centers' (Ward & Zahavi, 1973).

In addition to the proximate functions of nest building suggested above, we draw attention to a possibly ultimate function: the connection of nest building and sleeping behavior. Is it possible that nests have permitted a different and better quality of sleep and if so, what implications may that kind of sleep have had on the cognitive evolution of the hominoids?

First, nests may have improved sleep by providing better thermoregulation or by the potential for more relaxed and comfortable sleeping postures. Second, apes constructed nests at places less accessible to predators; thus nests may have provided safer sleep. With improved security during sleep, the need for vigilance decreased. Allison & Van Twyver (1970) demonstrated that the proportion of different characteristics of sleep change with the need for vigilance during sleep. 'Secure sleepers' spent more time in REM sleep (rapid eye movement) than 'insecure sleepers'. REM or active sleep is of great importance for maturation of the fetal nervous system. In adults REM sleep seems related to periods of quiet sleep (QS) (QS = SWS + LQS; SWS = slow wave sleep, LQS = light quiet sleep; Meddis, 1975). SWS is associated with metabolic processes such as the release of growth hormones. It accounts for about 25% of adult human sleep. Meddis (1975) contrasts traditional theories on the function of sleep, namely sleep as physiological recuperation, with his theory of 'sleep the immobiliser' (p. 680). Recent investigations on REM and SWS have shown that both phases of sleep combine to establish memory (Wilson & McNaughton, 1994; Karni *et al.*, 1994). Accordingly, REM sleep is thought to repeat practiced tasks, while SWS is responsible for the transfer of information to the cortex resulting in long-term memory. If we apply this to apes' sleep, we may better understand the increase in learning abilities that evolved from monkeys to apes.

Summarizing the previous paragraphs, we conclude: nest building is a phylogenetically conservative behavior that must have evolved in the Miocene. Since we do not have fossil nests from that epoch, the evolutionary scenario mentioned above remains highly speculative. Increase in body size and special patterns of food consumption have been considered as preadaptations for nest building. First, these feeding techniques led to the ability to monopolize food. They presented a tool to compete successfully with other frugivorous animals. Second, nests allowed apes, despite their increased body size, to sleep in places inaccessible to ground predators. Third, nest building had implications for facilitating the evolution of cognitive abilities within hominoids.

Nests were not only the first products of exploratory behavior with twigs and sticks, but also the foundation for all future tool use ability. Nest building may have promoted higher levels of tool use that helped to open up new food resources. Nests may thus have been the cradle for higher cognition, manipulation and technological skills, culminating in human abilities for environmental alteration. Thus, the nest served as the spring board for the great leap forward in hominid evolution.

ACKNOWLEDGMENTS

This paper was originally prepared for the Wenner-Gren conference 'The Great Apes Revisited,' held November 12–19, 1994 in Cabo San Lucas, Baja California, Mexico. BF thanks W. C. McGrew, T. Nishida and the Wenner-Gren Foundation for the opportunity to take part in the conference and the editors for the invitation to contribute to this volume. We thank I. Eibl-Eibesfeldt, G. Neuweiler and D. Ploog for constant support, encouragement and advice. Special thanks to all respondents to the questonnaire: R. Alp, C. Boesch, A. Fletcher, K. McFarland, W. McGrew, J. Moore, T. Kano, T. Nishida, J.-M. Pavy, J. Sabater Pi, Y. Sugiyama, C. Tutin, E. Williamson, R. Wrangham, D. Wrogemann, and J. Yamagiwa. L. Marchant, W. McGrew, J. Moore and A. Zihlman made thoughtful suggestions and comments on a previous draft. LFM and WCM improved the manuscript with their careful editing; all their help is gratefully acknowledged. Fieldwork in Zaïre was enabled by the German Embassy (Kinshasa), the Catholique Mission, the 'Ministère de la Recherche Scientifique et Technologie,' Kinshasa, and the 'Centre de Recherche en Sciences Naturelles,' Lwiro. Financial support was provided by the German Academic Exchange Service (DAAD), the German Research Council (DFG), the Max-Planck-Society (MPG) and a private donor. Thanks to T. Parish, F. Salter, T. Van't Hof, and P. Wiessner for correcting the English text.

REFERENCES

Alcock, J. 1972. The evolution of the use of tools by feeding animals. *Evolution*, **26**, 464–73.

Allison, T. & Van Twyver, H. 1970. The evolution of sleep. *Natural History*, **79**, 56–65.

Anderson, J. R., Williamson, E. A. & Carter, J. 1983. Chimpanzees of Sapo Forest, Liberia: density, nests, tools and meat-eating. *Primates*, **24**, 594–601.

Andrews, P. 1987. Species diversity and diet in monkeys and apes during the Miocene. In: *Primate Evolution and Human Origins*, ed. R. L. Ciochon & J. G. Fleagle, pp. 194–204. New York: Aldine de Gruyter.

Badrian, A. & Badrian, N. 1977. Pygmy chimpanzees. *Oryx*, **13**, 463–8.

Baldwin, P. J., Sabater Pi, J., McGrew, W. C. & Tutin C. E. G. 1981. Comparisons of nests made by different populations of chimpanzees. *Primates*, **22**, 474–86.

Beck, B. B. 1980. *Animal Tool Behavior*. New York: Garland STPM.

Casimir, M. J. 1979. An analysis of gorilla nesting sites of the Mt Kahuzi region (Zaïre). *Folia Primatologica*, **32**, 290–308.

Davenport, R. K. 1967. The orangutan in Sabah. *Folia Primatologica*, **5**, 247–63.

Di Fiore, A. & Rendall, D. 1994. Evolution of social organization: a reappraisal for primates by using phylogenetic methods. *Proceedings of the National Academy of Sciences of the United States of America*, **91**, 9941–5.

Donisthorpe, J. H. 1958. A pilot study of the mountain gorilla (*Gorilla gorilla beringei*) in south west Uganda, February to September, 1957. *South African Journal of Science*, 195–217.

Du Chaillu, P. 1861. *Explorations and Adventures in Equatorial Africa*. London: John Murray.

Fruth, B. & Hohmann, G. 1993. Ecological and behavioral aspects of nest building in wild bonobos (*Pan paniscus*). *Ethology*, **94**, 113–26.

Fruth, B. & Hohmann, G. 1994. Comparative analyses of nest building behavior in bonobos (*Pan paniscus*) and chimpanzees (*Pan troglodytes*). In: *Chimpanzee Cultures*, ed. R. W. Wrangham, W. C. McGrew, F. B. M. de Waal & P. G. Heltne, pp. 109–28. Cambridge, MA: Harvard University Press.

Galdikas, B. M. F. 1975. Orangutans, Indonesia's 'people of the forest'. *National Geographic*, **148**, 444–73.

Galdikas, B. M. F. 1982. Orang-utan tool use at Tanjung Puting reserve, central Indonesian Borneo (Kalimantan Tengah). *Journal of Human Evolution*, **10**, 19–33.

Garner, R. L. 1896. *Gorillas and Chimpanzees*. London: Harpers.

Ghiglieri, M. P. 1984. *The Chimpanzees of Kibale Forest: A Field Study of Ecology and Social Structure*. New York: Columbia University Press.

Ghiglieri, M. P. 1987. Sociobiology of the great apes and the hominid ancestor. *Journal of Human Evolution*, **16**, 319–57.

Goodall, J. 1968. The behaviour of free-living chimpanzees in the Gombe Stream Reserve. *Animal Behaviour Monographs*, **1**, 163–311.

Groves, C. P. & Sabater Pi, J. 1985. From ape's nest to human fix-point. *Man*, **20**, 22–47.

Hiraiwa-Hasegawa, M. 1986. Sex differences in the behavioral development of chimpanzees at Mahale. In: *Understanding Chimpanzees*, ed. P. G. Heltne & L. A. Marquardt, pp. 104–15. Cambridge, MA: Harvard University Press.

Hohmann, G. & Fruth, B. 1994. Structure and use of distance calls in wild bonobos (*Pan paniscus*). *International Journal of Primatology*, **15**, 767–82.

Horn, A. D. 1980. Some observations on the ecology of the bonobo chimpanzee (*Pan paniscus*, Schwarz 1929)

near Lake Tumba, Zaïre. *Folia Primatologica*, **34**, 145–69.

Hornaday, W. T. 1879. On the species of bornean orangs, with notes on their habits. *Proceedings of American Association for the Advancement of Science*, **28**, 438–55.

Hunt, K. D. 1994. The evolution of human bipedality: ecology and functional morphology. *Journal of Human Evolution*, **26**, 183–202.

Itani, J. 1979. Distribution and adaptation of chimpanzees in an arid area. In: *The Great Apes*, ed. D. A. Hamburg and E. R. McCown, pp. 54–71. Menlo Park, CA: Benjamin/Cummings.

Kano, T. 1983. An ecological study of the pygmy chimpanzees (*Pan paniscus*) of Yalosidi, Republic of Zaïre. *International Journal of Primatology*, **4**, 1–31.

Kano, T. 1992. *The Last Ape*. Stanford, CA: Stanford University Press.

Karni, A., Tanne, D., Rubenstein, B. S., Askenasy, J. J. M. & Sagi D. 1994. Dependence on REM sleep of overnight improvement of a perceptual skill. *Science*, **265**, 679–82.

Kawai, M. & Mizuhara, H. 1959. An ecological study on the wild mountain gorilla (*Gorilla gorilla beringei*). *Primates*, **2**, 1–42.

MacKinnon, J. 1974. The behaviour and ecology of wild orang-utans (*Pongo pygmaeus*). *Animal Behaviour*, **22**, 3–74.

MacKinnon, J. 1977. A comparative ecology of Asian apes. *Primates*, **18**, 747–72.

Marchesi, P., Marchesi, N., Fruth, B. & Boesch, C. 1995. Census and distribution of chimpanzees in Côte D'Ivoire. *Primates*, **36**, 591–607.

Martin, R. D. 1990. *Primate Origins and Evolution. A Phylogenetic Reconstruction*. London: Chapman & Hall.

McGrew, W. C. 1992. *Chimpanzee Material Culture: Implications for Human Evolution*. Cambridge: Cambridge University Press.

Meddis, R. 1975. On the function of sleep. *Animal Behaviour*, **23**, 676–91.

Mwanza, N., Yamagiwa, J., Yumoto, T. & Maruhashi, T. 1992. Distribution and range utilization of eastern lowland gorillas. In: *Topics in Primatology, Vol. 2, Behavior, Ecology, and Conservation*, ed. N. Itoigawa, Y. Sugiyama, G. P. Sackett & R. K. R. Thompson, pp. 283–300. Tokyo: University of Tokyo Press.

Remis, M. J. 1993. Nesting behavior of lowland gorillas in the Dzanga-Sangha Reserve, Central African Republic: implications for population estimates and understandings of group dynamics. *Tropics*, **2**, 245–55.

Rjiksen, H. D. 1978. *Sumatran Orang Utans* (Pongo pygmaeus abelii *Lesson 1827) Ecology, Behaviour and Conservation*. Wageningen: H. Veenman and B. V. Zonen.

Rodman, P. S. 1979. Individual activity patterns and the solitary nature of orangutans. In: *The Great Apes*, ed. D. A. Hamburg & E. R. McCown, pp. 234–55. Menlo Park, CA: Benjamin/Cummings.

Sabater Pi, J. & Veá, J. J. 1990. Estudio Eto-Etologico del Chimpance Bonobo (*Pan paniscus*) de la Region de Lokofe-Lilungu-Ikomaloki (Dist. de Ikela) Zaïre. Informe Preliminar. Universitat de Barcelona.

Savage, T. S. & Wyman, J. 1843. Observations on the external characters and habits of the *Troglodytes niger*, Geoff. and on its organization. *Boston Journal of Natural History*, 4, 362–86.

van Schaik, C. P. 1983. Why are diurnal primates living in groups? *Behaviour*, 87, 120–44.

Schaller, G. B. 1961. The orang-utan in Sarawak. *Zoologica*, **46**, 73–82.

Schaller, G. B. 1963. *The Mountain Gorilla*. Chicago: University of Chicago Press.

Sugardjito, J. 1983. Selecting nest sites of Sumatran orang-utans, *Pongo pygmaeus abelii*, in the Gunung Leuser National Park, Indonesia. *Primates*, **24**, 467–74.

Tutin, C. E. G. & Fernandez M. 1984. Nationwide census of gorilla (*Gorilla g. gorilla*) and chimpanzee (*Pan t. troglodytes*) populations in Gabon. *American Journal of Primatology*, **6**, 313–36.

Tutin, C. E. G., Parnell, R. J., White, L. J. T. & Fernandez, M. 1995. Nest building by lowland gorillas in the Lopé Reserve, Gabon: environmental influences and implications for censusing. *International Journal of Primatology*, **16**, 53–76.

Ward, P. & Zahavi, A. 1973. The importance of certain assemblages of birds as 'information-centres' for food-finding. *Ibis*, **115**, 517–34.

Williamson, E. A. 1988. Behavioural ecology of western lowland gorillas in Gabon. Ph.D. thesis, University of Stirling.

Wilson, M. A. & McNaughton, B. L. 1994. Reactivation of hippocampal ensemble memories during sleep. *Science*, **265**, 676–9.

Wrangham, R. W. 1975. The behavioural ecology of chimpanzees in Gombe National Park, Tanzania. Ph.D. thesis, University of Cambridge.

Wrangham, R. W. 1980. An ecological model of female-bonded primate groups. *Behaviour*, **75**, 262–300.

Wrangham, R. W. 1983. Ultimate factors determining social structure. In: *Primate Social Relationships*, ed. R. A. Hinde, pp. 255–62. Oxford: Blackwell Scientific Publications.

Wrangham, R. W. 1987. The significance of African apes for reconstructing human social evolution. In: *The Evolution of Human Behavior: Primate Models*, ed. W. G. Kinzey, pp. 51–71. Albany, NY: State University of New York Press.

Wrogemann, D. 1992. Wild chimpanzees in Lopé, Gabon: census method and habitat use. Ph.D. thesis, University of Bremen.

Yamagiwa, J. 1983. Diachronic changes in two eastern lowland gorilla groups (*Gorilla gorilla graueri*) in the Mt Kahuzi region, Zaïre. *Primates*, **24**, 174–83.

Yerkes, R. M. & Yerkes, A. W. 1929. *The Great Apes*. New Haven, CT: Yale University Press.

Yoshiba, K. 1964. Report of the preliminary survey on the orang-utan in North-Borneo. *Primates*, **5**, 11–26.

18 • Comparative studies of African ape vocal behavior

JOHN C. MITANI

INTRODUCTION

The close phylogenetic relationship between the African apes and humans has been known to Western science for well over a century. In comparing the differences between the African apes and humans, Thomas Henry Huxley (1863, p. 123) commented:

> It would be no less wrong than absurd to deny the existence of this chasm, but it is at least equally wrong and absurd to exaggerate its magnitude, and, resting on the admitted fact of its existence, refuse to inquire whether it is wide or narrow.

Since the last Wenner-Gren conference on the great apes, the chasm to which Huxley referred has narrowed considerably. Ongoing research in captivity and in the field has progressively reduced a previously long list of traits that could be employed to differentiate the African apes from humans, and with this research, it has become increasingly clear that humankind's uniqueness may depend on a single characteristic, namely our ability to use speech and language.

While the proposition that speech is special is arguably true, it is important to note that this claim is based on very few data regarding the vocal behavior of our closest living relatives in the wild. Pioneering field work by Schaller (1963), Goodall (1968a), Marler (1969, 1976), Fossey (1972), Marler & Hobbett (1975) and Marler & Tenaza (1977) gave us an early, but admittedly rudimentary glimpse into the vocal repertoires of the African apes. The past few years have witnessed renewed efforts to explore the natural vocal communication systems of these animals, and in this contribution I summarize some recent findings, focusing specifically on two questions concerning the manner in which apes produce and pronounce vocal sig-

nals (production) and give calls in specific situations (usage). First, what are the similarities and differences in vocal behavior between species? Second, what are the sources of vocal variability within species? I conclude with a brief review of our current state of knowledge regarding the vocal behavior of our closest living relatives, comment on the limits of vocal signaling by wild apes and suggest some future lines of inquiry.

A COMPARATIVE EVOLUTIONARY FRAMEWORK

Early ethological studies focused on species as the unit of analysis and emphasized the importance of interpreting similarities and differences in their behavior within a comparative phylogenetic framework. In a seminal paper, Tinbergen (1963, p. 421) summarized the comparative method:

> Through comparison he notices both similarities between species and differences between them. Either of these can be due to one of two sources. Similarity can be due to affinity, to common descent; or it can be due to convergent evolution. It is the convergences which call his attention to functional problems. . . . The differences between species can be due to lack of affinity, or they can be found in closely related species. The student of survival value concentrates on the latter differences, because they must be due to recent adaptive radiation.

To the early ethologists, intraspecific variability in behavior was sometimes viewed as noise that obscured patterns of species-typical behavior. Goodall (1968b) was one of the first to emphasize the degree to which individuals differ in their behavior, and intraspecific behavioral variability has become a major focus of current ethological investigation. The same evolutionary approach promoted by Tinbergen has proved useful for understanding the functional significance of individual differences in behavior, and today behavioral ecologists typically employ cost-benefit analyses to explain adaptive behavioral differences between members of the same species (Krebs & Davies, 1993).

In the following, I adopt a comparative, evolutionary framework to interpret patterns of interspecific similarities and differences and intraspecific variation in vocal behavior among the African apes. Interspecific similarities in vocal behavior reflect these animals' close phylogenetic relationship, while identification of the selective fitness costs and benefits of performing a behavior provides insights into the functional significance of inter- and intraspecific behavioral variation.

INTERSPECIFIC SIMILARITIES IN VOCAL PRODUCTION

There is a close correspondence between the acoustic structure of calls produced by each species

Given the close phylogenetic relationship of the three species of African ape, it comes as no surprise to find similarities in the acoustic structure of the calls that they produce. In the first systematic study of the vocal behavior of an African ape, Marler (1969) noted that the call repertoire of the chimpanzee comprises variations of grunts, barks, screams and hoots. Subsequent investigations of captive bonobos and wild mountain gorillas confirmed that these species produce calls that do not differ in their gross acoustic morphology from those of the chimpanzee (Fossey, 1972; de Waal, 1988; Harcourt et al., 1993). Figure 18.1 illustrates the close correspondence in the acoustic structure of some calls made by each of the three species.

Signal repertoire sizes do not vary markedly between species

In the only major review of ape vocal behavior, Marler & Tenaza (1977) noted the remarkable similarity in the number of elements in the vocal repertoires of each species. Subsequent studies have generally confirmed this early finding (Goodall, 1986; de Waal, 1988; Harcourt et

al., 1993). Estimates of vocal repertoire sizes vary among studies, and differences in assessments can be ascribed largely to the length and focus of research. In the absence of psychophysical (e.g. Zoloth et al., 1979) or field playback studies (e.g. Mitani, 1987) that empirically evaluate the animal's perception of call types, it is difficult to resolve these different appraisals of repertoire sizes. Irrespective of classification, current estimates of the number of calls uttered by each ape species do not appear to vary considerably from those made for other non-human primate species (e. g. vervets, *Cercopithecus aethiops* > 36 calls, Struhsaker, 1967; cotton-top tamarins, *Saguinus oedipus* > 13 calls, Cleveland & Snowdon, 1982).

Adult males are the most vocal age–sex class among chimpanzees and gorillas

Early field research by Marler (1976) on chimpanzees at the Gombe provisioning area indicated that all age–sex classes produce calls with equal frequency. In contrast, observations of wild mountain gorillas have consistently shown that silverback males are the most vocal members of this species (Fossey, 1972; Marler, 1976; Harcourt et al., 1993; Seyfarth et al., 1994). These data led Marler (1976) to speculate that patterns of vocal production differed significantly between chimpanzees and gorillas. More recent field observations by Clark (1993) on chimpanzees living in a more diverse and species-typical set of social situations than those that prevailed at the Gombe provisioning site indicated that adult females, subadult males and low-ranking males were generally quiet except when foraging in large parties of mixed composition. Like Marler (1976), Clark (1993) found that the long-distance pant-hoot was the dominant element in the call repertoire of adult male chimpanzees, and, noting this, he interpreted age–sex class differences in vocal production in terms of divergent selective costs and benefits of calling by different individuals. Adult females and low-ranking males incur high costs in feeding competition and are vulnerable to lethal aggression from males of other groups, and their relative reluctance to call may reflect the high costs they would suffer by revealing their location to conspecifics. In contrast, high-ranking adult males appear to gain important selective benefits from indicating their positions to others. These benefits include: protection against extra-group males, maintaining within-group alliances and attracting estrous

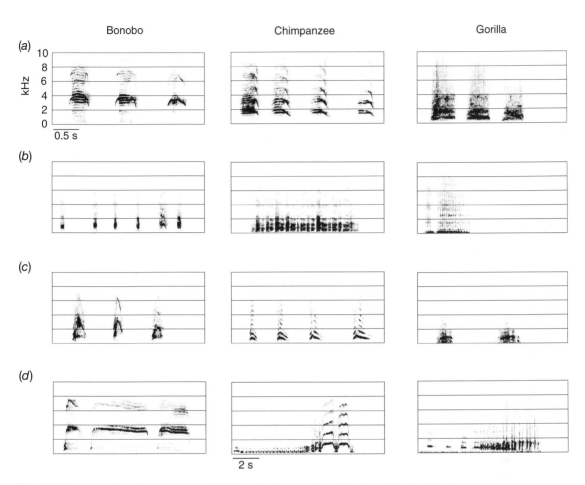

Fig. 18.1. Representative audiospectrograms of calls produced by bonobos, chimpanzees, and gorillas. (*a*): Adult female screams; (*b*): adult male grunts; (*c*): adult male barks; (*d*): adult male hoots. Spectrograms were produced using SoundEdit Pro software for the Macintosh. Analysis range = 11 kHz, frequency resolution = 86 Hz, except for the gorilla hoots in (*d*): Analysis range = 5 kHz, frequency resolution = 39 Hz.

females (Wrangham, 1977; Mitani & Nishida, 1993). Similar benefits may be accrued by silverback males who are the focus of group activity and the most voluble members of gorilla societies (Harcourt *et al.*, 1993).

INTERSPECIFIC SIMILARITIES IN VOCAL USAGE

There is considerable homology in the use of vocal signals

Currently available data suggest that the African apes show additional similarities in the manner in which they use calls. Marler (1976) again was the first to note the striking consistency with which chimpanzees and gorillas used acoustically homologous calls. More recent observations of captive bonobos and chimpanzees led de Waal (1988) to a similar conclusion: both species employ many of the same vocal elements in similar fashions. Specific examples of common usage include screams produced by subordinate individuals who receive aggression (see Fig. 18.1), and laughter by immatures during bouts of play. Despite these common patterns of vocal usage among bonobos, chimpanzees and gorillas, the three species appear to share fewer similarities in usage than in production (see below).

INTERSPECIFIC DIFFERENCES IN VOCAL PRODUCTION

There are consistent species differences in the acoustic structure of calls

Despite their homology, calls produced by the African apes show consistent acoustic differences between species. Animals frequently use calls to attract mates or maintain characteristic patterns of social dispersion (e.g. Marler, 1957; Payne, 1986), thus creating strong selective pressures for vocal differences between those species living in sympatry. The African apes are often found allopatrically, however, and under these conditions, strong selection for species-specificity may be relaxed.

If selective factors operate only minimally to differentiate calls, what other sources may account for the acoustic variability found between homologous calls produced by the African apes? We recently documented consistent acoustic differences between the screams made by bonobos and chimpanzees (Mitani & Gros-Louis, 1995). The calls given by chimpanzees, emitted at relatively low frequencies, were readily distinguishable from the higher-pitched calls uttered by bonobos (see Fig. 18.1). We attribute these vocal differences to interspecific variability in body size. Bonobos display neotenous characteristics, especially in their craniofacial morphology and in regions surrounding the basicranium (reviews in Susman, 1984), and as a result are likely to possess smaller vocal tracts than chimpanzees. This size difference presumably gives rise to related differences in laryngeal and vocal tract production mechanisms, leading chimpanzees to produce calls that are typically lower in frequency than those made by bonobos.

While studies such as these routinely reveal species-specific patterns in the calls of apes and other primates (review in Snowdon, 1986), it is unclear why calls differ in the precise acoustic manner that they do. Traditional ethological studies emphasize the relative roles that acoustic differentiation and variability play in species' call discrimination and recognition. One hypothesis considers the discrimination task within the context of a community of multiple callers and proposes that those acoustic characteristics that differ significantly among species will provide the most reliable cues in recognition (e.g. Marler, 1960). In contrast, an alternate hypothesis suggests that individuals may use acoustic features that differ little within species to discriminate conspecific calls from those of heterospecifics (e.g. Emlen, 1972).

We used our recently completed analysis of species differences in the screams of bonobos and chimpanzees to investigate these hypotheses (Mitani & Gros-Louis, 1995). Measurement of 11 acoustic features revealed consistent and reliable differences between the calls produced by the two species (see above), and the effectiveness with which an acoustic feature could be used to discriminate the two species was negatively associated with its relative variability in both bonobos and chimpanzees (see Fig. 18.2). These data are consistent with the hypothesis that optimal signals for species discrimination vary little within species but differ widely between species.

In addition to the calls that have homologous equivalents across species, each of the African apes appear to emit a few unique calls. These calls include high-pitched 'peeps' given by bonobos in the context of food, 'wraagh-like' alarms produced by chimpanzees, and 'singing' or 'humming' by gorillas (see Fig. 18.3).

Overall rates of vocal production differ between species

Marler (1976) originally suggested that chimpanzees and gorillas are equally voluble; early observations of provisioned chimpanzees at Gombe and newly habituated gorillas at Karisoke indicated that both species called more than ten times per hour. More recent research casts doubt on this early assessment of chimpanzee and gorilla vocal production. Observations of wild, unprovisioned chimpanzees at the Kibale forest by Clark (1993) revealed that adults delivered only two to four calls per hour. In contrast, adults gave more calls in larger mixed-sex parties, socio-ecological situations that approximated those of Marler's (1976) at the Gombe feeding station. In addition, long-term observations of habituated gorillas showed that individuals called less frequently than previously reported, albeit more than habituated chimpanzees, at a rate of about eight times per hour (Harcourt *et al.*, 1993). My observations of male gorillas at Karisoke and male chimpanzees at Mahale are consistent with the results of these more recent studies. Gorillas appear to call significantly more often than chimpanzees (see Fig. 18.4).

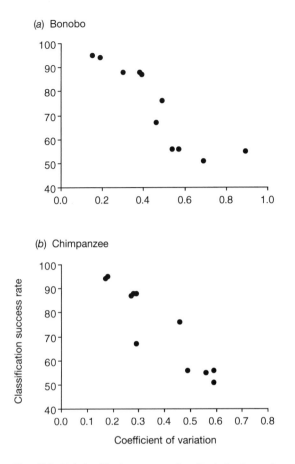

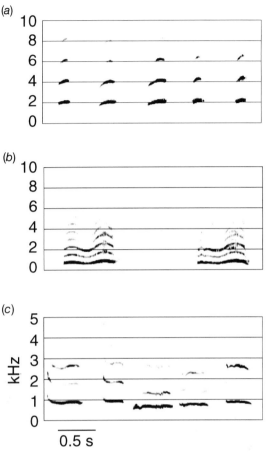

Fig. 18.2. Relationship between species discrimination and acoustic feature variability in the screams of bonobos and chimpanzees. Each plot shows the accuracy with which each of 13 acoustic variables classified calls in a discriminant analysis versus each acoustic feature's relative variability as measured by its coefficient of variation. Relatively invariant features discriminated between species more successfully than more variable features. (a): bonobos; (b): chimpanzees.

Fig. 18.3. Representative audiospectrograms of species-specific calls produced by bonobos, chimpanzees, and gorillas. (a): Bonobo peeps; (b): chimpanzee alarm calls; (c): gorilla hums. Spectrograms were produced as in Fig. 18.1, with the analysis range for gorilla humming similar to the hooting series in 18.1(d).

What is responsible for this species difference in vocal production? The maintenance of social relationships is a primary function of vocal communication, and insofar as communication processes mediate these relationships, social factors should have an important influence on patterns of vocal behavior. Recent studies indicate that variations in social situations affect vocal production by animals. For example, several birds and mammals vary their alarm calling behavior as a function of the presence or absence of specific individuals (review in Cheney &

Seyfarth, 1990). Gorillas live in groups that are relatively stable in size and composition (Watts, Chapter 2). In contrast, chimpanzees form loosely organized unit-groups or communities, whose members associate in ephemeral parties and are often widely dispersed (Goodall, 1986; Nishida, 1990). In the case of chimpanzees, the absence of reliable signal recipients from whom one may gain fitness benefits via calling appears to be associated with a corresponding reduction in vocal activity. These observations highlight a transparent but nonetheless widely overlooked generalization that living

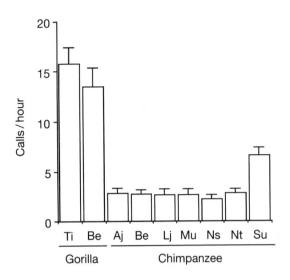

Fig. 18.4. Frequency of call production by adult male gorillas and chimpanzees. Two silverback male gorillas are from a typical bisexual group (Beetsme's) at the Karisoke Research Centre in 1992, while chimpanzees represent seven of the ten adult males of the M-group chimpanzees at the Mahale Mountains National Park in 1990. Means ± one SE are shown. Estimates are based on 25 one-hour observation sessions of each individual, with the exception of the gorilla Be, who was observed during 20 hourly samples. Gorillas called significantly more often than chimpanzees, Mann–Whitney U test, $N_1 = 2$, $N_2 = 7$, $p < 0.05$.

in a relatively stable social group places a selective premium on vocal communication. The near incessant vocal activity that takes place among other socially living primates is impressive to even the casual observer, and the importance of maintaining long-term social relationships within the context of a stable group may prove to be a primary determinant of the frequency with which individuals call.

INTERSPECIFIC DIFFERENCES IN VOCAL USAGE

There is disproportionate use of short-range calls by gorillas and long-range calls by chimpanzees

Chimpanzees and gorillas do not use acoustically homologous calls in their vocal repertoires with equal fre-

quency. Field studies consistently show that the long-distance pant-hoot is the most frequently emitted call by adult male chimpanzees (Marler, 1976; Goodall, 1986; Clark, 1993). In contrast, vocal output by gorillas is dominated by short-range grunts (Marler, 1976; Harcourt et al., 1993). My observations of adult male chimpanzees and gorillas provide quantitative support to these findings (see Fig. 18.5).

In the same manner that social factors influence overall rates of vocal production, vocal usage appears to depend critically on the social signaling environment. It is not surprising to find that gorillas, whose group members usually stay together, predominantly use calls during short-range communication. Despite recent study, the functional significance of many of these calls remains obscure (Harcourt et al., 1993; Stewart & Harcourt, 1994). In contrast, chimpanzees, whose group members are often dispersed and out of sight of each other, depend heavily on a long-distance call to communicate. The selective benefits of using this call have been the focus of our recent field research; our observations suggest that males may use this call selectively to recruit the company and support of allies and associates (Mitani & Nishida, 1993).

INTRASPECIFIC VARIABILITY IN VOCAL PRODUCTION

Populational differences exist in the acoustic structure of calls produced by chimpanzees

One hallmark of human speech is that it is a learned motor skill. Vocal learning combined with limited dispersal from natal areas after acquisition of the vocal repertoire commonly leads to the formation of local dialects, and learning among animals is frequently inferred from the existence of acoustic variability in the calls of individuals living in different populations (Kroodsma, 1982).

The preceding considerations set the stage for a recent study of acoustic variation between calls produced by males from the two well-studied chimpanzees populations of Gombe and Mahale. These analyses provide some of the first evidence of populational variability in the calls of non-human primates (Mitani et al., 1992). More recently completed analyses have confirmed the earlier finding that males from the two populations pro-

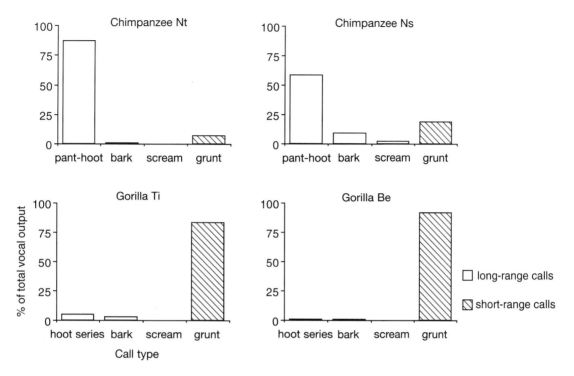

Fig. 18.5. Differential usage of short- and long-distance calls by chimpanzees and gorillas. Frequency of use of calls, expressed as a percentage of total vocal output, is shown for each of four classes of calls produced by two adult males of each species. Gorilla barks include hiccup and question barks (*sensu* Fossey, 1972), while grunts comprise double, cough and train grunts

(*sensu* Harcourt *et al.*, 1993). Chimpanzee grunts include rough and pant-grunts (*sensu* Marler & Tenaza, 1977). Figs. are based on 25 hours of observation of each subject, with the exception of the gorilla Be, who was followed during 20 hourly sessions. *N* total calls: Nt = 70, Ns = 53, Ti = 397, Be = 282.

duce acoustically distinguishable calls (Mitani & Brandt, 1994). Although alternative interpretations exist, we have tentatively ascribed these populational differences to vocal learning (Mitani *et al.*, 1992).

One perplexing problem in current ethological research is that to date, studies of male birds provide the only compelling evidence for the vocal learning process in non-human animals (review in Marler, 1991). Our understanding of the evolution of vocal learning and its apparent taxonomically restricted distribution is currently hindered by the lack of a theory specifying why and under what conditions such developmental plasticity should take place. The development of such a theory will be critically dependent upon incorporating considerations of the selective costs and benefits of learning, and in this context the social system of chimpanzees may have created an appropriate selective mileau favoring the

evolution of vocal learning (Mitani *et al.*, 1992). Male chimpanzees of neighboring unit groups are unusually xenophobic and aggressive toward each other (Goodall *et al.*, 1979). Given these circumstances, it would be selectively advantageous for chimpanzees to have the ability to recognize friend from foe at a distance. Acoustic similarities between the calls of males from the same group may provide a means for such recognition, with vocal learning the mechanism by which these similarities arise.

There are individual differences in vocal production

Individual differences often exist among the calls produced by primates (review in Snowdon, 1986). Most primates live within complex social groups and interact with several individuals who vary in age, sex, kinship and rank. Since calls frequently mediate these interactions, there must be strong selective pressures placed on the ability to identify the calls of different individuals.

Marler & Hobbett (1975) were the first to document acoustic differences in the calls produced by individual chimpanzees. Their acoustic analyses revealed that variations in the frequency and temporal structure of calls could be used to identify the long-distance calls made by different individuals. More recent analyses have confirmed that short-range calls uttered by mountain gorillas are individually distinctive (Seyfarth et al., 1994). Additional research indicates that an individual's age, sex and rank affect the acoustic structure and frequency of production of several calls made by bonobos, chimpanzees and gorillas (Fossey, 1972; Marler & Hobbett, 1975; Pusey, 1990; Harcourt et al., 1993; Mitani & Nishida, 1993; Clark & Wrangham, 1994; Hohmann & Fruth, 1994; Mitani & Gros-Louis, 1995).

While the pressures created by social life undoubtedly form a strong impetus for selection to encode individuality into the voices of animals, is there any reason to suspect that such selection varies in different situations? Recent studies suggest that under certain conditions, natural selection may favor the elaboration of acoustic features that differentiate individuals. For example, the calls of animals living in large social groups, where recognition is hindered, appear to contain more individually distinctive acoustic characteristics than those living in smaller groups (Beecher & Stoddard, 1990). We have recently extended this line of reasoning by investigating whether selection acts to encode individually distinctive features differentially in signaling systems. Field studies routinely reveal that chimpanzees live in fluid unit-groups whose members associate in temporary parties of changing size and composition (see above). Rank relationships play a significant role in the lives of chimpanzees, especially males, most of whom jockey for higher rank on a near-continuous basis (Riss & Goodall, 1977; Nishida, 1983). As a result, the two dominant components of the vocal repertoires of chimpanzees are calls used to regulate spacing between widely separated individuals and dominance relationships between animals. These vocalizations are the long-distance pant-hoot, and short-range pant-grunt, respectively.

Consider the recognition tasks involved during the perception of these two calls. Pant-hoots are long-distance signals typically delivered to potential recipients who may be as far away as several hundred meters. In contrast, pant-grunts are commonly uttered within close visual range of call recipients. Given their markedly different usage, we hypothesized that selection will act differentially on the two signaling systems by encoding a greater degree of individuality in the long-range pant-hoot compared with the pant-grunt. Selection for individuality in the short-range pant grunt would presumably be relaxed given the ability of recipients to identify signalers visually.

In keeping with results of our acoustic analyses that revealed absolute differentiation and relative variability play important roles in encoding species-specificity (see above), we predicted that when compared with pant-grunts, pant-hoots would vary little within-individuals but differ markedly between individuals (Mitani & Macedonia, unpublished data). Results of our analyses confirmed both predictions (see Table 18.1), thus providing the first empirical demonstration that selection may act to encode varying degrees of individuality in different components of the vocal repertoire of a single species.

Audience effects reveal within-individual differences in vocal production

Early ethological research suggested that animal signals were frequently elicited as fixed responses to specific external stimuli (e.g. Lorenz, 1935). As noted above, however, recent research shows that the signaling context influences vocal production by animals. An important implication of this research is that animals may be able to control their vocal output voluntarily. Goodall (1986) described chimpanzees who altered their vocal output in different social situations. Her reports of a young male suppressing calls at the banana provisioning station and of vocal restraint by adult males during boundary patrols provided anecdotal support for the hypothesis that call production is under the volitional control of chimpanzees. More recent research adds quantitative support to Goodall's observations by showing the effects of social context on call production. Field observations of chimpanzees at the Kibale Forest indicated that adult females called more often in large parties than when alone (Clark, 1993), and that subadult males pant-hoot upon arriving

Table 18.1. *Within- and between-individual acoustic variability in male chimpanzee pant-hoots and pant-grunts*

(a) Mean coefficients of variation for each of 5 acoustic variables*[a]

Acoustic variable	Pant-hoots (mean c.v.)	Pant-grunts (mean c.v.)
Duration	30	31
Minimum frequency	21	91
Maximum frequency	15	30
Frequency range	23	27
Average frequency	17	45

(b) Variance components attributable to call elements, call bouts and individuals for each of three principal components generated using the five original acoustic variables in (a)

	Principal component	% variance due to:		
		Call elements	Call bouts	Individuals
Pant-hoots				
	1	44	42***	14*
	2	54	11	35***
	3	39	24***	36***
Pant-grunts				
	1	69	24**	7
	2	71	22**	7
	3	60	26***	13

Notes: [a] $N = 7$ males Pant-grunts are significantly more variable than pant-hoots. Combined probability test, $Chi^2 = 33.51$, 1 tailed, $p < 0.001$. [b] Variance components estimated from a nested analysis-of-variance computed on elements within bouts nested within individuals. Pant-hoots showed significant variability between individuals, while pant-grunts differed only between bouts produced by the same caller. *$P < 0.05$; **$P < 0.005$; ***$P < 0.001$.

at a large fruiting tree only if high-ranking males are present (Clark & Wrangham, 1994). Recent studies of adult male chimpanzees at Mahale revealed that males vary their production of pant-hoots as a function of the presence or absence of specific individuals, in particular allies and close associates (Mitani & Nishida, 1993).

Context-specific acoustic variation and representational signaling have been surprisingly difficult to document

Research on the vocal behavior of non-human primates reveals unexpectedly subtle acoustic variability in calls produced by individuals in different behavioral and social contexts. Early studies by Green (1975), for example, suggested that Japanese macaques (*Macaca fuscata*) use seven variants of their 'coo' call in ten differ-

ent social contexts. Additional research has revealed similar context-specific phonetic variation in the vocal repertoires of non-human primates, and that some of these acoustic variants are employed in a referential fashion to designate objects and events in the external world in a manner similar to the way humans use words (review in Cheney & Seyfarth, 1990).

Goodall (1986) was the first to hypothesize that chimpanzees produce calls that vary in subtle acoustic fashion in different situations. She suggested that males may utter distinct versions of two calls, their long-distance pant-hoots and screams. Following these suggestions by Goodall and independent observations by Wrangham (1977), Clark & Wrangham (1993) investigated whether or not male chimpanzees produce an acoustically distinct pant-hoot upon arriving at food sources. Their analysis revealed that a high percentage of pant-hoots delivered upon arriving at food sources contained one particular acoustic component, the 'let-down', which was not produced frequently in other contexts. No reliably consistent acoustic differences between pant-hoots produced in feeding and non-feeding situations could be documented, however.

A second attempt to uncover referential vocal signaling by chimpanzees followed the leads provided by previous studies of primates (Cheney & Seyfarth, 1985) and gallinaceous birds (Marler *et al.*, 1986). Chimpanzees utter distinctive calls in the presence of food (Marler, 1976; Goodall, 1986), and since these calls are given specifically in the context of food, Hauser (Hauser & Wrangham, 1987; Hauser *et al.*, 1993) has suggested that individuals might use this call in referential fashion. Possible referents include the amount and divisibility of food. Nonetheless, experiments with captive chimpanzees indicate that individuals vary their rate of calling in direct proportion to the amount of food discovered (Hauser & Wrangham, 1987; Hauser *et al.*, 1993), suggesting the more parsimonious interpretation that calls simply reflect the underlying affective states of the signalers (cf. Snowdon, 1990). Additional observations regarding the acoustic structure and social contexts of production are not consistent with the hypothesis that 'food' calls are functionally referential. Calls are produced at extremely low amplitudes and are frequently given within sight of potential recipients. Given these circumstances, it is unlikely that chimpanzees give these calls to transmit functionally relevant information about the location, availability or characteristics of food to others.

The general failure to show that chimpanzees produce referential signals has led others to seek other sources that may influence patterns of within-individual call variability. Field and laboratory work reveals a surprising amount of within-individual variability in the acoustic structure of pant-hoots; within-individual acoustic variability appears to exceed that found between individuals (Mitani & Brandt, 1994). Two factors may account for this pattern. First, male chimpanzees frequently pant-hoot with others, and second, males tend to match the call characteristics of others during these choruses (Mitani & Brandt, 1994; Fig. 18.6). One corollary derived from these observations is that those individuals who chorus most often with the largest number of different males should show the most variability in

Fig. 18.6. Three adult male chimpanzees bark in coalition at Mahale. (Photo by T. Nishida)

their calls, a prediction currently supported by our limited set of recordings (Mitani & Brandt, 1994). In addition to these social factors, endogenous variables related to the emotional states of signalers may contribute, in part, to the large amount of within-individual acoustic variation found in calls. Recent attempts to document referential signals in the acoustic repertoire of chimpanzees have tended to obscure traditional ethological studies of vocal communication that emphasize the manner in which call morphology depends on the motivational states of signalers (Darwin, 1872; Scherer, 1985). Nonetheless, an understanding of affect and motivation provides a framework within which to evaluate the causation of behavior. Vocal assays that furnish a non-invasive means of monitoring the internal states of individuals promise to provide novel glimpses of the world as viewed by the animals themselves.

REVIEW AND DISCUSSION

In keeping with the findings of earlier reviews (Marler, 1976; Marler & Tenaza, 1977), recent research confirms both similarities and differences in the vocal behavior of the African apes. Similarities include a close correspondence between the acoustic structure of calls, signal repertoire sizes and use of vocal signals. Differences comprise consistent variations in the acoustic structure of homologous calls and in the use of different call elements by gorillas and chimpanzees. Additional observations of fully habituated subjects occupying a wide range of social conditions have refined previous generalizations. While early data suggested that all age-sex classes of chimpanzees call with equal frequency, subsequent research indicates that adult males are the most vocal individuals within a community. In addition, a previous proposal that overall rates of vocal production do not differ between species has not been supported by more recent observations.

Our understanding of broad patterns of interspecific similarities and differences in vocal behavior has changed only slightly over the years (cf. Marler, 1976; Marler & Tenaza, 1977). In contrast, our knowledge of intraspecific variability in ape vocal behavior has progressed steadily. For example, we have now documented consistent population differences in the calls of chimpanzees. While early studies showed clear individual differences in the acoustic structure of calls, recent research suggests that selection may act to encode varying levels of individuality into different calls uttered by the same species. Current studies have revealed the importance of within-individual differences in call production, subtly influenced by a subject's audience or social companions. Additional research has somewhat curiously failed to replicate in apes one of the most startling and important findings of recent ethological research, namely referential signaling by non-human animals (e.g. Gouzoules et al., 1984). This last result, more than anything else, causes one to consider the limits of the vocal behavior of wild apes.

Studies in captivity reveal the remarkable cognitive abilities of the African apes and suggest that these animals possess the rudiments required for spoken language (Savage-Rumbaugh & Lewin, 1994). This research has led some to speculate that similar capabilities will be shown by their wild counterparts and that the apes will display qualitatively different vocal skills than monkeys (e.g. Cheney & Seyfarth, 1990). Until recently, virtually nothing was known about the vocal behavior of our closest living relatives in their natural environmental and social settings, and in the absence of these data such conjecture was possible. Recent studies summarized here,

however, do not support the proposition that the African apes use the vocal channel to communicate with conspecifics in a qualitatively different fashion than do other animals. To date, wild apes have given us no clear indication that they possess special vocal skills, and as a result their communication system does not appear to be extraordinary.

Efforts to draw parallels between human language and animal vocal communication have focused attention away from the very real differences that exist between them. While some studies have highlighted these parallels by demonstrating the remarkable vocal abilities of animals, strong differences are immediately apparent upon considering the manner in which the apes fail to use calls. Recent research suggests that while monkeys know much about themselves and their social worlds, they do not appear to know that others know (Cheney & Seyfarth, 1990). The selective advantages of possessing the ability to attribute knowledge to others is readily apparent. For example, humans use speech and attribution to exchange important, sometimes essential information between individuals. Do the apes give us any overt manifestation that they attribute knowledge to others through their vocal behavior? Consider the following scenario: envision yourself following a solitary chimpanzee who has just encountered a predator, say a lion. This individual rapidly races to join others in his group. Upon encountering his group-mates, what transpires? Far from there being any calm vocal dialogues during which our chimpanzee passes on his important new-found knowledge, these reunions are noisy, raucous, highly emotional affairs. This scenario does not explicitly deny the possibility that chimpanzees attribute knowledge to others (cf. Povinelli et al., 1992) nor that complex information exchange occurs (cf. Menzel, 1979), but it indicates that if such transfer occurs, it probably does not occur vocally.

CONCLUDING COMMENTS

The preceding review emphasizes that a paucity of data still exists on the vocal behavior of wild apes. We specifically lack information regarding the vocal repertoires of bonobos and gorillas, and as a result many of the preceding conclusions must be offered tentatively. It is possible that subsequent study will reveal that these animals possess more complex vocal skills than heretofore documented. In this context, calls used to regulate close-range social interactions clearly warrant further investigation (cf. Harcourt *et al.*, 1993; Stewart & Harcourt, 1994). In addition, new empirical and theoretical perspectives may enrich our understanding of the complexities of ape communication in the wild. For example, the reader may find it curious that those who study animal vocal communication do not treat it as a dynamic process. Vocal communication clearly entails a signaler, the signal and a recipient-respondent, yet with few exceptions, ethological studies focus on only one side of the communicative channel at a time. Reasons for this are both methodological and empirical. Methodologically, it is inherently difficult for a single observer to follow activities occurring on both sides of the communicatory channel simultaneously. Clearly, new techniques need to be devised to deal with this problem. Experimental playbacks of pre-recorded calls, used successfully to investigate the function and perception of calls by animals (e.g. Waser, 1975; Zoloth *et al.*, 1979), have not yet been applied to the study of ape vocal signaling in the wild. Despite the inevitable logistic and often overlooked ethical problems associated with employing this technique, future advances in our understanding are likely to depend on experimental control of the signaling system and environment.

Empirically, vocal communicative exchanges appear to occur astonishingly infrequently among the African apes. Two notable exceptions are double grunt or belch exchanges between gorillas (Seyfarth *et al.*, 1994), and hooting exchanges that occur between bonobos prior to nesting at night (Hohmann & Fruth, 1994). These call systems provide tempting possibilities to investigate vocal communication as a dynamic process involving signaler and recipient simultaneously. Nonetheless, the noticeable lack of vocal exchanges among apes once again underscores the limits of African ape vocal complexity.

Finally, there is also a pressing need for construction of a robust theory that can be used to generate *a priori* predictions about what animals must talk about in order to survive and to reproduce successfully in their natural habitats. Previous studies of primate vocal communication have often been motivated by the search for parallels between animal calls and human speech, but it is only rarely asked why similarities should exist in the first place. I have suggested earlier that a comparative ethological framework provides the basis for understanding patterns of inter- and intraspecific similarities and differences in behavior. Viewed from this perspective, description of the similarities between speech and animal calls is subsumed under a research protocol that attempts to identify the factors that account for differences as well as similarities. Here two lines of inquiry will be necessary, one that stresses an integration of a phylogenetic perspective into comparative ethological studies (Harvey & Pagel, 1991), and a second that specifies the fitness costs and benefits of vocal behavior performed by animals in their natural social and environmental settings. Questions such as these will provide fertile ground for functionally-based studies of primate vocal communication in the wild, and serve as the foundation for interpreting the diversity in animal signaling systems.

ACKNOWLEDGMENTS

This paper was originally prepared for the Wenner-Gren conference, 'The Great Apes Revisited,' held November 12–19, 1994 in Cabo San Lucas, Baja, Mexico. I thank Bill McGrew, Toshisada Nishida and the Wenner-Gren Foundation for giving me the opportunity to participate and contribute this paper. My research in Africa has been sponsored by the Tanzanian Commission for Science and Technology, the Serengeti Wildlife Research Institute, the Tanzanian National Parks, the Rwanda Office of Tourism and National Parks, and the Zaïre Center of Natural Science Research. I am indebted to my scientific colleagues and the staffs at the Mahale Mountains National Park, Karisoke Research Centre and Scientific Reserve of the Luo for logistic support in the field. J. Gros-Louis provided indispensable assistance preparing the manuscript. I thank A. Clark-Arcadi, L. Marchant, P. Marler, W. McGrew, T. Nishida and J. Pepper for comments on the manuscript. My research is currently supported by a NSF Presidential Faculty Fellows Award, DBS-9253590. I dedicate this contribution to Peter Marler, pioneering ethologist and mentor, who paved the way and ushered me along the path.

REFERENCES

Beecher, M. & Stoddard, P. 1990. The role of bird song and calls in individual recognition: contrasting field and laboratory perspectives. In: *Comparative Perception, Vol. II, Complex Signals*, ed. W. Stebbins & M. Berkley, pp. 375–408. New York: John Wiley.

Cheney, D. & Seyfarth, R. 1985. Vervet monkey alarm calls: manipulation through shared information. *Behaviour*, 94, 150–66.

Cheney, D. & Seyfarth, R. 1990. *How Monkeys See the World*. Chicago: University of Chicago Press.

Clark, A. 1993. Rank differences in the vocal production of Kibale Forest chimpanzees as a function of social context. *American Journal of Primatology*, 31, 159–79.

Clark, A. & Wrangham, R. 1993. Acoustic analysis of wild chimpanzee pant hoots: do Kibale Forest chimpanzees have an acoustically distinct food arrival pant hoot? *American Journal of Primatology*, 31, 99–109.

Clark, A. & Wrangham, R. 1994. Chimpanzee arrival pant hoots: do they signify food or status? *International Journal of Primatology*, 15, 185–206.

Cleveland, J. & Snowdon C. 1982. The complex vocal repertoire of the adult cotton-top tamarin (*Saguinus oedipus oedipus*). *Zeitschrift für Tierpsychologie*, 58, 231–70.

Darwin, C. 1872. *The Expression of the Emotions in Man and Animals*. London: John Murray.

Emlen, S. 1972. An experimental analysis of the parameters of bird song eliciting species recognition. *Behaviour*, 41, 130–71.

Fossey, D. 1972. Vocalizations of the mountain gorilla. *Animal Behaviour*, 20, 36–53.

Goodall, J. 1968a. Expressive movements and communication in the Gombe Stream chimpanzees. In: *Primates: Studies in Adaptation and Variability*, ed. P. Jay, pp. 313–74. New York: Holt, Rinehart & Winston.

Goodall, J. 1968b. The behaviour of free-living chimpanzees in the Gombe Stream Reserve. *Animal Behaviour Monographs*, 1, 161–311.

Goodall, J. 1986. *The Chimpanzees of Gombe*. Cambridge, MA: Harvard University Press.

Goodall, J., Bandora, A., Bergmann, E., Busse, C., Matama, H., Mpongo, E., Pierce, A. & Riss, D. 1979. Intercommunity interactions in the chimpanzee population of the Gombe National Park. In: *The Great Apes*, ed. D. A. Hamburg & E. R. McCown, pp. 13–54. Menlo Park, CA: Benjamin/Cummings.

Gouzoules, H., Gouzoules, S. & Marler, P. 1984. Rhesus monkey (*Macaca mulatta*) screams: representational

signalling in the recruitment of agonistic aid. *Animal Behaviour*, 32, 182–93.

Green, S. 1975. Variation of vocal pattern with social situation in the Japanese monkey (*Macaca fuscata*): a field study. In: *Primate Behavior*, ed. L. A. Rosenblum, pp. 1–102. New York: Academic Press.

Harcourt, A., Stewart, K. & Hauser, M. 1993. Functions of wild gorilla 'close' calls. I. Repertoire, context and interspecific comparison. *Behaviour*, 124, 89–121.

Harvey, P. & Pagel, M. 1991. *The Comparative Method in Evolutionary Biology*. Oxford: Oxford University Press.

Hauser, M. & Wrangham, R. 1987. Manipulation of food calls in captive chimpanzees. A preliminary report. *Folia Primatologica*, 48, 207–10.

Hauser, M., Teixidor, P., Field, L. & Flaherty, R. 1993. Food-elicited calls in chimpanzees: effects of food quantity and divisibility. *Animal Behaviour*, 45, 817–19.

Hohmann, G. & Fruth, B. 1994. Structure and use of distance calls in wild bonobos (*Pan paniscus*). *International Journal of Primatology*, 15, 767–82.

Huxley, T. H. 1863. *Evidence as to Man's Place in Nature*. New York: D. Appleton.

Krebs, J. & Davies, N. 1993. *An Introduction to Behavioural Ecology*, 3rd edn. Oxford: Blackwell Scientific Publications.

Kroodsma, D. 1982. Learning and the ontogeny of sound signals in birds. In: *Acoustic Communication in Birds*, ed. D. Kroodsma & E. Miller, pp. 1–23. New York: Academic Press.

Lorenz, K. 1935. Der Kumpan in der Umwelt des Vogels. *Journal of Ornithology*, 83, 137–213.

Marler, P. 1957. Specific distinctiveness in the communication signals of birds. *Behaviour*, 11, 11–39.

Marler, P. 1960. Bird songs and mate selection. In: *Animal Sounds and Communication*, ed. W. Lanyon & W. Tavolga, pp. 348–67. Washington, DC: American Institute of Biological Sciences.

Marler, P. 1969. Vocalizations of wild chimpanzees: an introduction. In: *Proceedings of the Second International Congress of Primatology*, ed. C. R. Carpenter, pp. 94–100. Basel: Karger.

Marler, P. 1976. Social organization, communication, and graded signals: the chimpanzee and gorilla. In: *Growing Points in Ethology*, ed. P. G. Bateson & R. A. Hinde, pp. 239–80. Cambridge: Cambridge University Press.

Marler, P. 1991. Song-learning behavior: the interface

with neuroethology. *Trends in Neuroscience*, **14**, 199–205.

Marler, P. & Hobbett, L. 1975. Individuality in a long-range vocalization of wild chimpanzees. *Zeitschrift für Tierpsychologie*, **38**, 97–109.

Marler, P. & Tenaza, R. 1977. Signaling behavior of apes with special reference to vocalization. In: *How Animals Communicate*, ed. T. Sebeok, pp. 965–1003. Bloomington: Indiana University Press.

Marler, P., Dufty, A. & Pickert, R. 1986. Vocal communication in the domestic chicken: I. Does a sender communicate information about the quality of a food referent to a receiver? *Animal Behaviour*, **34**, 188–93.

Menzel, E. 1979. Communication of object-locations in a group of young chimpanzees. In: *The Great Apes*, ed. D. A. Hamburg & E. R. McCown, pp. 359–71. Menlo Park, CA: Benjamin/Cummings.

Mitani, J. 1987. Species discrimination of male song in gibbons. *American Journal of Primatology*, **13**, 413–23.

Mitani, J. & Brandt, K. 1994. Social factors influence the acoustic variability in the long-distance calls of male chimpanzees. *Ethology*, **96**, 233–52.

Mitani, J. & Gros-Louis, J. 1995. Species and sex differences in the screams of chimpanzees and bonobos. *International Journal of Primatology*, **16**, 393–411.

Mitani, J. & Nishida, T. 1993. Contexts and social correlates of long distance calling by male chimpanzees. *Animal Behaviour*, **45**, 735–46.

Mitani, J., Hasegawa, T., Gros-Louis, J., Marler, P. & Byrne, R. 1992. Dialects in wild chimpanzees? *American Journal of Primatology*, **27**, 233–43.

Nishida, T. 1983. Alpha status and agonistic alliance in chimpanzees. *Primates*, **24**, 318–36.

Nishida, T. (ed.) 1990. *The Chimpanzees of the Mahale Mountains*. Tokyo: University of Tokyo Press.

Payne, R. 1986. Bird songs and avian systematics. In: *Current Ornithology*, ed. R. Johnston, pp. 87–126. New York: Plenum Press.

Povinelli, D., Nelson, K. & Boysen, S. 1992. Comprehension of role reversal in chimpanzees: evidence of empathy? *Animal Behaviour*, **43**, 633–40.

Pusey, A. 1990. Behavioural changes at adolescence in chimpanzees. *Behaviour*, **115**, 203–46.

Riss, D. & Goodall, J. 1977. The recent rise to the alpha rank in a population of free-living chimpanzees. *Folia Primatologica*, **27**, 134–51.

Savage-Rumbaugh, E. S. & Lewin, R. 1994. *Kanzi: The Ape at the Brink of the Human Mind*. New York: John Wiley.

Schaller, G. 1963. *The Mountain Gorilla*. Chicago: University of Chicago Press.

Scherer, K. 1985. Vocal affect signalling: a comparative approach. In: *Advances in the Study of Behavior, Vol. 15*, ed. J. Rosenblatt, C. Beer & M. Busnel, pp. 189–244. New York: Academic Press.

Seyfarth, R., Cheney, D., Harcourt, A. & Stewart, K. 1994. The acoustic features of gorilla double grunts and their relation to behavior. *American Journal of Primatology*, **33**, 31–50.

Snowdon, C. 1986. Vocal communication. In: *Comparative Primate Biology*, ed. G. Mitchell & J. Erwin, pp. 495–530. New York: Alan R. Liss.

Snowdon, C. 1990. Language capacities of nonhuman animals. *Yearbook of Physical Anthropology*, **33**, 215–43.

Stewart, K. & Harcourt, A. 1994. Gorillas vocalizations during rest periods: signals of impending departure? *Behaviour*, **130**, 29–40.

Susman, R. (ed.) 1984. *The Pygmy Chimpanzee: Evolutionary Biology and Behavior*. New York: Plenum Press.

Struhsaker, T. 1967. Auditory communication among vervet monkeys (*Cercopithecus aethiops*). In: *Social Communication among Primates*, ed. S. A. Altmann, pp. 281–324. Chicago: University of Chicago Press.

Tinbergen, N. 1963. On the aims and methods of ethology. *Zeitschrift für Tierpsychologie*, **20**, 410–33.

Waser, P. 1975. Experimental playbacks show vocal mediation of intergroup avoidance in a forest monkey. *Nature*, **255**, 56–8.

de Waal, F. 1988. The communicative repertoire of captive bonobos (*Pan paniscus*) compared to that of chimpanzees. *Behaviour*, **106**, 183–251.

Wrangham, R. 1977. Feeding behaviour of chimpanzees in Gombe National Park, Tanzania. In: *Primate Ecology*, ed. T. H. Clutton-Brock, pp. 503–38. London: Academic Press.

Zoloth, S., Petersen, M., Beecher, M., Green, S., Marler, P., Moody, D. & Stebbins, W. 1979. Species-specific perceptual processing of vocal sounds by monkeys. *Science*, **204**, 870–2.

19 • On which side of the apes? Ethological study of laterality of hand use

WILLIAM C. MCGREW AND LINDA F. MARCHANT

Of particular interest is the study of Shafer (1987) in which, for the first time in other primates, a significant humanlike pattern consisting of a greater number of animals [i.e. *Gorilla gorilla*] which preferred the right hand *for all acts* was observed.

MacNeilage, Studdert-Kennedy & Lindblom, 1991, p. 344

Overall, our findings suggest that functional motor asymmetries are present in great apes. A left-hand population preference was found for carrying, while right-hand population preferences were found for object manipulation and leading limbs in locomotion.

Hopkins & Morris, 1993, p. 20

INTRODUCTION

Until 1987, the received wisdom in psychobiology was that laterality of function in *Homo sapiens* was unique, and so qualitatively different from all other species, including even our nearest relations, the great apes. Humans were said to be overwhelmingly right-handed at species (or population) level, that is, about 90% of persons show right-hand dominance for virtually all kinds of hand-use. In contrast, other hominoids were said to be randomly lateralized, either showing no consistent, overall preference for left or right, or being individually lateralized to either the left or right side in about equal numbers.

The published data available on apes at the time seemed to support these views of 50 : 50 randomness, but findings were sparse. Finch's (1941) benchmark study showed a population of 30 chimpanzees in which all but one were lateralized, but to an equal extent for left and right. Marchant's (1983) study of 27 chimpanzees showed similar individual lateralization for some

behavioral patterns (e.g. reach) but apparent randomness for others (e.g. groom). However, all other studies were flawed by basic methodological shortcomings such as too few subjects, and no studies of apes other than *Pan troglodytes* met minimal criteria for reliability and validity (Marchant & McGrew, 1991).

In 1987, two events changed the picture: MacNeilage *et al.* (1987) published their provocative reassessment in which they claimed that non-human primates *do* show species- or population-level handedness. They provided a theoretical framework, 'postural origins theory,' to explain the evolutionary origins of right-handedness in the Order Primates. Needless to say, their assault on a bastion of human uniqueness sparked controversy (MacNeilage *et al.*, 1988, 1991). Shafer (1987) produced a monumental master's thesis in which 47 *Gorilla gorilla* from 5 zoos were seen to perform more than 53 000 hand movements, and almost three-quarters of the apes did so more often with their right hands. Handedness research was revitalized, so that in only a few years quotations such as those above began to appear, as new data poured in (e.g. Ward & Hopkins, 1993). To what extent are the emerging claims of the 1990s justified by the data?

The aim of this paper is to scrutinize and to synthesize the data on laterality of hand-use in the great apes, using an ethological viewpoint. In doing so, we will present new findings from captivity and from nature, including the first comprehensive ethological study of any species of non-human primate in the field (Marchant & McGrew, 1996). In tackling this, we are forced to confront persistent methodological issues that vex handedness research. We propose a new conceptual framework and new analytical criteria for cross-species comparison. Finally, we attempt tentative comparisons across the three hominoid species for which there are enough data: *Pan troglodytes*, *Gorilla gorilla* and *Homo sapiens*.

However, before proceeding, it is worth saying what this paper does *not* seek to address: with only a few exceptions, we deal just with hand-use, and not with laterality of function in other paired organs, i.e. footedness, eyedness, earedness, toothedness, etc. This is mostly a reflection of shortage of data, at least for apes. We deal only with *behavioral* function, and not with anatomical or physiological function, so questions of asymmetry of structure, especially cerebral, are ignored. This is a matter of limited space availability. Similar constraints dictate that our coverage is restricted to major studies, that is, those with 20 or more subjects. Like all other areas in behavioral primatology, research on laterality has suffered from small sample effects, and these we seek to avoid. For simplicity's sake, we stick to one-handed acts, rather than two-handed ones, although acknowledging that bimanuality is likely to prove more interesting and informative in some cases, e.g. when the two hands work simultaneously but complementarily on the same task. There are as yet few data available on this (but see Byrne & Byrne, 1991, 1993; Hopkins, 1994, for exemplary exceptions). Finally, this paper does not tackle questions of causation, adaptation or ontogeny, on the assumption that we must have confidence in what exactly there is to explain, before seeking to do so.

METHODOLOGY AND METHODS

In principle, studying sagittal laterality is simple: the dependent variable is binary (left or right) and the null-hypothesis is symmetrical randomness (50 : 50). In practice, there are surprisingly many pitfalls. Different terms are used for the same phenomenon, and vice versa. Table 19.1 presents a standardized 2×2 matrix framework that copes exhaustively with the four logical possibilities generated by two key factors: individual versus population, and single versus multiple tasks. We believe that *handedness* should be reserved for the most robust and comprehensive case, when most subjects show consistency for many types of hand use. At the other extreme, when only one subject shows consistency for only one task, this is a case of *hand preference*. In the heterogenous con-

Table 19.1. *Laterality of hand function in terms of tasks and subjects*

	Subjects(s)	
	Within	Across
Task(s)		
Within	Hand preference	Task specialization
Across	Hand specialization	Handedness

ditions, *hand specialization* refers to the individual's same hand being used for many tasks, while *task specialization* refers to many subjects using the same-sided hand for some key task. Of course, handedness is just the sum of enough hand specializations to be persuasive, whether at group, population, or species level – here we arbitrarily set the threshold at no fewer than 20 subjects with enough instances for statistical testing ($N \geqslant 6$).

One source of confusion in laterality research is statistical testing. The test of choice is the binomial test on the relative numbers of left and right independent responses within a subject, and the standard measure is the z-score (Siegel, 1956). By testing within individuals rather than combining their scores, one avoids the problems of pooling (Machlis *et al.*, 1985). With regard to probability, the more conservative 2-tailed values are preferable, given the current contradictory theory and findings. Non-parametric statistical tests are preferred for the same reasons. The key issue is that of independence of data-points: if a subject performs a long sequence of identical responses without interruption, such as picking up one piece after another of fruit off the floor, then data-sets quickly become large. However, such mechanical repetitions are hardly independent of one another, and the result is an inflated sample size. The binomial test is sensitive to this: a ratio of 30L : 20R is the same as 300L : 200R, but the former produces a non-significant score of $z = 1.27$ ($p = 0.21$), while the latter produces a highly significant difference of $z = 4.42$ ($p < 0.0000$). Such complications are well-known in experimental psychology, where perseveration in responding based on a 'win–stay, lose–shift' strategy will lead to position preferences in a standard test setting (Chance, 1994).

To enhance independence of data-points requires

using *bouts* of such events, rather than counting each response separately, but this raises the issue of what distinguishes two bouts? (See Martin & Bateson, 1993, for general discussion of these issues and McGrew & Marchant, 1992, for specific application to laterality). A change of hands by definition means a new bout, but what about runs of same-handed events? Pauses between such bouts, even lengthy ones, are not enough, as hand position relative to the goal-object may be maintained; a better criterion is that the subject switches posture or position between bouts of hand use. Even better is that the hand performs an intervening activity, e.g. between bouts of eating, the hand is used to self-groom. Studies that count every response (or fail to mention addressing these issues) and comprise a few, large blocks of data-points are highly likely to produce statistically significant but false–positive differences.

Another set of issues has to do with choice of variables. Dependent variables (that is, recorded behavioral patterns) may be problematic: testing arboreal apes by having them pick up objects from flat, horizontal surfaces (e.g. floor) is unnatural; scoring 'Touch Other' as a category when such contact can range from a slap to a groom is too imprecise. Independent variables (that is, determining or influencing factors) can also be troublesome: testing quadrupedal apes in bipedal postures may reveal trainability but is hardly normal behavior. In one-handed tasks, recording what the other hand is doing may be crucial: is it idle and so potentially available as an alternative, or is it committed to maintaining positional stability? (This is especially important for arboreally evolved species like great apes.) Another missing variable of special interest to studies of human-reared apes is the handedness of their human companion(s); given that rats can learn laterality by observing others (Heyes & Dawson, 1990), it seems likely that captive apes can learn it from their surrogate caretakers. This may explain why zoo-born gorillas (many of them human-reared, presumably) are significantly more right-handed than their wild-born counterparts (Shafer, 1987). Finally, how dependent variables are selected from the total repertoire of the species or population is important; the fewer chosen, the higher the chance of bias. Unfortunately, the median number of measures per study of laterality in apes is only three (Marchant & McGrew, 1991). The safest (but most time-consuming!) approach is to record all common hand-use activities and to preclude such bias.

Many of the problems of methodology are dealt with by using ethological methods. Ethologists tend to record spontaneous (as opposed to induced), rigorously defined motor patterns as performed in natural or naturalistic (as opposed to contrived, artificial) settings. They let the organism evince its own behavioral categories, rather than trying to elicit tailored actions from the constrained subject. Ethological study is tiring, frustrating and procedurally messy, as it entails surrendering much or all researcher control. But it can be done in captivity as well as in the wild and its overriding benefit is high validity. Ethologists measure traits expressed as directly as possible from a history of natural selection, in settings that approximate as closely as possible the environment of evolutionary adaptedness. This is bound to be helpful.

APE LATERALITY IN NATURE AND CAPTIVITY

Most major studies of laterality of hand use have been done with chimpanzees (see Table 19.2), being more than twice as many as the other three species of great ape combined (see Table 19.3). The lack of laterality research on orangutans, *Pongo pygmaeus*, is disappointing, although a major study of 35 free-ranging but released subjects has been presented (Rogers & Kaplan, 1992). The situation for bonobos, *Pan paniscus*, is similar with the most extensive studies being of 11 subjects by Hopkins *et al.* (1993b) and of 10 subjects by Hopkins & de Waal (1995). More studies have been done in captivity than in the wild for both chimpanzees and gorillas, but more importantly, most field studies have focused on skilled object manipulation such as tool use, while most captive studies have recorded simpler motor patterns such as reaching. Methodologically, most studies have used subjects of all ages and both sexes and have provided raw data on individual scores (exceptions are noted below). The 15 studies of chimpanzees cover a range of tasks from the simplest touching to complex tool use, but only one has attempted comprehensive coverage of manual activities (Marchant & McGrew, 1996). For gorillas, the only wild data focus on herbaceous plant food processing by mountain gorillas, *Gorilla gorilla*

Table 19.2. *Studies of laterality of hand use with about 20 or more subjects, in* Pan troglodytes: *tool use in italic*

No. of Subjects	Tasks	Age of subjects	Captive/Wild	Individual scores	Reference
27	Touch face or body or other or environment	All ages	C	No	Aruguete *et al.*, 1992
31	Reach, drink, swirl, *throw*	All ages	C	Yes & No	Colell *et al.*, 1995
68	*Hammer, dip*, reach, groom	All ages	W	Yes	Boesch, 1991
30	Obtain fruit (4 types)	Mostly adult	C	Yes & No	Finch, 1941
20	Walk–run, eat, sit–support, climb, hang, carry–hold, groom, touch, manipulate, brachiate, scratch, *hit-throw*	All ages	C	Yes	Heestand, 1986
40	Reach	All ages	C	Yes	Hopkins, 1993
140	Bimanual feed	All ages	C	Yes	Hopkins, 1994
51	Reach	Juveniles	C	Yes	Hopkins, 1994
110	Bimanual finger-feed	All ages	C	No	Hopkins, 1995
36	Throw, cradle	All ages	C	Yes	Hopkins *et al.*, 1993
27	Feed, reach, social reach, hold, carry, groom, *throw*	All ages	C	Yes	Marchant, 1983
42	43 categories	All ages	W	Yes	Marchant & McGrew, 1995
20	*Ant-fish*	All ages	W	Yes	Nishida & Hiraiwa, 1982
18–19	Pick food, *hammer*	All ages	W	Yes	Sugiyama *et al.*, 1993
80	Reach	All ages	C	Yes	Tonooka & Matsuzawa, 1995

Table 19.3. *Studies of laterality of hand use with 20 or more subjects, in other great apes: tool use in italic*

No. of Subjects	Tasks	Age of Subjects	Captive/Wild	Individual Scores	Reference
Gorilla gorilla					
31	Eat	All ages	C	Yes	Annett & Annett, 1991
38	Food process (6 types)	All ages	W	Yes	Byrne & Byrne, 1991
38	Leaf process (28 elements)	All ages	W	Yes	Byrne & Byrne, 1993
29	Walk–run, eat, sit–support, climb, hang, carry–hold, groom, touch, manipulate, brachiate, scratch, *hit-throw*	All ages	C	Yes	Heestand, 1986
47	Touch self or other, hit, *throw*, eat, manipulate object (large or small), dig, gesture, misc.	All ages	C	Yes	Shafer, 1987
Pongo pygmaeus					
35	Touch face or body; feed	All ages	C/W	?	Rogers & Kaplan, 1992
Pan paniscus					
None					

beringei (Byrne & Byrne, 1991, 1993). Such concentrated folivory is not typical of the much more common western lowland gorilla, *Gorilla gorilla gorilla* (Tutin *et al.*, 1991), upon whom all captive studies have had to depend. Only one major study has systematically compared species of apes: Heestand (1986) presented data on chimpanzees ($N = 20$), gorillas ($N = 29$), orangutans ($N = 8$), and siamangs ($N = 11$) from 5 zoos, using 12 categories of behavior.

Table 19.4 presents data on laterality of hand-use in

Table 19.4. *Laterality of hand use in wild chimpanzees, by individual: tool use in italic*

Behavioral pattern	All left	Signi. left	Ambilateral	Signi. right	All right	Ss Total
Boesch, 1991						
Reach/Pick up	1	2	13	3	1	20
Groom (social)	0	0	10	4	1	15
Dip sponge	3	1	3	0	9	16
Hammer nuts	8	10	5	8	10	41
Sugiyama et al., 1993						
Reach/pick	0	1	18	0	0	19
Hammer nuts	5	0	1	2	5	13
Marchant & McGrew, 1995						
Pluck food	0	0	32	0	0	32
Scratch self	0	1	34	1	0	36
Eat	0	1	31	3	0	35
Groom social	0	1	28	2	0	31
Termite–Fish	12	4	9	5	6	36

Note: Statistical significance when $z \geqslant 1.97$, $p < 0.05$, 2-tailed, $N \geqslant 6$.

wild chimpanzees; each individual giving enough data is classified on a 5-point scale. Categories 'all left' and 'all right' refer to *exclusive* use of that hand for a given task; 'signi. left' and 'signi. right' refer to a statistically significant *preference* for that hand in that task; 'ambilateral' refers to the interchangeability of the two hands, that is, no statistically significant difference between left and right.

For Taï in Ivory Coast, Boesch (1991) presented raw data for adults only ($Ns = 15$–41) on four tasks of apparently increasing motor complexity. Reaching to pick up fruit on the forest-floor was mostly unlateralized (13/20), as was social grooming (10/15) (see Fig. 19.1). The two forms of tool use showed a very different picture: in using wadges as sponges to dip drinking water from puddles and in using wooden or stone hammers to crack open nuts, few individuals were ambilateral and most were highly lateralized. For dipping, 12 of 16 used only one hand, as did 18 of 41 for hammering (see Fig. 19.2). However, there was no evidence of overall dominance to the right or left.

In the smaller community at Bossou in Guinea, Sugiyama *et al.* (1993) compared two tasks: picking fruits and leaves from branches and using stone hammers to crack oil palm nuts on stone anvils. Like Taï, Bossou's chimpanzees were unlateralized for bouts of simple reaching

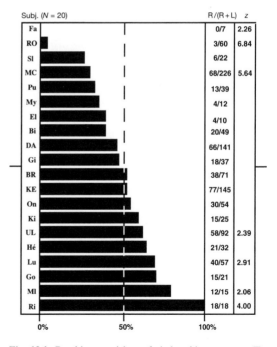

Fig. 19.1. Reaching to pick up fruit by chimpanzees at Taï, percentage of right-handed responses (from Boesch, 1991). Double upper-case codes are males; single upper and lower-case codes are females. This convention applies to Figs 19.1–19.5, 19.7 and 19.11. Only statistically significant z-scores are given.

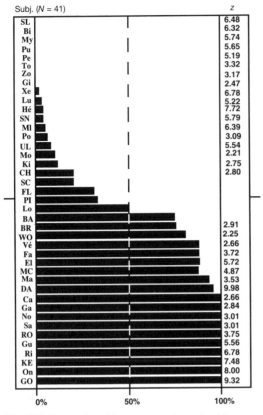

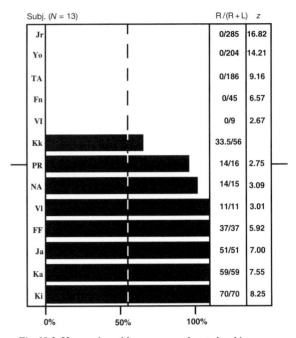

Fig. 19.2. Hammering with stone to crack nuts by chimpanzees at Taï, percentage of right-handed responses (from Boesch, 1991).

Fig. 19.3. Hammering with stone to crack nuts by chimpanzees at Bossou, percentage of right-handed responses (from Sugiyama *et al.*, 1993).

(18/19) but were highly lateralized for complex tool-use: 10 of 13 individuals were exclusively left- or right-preferent (see Fig. 19.3; also Matsuzawa, Chapter 15). Again, there was no overall bias to either side.

At Gombe in Tanzania, Marchant & McGrew (1996) recorded almost 10 000 bouts of hand-use; the four most common, non-tool-use categories are given in Table 19.4. Scratching oneself, eating, social grooming and reaching to pluck attached plant foods all showed a very similar pattern of random use of either hand. (This was consistent throughout all non-tool-use categories of behavior for which sufficient numbers of bouts were recorded on enough subjects; see Fig. 19.4.) In contrast, for the hand used to insert the probe in termite fishing, significant lateralization emerged: half (18 of 36) of individuals showed exclusive use of either left or right hand (see Fig. 19.5). (An earlier pilot study on only 15 subjects showed a similar pattern; McGrew & Marchant, 1992.)

For Mahale in Tanzania, Nishida & Hiraiwa (1982) presented data on fishing for arboreal, wood-boring ants, with 68 clearly independent bouts for 20 subjects. Nishida (personal communication) reports that additional data show most individuals to be ambilateral for this arboreal form of tool use, but no data on non-tool-use tasks are yet available.

To summarize, results from four field sites show the following pattern of laterality of hand-use at population level: only terrestrial tool-use is lateralized, but it takes the form of the strongest individual preferences, rather than task specialization. There is no evidence of handedness for any non-tool-use activity, and it appears that for a variety of tasks, chimpanzees randomly use whichever hand is convenient at the time.

Table 19.5 presents results on laterality of hand-use in captive chimpanzees. Finch's (1941) pioneering study at Orange Park, Florida, used four tasks of acquiring fruit, ranging from simple picking-up of pieces of food to a demanding string-pulling task. Each subject got 8 blocks of 100 trials, and the data were pooled into a single dependent variable with 800, apparently non-independent events. Of the 30 subjects, 26 were adults

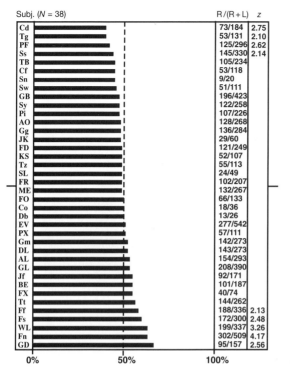

Fig. 19.4. Pooled hand use, except tool use, by chimpanzees at Gombe, percentage of right-handed responses (from Marchant & McGrew, 1995).

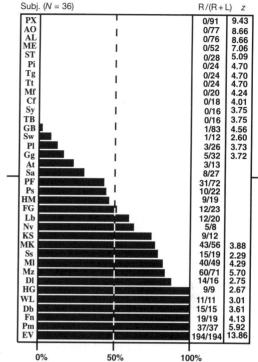

Fig. 19.5. Fishing for termites with a probe by chimpanzees at Gombe, percentage of right-handed responses (from McGrew & Marchant, in preparation).

and 25 were females. The result was almost total (29/30) but incomplete lateralization: only one subject was unlateralized, but only one subject used one hand *exclusively* for obtaining pieces of fruit (see Table 19.5 and Fig. 19.6).

Marchant's (1983) 27 subjects lived on islands in Lion Country Safari Park, Florida, and were observed ethologically (and a subset was tested experimentally). For five of the six behavioral patterns (hold object, social groom, carry object, feed, throw object) at least half of the subjects were unlateralized, albeit to varying extents: 17 of 20 for groom, but only 9 of 18 for the only tool use pattern recorded, throwing. The exception was non-social reaching for food or another inanimate object, or for another chimpanzee, for which 21 of 26 subjects were lateralized (see Fig. 19.7). No subject showed exclusive use of either hand for any category, and there was no significant skew to left or right either within or across tasks.

Heestand (1986) studied 20 chimpanzees in 3 zoos, recording 12 compound categories of behavior; separate

bouts of behavior were demarcated only by a pause of at least 3 seconds. She presented results for only five tasks: *walk–run*, refers to the leading arm or leg in quadrupedal or bipedal walking or running; *sit–support* refers to the arm or leg leaned against while sitting; *climb* refers to the leading arm or leg in vertical ascent or descent. Thus, these three were heterogenous categories, not specified as to upper or lower limb. Two categories of *eat–manipulate* refer to any object manipulated or eaten, using either fingers (precision grip) or whole hand (power grip). Four of the five tasks showed predominant lack of laterality, ranging from 14–17 of 20 subjects. *Walk–run*, on the other hand, was lateralized, and the trend toward right-sided task specialization of 13 of 17 subjects almost reached statistical significance ($z = 1.94$, $p = 0.052$). No subject showed exclusive use of one hand (or foot) for any activity.

In a series of studies, Hopkins and his colleagues have made use of the big population of *Pan troglodytes* at the Yerkes Primate Center, Georgia, where breeding

Table 19.5. *Laterality of hand use in captive chimpanzees by individual: tool use in italic*

Behavioral pattern	All left	Signi. left	Ambilateral	Signi. right	All right	Ss Total
Finch, 1941						
Obtain fruit	0	15	1	13	1	30
Marchant, 1983						
Groom	0	2	17	1	0	20
Hold	0	4	21	1	0	26
Carry	0	7	15	4	0	26
Feed	0	9	13	4	0	26
Throw	0	4	9	5	0	18
Reach	0	11	5	10	0	26
Heestand, 1986						
Eat/manip. (precision)	0	11	7	2	0	20
Sit–support	0	41	6	0	0	20
Eat/manip. (power)	0	3	15	2	0	20
Climb	0	0	14	6	0	20
Walk–run	0	4	3	13	0	20
Hopkins, 1993						
Reach	0	7	19	14	0	40
Hopkins et al., 1993a						
Throw	3	1	6	2	9	21
Hopkins, 1994a						
Bimanual feeding	0	25	66	49	0	140
Hopkins, 1994b						
Reach	5	14	5	23	4	51
Tonooka & Matsuzawa, 1995						
Reach	2	20	41	17	0	80
Colell et al., 1995						
Reach	0	10	6	15	0	31

Note: Statistical significance when $z \geqslant 1.97$, $p < 0.05$, 2-tailed, $N \geqslant 6$.

involves both mother-rearing and hand-rearing by humans. In a study of 40 mostly captive-born (33/40) chimpanzees, Hopkins (1993) recorded reaching to pick up peanuts or raisins while the subject was quadrupedal or bipedal. The criterion for demarcation of trials was postural repositioning (unspecified). About half (19/40) the subjects showed ambilaterality, and the other half showed a non-significant trend (7L : 14R, $z = 1.53$, $p = 0.12$) toward right-sided task specialization. Hopkins (1995a) also studied 51 juvenile chimpanzees (2–5 years old) reaching to pick up peanuts placed just outside a meshed barrier. The criterion for separate bouts of reaching was intervening postural readjustment. (This task closely approximated the simplest of Finch's, 1941,

four measures of food procurement.) Like Finch, the results showed almost all (46/51) subjects to be lateralized, and 9 of the 51 showed exclusive use of one hand. However, there was no overall bias to left or right side.

Hopkins *et al.* (1993a) recorded underarm or overarm throwing from either quadrupedal or bipedal postures in 36 chimpanzees. No criteria for delimiting bouts were mentioned. Most (12/21) of the subjects were highly lateralized, showing exclusive use of either left or right hand, while only six were unlateralized. Again, there was a trend toward right-sided task specialization, but this did not reach statistical significance (4L : 11R, $z = 1.55$, $p = 0.12$).

Hopkins' (1994) monumental study of 140 chimpan-

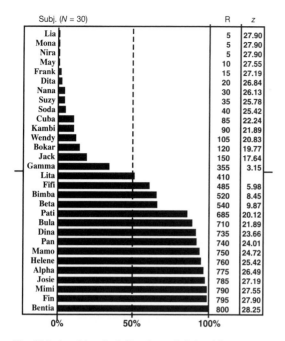

Fig. 19.6. Acquiring food (4 tasks pooled) by chimpanzees at Orange Park, percentage of right-handed responses (from Finch, 1941). R + L = 800.

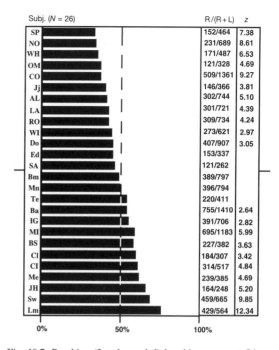

Fig. 19.7. Reaching (2 tasks pooled) by chimpanzees at Lion Country, percentage of right-handed responses (from Marchant, 1983).

zees eating with one hand while holding other food with the other hand (bimanual feeding) showed about half ($N = 66$) the subjects to be unlateralized, while of those that were lateralized ($N = 74$), about twice as many were right-sided as left-sided (25L : 49R, $z = 2.67$, $p = 0.007$). The right-sided predominance held for both mother- and nursery-reared males and females (see Fig. 19.8).

Colell *et al.* (1995) observed 31 chimpanzees, 2 bonobos and 3 orangutans in 3 zoos; they recorded four behavioral patterns: reaching for food, scooping up water in the cupped hand, swirling water in the moat and throwing. Only reaching by chimpanzees occurred often enough for inclusion here: only six subjects were unlateralized, and of those that were, there was no statistically significant bias to either left or right (10L : 15R, $z = 0.80$, $p = 0.42$).

Tonooka & Matsuzawa (1995) studied 80 chimpanzees in 3 laboratories in Japan and recorded the hand and grip used to pick up raisins from the floor. No criteria for bouts were given. About half the subjects were unlateralized and the remainder were split evenly between left and right (22L : 17R, $z = 0.94$, $p = 0.34$).

Aruguete *et al.*'s (1992) study at Tempe of 27 chim-

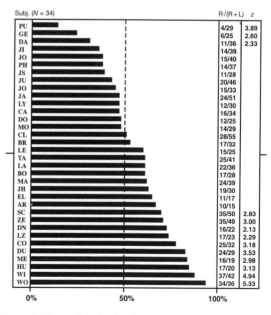

Fig. 19.8. Bimanual feeding by chimpanzees at Yerkes, percentage of right-handed responses (from Hopkins, 1994). Partial results only: nursery-reared males.

panzees showing four kinds of touching did not supply raw data. They defined bouts as uninterrupted sequences with no significant postural reorientation between responses. Without raw scores, these studies unfortunately cannot be compared systematically here. The same limitation applies to Hopkins' (1995b) study of 110 chimpanzees showing bimanual manipulation of a tube containing peanut butter; only summary statistics are provided and demarcation of bouts is not mentioned.

In summary, results on chimpanzee laterality from laboratories and zoos show a more heterogeneous picture than do those from nature. Reaching to pick up food-items may be unlateralized (Heestand, 1986), mildly lateralized (Marchant, 1983; Hopkins, 1993; Tonooka & Matsuzawa, 1995), highly lateralized but not exclusively so (Finch, 1941; Hopkins, 1994; Colell *et al.*, 1995), or highly lateralized, including some (18%) exclusive hand-use (Hopkins, 1995a), depending on the population studied.

Other patterns of non-tool use show varying degrees of lateralization, but only tool use in throwing shows exclusivity of one hand to be the norm (Hopkins *et al.*, 1993a). Even that is not matched by throwing in Marchant's (1983) study, whose subjects were only mildly lateralized.

Table 19.6 shows results for laterality of hand function in gorillas in the wild and in captivity. Annett & Annett (1991) recorded reaching to pick up from the substrate and to eat food, using 31 subjects (presumably, western lowland) from 5 European zoos. No criteria to distinguish bouts were given and large numbers of data-points were swiftly accrued (mean of 136 reaches in mean of 3.4 sessions). The modal position was ambilaterality, but overall 61% of individuals showed significant hand preferences to left or right. None showed exclusive use of either hand.

Working at Karisoke in the Virunga Volcanoes of Rwanda, Byrne & Byrne (1991) recorded hand use in the bimanual processing of foliage for food for 38 mountain gorillas. They presented raw scores for four of these: nettle leaves, thistle leaves, celery pith and *Galium* stems and leaves (see Fig. 19.9). For all four plant-food-types, 87–92% of individuals were lateralized for processing, and for three of the four foods, most of the subjects used one dominant hand exclusively. (The exception was *Galium*, for which only 24% of subjects showed exclusivity in hand use.)

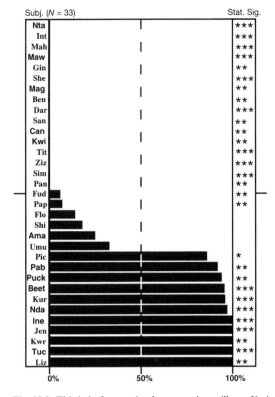

Fig. 19.9. Thistle leaf processing by mountain gorillas at Karisoke, percentage of right-handed responses (from Byrne & Byrne, 1991). * = $p \leqslant 0.05$, ** = $p \leqslant 0.01$; *** = $p \leqslant 0.001$.

In a follow-up article on the same subjects, Byrne & Byrne (1993) looked at the 28 component elements of the processing techniques for the leaves of nettle and *Galium*. They lumped these into the three stages of processing: procurement, holding or accumulating and eating. For all six tasks, the majority of subjects were significantly lateralized for the dominant hand, ranging from 74–97% of individuals. However, a smaller but highly variable proportion of individuals used exclusively one hand: 0% for *Galium* procurement to 50% for nettle eating. In neither study was there any indication of an overall bias to the left or right side.

Heestand's (1986) results for gorillas in five zoos closely resembled hers for chimpanzees. The same four categories of behavior were barely lateralized: sit-support, climb, eat/manipulate (precision grip), eat/manipulate (power grip). The same fifth category, walk-run, was highly (21/29) lateralized, and even more

Table 19.6. *Laterality of hand use in gorillas, by individual*

Behavioral pattern	All left	Signi. left	Ambilateral	Signi. right	All right	Ss Total
Annett & Annett, 1991						
Eat	0	8	12	11	0	31
Byrne & Byrne, 1991						
Galium	8	13	3	13	1	38
Nettle	12	7	4	4	9	36
Thistle leaf	16	2	4	6	5	33
Celery	11	7	5	3	12	38
Byrne & Byrne, 1993						
Procure (N)	1	9	10	16	2	38
Procure (G)	0	20	5	13	0	38
Accumul. (G)	2	19	4	12	1	38
Hold (N)	3	11	3	19	2	38
Eat (G)	7	16	2	12	1	38
Eat (N)	11	11	1	7	8	38
Heestand, 1986						
Eat/manip. (power)	0	1	25	3	0	29
Eat/manip. (precision)	0	2	23	4	0	29
Climb	0	1	20	4	0	25
Sit–support	0	4	20	5	0	29
Walk–run	0	2	8	19	0	29
Shafer, 1987						
Touch other	0	3	27	6	0	36
Hit/slap	0	2	24	7	1	33
Gesture	0	1	13	3	1	18
Handle large object	0	3	32	10	0	45
Touch self	0	4	25	16	0	47
Miscellaneous	0	4	28	15	0	47
Handle small object	0	5	25	16	0	46
Eat	0	10	17	20	0	47

Note: Statistical significance when $z \geq 1.97$, < 0.05, 2-tailed, $N \geq 6$.

skewed to the right in the gorillas than in the chimpanzees (see Fig. 19.10). Walk–run's proportions of 2L : 19R are highly statistically significantly different ($z = 3.45$, $p < 0.0000$), but no individual made exclusive use of one limb for any task.

Shafer's (1987) study of 47 gorillas from 5 zoos using 10 categories is the most extensive research on laterality done with any species of ape. Only one category, dig/sift, failed to occur often enough to be useful here, though the miscellaneous category included everything from catching feces to chest-beating. No indication was given of use of discrete bouts, rather than every event in a sequence, and the sample sizes are massive, e.g. 20 156 cases of eat, giving a mean of 429 data-points per subject. For seven of the eight categories, most individuals were unlateralized; the exception was eating in which 64% (30/47) of the gorillas were significantly lateralized. All eight measures are skewed to the right side, and for three of these the right-sidedness was statistically significant: touch self (16/20, $z = 2.46$, $p = 0.01$); miscellaneous (15/19, $z = 2.29$, $p = 0.02$); manipulate small object (16/21, $z = 2.18$, $p = 0.03$, see Fig. 19.11). However, in only 2 of the 319 possible combinations of subject and behavioral category did an individual make exclusive use of one hand to perform the task.

To summarize, most captive lowland gorillas are unlateralized for hand-use, but when a behavior pattern is skewed it tends to be to the right side, though this

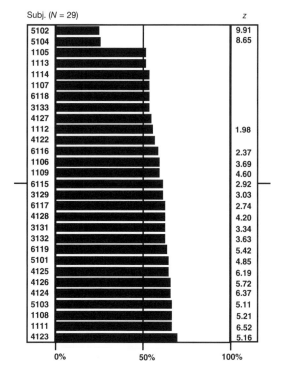

Fig. 19.10. Leading limb in walk-run by western lowland gorillas in five zoos, percentage of right-limbed responses (from Heestand, 1986).

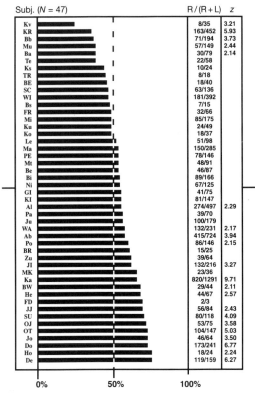

Fig. 19.11. Manipulation of small objects by western lowland gorillas in five zoos, percentage of right-handed responses (from Shafer, 1987).

reaches statistical significance in only a few cases. The data from wild mountain gorillas are notably different, being on very specialized activities, and these show consistently high levels of lateralization, with many cases of exclusive use of one hand, but no overall tendency toward left or right.

LEVELS OF LATERALITY?

There seem to be at least five levels of lateralization of hand use shown by great apes (see Fig. 19.12). Level 1 is the simplest, in which virtually all members of a population show no hand preferences, that is, when the norm for individuals is ambilaterality. (The few exceptions then are likely to be statistical anomalies; for every 20 statistical tests done when $\alpha = 0.05$, one is likely to be a spurious false-positive result.) Level 2 is when a significant proportion of individuals are significantly lateralized

to either side, but overall the distribution remains symmetrical. Level 3 is when a significant proportion of a population show not only laterality, but exclusive use of one hand or the other while still retaining symmetry. Through this level the overall distribution is still one of hand preference, as there is no sign of a skew to left or right. Level 4 is when the distribution of lateralized individuals is asymmetrical, that is there are significantly more right- than left-preferent members of a population, or vice-versa. Finally, Level 5 is Level 4 carried a step further, in which the distribution of exclusive users is asymmetrical, either biased to left or right.

So, what does the laterality of great apes look like in this framework? Level 1 is the norm for wild chimpanzee hand-use, except for tool use. It also applies to a wide variety of behavioral patterns in some captive chimpanzees (Marchant, 1983; Heestand, 1986; Fig. 19.13) and

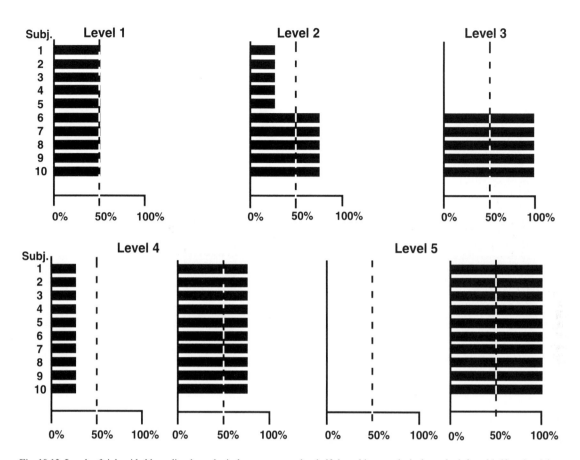

Fig. 19.12. Levels of right-sided laterality: hypothetical, extreme distribution of ten subjects through five levels of lateralization. Level 1 has all subjects ambilateral; Level 2 has half the subjects significantly lateralized to the left and half to the right; Level 3 has half the subjects exclusively to the left and half to the right; Level 4 has either all subjects significantly to the left or to the right; Level 5 has either all subjects exclusively to the left or to the right.

some captive gorillas (Heestand, 1986; Shafer, 1987). This may be the baseline condition for an arboreally-derived frugivore–folivore who forages in a 3-dimensional world in which food-items are equally available in all directions, and the same convenience factor may also operate in the more 2-dimensional world of terrestriality.

Level 2 laterality occurs exceptionally for a few behavioral categories in captive chimpanzees: reach in Marchant (1983) and in Colell et al. (1995), walk–run in Heestand (1986). It is the norm for some single-variable studies: acquire fruit in Finch (1941), reach for both adults and juveniles in Hopkins (1993, 1995a) and in Tonooka & Matsuzawa (1995). A similar picture emerges for captive gorillas: eat in Annett & Annett (1991), walk–

run in Heestand (1986), eat in Shafer (1987). We suspect that many of these results are methodological artifacts. Walk–run as recorded by Heestand lumps type of locomotion as well as type of limb. Finch (1941) combined four measures of very different complexity; we suspect that taken separately his measures might have shown a range from Level 1 for picking up fruit-pieces to Level 3 for string-pulling. Finally, some results (e.g. Annett & Annett, 1991) are likely to be statistical artifacts of inflated sample size that enhanced normal variance to statistically significant levels. However, Hopkins' (1993, 1995a) results on reaching cannot be so impeached. One is tempted to suspect that the high proportion of human-reared chimpanzees at Yerkes might account for the

Fig. 19.13. Chimpanzees at Lion Country Safari eating. *Left to right*: adult female holds in left hand; adult male eats with both hands; adult male reaches to pick up food with right hand. (Photo by L. Marchant)

result, given Shafer's (1987) finding on captive-born versus zoo-born gorillas cited above, but Hopkins (1993, p. 165) specifically excludes this.

Level 3 laterality is more circumscribed: it is characteristic of most tool use and skillful food processing. For wild chimpanzees cracking nuts, fishing for termites or sponging for water, exclusive use of one hand or the other is the norm (Boesch, 1991; McGrew & Marchant, 1992; Sugiyama *et al.*, 1993; Matsuzawa, Chapter 15; Fig. 19.14). For captive chimpanzees throwing objects, exclusive hand use is the norm in one population (Hopkins *et al.*, 1993a) but not another (Marchant, 1983). Notably missing are data from captive apes of any species on hand preference in tool use along a gradient of complexity. Level 3 laterality also characterizes many of Byrne & Byrne's (1991, 1993) data on wild gorillas, though Level 2 patterns are also present. Their carefully documented sequential analyses of plant food processing show it to demand equivalent motor skills to tool use, at least in some components. Notably the least exclusively lateralized elements are the initial ones of procurement, which entails reaching, yanking, picking, etc. vegetation (Byrne & Byrne, 1993, p. 245). Comparable

Fig. 19.14. Adult female chimpanzee uses hammer stone to crack oil palm nut on bough anvil at Bassa Islands, Liberia. (Photo by A. Hannah)

studies of skilled food processing in captive apes are frustratingly absent, as most eating in captivity consists of little more than putting pre-processed items into the mouth.

Level 4 laterality emerges strongly only in Shafer's (1987) and Heestand's (1986) studies of gorillas and in Hopkins' (1994) study of chimpanzees. For Shafer, even if only three of the eight behavioral patterns show statistically significant right-sidedness, the fact that all eight patterns show rightward tendencies is itself a significant result ($z = 2.48$, $p = 0.01$). For Heestand, only one of five patterns, walk–run, is so skewed: 2L : 19R, $z = 3.49$, $p = 0.0004$. Comparing Shafer's and Heestand's results should be informative, since three of the five zoos studied by each were the same, but there is little overlap in the categories of behavior recorded. In Heestand's (1986) study, few gorillas were lateralized for eat-manipulate; in Shafer's (1987) study, when gorillas were lateralized for eat ($p = 0.10$), handle large objects ($p = 0.09$) and handle small objects ($p = 0.03$), it was always to the right side. It may be that huge and probably invalid sample sizes produced spurious near-significant trends and significant differences. Thus, MacNeilage et al.'s (1991) quotation, which opened this chapter, should be taken with caution (see also McGrew & Marchant, 1993).

For Hopkins' (1994) study of which hand is used in eating while the other hand holds food (bimanual feeding), the right hand was preferred over the left, though only in 49 of 140 subjects. In another bimanual study, in which one hand holds a tube while the other hand pokes out food, a similar right-sided task specialization emerged: 30L : 54R, $z = 2.51$, $p = 0.01$ (Hopkins, 1995b, personal communication).

Level 5 laterality occurs only for processing of thistle leaves by wild mountain gorillas (Byrne & Byrne, 1991), and the significant bias is to the left: 16L : 5R, $z = 2.18$, $p = 0.03$ (Fig. 19.15). It is hard to know what to make of a single result and until more persuasive data emerge, the presence of Level 5 laterality in great apes remains unproven.

What, if anything, can be said about laterality of hand use in the other two species of great ape? For orangutans, the most extensive study was that of Olson et al. (1990), who tested 13 orangutans at Yerkes on three tasks: picking up small fruits off the floor ($N = 12$), through chain-link fencing ($N = 12$), and from a WGTA testing

Fig. 19.15. Adult female mountain gorilla processes terrestrial herbaceous vegetation in the Virunga Volcanoes. (Photo by D. Watts)

device ($N = 8$). In these small samples there was Level 2 laterality but much inconsistency across tests.

For bonobos, the only systematic research is by Hopkins et al. (1993b) on 11 bonobos at Yerkes and by Hopkins & de Waal (1995) on 10 bonobos at San Diego Zoo. They scored seven behavioral categories from videotape: feed, carry, touch self, touch face, leading limb, gesture, reach. Bouts were demarcated by a change of hand activity for at least 3 seconds. Raw scores were presented for all but feed, and all but one of these showed only Level 1 lateralization. The exception was leading limb, for 5 of 11 individuals in Hopkins et al. (1993b) and 7 of 10 in Hopkins & de Waal (1995) showed significant right preference, although it was not specified for hand or foot. It goes without saying that more data are needed.

COMPARISONS WITH HUMANS

> The vast majority of human accomplishments have resulted perhaps not from the hands, broadly defined, but from one hand in particular: the right one. Roughly 90% of the world's population is right-handed: such people tend to show more strength and skill with the right hand than with the left, and they typically and unconsciously employ it for most tasks.
>
> *MacNeilage, Studdert-Kennedy & Lindblom, 1993, p. 32*

The standard story from textbooks (Springer & Deutsch, 1989) and scholarly treatises (Bradshaw & Rogers, 1993) is as given in the above quotation: that right-handedness is universally predominant. That is, the typical *Homo sapiens* shows Level 5 right-handedness. In fact, there is very little cross-cultural evidence to support this conclusion. In a recent meta-analysis, McGrew & Marchant (1994) showed that there are few valid and reliable empirical studies, and that the number of reported left-handers in a given population may vary from 0–23%.

Within Western populations, hand-use is not simply dichotomized. Putting aside the small proportion of truly ambidextrous individuals, the rest of a typical population shows not a simple gradient from totally left-handed to totally right-handed (Healey *et al.*, 1986), but instead groups of varying intermediate status (Peters & Murphy, 1992). These data typically come from inventories administered as questionnaires, in which for each task or activity an individual is asked to score it as being done (1) always with left hand, (2) usually left, (3) either left or right, (4) usually right, or (5) always right. These inventories are not random or representative samples of common or everyday manual activities, but are heavily biased toward precision tool use (Marchant *et al.*, 1995).

Marchant *et al.* (1995) coded ongoing behavior in daily life from cinematic archives of three traditional societies: G/wi gatherer–hunters of Botswana, Himba pastoralists of Namibia and Yanomamö horticulturalists of Venezuela. This ethological study found in all three cultures that most common non-tool-use activities were unlateralized, that is showed only Level 1 lateralization. This was true for both object manipulation (e.g. eat, handle, carry) and for non-object manipulation (e.g. touch, gesture, groom) in each culture. In contrast, gross or less skilled tool-use with a power grip showed Level

4 laterality, and only skilled tool use with precision grip was at Level 5: of 58 individuals combined from the three cultures, 32 used only the right hand, 1 used only the left hand and 22 used both hands.

In conclusion, ethological analyses of what hominoids, either apes or humans, actually do spontaneously in real life (as opposed to what they can be induced to do artificially) shows them to be less lateralized than previously thought. At least with tool use, apes go beyond hand preference to committed individual lateralization, that is exclusive use of one hand, while humans take this a step further to species-level handedness.

ACKNOWLEDGMENTS

This paper was originally prepared for the Wenner-Gren conference 'The Great Apes Revisited,' held November 12–19, 1994 in Cabo San Lucas, Baja, California. We thank Bill Hopkins for sharing unpublished data, Carol Kist for manuscript preparation, and Martha Holder, Bill Hopkins and Toshisada Nishida for critical comments on the manuscript.

REFERENCES

Annett, M. & Annett, J. 1991. Handedness for eating in gorillas. *Cortex*, **27**, 269–75.

Aruguete, M. S., Ely, E. A. & King, J. E. 1992. Laterality in spontaneous motor activity of chimpanzees and squirrel monkeys. *American Journal of Primatology*, **27**, 177–88.

Boesch, C. 1991. Handedness in wild chimpanzees. *International Journal of Primatology*, **12**, 541–58.

Bradshaw, J. L. & Rogers, L. J. 1993. *The Evolution of Lateral Asymmetries, Language, Tool Use, and Intellect*. San Diego, CA: Academic Press.

Byrne, R. W. & Byrne, J. M. 1991. Hand preferences in the skilled gathering tasks of mountain gorillas (*Gorilla g. beringei*). *Cortex*, **27**, 521–46.

Byrne, R. W. & Byrne, J. M. 1993. Complex leaf gathering skills of mountain gorillas (*Gorilla g. beringei*): Variability and standardization. *American Journal of Primatology*, **31**, 241–61.

Chance, P. 1994. *Learning and Behavior*. 3rd edn. Pacific Grove, CA: Brooks/Cole.

Colell, M., Segarra, M. D. & Sabater Pi, J. 1995. Hand preferences in chimpanzees (*Pan troglodytes*), bonobos (*Pan paniscus*) and orangutans (*Pongo pygmaeus*) in food-reaching and other daily activities. *International Journal of Primatology*, **16**, 413–34.

Finch, G. 1941. Chimpanzee handedness. *Science*, **94**, 117–18.

Healey, J. M., Liederman, J. & Geschwind, N. 1986. Handedness is not a unidimensional trait. *Cortex*, **22**, 33–53.

Heestand, J. E. 1986. *Behavioral Lateralization in Four Species of Apes?* Ann Arbor, MI: University Microfilms International.

Heyes, C. M. & Dawson, G. R. 1990. A demonstration of observational learning in rats using a bidirectional control. *Quarterly Journal of Experimental Psychology*, **42B**, 59–71.

Hopkins, W. D. 1993. Posture and reaching in chimpanzees (*Pan troglodytes*) and orangutans (*Pongo pygmaeus*). *Journal of Comparative Psychology*, **107**, 162–8.

Hopkins, W. D. 1994. Hand preferences for bimanual feeding in 140 captive chimpanzees (*Pan troglodytes*): rearing and ontogenetic determinants. *Developmental Psychobiology*, **217**, 395–407.

Hopkins, W. D. 1995a. Hand preferences for simple reaching in juvenile chimpanzees (*Pan troglodytes*): continuity in development. *Developmental Psychology*, **31**, 619–25.

Hopkins, W. D. 1995b. Hand preferences for a coordinated bimanual task in 110 chimpanzees (*Pan troglodytes*): cross-sectional analysis. *Journal of Comparative Psychology*, **109**, 291–7.

Hopkins, W. D. & Morris, R. D. 1993. Handedness in great apes: a review of findings. *International Journal of Primatology*, **14**, 1–25.

Hopkins, W. D. & de Waal, F. B. M. 1995. Behavioral laterality in captive bonobos (*Pan paniscus*): Replication and extension. *International Journal of Primatology*, **16**, 261–76.

Hopkins, W. D., Bard, K. A., Jones, A. & Bales, S. L. 1993a. Chimpanzee hand preference in throwing and infant cradling: implications for the origin of human handedness. *Current Anthropology*, **34**, 786–90.

Hopkins, W. D., Bennett, A. J., Bales, S. L., Lee, J. & Ward, J. 1993b. Behavioral laterality in captive bonobos (*Pan paniscus*). *Journal of Comparative Psychology*, **107**, 403–10.

Machlis, L., Dodd, P. W. D. & Fentress, J. C. 1985. The pooling fallacy – problems arising when individuals contribute more than one observation to the data set. *Zeitschrift für Tierpsychologie*, **68**, 201–14.

MacNeilage, P. F., Studdert-Kennedy, M. G. &

Lindblom, B. 1987. Primate handedness reconsidered. *Behavioral and Brain Sciences*, **10**, 247–303.

MacNeilage, P. F., Studdert-Kennedy, M. G. & Lindblom, B. 1988. Primate handedness: a foot in the door. *Behavioral and Brain Sciences*, **11**, 720–46.

MacNeilage, P. F., Studdert-Kennedy, M. G. & Lindblom, B. 1991. Primate handedness: the other theory, the other hand and the other attitude. *Behavioral and Brain Sciences*, **14**, 338–49.

MacNeilage, P. F., Studdert-Kennedy, M. G. & Lindblom, B. 1993. Hand signals. Right side, left brain and the origins of language. *The Sciences*, **Jan.–Feb.**, pp. 32–7.

Marchant, L. F. 1983. *Hand Preference among Captive Island Groups of Chimpanzees (Pan troglodytes)*. Ann Arbor, MI: University Microfilms International.

Marchant, L. F. & McGrew, W. C. 1991. Laterality of function in apes: a meta-analysis of methods. *Journal of Human Evolution*, **21**, 425–38.

Marchant, L. F. & McGrew, W. C. 1996. Laterality of limb function in wild chimpanzees of Gombe National Park: comprehensive study of spontaneous activities. *Journal of Human Evolution* (in press).

Marchant, L. F., McGrew, W. C. & Eibl-Eibesfeldt, I. 1995. Is human handedness universal? ethological analyses from three traditional cultures. *Ethology*, **101**, 239–58.

Martin, P. & Bateson, P. 1993. *Measuring Behaviour*, 2nd edn. Cambridge: Cambridge University Press.

McGrew, W. C. & Marchant, L. F. 1992. Chimpanzees, tools, and termites: hand preference or handedness? *Current Anthropology*, **33**, 114–19.

McGrew, W. C. & Marchant, L. F. 1993. Are gorillas right-handed or not? *Human Evolution*, **8**, 17–23.

McGrew, W. C. & Marchant, L. F. 1994. Primate ethology: a perspective on human and nonhuman handedness. In: *Handbook of Psychological Anthropology*, ed. P. K. Bock, pp. 171–84. Westport, CT: Greenwood Press.

Nishida, T. & Hiraiwa, M. 1982. Natural history of a tool-using behavior by wild chimpanzees in feeding upon wood-boring ants. *Journal of Human Evolution*, **11**, 73–99.

Olson, D. A., Ellis, J. E. & Nadler, R. D. 1990. Hand preferences in captive gorillas, orang-utans and gibbons. *American Journal of Primatology*, **20**, 83–94.

Peters, M. & Murphy, K. 1992. Cluster analysis reveals at least three, and possibly five distinct handedness groups. *Neuropsychologia*, **30**, 373–80.

Rogers, L. & Kaplan, G. 1992. Lateralization of hand-use in orangutans. Abstract, 6th Biennial Meeting of International Society for Comparative Psychology, Brussels.

Shafer, D. D. 1987. Patterns of hand preference among captive gorillas. MA thesis, San Francisco State University, 142 pp.

Siegel, S. 1956. *Nonparametric Statistics for the Behavioral Sciences*. New York: McGraw-Hill.

Springer, S. P. & Deutsch, G. 1989. *Left Brain, Right Brain*, 3rd edn. New York: W. H. Freeman.

Sugiyama, Y., Fushimi, T., Sakura, O. & Matsuzawa, T. 1993. Hand preference and tool use in wild chimpanzees. *Primates*, **34**, 151–59.

Tonooka, R. & Matsuzawa, T. 1995. Hand preference of captive chimpanzees (*Pan troglodytes*) in simple reaching for food. *International Journal of Primatology*, **16**, 17–35.

Tutin, C. E. G., Fernandez, M., Rogers, M. E., Williamson, E. A. & McGrew, W. C. 1991. Foraging profiles of sympatric lowland gorillas and chimpanzees in the Lopé Reserve, Gabon. *Philosophical Transactions of the Royal Society, Series B*, **334**, 179–86.

Ward, J. P. & Hopkins, W. D. (eds.) 1993. *Primate Laterality: Current Behavioral Evidence of Primate Asymmetries*. New York: Springer.

20 • Savanna chimpanzees, referential models and the last common ancestor

JIM MOORE

INTRODUCTION

Many a great ape grant proposal has waxed on about insights that the proposed research would provide into the behavioral ecology of our early ancestors (see Zihlman, Chapter 21), and the term 'model' is ubiquitous in the resulting literature (e.g. Jolly, 1970; Kinzey, 1987). What is a model in this context? It is important to know, because (justified or not) people often draw inferences about modern humans from such models. Skybreak (1984) describes the referential (i.e. analogical with an existing species) baboon model:

> Implicitly or explicitly, savanna baboons have been repeatedly proposed as models for our own earliest origins or as a basis for understanding human aggression, xenophobia, class structure, and the domination of women, all seen as rooted in biological evolution (and by implication, thus very resistant to change).
>
> *Skybreak, 1984, p.84*

One might think that she would reject all such referential models, with their putative biological determinism, but instead she supports Tanner's (1981) chimpanzee model. One of the many meanings of the word 'model' is 'something held up before one for imitation or guidance' (*Webster's New Collegiate Dictionary*, 1951), and one wonders whether there might be some cognitive spillover among usages. It would be understandable; the word is remarkably protean, as illustrated by this imaginary dialogue (italicized words have been used as approximate synonyms of 'model' in the literature):

'Your so-called *hypothesis* is nothing but an untestable *scenario* based loosely on some vague *metaphor*. It has no *theory* supporting it, leaving it no more than a crude *sketch*, an *overview* without substance.'

'On the contrary! It is a detailed *reconstruction* based on the best available *analogy* for the phenomenon, suggesting a reasonable *prototype* for consideration as a *precursor* to hominids.'

In addition to argument over the best referential model (i.e. species) to use, there seems to be much uncertainty about what a model *is* and what *type* of model (if any) is useful in scientific inquiry. I begin by addressing these two issues, and then discuss the potential for use of savanna-dwelling chimpanzees in modeling early hominids and the last common ancestor (LCA) of humans and the African apes. My goal is not to construct a particular scenario, but to pose a series of specific questions that we might ask of a chimpanzee model. By doing so I hope to suggest directions for paleoanthropologically relevant research on modern chimpanzees, and to stimulate discussion regarding the utility of such referential models in research on hominid behavioral ecology.

WHAT IS A MODEL?

A model can be a world view ('the scientific model'), a body of theory (saltational models of evolution), a strict set of formal instructions (a mathematical model), a set of ideas related to a central conceptual focus ('the hunting model'), or a direct referential analogy ('the chimpanzee model'). At first these usages seem divisible into two classes that are sets of linked ideas and concrete physical analogies, but the line is blurred, e.g. the 'social carnivore model.' Table 20.1 presents an attempt to sort out the relevant terminology. It is offered not in any naive hope that the discipline will adopt these distinctions, but to encourage others at least to be explicit about how they use the terms.

Table 20.1. *An idiosyncratic thesaurus for hominid modelers*

Analogy. A relation based on similarity between specific attributes of two things, a model and its referent. The similar properties can be thought of as the positive analogy of the model and dissimilar properties the negative analogy; finally, properties of the model which may or may not be similar to the referent are the neutral analogy – and it is these properties that are of interest (Hesse, 1966, pp. 8–10). Formally, analogy refers to similarity due to convergence rather than common descent (i.e., homology); in the present context the two seem almost inextricably interwoven.

Hypothesis. A specific tentative idea that is in principle amenable to qualitative or quantitative testing; distinguished from scenario in being more narrowly circumscribed.

'Just-so' story. Someone else's model (*sensu lato*), with which one disagrees.

Metaphor. A figure of speech intended to invoke a feeling of similarity without providing a formal hook upon which to hang it; useful rhetorically but not analytically.

Model. (noun) The known member of an analogous pair. (verb) The process of constructing a scenario of or about the unknown member of an analogous pair (the referent).

Overview. Similar to 'sketch,' with little power to mislead.

Precursor. Outside of science fiction, we cannot make a literal precursor of an extinct species; the very concept of an evolutionary 'precursor' is uncomfortably teleological. To be avoided.

Prototype. As with 'precursor,' only more so; carries the additional danger of invoking the image of a Platonic 'ideal type' and so implicitly downplaying variation (ecological, genetic, sexual) among both model and target populations.

Reconstruction. This is not literally possible (except as applied to fossils); as commonly used, it is synonymous with 'scenario' but sounds more precise. Precisely for that reason, the noun should be avoided. As a verb it is more useful; one can use an analogy to help reconstruct (model) a facet of the past for use *in* a scenario.

Scenario. A broad, internally consistent 'story' about the behavior and ecology of a specific extinct taxon, e.g. 'the hunting scenario for *Homo habilis*' (which generates a number of specific hypotheses, e.g. 'if hunting was important, butchery marks on bones should be distributed in X fashion').

Sketch. Deliberately self-effacing, the term cannot do much damage however it is used.

Theory. Too often used as a formal-sounding synonym for 'idea' or 'suggestion,' which is the sense in which it has overlapped with 'model' in the evolutionary literature.

In this paper I will argue for the utility of a chimpanzee analogical model in trying to construct testable scenarios describing the behavioral ecology of the LCA and earliest hominids. Before doing so, the danger of 'mental spillover' between theory and model must be emphasized. It is all too easy to see in the data what theory predicts, or begin incorporating into theory what is really only an aspect of the analogical model (Fedigan, 1992, 16–21). For example, a considerable amount of primatological 'theory' may owe its origins more to the 'terrestrial cercopithecine model' than anything else, given unavoidable biases against arboreal and New World monkeys

(Moore, 1993; Strier, 1994). (Note: there is little difference between referentially using one species or a dozen; the plural of example is examples, not 'law', at least until such time as the sample is exhaustive.)

ARE ANALOGICAL OR REFERENTIAL MODELS USEFUL IN SCIENCE?

There are two axes to this question. First, are referential models anything more than heuristic devices; second, if we accept referential models as useful, what are the roles of analogy and homology in choosing a given model?

Models: mere heuristics, or essential components of understanding?

Hesse (1966) discusses analogies and models using the arguments of physicists Pierre Duhem and N. R. Campbell to characterize two major positions. To Duhem, analogical models may be useful heuristic devices in thinking about a problem, but ideally are supplanted by the final static logical or mathematical theory that fully specifies the solution. Campbell believes that theories are dynamic and depend for growth and predictive ability in new domains upon analogies with existing models; secondarily, to be successful a theory must be intellectually satisfying and this often depends upon being intelligible in terms of a model (Hesse, 1966, pp. 1–5).

Tooby & DeVore (1987) open a similar debate in paleoanthropology. They argue that 'referential models' based on analogy are inherently inferior to 'conceptual/strategic models' derived from evolutionary theory. Without rules for determining which traits are shared by model and referent, choice of a model is arbitrary; furthermore, without some guiding theory there is no way to know which differences are important or what their effects will be. In the absence of such rules and theory, referential modelers are forced to emphasize similarities between model and referent – when the ultimate goal of paleoanthropologists is to understand what makes humans different. Therefore, Tooby & DeVore say, we should focus on developing conceptual rules and theories; individual taxa should be treated as sources of data points for comparative studies, not as models.

It is likely a mistake to try to opt absolutely in favor of either conceptual or referential modeling in paleoanthropology. It is more productive to see each as contributing to overall understanding through a cyclical process of refinement (see Rubinstein et al., 1984, pp. 131–9). The essential step is to attempt to use a combination of referential and conceptual modeling to generate hypotheses that are or may become testable using paleoanthropological or comparative data.

For example, in discussing possible sex differences in hominid food sharing, Tooby & DeVore (1987, p. 218) rely on the theoretically-based assumption that emigration is essential in order to avoid inbreeding depression. Theoretical debates notwithstanding (Moore, 1993), evidence is mounting that in some primates and delphinids, *neither* sex consistently emigrates from its natal social group (Moore, 1992c; Amos et al., 1993). Thus 'referential models' – here, Barbary macaques (*Macaca sylvanus*), pilot whales (*Globicephala melas*) and killer whales (*Orcinus orca*) – can demonstrate the inadequacy of existing theory for conceptual/strategic modeling of a particular unique species: when there are three exceptions to a pattern, there may be more. Clearly humans are not killer whales; these exceptions show that Tooby & DeVore were *premature* in their conclusion but they do not show that the conclusion was *wrong*. Only after discovering the underlying basis for the distribution of natal philopatry by both sexes can we hope to say that early hominid food sharing likely was or was not constrained by emigration patterns. It is the continuing cycle of incorporating insights derived from natural history and referential modeling into the theories used by strategic models that is important.

Analogical or metaphorical models can be important ways of representing phenomena. To explore this issue would lead into philosophy of science on the one hand and into cognitive science and linguistics on the other (see Merrell, 1991; D'Andrade, 1992; Wharton et al., 1994; Holyoak & Thagard, 1995, for starting points). To the extent that (some) people think metaphorically, models of one sort or another are simply unavoidable; as stories they are remarkably common in paleoanthropology (Landau, 1991).

Models: the proper use of homology and analogy

A referential model may be based on homology (similarity due to common descent, e.g. related chimpanzee), or analogy (similarity due to common adaptation; e.g., terrestrial baboon). The importance of distinguishing such bases is illustrated in the controversy over the 'pygmy chimpanzee model' of Zihlman et al. (1978). Responding to Johnson (1981) and to Latimer et al. (1981), Cronin (1983) writes:

> . . . objections to our model center around the observation that *Australopithecus* sp. . . . do not look like *P. paniscus*. This constitutes an absurd mastery of the obvious. . . . We did not say that *P. paniscus* would be identical to the *Australopithecus* sp., but instead that the earliest hominids

would resemble *P. paniscus* more than the other living African hominoids.

Cronin, 1983, p.132.

That so many capable researchers end up arguing at such cross-purposes is more than a little worrying. I believe that there was confusion over the *purpose* of the model, which in principle may have been intended either to: (a) suggest a morphotype of the LCA based on homology, in order to facilitate recognition of relevant fossils; or (b) help illuminate the general characteristics of the LCA, in order to function as an analogy – in which case the key thing is *not* to resemble the target too closely, since the information sought may lie as much in the difference as in the similarity (Hesse, 1966; Hinde, 1976).

In reading the exchange, it seems that neither Zihlman *et al.* (1978) nor Latimer *et al.* (1981) have a very clear idea of which of these was intended. The former cite pygmy chimpanzee sexual and communicative behavior, irrelevant to an osteological morphotype; the latter dismiss the model because of lack of identity – which clearly misses the point, as Cronin (1983) shows. Explicit attention needs to be paid to the purpose as well as to the structure of referential models. Further discussion of analogy and homology can be found in von Cranach (1976), Cartmill (1990, 1994) and Gifford-Gonzalez (1991).

Are referential models useful in paleoanthropology?

A referential model potentially can prove useful in several ways (see Potts, 1987 and McGrew, 1992, for further discussion):

> It can show what is possible, e.g. the hypothesis that significant rates of hominid hunting would have required weapons and conscious coordination of hunters is disproven by chimpanzees (Stanford, 1996).

> As part of a narrow-focus comparative approach, it provides the only possible method of incorporating certain demographic parameters such as dispersal into our scenarios (cf. Wrangham, 1987).

> It can suggest important new ideas, such as the realization that sex differences in chimpanzee

insectivory might illuminate origins of sexual division of labor in hominids (McGrew, 1979, 1992).

> It can generate a detailed scenario from which testable predictions can be derived ('if the LCA was just like a baboon except for [insert fossil data], then we should see X in the paleontological record of the LCA'; see below). While this approach can be useful, it is the most dangerous because it is here that confusion of model and referent is most likely to occur (cf. Tooby & DeVore, 1987).

Referential models have a number of advantages as well as dangers. At the least they provide a heuristic framework for thinking about a problem, and at best can be used to set up middle-range tests of hypotheses (see below). Their greatest danger is that models are easily mistaken for that which they model; critically, by creating a mental prototype they can mask variation and so block use of the comparative method for testing hypotheses (Hinde, 1976; Strum & Mitchell, 1987; Wrangham, 1987; Potts, 1994). This danger is especially serious in hominid modeling; normally the information provided by an analogy lies entirely in the difference between model and target (Hesse, 1966; Hinde, 1976), but in paleoanthropology this is too strict – a living hominoid identical to the LCA would be extremely informative. This difference in the nature and utility of paleontological analogy versus analogies in physics or human biology may contribute to the confusion over appropriate use of models in the field. Every sort of model has its hazards.

CHIMPANZEES AND THE LCA: SIMILARITY AND DIFFERENCE

Before any living ape can serve as a referential model for early hominids and the LCA, it is important to be as clear as possible regarding similarities and differences between the model and the referent. My thesis is that:

1. The LCA was roughly similar to modern chimpanzees in size, encephalization, habitat, diet (but with important differences) and perhaps locomotion. This thesis is supported by two lines of evidence: the fossils themselves and

molecular evidence suggesting that the African ape lineage is morphologically conservative (see Appendix to this chapter and also Zihlman, Chapter 21). It simply is not true that chimpanzees and modern hunter–gatherers are equally useful (or useless) as models of the LCA (*pace* Cameron, 1993).

2. These similarities suggest that the LCA and early hominids would have been subject to selective forces similar to those affecting modern chimpanzees. Further, the transition from forest to savanna that occurred in the hominid lineage may therefore have involved ecological and social adaptations similar to those exhibited by a comparison of forest versus savanna chimpanzees (Collins & McGrew, 1988).

   However, the LCA and early australopithecines may have been very unlike modern chimpanzees in being highly sexually dimorphic (Kelley, 1993; Kimbel *et al.*, 1994). It is therefore necessary in effect to consider two chimpanzee models. The first is a direct analogy in which the LCA and early hominids are considered similar in dimorphism to modern chimpanzees (cf. Zihlman, 1985). In the second, they are considered highly dimorphic and the principles of strategic modeling (*sensu* Tooby & DeVore, 1987) are used to 'adjust' the analogy accordingly.

3. The initial hominid adaptations to a 'savanna'-type niche took place in savanna woodland, rather than in denser forest or open shortgrass plains (Suzuki, 1969; Moore, 1992a). For now, this is merely an assumption of the model to be tested against future paleo-environmental reconstructions. Recent analyses support a woodland habitat for Plio-Pleistocene hominids, so the assumption is not ungrounded (e.g. Sikes, 1994).

4. The utility of the analogy will depend on the ability to test elements of it against paleontological data; with increasing correspondence between discoverable elements will come increasing confidence in nonverifiable aspects of the analogy. Conversely, to the extent that elements of the analogy are untestable or

are disconfirmed by data, the usefulness of the general 'chimpanzee model' is questioned.

5. Even if the analogy seems to work well, it is important to remember the limits of analogy. The analogy between waves of sound and light was useful in the development of optical theory, but did result in the erroneous postulation of 'ether' (Hesse, 1966).

Together these assumptions constitute the basis for a referential model that can be used to generate specific hypotheses to be tested against either fossil or archaeological data (e.g. 'savanna chimpanzees prefer hillsides, so the distribution of early hominid sites should be biased toward hillsides') or modern forest versus savanna comparisons (e.g., 'early australopithecines showed a trend toward postcanine megadonty suggesting exploitation of rough or hard foods, so savanna chimpanzees should incorporate more such foods in their diets'). These two examples may seem hopeless or naive: how many *in situ* early sites of demonstrable topography can we hope to find? Why assume a connection between expanded diet breadth in modern apes and the highly specialized megadonty of australopithecines? Here I hope to push this chimpanzee model to its limit; if the result is uninformative, then skepticism toward chimpanzee models in general is supported.

THE SAVANNA CHIMPANZEE MODEL

What about humans would we like a model to help explain? Several authors have generated lists of 'human' attributes (Alexander & Noonan, 1979; Tooby & DeVore, 1987) (see Table 20.2). Human attributes we might expect to have existed in the LCA fall into two categories: those we are confident appeared near the time of the LCA (column 1) and those that might have appeared at virtually any time in our subsequent evolution (column 2). A number of these attributes leave no direct fossil or archeological evidence. Alternative approaches that have been used to build explanatory scenarios for their evolution include:

1. Direct referential models (e.g., McGrew, 1981; Tanner, 1981, 1987).

2. Cladistic analysis (Cameron, 1993; Foley, 1989; Wrangham, 1987).

Table 20.2 *Attributes of the LCA and hominids addressed by hominid modeling*

Likely characteristics of the LCA	Novel attributes of early australopithecines	Novel attributes arising sometime post-LCA	Novel attributes arising with *Homo*
Closed social network	?Increased terrestriality	Increased meat eating	Increased encephalization
Female dispersal	Bipedality	Hunting of prey larger than self	Fire
Not monogamous	?Postcanine megadonty	Increased infant helplessness	Language[a]
Hostile intergroup relations		Increased male parental investment	Extensive reliance on (stone) tools
Male stalk and attack strangers		Concealed ovulation	Loss of megadonty
Extreme sexual dimorphism[b]		Prolonged female sexual receptivity	Reduction in sexual dimorphism[b]
		Enlarged penis	Adolescent growth spurt [females only?][c]
		Hairlessness	Food transport and storage
		Increased sexual division of labor	Kin involvement in mate selection
		Large structured male coalitions	

Source: Based primarily on Alexander & Noonan (1979), Tooby & DeVore (1987) and Wrangham (1987).
[a]Timing speculative, but probably postdates the origin of *Homo*.
[b]Reality of the attributes depends on particular taxonomic decisions.
[c]It is unclear whether or not the adolescent growth spurt is unique to humans, when it appeared, and what relation (if any) it might have to changes in (especially female) stature with *H. erectus* (Leigh, 1992; McHenry, 1994).

3. Strategic modeling (Tooby & DeVore, 1987).
4. The use of intraspecific variability in a referent to model diachronic change in the hominid lineage (Susman, 1987); to focus on process and not stasis (Potts, 1987).

In this paper I use the fourth approach, focusing on comparison of savanna with forest populations of chimpanzees and building on the work of Suzuki (1969), McGrew *et al.* (1981); Kortlandt (1983, 1984) and Laporte & Zihlman (1983); see Moore (1992a) for a definition of 'savanna chimpanzees'. I begin by asking what, from a hominoid's point of view, is different about savanna woodland?

Climate

Relative to forests, savanna woodlands are drier and rainfall is more seasonal (McGrew *et al.*, 1981; Moore, 1992a; Fig. 20.1). Water may be seasonally hard to locate and thermoregulation may be demanding.

Food

Greater aridity should be most important via effects on the supply of food plants. Taxonomic differences in vegetation are presumably not informative with regard to modeling the LCA, but the following general features of savanna plants may be:

1. *Types.* Many savanna plants have hard-shelled seeds or nuts, and many have underground storage organs (USOs) such as tubers; both food-types have been proposed as key resources for early hominids (Hatley & Kappelman, 1980; Peters, 1987).
2. *Seasonality.* The greater seasonality in rainfall experienced at savanna sites (Bourliére & Hadley, 1983) is likely to result in greater seasonal variation in relevant resources, though this needs to be confirmed empirically.
3. *Dispersion.* Without scale-independent measures of patchiness for savanna and forest sites it is difficult to know whether hominoid resources in savannas are more patchily distributed or simply more thinly distributed. Chimpanzees range more widely in savannas (Suzuki, 1969; Kano, 1972; Baldwin *et al.*, 1982), *ipso facto* evidence that important resources are more dispersed in such habitats. Assuming comparability of diets, a savanna hominoid must travel farther on at least a seasonal basis than a forest hominoid.

Fig. 20.1. Savanna habitat inhabited by chimpanzees in dry season at Ugalla. (Photo by J. Moore)

Predation

Relative to forest, savanna woodland has a greater variety of predators – lion (*Panthera leo*), hyena (*Crocuta crocuta*) and wild dog (*Lycaon pictus*) are more common, and leopard (*Panthera pardus*) appear comparably abundant. However, savanna woodland is importantly different from open shortgrass savanna in that relative safety in the trees is generally only a few meters away (Moore, 1992a). It is difficult to judge the actual risk of predation, since greater time spent on the ground (hence vulnerable) may be balanced by the greater visibility at ground level and hence reduced risk of ambush. At Ugalla near-ground visibility is usually > 20 m, while at Gombe and Mahale near-ground visibility is usually < 5 m (personal observations and unpublished data). Two caveats apply here: first, while chimpanzees at Ugalla utilize open woodland, they almost certainly bias their time toward denser vegetation. Second, the potential for ambush is still very real, as I realized when my assistant and I accidentally flushed a lion from behind a rock about 6 m away, where he was evidently stalking some warthogs (*Phacochoerus aethiopicus*) (see also Itani, 1979).

Competition

Relative to the forest, there are fewer species of primates and birds for chimpanzees to compete with. Whether competition is actually reduced or not is unknown, however, as elephants (*Loxodonta africana*), perhaps some browsing antelope (e.g. *Tragelaphus scriptus*), perhaps vervet monkeys (*Cercopithecus aethiops*) and especially baboons (*Papio* spp.) overlap with chimpanzees in diet (Peters & O'Brien, 1981; Collins & McGrew, 1988). McGrew *et al.* (1982) note that dietary overlap between Guinea baboons (*Papio papio*) and chimpanzees at the savanna site of Mt Assirik is about five times greater than the generic *Pan–Papio* overlap estimated by Peters & O'Brien, suggesting a possible increase in competition in arid habitats.

These differences between savanna woodland and more forested habitats may promote morphological, social or behavioral differentiation between savanna and forest chimpanzees (or facultative shifts in behavior among chimpanzees according to habitat at mosaic sites such as Gombe or Mahale). When such potential differentiation bears on traits considered important in human evolution, understanding of the bases for the differences – or their absence – can shed light on our scenarios of hominid behavioral ecology. The following questions, grouped by topic, are intended to suggest an approach to the use of chimpanzee referential models in paleoanthropology. The list is not exhaustive, and the questions vary markedly in both potential ease of answering and in the degree to which, when answered, they disconfirm hypotheses about early hominids and the LCA.

Bipedalism

Hunt (1994) has recently evaluated leading hypotheses for the origin of hominid bipedalism, which include vertical climbing (Fleagle et al., 1981; Prost, 1985; Doran, Chapter 16), a combination of vertical climbing and above-branch bipedal locomotion (Tuttle, 1981), the need to increase terrestrial travel in order to reach scattered arboreal resources (Rodman & McHenry, 1980), or thermoregulation in a dry, open habitat (Wheeler, 1994). Hunt argues that bipedal posture was initially favored in the context of feeding upon small fruits in closely-spaced, rapidly depleted patches; chimpanzees exploit such fruits by shuffling bipedally while harvesting with their hands.

Do savanna chimpanzees make greater use of large vertical trunks?

Hunt (1994) also points out that vertical climbing of large-trunked trees is rare at Gombe and Mahale, with apes preferring to enter trees via lianas or smaller-trunked neighbors (but at other forest sites vertical climbing may be common: Kortlandt, 1986; Doran, Chapter 16). At Ugalla, lianas are only locally abundant and many trees used for nesting appear to require vertical climbing. While large-stratum vertical climbing may not be the chimpanzee's primary adaptation (Hunt, 1994), greater forced reliance upon this niche in savanna chimpanzees would be consistent with Tuttle's (1981) and Prost's (1985) hypotheses regarding the LCA. While discovery of significantly more vertical climbing in woodlands could not disprove Hunt's shuffling-forager hypothesis, lesser amounts of vertical climbing in woodland seems inconsistent with the vertical-climber hypotheses.

Do savanna chimpanzees have larger day ranges?

We do not yet know whether they exploit their large home ranges by increasing average or maximum day ranges (cf. the Rodman & McHenry, 1980, hypothesis) or by steadily traveling through the annual range, with minimal backtracking, in increments similar to day ranges at more forested sites (e.g. Gombe: 3–5 km; Goodall, 1986). If the latter, it would demonstrate that increased locomotion is not a necessary consequence of lower resource density and greater home-range size, and so cannot be assumed for the LCA facing shrinking or mosaic forests.

Do savanna chimpanzees do more terrestrial shuffle-foraging?

In conjunction with development of paleoecological analogs (Sept, 1994), observations of notable reliance on shuffle-foraging at savanna sites would provide strong support for Hunt's hypothesis.

Are savanna chimpanzees constrained by thermoregulation and water balance?

Failure of savanna-living chimpanzees to exhibit signs of thermoregulatory constraint would be evidence against Wheeler's major hypothesis (Wheeler, 1994). While it has been suggested that chimpanzees at Mt Assirik are so constrained because they favor gallery forest in the hot season (McGrew et al., 1981), data to distinguish effects of temperature from seasonal resource distribution have not yet been presented.

Home bases and the division of labor

Much effort has gone into trying to discover the function of hominid archeological sites and the possible implications of such sites for social organization and behavior (e.g. Isaac, 1989; Clark, 1993). It has been argued that regular use of specific sleeping or feeding places is a necessary and sufficient indication of a division of labor in which males (?) travel long distances hunting or scavenging for meat while females (?) gather plant foods; resources are then shared at the 'home base' (Isaac, 1978; reviewed in Potts, 1994). Sept (1992) pointed out that creation of 'sites' by chimpanzees would invalidate the sufficient portion of the above argument. (Here, 'site' means to re-use a specific grove or location, on the order of a few tens of meters in diameter at most.)

Do savanna chimpanzees create 'sites' by repeated returns to the same location?

Data on nest distributions indicate that the answer is 'yes' (Sept, 1992; Moore, 1992b and personal observations), but the relative importance of such site fidelity remains to be established. Knowing the prevalence of such sites is important for comparison with the archeological record of site-concentrated versus background scatter of artifacts (the 'scatter between the patches') (Isaac, 1981).

Do savanna chimpanzees re-use sites more than do forest chimpanzees?

'Yes' would provide support for the idea that adaptation to drier habitats played a role in accentuating the importance of sites. 'No' suggests that the explanation for eventual adoption of home-base sites in hominids should be sought in some other aspect of hominid behavior and ecology.

What are the environmental correlates of chimpanzee sites?

Preliminary evidence shows that savanna chimpanzee nesting areas are non-randomly placed at least with respect to topography (Baldwin *et al.*, 1982; Kortlandt, 1992; Moore, 1992b; Sept, 1992; Fruth & Hohmann, Chapter 17; see Fig. 20.2). If hominid sites were placed similarly, *ceteris paribus* we could conclude that they need not indicate anything other than chimpanzee-like social organization.

Even in the absence of specific 'sites', the increased patchiness of savanna habitats may promote return use of certain areas (e.g. riverine strip or valley, on the order of 1 km² or so) and thus facilitate reunions of parties through vocal contact; ease of reunion could promote increased sex differences in ranging patterns (cf. McGrew, 1981, p.60). Thus, adaptation to savanna might promote increased sexual differentiation of diet and behavior without the exchange or provisioning envisioned in home-base scenarios. The question then is, does savanna habitat promote increased ecological segre-

Fig. 20.2. Nest-site in savanna with herbaceous understory at Ugalla: two nests visible at top center. (Photo by J. Moore)

gation of the sexes in chimpanzees? Such differentiation, a stable adaptation in itself, would be an obvious pre-adaptation for later division of labor.

Are sex differences in diet greater for savanna chimpanzees?

Without detailed behavioral data on habituated apes, answering this question requires an indirect approach. Based on stable isotope ratios in hairs found in nests, Schoeninger (personal communication) has recently found indications of high within-group dietary variation among savanna chimpanzees; the range of $\delta^{15}N$ values in a small sample was twice that found in controlled diet studies (Fitzer *et al.*, 1993). Further analysis and correlation with sex (determined cytogenetically from hair follicles, a procedure just moving from experimental to practical) should answer the question.

Increased sexual dimorphism in body size might correlate with increased sex differences in diet and ranging, either because males and females have different optimal sizes due to different energetic or social requirements, or through intersexual character displacement driven by competition for similar resources (Slatkin, 1984). Alternatively, increased sexual dimorphism might be favored by sexual selection if parties become more stable in savanna habitats, shifting toward a hamadryas-like social pattern (K. Hunt, personal communication). Given the extreme sexual dimorphism reported for early australopithecines, any insights chimpanzees could offer to our understanding of the ecological basis of sexual dimorphism would be important (see also McGrew, 1992, Chapter 5).

Do savanna chimpanzees exhibit greater size dimorphism?

It seems unlikely that any savanna chimpanzee populations have been isolated long enough for such evolution to occur (Morin et al., 1994), but since Groves, Shea and others are already studying skeletal variation in *Pan* (e.g. Shea et al., 1993), asking this question may be justified by its low cost and high potential gain.

Are savanna chimpanzee parties more harem-like in stability and composition?

There is some evidence of greater mixed-party stability at Mt Assirik (Tutin et al., 1983). In principle, party stability and its relationship with paternity can be assessed using DNA from shed hairs (cf. Morin et al., 1994).

Diet and tool use

The scarcity of reported scavenging by chimpanzees and other primates has been cited as evidence that scavenging was unimportant to early hominids (McGrew, 1981; Tooby & DeVore, 1987; but see Hasegawa et al., 1983; Nishida, 1994). It is worth considering, however, that Gombe National Park has few large predators and that the combination of moister habitat and denser forest at Mahale may reduce the effective 'shelf-life' of predator kills while at the same time making them harder to locate. If savanna woodland is at an ecological extreme for chimpanzees, one would expect their diet breadth to broaden and thus scavenging might become more

important; the same logic predicts increased hunting as well. Another aspect of savanna diet expansion might be an increased reliance on tools for extracting or processing foods such as hard seeds or tubers (McGrew, 1981; but see Boesch-Achermann & Boesch, 1994; Nishida, 1973, regarding the significance of tool-use in the forest). Finally, increased reliance on nuts, tubers or grass seeds by savanna chimpanzees would provide support for various explanations of postcanine megadonty and enamel thickness in australopithecines (Moore, 1992a).

Are savanna chimpanzees more likely to hunt or to scavenge?

This is an obvious question, but one both difficult to answer (McGrew, 1992, p.150–3) and ambiguous in its implications since scavenging niches have changed drastically since the loss of Pliocene megafauna and their specialized sabertooth predators. The sabertooths probably could not chew close to the bone and so likely left both relatively and absolutely more scavengable material at their kills (Marean, 1989).

Do savanna chimpanzees have more robust mandibles?

Greater robusticity would be a sensitive indication that savanna diets place greater demands on the chewing apparatus. Such robusticity in savanna chimpanzees would likely be purely developmental rather than phylogenetic.

Do savanna chimpanzees use tools more often or in more ways?

There is no indication that they do, but since observations of tool use are correlated with length of observation and degree of habituation (McGrew, 1992), this is not surprising or meaningful. 'Yes' would provide support for the general 'out of the trees, onto the ground, pick up tools' scenario, but 'No' would not be informative, since this failure could be seen as the key point at which hominids and hominoids diverged.

Intergroup relations

Intercommunity relationships in chimpanzees appear to be generally hostile, and similarities with patterns of human aggression have led Wrangham to suggest that the LCA would also have exhibited intercommunity hostility by males, with stalking and fatal attacks probable (Wrangham, 1987; Manson & Wrangham, 1991). At the same time, it seems plausible that lower population densities in savanna hominoids would militate against the fission-fusion system at the core of forest chimpanzees' intergroup relations (Moore, 1992a). Also, where resource patches are dispersed but non-depleting (e.g. dry-season waterholes) communities might be forced into periods of coexistence (McGrew, 1981) (or outright battle, as in the opening scene of the film *2001: A Space Odyssey*).

Do savanna chimpanzees exhibit intergroup hostility, with stalking attacks by males?

There is some evidence that they do (Goodall, 1986, p. 521). However, this incident involves reactions of wild individuals to tame ones, and the possibility remains that the attacks were stimulated by some aspect of the human presence or relationship with the rehabilitants. A negative answer to this question would support a 'U-shaped' trajectory for intergroup violence in human evolution, analogous to that proposed by Knauft (1991), in which forest chimpanzee-like hostility gave way to relatively more peaceful relationships in early, low-density hominids only to return later, with the emergence of *Homo* or of the state.

Is female-only dispersal the pattern for savanna chimpanzees?

If the fission–fusion social organization of forest chimpanzees breaks down at low densities, it may be that the female-only dispersal pattern may not apply to savanna-living populations (cf. Sugiyama *et al.*, 1993). If so, we might see the origin of the mixed dispersal patterns exhibited by human societies in the low densities and large ranges of savanna-dwelling early hominids.

Ecology and brains

Given their large home ranges, it seems reasonable to postulate that misremembering the location of a waterhole or the phenological cycle of a particular fig tree might be more costly to a savanna chimpanzee (McGrew,

1981, p.59). If savanna chimpanzees are living 'at the edge,' it may be necessary to adopt more complex or more secure resource acquisition tactics. Either or both of these possibilities suggests that savannas may be ecologically more challenging for chimpanzees than are forests. Might comparable complexity have played a role in the evolution of hominid encephalization?

CONCLUSION

The great apes are fascinating beings in their own rights, and richly deserving of long-term study for what we can learn about other ways of being intelligent, long-lived social animals; so are elephants, dolphins (Odontoceti) and perhaps others such as parrots (Psittidae). The apes' particular utility for illuminating the behavioral ecology of our extinct ancestors has, however, been questioned. Cladistic methods such as pioneered by Wrangham (1987) avoid the pitfalls of assuming that an extinct species could not have been uniquely different from any modern analog. Unfortunately such cladistic conclusions are limited to those traits that are shared by all the modern taxa used in the analysis; while we have some confidence in the picture of the LCA so created, it is missing precisely the bits that might explain differences between hominids and apes.

Strategic modeling relies upon our understanding of the principles of behavioral ecology. It is the only approach that can hope to address the significance of certain early hominid traits such as the extreme dimorphism of *Australopithecus afarensis*. However, as long as we continue to be surprised by the natural history of modern taxa, we need to keep in mind that our strategically-constructed scenarios concerning the past are based on a young, developing and still uncertain field (Wrangham, 1987).

In this paper I have advocated the use of a referential approach in which the model is not a single typological modern species *per se*, but the set of differences observed between populations of that highly variable species (in archeological terms, a *relational* as opposed to *formal* analogy; Gifford-Gonzalez, 1991). I believe such an approach has great potential, though it will be difficult and expensive to collect all the relevant data. I do not believe the method can stand on its own, though, any more than can the simple analogical, cladistic, or strategic modeling approaches; and none of these approaches to behavioral scenarios will get far without careful attention to the fossil and archeological evidence. As Glynn Isaac (1980, p. 227) put it in a not-too-different context, 'the varied company of scientists inquiring into human origins must cast their net of investigation more widely.'

ACKNOWLEDGMENTS

This paper was originally prepared for the Wenner-Gren conference 'The Great Apes Revisited,' held 12–19 November, 1994 in Cabo San Lucas, Baja, Mexico. My thanks to the Wenner-Gren Foundation and the organizers and participants of the conference that set the rhythm for this contribution, especially to Bill McGrew for two decades of good advice. I am grateful to Kenna Barret, Nancy Friedlander, Kevin Hunt, Jay Kelley, Curtis Marean, Bill McGrew, Toshisada Nishida and David Pilbeam for their comments and encouragement, even if I haven't convinced them all. Major thanks to Jeanne Sept for her intellectual and practical contributions to my work in Ugalla. I thank Moshi Rajabu and Ramadhani Bilali for their help in the field, Toshisada Nishida for his advice, the Government of Tanzania, COSTECH and SWRI for permission to work at Ugalla, and the L. S. B. Leakey Foundation and the University of California Committee on Research for financial support.

APPENDIX: FOSSIL AND MOLECULAR EVIDENCE FOR A CHIMPANZEE-LCA ANALOGY

Fossils

The fossil record provides little evidence regarding the LCA (Ciochon, 1983; Hill & Ward, 1988), and so physical reconstruction of the LCA must be based primarily on the earliest known hominids, *Australopithecus afarensis*, present as early as 3.4–3.9 mya (Kimbel *et al.*, 1994), the recently described *Australopithecus ramidus*, at *c.* 4.4 mya (White *et al.*, 1994), and on the Miocene hominoids (Pilbeam, 1985). Given an ape-hominid divergence of about 5 mya, *A. afarensis* is by far the temporally closest, reasonably well-characterized taxon. Everything currently known of *A. ramidus* suggests that it was considerably more chimpanzee-like than *A. afarensis* (White *et al.*, 1994), rendering the following comparisons conservative with respect to a chimpanzee or bonobo analogy.

Size

Australopithecus afarensis weighed about 30–45 kg, within the range of chimpanzees, but perhaps slightly on the small side (McHenry, 1994), though Hartwig-Scherer (1993) estimates up to *c.* 75 kg or more for some specimens. *A. afarensis* were built like chimpanzees in terms of weight-for-stature (Aiello, 1992). There is a general evolutionary trend for body-size to increase within lineages (Cope's Law; see Maurer *et al.*, 1992), so it would be unlikely for the LCA to have been larger than *A. afarensis* or modern chimpanzees; it may have been smaller, in the 15–25 kg range.

Encephalization

Relative and absolute brain sizes of the earliest australopithecines were no different from modern chimpanzees (Foley & Lee, 1991; Hartwig-Scherer, 1993). Much debate exists about the organization of their brains (Falk, 1987).

Habitat

Considering the range of habitats used by modern chimpanzees, only if the LCA was found in habitats at least as dry or open as bushland or *Acacia* woodland would they differ (terms from Pratt *et al.*, 1966). Several recent studies have concluded that habitats of the earliest hominids were well-wooded (Cerling, 1992; Kingston *et al.*, 1994), and arboreal adaptations of *Australopithecus afarensis* suggest significant time spent in trees (Susman, 1987). Based on remains from a single site, *Australopithecus ramidus* is thought to have lived in a 'wooded' habitat (WoldeGabriel *et al.*, 1994).

Assuming that hominids in some sense moved out of the trees, the LCA would likely have lived in forest-woodland or wetter habitats (e.g. Bernor, 1983). The LCA may have preferred drier habitats than those in which chimpanzees are found (500–800 mm rain per year; Pickford, 1983) but the inferred habitat of *A. ramidus* makes this unlikely.

Diet

The resolution of dental morphology and microwear studies is not fine enough to determine *Australopithecus afarensis'* diet with any precision. Research suggests broad dietary similarity with chimpanzees (Grine & Kay, 1988; Andrews & Martin, 1991), or gorillas, but with seasonal reliance on hard gritty foods (Ryan & Johanson, 1989). It is worth noting that Ryan & Johanson based their comparison on western lowland gorillas which appear to have diets highly overlapping with chimpanzees' (Tutin & Fernandez, 1993). Molars of many Miocene hominoids as well as of early australopithecines were much larger, and their enamel thicker, than those of the African apes (Andrews & Martin, 1991). This suggests that the LCA's diet differed in some important ways from modern chimpanzees: at least seasonally, the LCA and earliest hominids probably relied on gritty or hard foods (Peters, 1987; Ryan & Johanson, 1989). *Australopithecus ramidus* molar enamel is thin, comparable to chimpanzees' (White *et al.*, 1994). While strengthening a chimpanzee analogy with the LCA, this finding complicates the overall picture of enamel evolution in the hominoid and hominid lineages.

Locomotion

A knuckle-walking LCA is possible (Zihlman, Chapter 21; though the available *Australopithecus ramidus* material shows no signs of knuckle-walking adaptations). *Australopithecus afarensis* was at least capable of exploiting the arboreal environment (Susman *et al.*, 1985; Tuttle, 1981). Since the energetic efficiency of locomotion may have been important in the evolution of hominoid social structure (Rodman & McHenry, 1980; Wrangham, 1979), locomotor mode remains an important loose end in models of the LCA.

Molecular evidence

Chimpanzees and bonobos diverged about 2–3 mya (Cronin, 1983; Goodman *et al.*, 1994). At that time our ancestors (early *Homo* sp.) were very different from modern humans in size, dentition and brain size (Wood, 1992); by contrast, chimpanzees and bonobos are easily mistaken for each other by the casual observer and hybridization in captivity has been reported (Vervaecke & van Elsacker, 1992). More recently, on the basis of mtDNA divergence, Morin *et al.* (1994) have concluded that *Pan troglodytes verus* and the other two chimpanzee subspecies diverged about 1.6 mya, roughly contemporaneously with the origin of *Homo erectus*; *P. t. verus* cannot be distinguished reliably from *Pan troglodytes*

troglodytes or *Pan troglodytes schweinfurthii* on morphological grounds (Shea *et al.*, 1993). Mountain and lowland gorillas differ in mitochondrial COII sequences more than do chimpanzees and bonobos (Ruvolo *et al.*, 1994), suggesting a divergence date of almost 3 mya (Morell, 1994); again, morphological differences are slight. Finally, lowland gorillas and chimpanzees (molecular divergence date about 5.5–7 mya; Hasegawa, 1992; see discussion in Goodman *et al.*, 1994, and Zihlman, Chapter 21) are similar in shape and diet, and gorillas can be viewed as essentially allometrically scaled-up versions of chimpanzees (Shea, 1990; Doran, Chapter 16). All of these comparisons indirectly suggest much morphological conservatism in the African pongid lineage, and therefore support the notion that the LCA would have resembled a chimpanzee.

REFERENCES

Aiello, L. C. 1992. Allometry and the analysis of size and shape in human evolution. *Journal of Human Evolution*, **22**, 127–47.

Alexander, R. D. & Noonan, K. M. 1979. Concealment of ovulation, parental care, and human social evolution. In: *Evolutionary Biology and Human Social Behavior*, ed. N. A. Chagnon & W. Irons, pp. 436–53. North Scituate, MA: Duxbury.

Amos, B., Schlötterer, C. & Tautz, D. 1993. Social structure of pilot whales revealed by analytical DNA profiling. *Science*, **260**, 670–72.

Andrews, P. & Martin, L. 1991. Hominoid dietary evolution. *Philosophical Transactions of the Royal Society of London, Series B*, **334**, 199–209.

Baldwin, P. J., McGrew, W. C. & Tutin, C. E. G. 1982. Wide-ranging chimpanzees at Mt Assirik, Senegal. *International Journal of Primatology*, **3**, 367–85.

Bernor, R. L. 1983. Geochronology and zoogeographic relationships of Miocene hominoidea. In: *New Interpretations of Ape and Human Ancestry*, ed. R. L. Ciochon & R. S. Corruccini, pp. 21–64. New York: Plenum.

Boesch-Achermann, H. & Boesch, C. 1994. Hominization in the rainforest: the chimpanzee's piece of the puzzle. *Evolutionary Anthropology*, **3**, 9–16.

Bourliére, F. & Hadley, M. 1983. Present-day savannas: an overview. In: *Ecosystems of the World, 13, Tropical Savannas*, ed. F. Bourliére, pp. 1–17. Amsterdam: Elsevier.

Cameron, D. W. 1993. The Pliocene hominid and

protochimpanzee behavioral morphotypes. *Journal of Anthropological Archaeology*, **12**, 386–414.

Cartmill, M. 1990. Human uniqueness and theoretical content in paleoanthropology. *International Journal of Primatology*, **11**, 173–92.

Cartmill, M. 1994. A critique of homology as a morphological concept. *American Journal of Physical Anthropology*, **94**, 115–23.

Cerling, T. E. 1992. Development of grasslands and savannas in East Africa during the Neogene. *Palaeogeography, Palaeoclimatology, and Palaeoecology*, **97**, 241–47.

Ciochon, R. L. 1983. Hominoid cladistics and the ancestry of modern apes and humans: a summary statement. In: *New Interpretations of Ape and Human Ancestry*, ed. R. L. Ciochon & R. S. Corruccini, pp. 783–843. New York: Plenum.

Clark, J. D. 1993. Distinguished lecture: coming into focus. *American Anthropologist*, **95**, 823–38.

Collins, D. A. & McGrew, W. C. 1988. Habitats of three groups of chimpanzees (*Pan troglodytes*) in western Tanzania compared. *Journal of Human Evolution*, **17**, 553–74.

von Cranach, M. 1976. *Methods of Inference from Animal to Human Behavior*. Chicago: Aldine.

Cronin, J. E. 1983. Apes, humans, and molecular clocks: a reappraisal. In: *New Interpretations of Ape and Human Ancestry*, ed. R. L. Ciochon & R. S. Corruccini, pp. 115–35. New York: Plenum.

D'Andrade, R. 1992. Schemas and motivation. In: *Human Motives and Cultural Models*, ed. R. D'Andrade & C. Strauss, pp. 23–44. Cambridge: Cambridge University Press.

Falk, D. 1987. Hominid paleoneurology. *Annual Review of Anthropology*, **16**, 13–30.

Fedigan, L. 1992. *Primate Paradigms: Sex Roles and Social Bonds*, 2nd edn. Chicago: University of Chicago Press.

Fitzer, P. W., Schoeninger, M. J. & Sept, J. M. 1993. Stable isotope variation in chimpanzees: implications for diet selectivity in early hominids. *American Journal of Physical Anthropology, Supplement*, **16**, 85–6.

Fleagle, J. G., Stern, J. T., Jungers, W. J. & Susman, R. L., Vangor, A. K. & Wells, J. P. 1981. Climbing: a biomechanical link with brachiation and with bipedalism. *Symposia of the Zoological Society of London*, **48**, 359–75.

Foley, R. A. 1989. The evolution of hominid social behaviour. In: *Comparative Socioecology: The Behavioural Ecology of Humans and Other Mammals*, ed. V. Standen & R. A. Foley, pp. 473–94. Oxford: Blackwell Scientific.

Foley, R. A. & Lee, P. C. 1991. Ecology and energetics of encephalization in hominid evolution. *Philosophical Transactions of the Royal Society of London, Series B*, **334**, 223–32.

Gifford-Gonzalez, D. 1991. Bones are not enough: analogues, knowledge, and interpretive strategies in zooarchaeology. *Journal of Anthropological Archaeology*, **10**, 215–54.

Goodall, J. 1986. *The Chimpanzees of Gombe: Patterns of Behavior*. Cambridge, MA: Harvard University Press.

Goodman, M., Bailey, W. J., Hayasaka, K., Stanhope, M. J., Slightom, J. & Czelusniak, J. 1994. Molecular evidence on primate phylogeny from DNA sequences. *American Journal of Physical Anthropology*, **94**, 3–24.

Grine, F. E. & Kay, R. F. 1988. Early hominid diets from quantitative image analysis of dental microwear. *Nature*, **333**, 765–8.

Hartwig-Scherer, S. 1993. Body weight prediction in early fossil hominids: towards a taxon-'independent' approach. *American Journal of Physical Anthropology*, **92**, 17–36.

Hasegawa, M. 1992. Evolution of hominoids as inferred from DNA sequences. In: *Topics in Primatology, Vol. I, Human Origins*, ed. T. Nishida, W. C. McGrew, P. Marler, M. Pickford & F. B. M. de Waal, pp. 347–57. Tokyo: University of Tokyo Press.

Hasegawa, T., Hiraiwa, M., Nishida, T. & Takasaki, H. 1983. New evidence on scavenging behavior in wild chimpanzees. *Current Anthropology*, **24**, 231–2.

Hatley, T. & Kappelman, J. 1980. Bears, pigs, and Plio–Pleistocene hominids: a case for the exploitation of belowground food resources. *Human Ecology*, **8**, 371–87.

Hesse, M. B. 1966. *Models and Analogies in Science*. South Bend, IN: University of Notre Dame Press.

Hill, A. & Ward, S. 1988. Origin of the Hominidae: the record of African large hominoid evolution between 14 My and 4 My. *Yearbook of Physical Anthropology*, **31**, 49–83.

Hinde, R. A. 1976. The use of differences and similarities in comparative psychopathology. In: *Animal Models in Human Psychobiology*, ed. G. Serban & A. Kling, pp. 187–202. New York: Plenum.

Holyoak, K. J. & Thagard, P. 1995. *Mental Leaps:*

Analogy in Creative Thought. Cambridge, MA: MIT Press.

Hunt, K. D. 1994. The evolution of human bipedality: ecology and functional morphology. *Journal of Human Evolution*, **26**, 183–202.

Isaac, B. (ed.) 1989. *The Archaeology of Human Origins: Papers by Glynn Isaac*. Cambridge: Cambridge University Press.

Isaac, G. L. 1978. The food-sharing behavior of protohuman hominids. *Scientific American*, **238–4**, 90–106.

Isaac, G. L. 1980. Casting the net wide. In: *Current Argument on Early Man*, ed. L. K. Konigsson, pp. 226–51. New York: Pergamon Press.

Isaac, G. L. 1981. Stone Age visiting cards: approaches to the study of early land-use patterns. In: *Patterns of the Past*, ed. I. Hodder, G. Isaac & N. Hammond, pp. 131–55. Cambridge: Cambridge University Press.

Itani, J. 1979. Distribution and adaptation of chimpanzees in an arid area. In: *The Great Apes*, ed. D. A. Hamburg & E. R. McCown, pp. 55–71. Menlo Park, CA: Benjamin/Cummings.

Johnson, S. C. 1981. Bonobos: generalized hominid prototypes or specialized insular dwarfs? *Current Anthropology*, **22**, 363–75.

Jolly, C. J. 1970. The seed-eaters: a new model of hominid differentiation based on a baboon analogy. *Man*, **5**, 5–26.

Kano, T. 1972. Distribution and adaptation of the chimpanzee on the eastern shore of Lake Tanganyika. *Kyoto University African Studies*, **7**, 37–129.

Kelley, J. 1993. Taxonomic implications of sexual dimorphism in *Lufengpithecus*. In: *Species, Species Concepts, and Primate Evolution*, ed. W. H. Kimbel & L. B. Martin, pp. 429–58. New York: Plenum.

Kimbel, W. H., Johanson, D. C. & Rak, Y. 1994. The first skull and other new discoveries of *Australopithecus afarensis* at Hadar, Ethiopia. *Nature*, **368**, 449–51.

Kingston, J. D., Marino, B. D. & Hill, A. 1994. Isotopic evidence for Neogene hominid paleoenvironments in the Kenya Rift Valley. *Science*, **264**, 955–59.

Kinzey, W. G. (ed.) 1987. *The Evolution of Human Behavior: Primate Models*. Albany: State University of New York Press.

Knauft, B. M. 1991. Violence and sociality in human evolution. *Current Anthropology*, **32**, 391–428.

Kortlandt, A. 1983. Marginal habitats of chimpanzees. *Journal of Human Evolution*, **12**, 231–78.

Kortlandt, A. 1984. Habitat richness, foraging range and diet in chimpanzees and some other primates. In: *Food Acquisition and Processing in Primates*, ed. D. A. Chivers, B. A. Wood & A. Bilsborough, pp. 119–59. New York: Plenum.

Kortlandt, A. 1986. Studying the treescape. In: *Primate Ecology and Conservation: Proceedings of the 10th Congress of the International Primatological Society*, ed. J. Else & P. C. Lee, pp. 263–75. Cambridge: Cambridge University Press.

Kortlandt, A. 1992. On chimpanzee dormitories and early hominid home sites. *Current Anthropology*, **33**, 399–401.

Landau, M. 1991. *Narratives of Human Evolution*. New Haven, CT: Yale University Press.

Laporte, L. F. & Zihlman, A. L. 1983. Plates, climate and hominoid evolution. *South African Journal of Science*, **79**, 96–110.

Latimer, B. M., White, T. D., Kimbel, W. H., Johanson, D. C. & Lovejoy, C. O. 1981. The pygmy chimpanzee is not a living missing link in human evolution. *Journal of Human Evolution*, **10**, 475–88.

Leigh, S. R. 1992. Patterns of variation in the ontogeny of primate body size dimorphism. *Journal of Human Evolution*, **23**, 27–50.

Manson, J. H. & Wrangham, R. W. 1991. Intergroup aggression in chimpanzees and humans. *Current Anthropology*, **32**, 369–90.

Marean, C. W. 1989. Sabertooth cats and their relevance for early hominid diet and evolution. *Journal of Human Evolution*, **18**, 559–82.

Maurer, B. A., Brown, J. H. & Rusler, R. D. 1992. The micro and macro in body size evolution. *Evolution*, **46**, 939–53.

McGrew, W. C. 1979. Evolutionary implications of sex differences in chimpanzee predation and tool use. In: *The Great Apes*, ed. D. A. Hamburg & E. R. McCown, pp. 441–63. Menlo Park, CA: Benjamin/Cummings.

McGrew, W. C. 1981. The female chimpanzee as a female evolutionary prototype. In: *Woman the Gatherer*, ed. F. Dahlberg, pp. 35–73. New Haven, CT: Yale University Press.

McGrew, W. C. 1992. *Chimpanzee Material Culture:*

Implications for Human Evolution. Cambridge: Cambridge University Press.

McGrew, W. C., Baldwin, P. J. & Tutin, C. E. G. 1981. Chimpanzees in a hot, dry and open habitat: Mt Assirik, Senegal, West Africa. *Journal of Human Evolution*, **10**, 227–44.

McGrew, W. C., Sharman, M. J., Baldwin, P. J. & Tutin, C. E. G. 1982. On early hominid plant-food niches. *Current Anthropology*, **23**, 213–14.

McHenry, H. M. 1994. Behavioral ecological implications of early hominid body size. *Journal of Human Evolution*, **27**, 77–87.

Merrell, F. 1991. Model, world, semiotic reality. In: *On Semiotic Modeling*, ed. M. Anderson & F. Merrell, pp. 247–83. Berlin: Mouton de Gruyter.

Moore, J. 1992a. 'Savanna' chimpanzees. In: *Topics in Primatology, Vol. I, Human Origins*, ed. T. Nishida, W. C. McGrew, P. Marler, M. Pickford & F. B. M. de Waal, pp. 99–118. Tokyo: University of Tokyo Press.

Moore, J. 1992b. Comment on 'Was there no place like home?' by J. Sept. *Current Anthropology, 33*, 198–9.

Moore, J. 1992c. Dispersal, nepotism, and primate social behavior. *International Journal of Primatology, 13*, 361–78.

Moore, J. 1993. Inbreeding and outbreeding in primates: What's wrong with 'the dispersing sex'? In: *The Natural History of Inbreeding and Outbreeding: Theoretical and Empirical Perspectives*, ed. N. W. Thornhill, pp. 392–426. Chicago: University of Chicago Press.

Morell, V. 1994. Will primate genetics split one gorilla into two? *Science, 265*, 1661.

Morin, P. A., Moore, J., Chakraborty, R., Jin, L., Goodall, J. & Woodruff, D. S. 1994. Kin selection, social structure, gene flow, and the evolution of chimpanzees. *Science, 265*, 1193–201.

Nishida, T. 1973. The ant-gathering behaviour by the use of tools among wild chimpanzees of the Mahali Mountains. *Journal of Human Evolution, 2*, 357–70.

Nishida, T. 1994. Afterword – Review of recent findings on Mahale chimpanzees: implications and future research directions. In: *Chimpanzee Cultures*, ed. R. W. Wrangham, W. C. McGrew, F. B. M. de Waal, P. G. Heltne & L. A. Marquardt, pp. 373–96. Cambridge, MA: Harvard University Press.

Peters, C. R. 1987. Nut-like oil seeds: Food for monkeys, chimpanzees, humans, and probably ape-men. *American Journal of Physical Anthropology, 73*, 333–63.

Peters, C. R. & O'Brien, E. M. 1981. The early hominid plant-food niche: insights from an analysis of plant exploitation by *Homo, Pan*, and *Papio* in eastern and southern Africa. *Current Anthropology, 22*, 127–40.

Pickford, M. 1983. Sequence and environments of the Lower and Middle Miocene hominoids of western Kenya. In: *New Interpretations of Ape and Human Ancestry*, ed. R. L. Ciochon & R. S. Corruccini, pp. 421–39. New York: Plenum.

Pilbeam, D. 1985. Patterns of hominoid evolution. In: *Ancestors: the Hard Evidence*, ed. E. Delson, pp. 51–9. New York: Alan R. Liss.

Potts, R. 1987. Reconstructions of early hominid socioecology: a critique of primate models. In: *The Evolution of Human Behavior: Primate Models*, ed. W. G. Kinzey, pp. 28–47. Albany: State University of New York Press.

Potts, R. 1994. Variables versus models of early Pleistocene hominid land use. *Journal of Human Evolution, 27*, 7–24.

Pratt, D. J., Greenway, P. J. & Gwynne, M. D. 1966. A classification of East African rangeland, with an appendix on terminology. *Journal of Applied Ecology, 3*, 369–82.

Prost, J. H. 1985. Chimpanzee behavior and models of hominization. In: *Primate Morphophysiology, Locomotor Analyses and Human Bipedalism*, ed. S. Kondo, pp. 289–303. Tokyo: University of Tokyo Press.

Rodman, P. S. & McHenry, H. 1980. Bioenergetics and the origin of hominid bipedalism. *American Journal of Physical Anthropology, 52*, 103–6.

Rubinstein, R. A., Laughlin Jr, C. D. & McManus, J. 1984. *Science as Cognitive Process: Toward an Empirical Philosophy of Science*. Philadelphia: University of Pennsylvania Press.

Ruvolo, M., Pan, D., Zehr, S., Goldberg, T., Disotell, T. & von Dornum, M. 1994. Gene trees and hominoid phylogeny. *Proceedings of the National Academy of Sciences, U.S.A., 91*, 8900–4.

Ryan, A. S. & Johanson, D. C. 1989. Anterior dental microwear in *Australopithecus afarensis*: comparisons with human and nonhuman primates. *Journal of Human Evolution, 18*, 235–68.

Sept, J. M. 1992. Was there no place like home? A new perspective on early hominid archaeological sites from the mapping of chimpanzee nests. *Current Anthropology*, **33**, 187–207.

Sept, J. M. 1994. Beyond bones: archaeological sites, early hominid subsistence, and the costs and benefits of exploiting wild plant foods in east African riverine landscapes. *Journal of Human Evolution*, **27**, 295–320.

Shea, B. T. 1990. Dynamic morphology: growth, life history, and ecology in primate evolution. In: *Primate Life History and Evolution*, ed. C. J. DeRousseau, pp. 325–52. New York: Wiley-Liss.

Shea, B. T., Leigh, S. R. & Groves, C. P. 1993. Multivariate craniometric variation in chimpanzees: implications for species identification in paleoanthropology. In: *Species, Species Concepts, and Primate Evolution*, ed. W. H. Kimbel & L. B. Martin, pp. 265–96. New York: Plenum.

Sikes, N. E. 1994. Early hominid habitat preferences in East Africa: paleosol carbon isotopic evidence. *Journal of Human Evolution*, **27**, 25–45.

Skybreak, A. 1984. *Of Primeval Steps & Future Leaps: An Essay on the Emergence of Human Beings, the Source of Women's Oppression, and the Road to Emancipation.* Chicago: Banner.

Slatkin, M. 1984. Ecological causes of sexual dimorphism. *Evolution*, **38**, 622–30.

Stanford, C. 1996. The hunting ecology of wild chimpanzees: implications for the behavioral ecology of Pliocene hominids. *American Anthropologist*, **98**, 1–18.

Strier, K. B. 1994. Myth of the typical primate. *Yearbook of Physical Anthropology*, **37**, 233–71.

Strum, S. C. & Mitchell, W. 1987. Baboon models and muddles. In: *The Evolution of Human Behavior: Primate Models*, ed. W. G. Kinzey, pp. 87–104. Albany: State University of New York Press.

Sugiyama, Y., Kawamoto, S., Takenaka, O., Kumazaki, K. & Miwa, N. 1993. Paternity discrimination and inter-group relationships of chimpanzees at Bossou. *Primates*, **34**, 545–52.

Susman, R. L. 1987. Pygmy chimpanzees and common chimpanzees: models for the behavioral ecology of the earliest hominids. In: *The Evolution of Human Behavior: Primate Models*, ed. W. G. Kinzey, pp. 72–86. Albany: State University of New York Press.

Susman, R. L., Stern Jr, J. T. & Jungers, W. L. 1985. Locomotor adaptations in the Hadar hominids. In: *Ancestors: the Hard Evidence*, ed. E. Delson, pp. 184–92. New York: Alan R. Liss.

Suzuki, A. 1969. An ecological study of chimpanzees in a savanna woodland. *Primates*, **10**, 103–48.

Tanner, N. M. 1981. *On Becoming Human.* Cambridge: Cambridge University Press.

Tanner, N. M. 1987. The chimpanzee model revisited and the gathering hypothesis [unpublished ms & published version]. In: *The Evolution of Human Behavior: Primate Models*, ed. W. G. Kinzey, pp. 3–27. Albany: State University of New York Press.

Tooby, J. & DeVore, I. 1987. The reconstruction of hominid behavioral evolution through strategic modelling. In: *The Evolution of Human Behavior: Primate Models*, ed. W. G. Kinzey, pp. 183–237. Albany: State University of New York Press.

Tutin, C. E. G., McGrew, W. C. & Baldwin, P. J. 1983. Social organization of savanna-dwelling chimpanzees, *Pan troglodytes verus*, at Mt Assirik, Senegal. *Primates*, **24**, 154–73.

Tutin, C. E. G. & Fernandez, M. 1993. Composition of the diet of chimpanzees and comparisons with that of sympatric lowland gorillas in the Lopé Reserve, Gabon. *American Journal of Primatology*, **30**, 195–211.

Tuttle, R. H. 1981. Evolution of hominid bipedalism and prehensile capabilities. *Philosophical Transactions of the Royal Society of London, Series B*, **292**, 89–94.

Vervaecke, H. & van Elsacker, L. 1992. Hybrids between common chimpanzees and pygmy chimpanzees in captivity. *Mammalia*, **56**, 667–9.

Wharton, C. M., Holyoak, K. J., Downing, P. E., Lange, T. E., Wickens, T. D. & Melz, E. R. 1994. Below the surface: analogical similarity and retrieval competition in reminding. *Cognitive Psychology*, **26**, 64–101.

Wheeler, P. E. 1994. The foraging times of bipedal and quadrupedal hominids in open equatorial environments (a reply to Chaplin, Jablonski & Cable, 1994). *Journal of Human Evolution*, **27**, 511–17.

White, T. D., Suwa, G. & Asfaw, B. 1994. *Australopithecus ramidus*, a new species of early hominid from Aramis, Ethiopia. *Nature*, **371**, 306–12.

WoldeGabriel, G., White, T. D., Suwa, G., Renne, P., de Heinzelin, J., Hart, W. K. & Heiken, G. 1994. Ecological and temporal placement of early Pliocene hominids at Aramis, Ethiopia. *Nature*, **371**, 330–3.

Wood, B. 1992. Origin and evolution of the genus *Homo*. *Nature*, **355**, 783–90.

Wrangham, R. W. 1979. On the evolution of ape social systems. *Social Science Information*, **18**, 335–68.

Wrangham, R. W. 1987. The significance of African apes for reconstructing human social evolution. In: *The*

Evolution of Human Behavior: Primate Models, ed. W. G. Kinzey, pp. 51–71. Albany: State University of New York Press.

Zihlman, A. L. 1985. *Australopithecus afarensis*: two sexes or two species? In: *Hominid Evolution*, ed. P. V. Tobias, pp. 213–20. New York: Alan R. Liss.

Zihlman, A. L., Cronin, J. B., Cramer, D. L. & Sarich, V. M. 1978. Pygmy chimpanzee as a possible prototype for the common ancestor of humans, chimpanzees, and gorillas. *Nature*, **275**, 744–6.

NOTE ADDED IN PROOF

Australophithecus ramidus has been placed in its own genus, *Ardipithecus*, in recognition of the belief that it may be a sister taxon of the hominids (White *et al.*, 1995). In 1995, *Australopithecus anamensis* was described, based on specimens found in Kenya at two sites that probably were riverine gallery-open dry woodland/bushland mosaics; this species dates from about 3.9–4.2 mya. It appears to have been bipedal and to have had dental enamel similar in thickness to *Australopithecus afarensis* (and much thicker than that of *A. ramidus*), features that are consistent with its being ancestral to *afarensis* (Leakey *et al.*, 1995). If *ramidus* is a sister taxon to the hominids and *anamensis* is ancestral to *afarensis*, the picture of dental enamel evolution in the early hominids would be simplified somewhat and the chimpanzee analogy with the LCA weakened (cf. p. 287).

Leakey, M. G., Feibel, C. S., McDougall, I. & Walker, A. 1995. New four-million-year-old hominid species from Kanapoi and Allia Bay, Kenya. *Nature*, **376**, 565–71.

White, T. D., Suwa, G. & Asfaw, B. 1995. Corrigendum. *Nature*, **375**, 88.

21 • Reconstructions reconsidered: chimpanzee models and human evolution

ADRIENNE ZIHLMAN

CHIMPANZEE MODELS: A LACK OF CONSENSUS

Until very recently, primatologists and paleoanthropologists did not agree on which apes, if any, were most relevant to reconstructing the common ape–human ancestor or how information was to be applied in speculating about human origins and the evolution of social behavior. Prior to the 1960s, little except anatomy was known of the great apes, but later, field observations and laboratory studies of cognition stimulated interest in ape models for hominid reconstructions. Fossil discoveries of extinct apes and early hominids multiplied, and molecular studies established the details of phylogenetic relationships among living primates. In the 1970s, information derived from laboratory and field research and from anatomical studies on the lesser-known pygmy chimpanzee (*Pan paniscus*) changed conceptions about the genus *Pan* and offered new ways to look at models for hominid origins. The combination of these different lines of evidence has led to a deeper and more complete understanding of the great apes and their relationship to human evolution.

CHIMPANZEES AND HUMAN EVOLUTION: NEW INFORMATION

Studies during the 1960s and 1970s of free-ranging primates, the growing fossil hominid record and molecular biology challenged previous notions about the evolutionary relationships of the apes and the timing of the ape–hominid divergence. These studies significantly influenced the development of physical anthropology over the next several decades.

Behavioral studies on free-ranging chimpanzees

The earliest chimpanzee field studies in the 1960s on *Pan troglodytes* revealed much new and surprising information about ape behavior. For example, Vernon Reynolds (1966) was struck by the ability of chimpanzees to assemble and to disperse with ease. Jane Goodall (1968) documented spontaneous use of probing tools, sponges and missiles, as well as meat eating (Fig. 21.1). Japanese primatologists detailed ranging patterns and group size and composition at Mahale, Tanzania (Nishida, 1968).

In the 1974 Great Apes conference volume, Hamburg & McCown (1979) noted in the introduction that there was no consensus about evolutionary relationships between humans and other primates. Several chapters implied that the apes provided a rich source of information for speculating about human evolution, but few authors addressed the issue directly. McGrew (1979), however, stated: 'it appears that the chimpanzee increasingly represents a better evolutionary model for early man' (p. 462).

Fossil discoveries

Prior to the discoveries of fossil hominids from Olduvai Gorge, Tanzania, the earliest human fossils – given the generic name of *Australopithecus* – came from South African caves. These deposits could not be dated accurately, and human origins were thought to be less than a million years ago. The East African discoveries by the Leakeys increased that time span significantly. New geochronological methods, such as potassium–argon dating, revealed a date of almost 2 million years for the base of Bed I, Olduvai Gorge. Koobi Fora on the eastern proportions intermediate between African apes and humans; pelvic and limb bones more humanlike than apelike; and footprints belonging to a bipedal hominid.

Paleontologists searched for a possible hominid ancestor in the fossil record. A leading candidate in the 1960s was *Ramapithecus,* first named by George Lewis (1934) for fossils from the Siwalik Hills in India, now Pakistan. Later, *Aegyptopithecus,* based on a fossil cranium from Egypt aged about 30 million years, was put forth as the putative ape ancestral to modern apes and humans (Simons, 1967). Few paleontologists were concerned with possible insights into hominid origins that might be derived from living apes.

Molecular findings

While *Ramapithecus* was gaining popularity as the ancestor to *Australopithecus,* a number of studies by molecular anthropologists were under way. By placing constraints on the time of the ape-human divergence, these studies critically challenged the supposition that *Ramapithecus* was a hominid.

The molecular biology revolution and the decipherment in 1960 of the genetic code – the translation of 3-base codons into individual amino acids, the building blocks of proteins – made it possible to establish genetic relationships among living species, indirectly through immunology or directly though sequencing DNA or proteins. The earliest methods applied to the study of primate relationships included protein immunological techniques (e.g. Goodman, 1963; Sarich & Wilson, 1967) and protein amino acid sequencing (e.g. Doolittle & Mross, 1970).

The molecular findings were surprising in several ways. First, they showed that, genetically, humans and African apes are very closely related, as close as the horse is to the zebra or the dog is to the fox. Second, the chimpanzee and gorilla are no closer to one another than either is to humans. This implies that morphological and genetic changes have occurred at different rates within the three lineages (e.g. King & Wilson, 1975). Finally, applying a molecular clock, Sarich & Wilson (1967) maintained that the chimpanzee, gorilla and human lineages separated only 5 million years ago, much more recently than the projected 15–30 million years. Estimates of the divergence of the African apes derived from

Fig. 21.1. Adult female chimpanzee eats red colobus monkey meat at Gombe. (Photo by C. Tutin)

shores of Lake Turkana in Kenya and the Omo River Valley in Ethiopia yielded fossils extending back to nearly 3 million years. Fossils from Hadar, Ethiopia, were more than 3 million years old, and footprints of bipedal hominids from Laetoli, Tanzania, along with jaws and teeth, pushed the known hominid record back to nearly 4 million years and over a wide geographic range in paleoenvironments, indicating a savanna mosaic (Curtis, 1981).

African hominids dated between 2 and 4 million years shared a number of morphological features: a brain about the size of a chimpanzee; very large and heavily worn molars and premolars; relatively small canine teeth; limb anatomical characteristics alone varied from 3 (Walker, 1976) to 15 million years (Simons, 1976).

The molecular clock

Zuckerkandl & Pauling (1962) first proposed the idea of a molecular clock. They observed that the number of amino acid differences between the hemoglobins of various vertebrates is roughly proportional to their divergence times estimated from the fossil record. The principle of the molecular clock is that the longer any two species have been separated, the more amino acid differences exist in homologous proteins, or in the case of DNA, in the nucleotide sequences.

Combining the immunological results from albumin and transferrin, Sarich & Cronin (1976) showed immunologically that African apes and humans differ by about 12% in their albumin and transferrin, African hominoids and Asian hominoids differ by about 25%, and African hominoids and cercopithecoids differ by about 56%. The Asian hominoids are roughly twice as different from the African hominoids, and Old World monkeys four to five times more different.

The relative amounts of difference can be linked to time by calibration with dates of divergence based on the geological record. For example, Sarich & Cronin used 35 million years for platyrrhine–catarrhine divergence, a geological time based on the presumed first appearance of fossil anthropoids. The cercopithecoid–hominoid divergence then is calculated to about 20 million years (35×0.56), and an African ape-human divergence to 4 million years (35×0.12). If 40 million years is used as the calibration point, then the cercopithecoid-hominoid split is 22 and the African ape–human split becomes 5 million years. The point is, one can alter the geological time used, but the set of quantitative relationships remains the same. Importantly, in order to arrive at a 15 million year split of African ape–human, the cercopithecoid–hominoid divergence would have to be 75 million years, a geological period which has no primate fossil record at all.

The rate of change, that is, the rate at which nucleotides mutate, differs among different proteins. Each protein has a specific rate of substitution of its amino acids, and the rate depends upon the function of the protein. Histones, for example, interact with DNA and their particular structure is critical to their function. This protein changes so slowly that its sequence is nearly identical in even distantly related species. In contrast, fibrinopeptides serve as spacers in blood clotting, so that the actual sequence of amino acids is unimportant for this function. Consequently, they mutate rapidly and provide a 'fast clock' for comparing closely related species like chimpanzees and humans. One does not assume an intrinsic molecular clock; time has to be calibrated using one or more points in the fossil record. It is possible that there is some intrinsic regularity in rate of base pair or amino acid substitution, but this is not yet established.

With molecular methods, species differences can then be measured directly and quantified, rather than being inferred from anatomical resemblances that are unavoidably subjective. Regardless of the method used – immunology (e.g. immunodiffusion, microcomplement fixation), DNA hybridization, or sequencing of proteins and DNA – the results agree, because the relative quantitative differences remain approximately the same between species (Wilson et al., 1977). Thus it became possible to estimate more specifically times of divergence between species.

THE LAST APE: *PAN PANISCUS*

By the middle 1970s, the field research on chimpanzees (*Pan*), the paleontological record and the molecular findings converged: (1) chimpanzees showed amazing similarities to humans in tool using, social relations and communication, behavior that parallels the similarities in anatomy between *Pan* and *Homo;* (2) the fossil hominids resembled chimpanzees more than modern humans; (3) the molecular data suggested a close relationship between humans and African apes but could not distinguish which pair, if either, was closer; and (4) the molecular clock suggested a very recent ape–human divergence. Despite the convergence of these different lines of evidence, most paleontologists were not convinced of the *Pan–Homo* connection.

During the 1970s, a growing interest in the other species of chimpanzee, *Pan paniscus,* was stimulated by preliminary field studies in Zaïre (Badrian & Badrian, 1977; Kano, 1979), laboratory studies on communication and sexual behavior (Savage-Rumbaugh & Wilkerson, 1978), molecular findings (Goodman et al., 1970) and anatomical studies (Cramer, 1977; Zihlman & Cramer, 1978). Integrating this new information, Zihlman *et al.* (1978) proposed that, of the living apes, *Pan paniscus* provided the best prototype for deriving both the earliest hominids and the other African apes.

Species distinctions in Pan

Prior to 1970, it was not clear how the two species of chimpanzees compared anatomically. Miller (1932) described muscle origin and insertion in a juvenile female *Pan paniscus* weighing only 8 kg. Coolidge (1933) compared one adult skeleton to *Pan troglodytes* and concluded that *Pan paniscus* has shorter upper limbs and a narrower chest. The distinguishing external features included facial shape and hair pattern, vocalizations and a tendency for bipedal behavior; much was made of the webbing between digits 2 and 3 on the foot.

Skull and dentition

Studies in the early 1970s on the dentition and craniofacial anatomy of the two species revealed that *Pan paniscus* had smaller dentition, most noticeably in the canines (Johanson, 1974). Although a number of cranio-facial dimensions overlap, the two species can be discriminated completely on the basis of mandibular length (Cramer, 1977). *Pan paniscus* skulls and dentition are notably gracile, this being the reason Coolidge called it 'pygmy' chimpanzee, which turned out to be misleading when applied to the postcranial anatomy. Cramer's sample of 60 skulls each of wild-caught *Pan paniscus* and *Pan troglodytes* showed that cranial capacities overlapped though the means differed significantly: *Pan paniscus*, 350 ml; *Pan troglodytes*, 390 ml.

Skeletal dimensions and limb proportions

Having studied the early hominid fossils and *Pan troglodytes*, I sought to learn more about morphological variation among living apes by studying the postcranial skeleton of *Pan paniscus*. From extensive research on apes, Adolph Schultz (personal communication) doubted that there were two distinct species. His small sample of *Pan paniscus* showed overlap with *Pan troglodytes* in body weights and in long bone lengths (Schultz, 1969).

Results from 21 adult *Pan paniscus* and 30 adult *Pan troglodytes* skeletons supported Coolidge's conclusions

based upon a single female skeleton. In the two groups of skeletons of both sexes, many of the measurements on bone lengths and breadths and joint sizes overlapped, but most (13 of 20) were statistically significantly different. Those differences that were statistically significant ($p < 0.01$) included: clavicle length; scapula length and breadth; humeral head diameter and biepicondylar breadth; innominate length, iliac and sacral breadths, acetabulum and femoral head diameters; and ratio of humerus to femur lengths (Zihlman & Cramer, 1978).

To address possible size and scaling effects, we also analyzed the measurements omitting male *Pan troglodytes* from the sample. Body weights from 18 free-ranging *Pan paniscus* (25–48 kg) showed extensive overlap with *Pan troglodytes* (25–60 +). Because male *Pan troglodytes* are much heavier than both female and male *Pan paniscus*, we compared the *paniscus* sample with only female *Pan troglodytes*. In this more comparable size comparison, we found that only clavicle length, humeral head diameter, iliac breadth, and femur and tibia lengths significantly distinguished the two species. The difference in humerus/femur ratio persisted in spite of more comparable body size (Zihlman & Cramer, 1978). Because *Pan paniscus* has relatively shorter upper limbs and longer lower limbs than *Pan troglodytes*, we concluded that *Pan paniscus* is not merely a scaled down version of *Pan troglodytes* but has distinct body proportions. Later studies confirmed the presence of two distinct patterns of body proportions (Jungers & Susman, 1984; Morbeck & Zihlman, 1989). Quantitative anatomical dissections also confirmed species differences in body proportions and demonstrated that *Pan paniscus* has more massive lower limbs (Zihlman, 1984).

Two patterns of sexual dimorphism

In evaluating the fossil record, sex can confound the identification of variation between species. Investigation of sex differences has revealed a distinct pattern of morphological variation within each of the two species of chimpanzee, a pattern not evident when looking at features singly. For example, *Pan troglodytes* males are heavier than females in average body weight and have larger average cranial capacity (male, 404 ml; female, 385 ml) (Cramer, 1977). Males are also larger in a number of dimensions including limb length, chest circumference and canine size (Schultz, 1956; Zihlman & Morbeck, unpublished data).

Pan paniscus exhibits a different pattern. On average, body weights show a similar degree of sex difference in both species: female *Pan paniscus* differ significantly from males (33 vs 46 kg), just as do female and male *Pan troglodytes* (40 vs 48 kg). However, the metric data show no significant differences between *Pan paniscus* females and males in cranial capacity, facial dimensions, tooth size, joint size, bone lengths or bone circumferences (Cramer & Zihlman, 1978).

The morphological differences between the two species are slight but important. Many of the statistically significant differences could not have been deduced without a sufficient sample of complete skulls and complete skeletons. These findings raise questions about the ability of paleoanthropologists to distinguish among species and between the sexes on the basis of incomplete individuals, small and incomplete sample sizes and non-existent population samples.

The pygmy chimpanzee proposal

Comparing humans, *Pan troglodytes* and *Pan paniscus*, the latter appears to be the most 'generalized' anatomically in its smaller body size, smaller facial and canine dimensions, and particularly in limb proportions. In addition, like humans, *Pan paniscus* is less sexually dimorphic in dentition. The existing reports of their tendency for bipedal behavior seem to follow from their having body proportions with more equal upper and lower limb lengths, and longer and heavier lower limbs. This predisposition for bipedal behavior and an inferred lower center of gravity suggest an easier transition from a quadrupedal ape to a bipedal hominid than previously imagined. For these reasons, Zihlman *et al.* (1978) proposed that *Pan paniscus* serves as a useful model of the prototypic ape that gave rise to the gorilla, chimpanzee and human lineages.

The pygmy chimpanzee model focused on several points. The model used chimpanzees as a point of reference for early hominid fossils. For example, instead of Le Gros Clark's gorilla skull (1955), a skull of *Pan paniscus* provided a more appropriate comparison with *Australopithecus*. Given the relatively generalized body proportions of *Pan paniscus*, we presumed that a bipedal hominid

could more easily be derived from it than from the larger gorilla or the more robust *Pan troglodytes*. This model also challenged the prevailing assumption that the hominid ancestor was highly sexually dimorphic. Lacking sex differences in canine teeth, cranial capacity and limb proportions, *Pan paniscus* provided a viable alternative to other primate models.

Finally, *Pan paniscus* drew renewed attention to the social behavioral dimension of chimpanzees (*Pan* spp.) for reconstructing early hominid behavior (e.g. Goodall & Hamburg, 1975; Tanner & Zihlman, 1976; McGrew, 1979, 1981) and increased the parameters of possible behavior. Their small canine teeth and less aggressive nature, their food sharing, social bonding and female dominance provided alternatives to the pervasive emphasis on male dominance and aggression (e.g. Kuroda, 1979; Kano, 1980; Zihlman, 1981).

Hypothesis and prediction

The strength of any hypothesis depends on its compatibility with all lines of evidence, ability to withstand falsification and predictive power. The *Pan paniscus* model predicted that, as hominid fossils were discovered older than 3–4 million years, they 'should continue to converge on the pygmy-chimpanzee-like condition, cranially and postcranially as they seem to do dentally' (Zihlman *et al.*, 1978, p. 746). This prediction is compatible with the new fossil evidence on *Australopithecus ramidus* from Aramis, Ethiopia (White *et al.*, 1994). *Pan paniscus* is now recognized as offering useful perspectives on early hominid anatomy and providing a broader base from which to speculate about social behavior.

THE CHIMPANZEE–HUMAN CONNECTION: TESTING THE HYPOTHESIS

Molecular advances: chimpanzees closer to humans than gorillas

Molecular techniques of the 1960s and 1970s were not sensitive enough to show with reasonable certainty which pair among humans, chimpanzees (*Pan*) and gorillas was most closely related. During the 1980s, hybridization and sequencing of DNA resolved the triad in favor of human-chimpanzee. Nuclear DNA provides about ten times more information than does protein sequencing – the previous molecular tool – because of the presence of introns, pseudogenes and the redundancy of the codons (Bailey *et al.*, 1992). A technique of hybridizing DNA (Sibley & Ahlquist, 1984) first demonstrated the closer chimpanzee–human relationship. Charles Sibley, a Yale ornithologist, had previously used this method to elucidate bird phylogeny, and anthropologist David Pilbeam suggested that he apply it to hominoid relationships. Although Sibley and Ahlquist's DNA hybridization data analysis was strongly attacked (e.g. Marks *et al.*, 1988), several independent studies now support their original findings, e.g. Caccone & Powell (1989).

Another significant source of information, mitochrondrial DNA, supports and refines the hominoid phylogeny (Hayasaka *et al.*, 1988; Ruvolo *et al.*, 1991; Horai *et al.*, 1992). Residing in the cytoplasm, mitochondria are passed through the generations solely through the female line and have been found to evolve at about ten times the average rate of nuclear DNA. Thus mitochondrial DNA is ideal for studying closely related species or populations within species.

In analyzing currently available molecular data, Bailey *et al.* (1992) concluded that these studies taken cumulatively support chimpanzee–human monophyly calculated to be statistically significant at the $p = 0.002$ level.

As noted previously, 'molecular clocks' must be calibrated to fossil divergences. In Sibley and Ahlquist's original study, they chose 16 million years (my) as a calibration point for orangutan–human separation, based upon Pilbeam's (1982) interpretation of *Sivapithecus* as an ancestral orangutan. This gave the human–chimpanzee divergence a date of 6.3–7.7 my, with gorillas separating at 8–9.9 my (Sibley & Ahlquist, 1984). In an 'added note' they recalculated their dates based on new fossils dated at 17 my from Buluk, Kenya, that were claimed to represent the ancestor of the great apes. Their revised divergence dates are: chimpanzee-human 5.9 my; gorilla 7.5 my; and orangutan 12 my. Although researchers vary the fossil dates for calibrating the clock, the data are consistent in supporting a human and chimpanzee branching.

The best approximate dates remain somewhere around 5–6 my for the chimpanzee–human divergence with the two chimpanzee species separating somewhere between 1.5 and 2.5 my (Ruvolo, 1994). The two chimpanzee species are distinct from each other genetically and are equally closely related to humans. Recent DNA research indicates that the western population of *Pan troglodytes* is so distinct genetically from the eastern population as to raise the question of whether or not they are different species (Morin *et al.*, 1994). Furthermore, new research on gorillas has shown that the western lowland, the eastern lowland and mountain gorilla populations differ more than do the two chimpanzee species, suggesting an older time of separation for gorilla populations than previously thought (Ruvolo *et al.*, 1994).

Anatomical studies: further distinctions within and between Pan species

Reacting to the *Pan paniscus* model, Johnson (1981) claimed that pygmy chimpanzees were insular dwarfs, and several authors agreed that *Pan paniscus* was generalized (see commentaries in Johnson, 1981). Because of overlap in linear measurements and body weights, and because the differences in body proportions were small (but significant), some taxonomists did not agree that *Pan paniscus* was a separate species. Groves (1986), for example, did not rule out the possibility that a population of small-bodied *Pan troglodytes* would metrically converge with *Pan paniscus*.

A 'natural experiment' refuted Groves' conjecture that a smaller *Pan troglodytes* would be more like *Pan paniscus*. Some Gombe chimpanzees (*Pan troglodytes schweinfurthii*) of known age and sex have body weights less than other *Pan troglodytes* populations. Gombe chimpanzees nonetheless conform to the *Pan troglodytes* pattern in cranial capacity and tooth size and, most significantly, have the same limb proportions. Compared with *Pan paniscus*, Gombe chimpanzees have lower body weight, higher cranial capacity, larger tooth size, broader trunk, shorter forelimb and hindlimb lengths and limb proportions like those of other *Pan troglodytes* and distinct from those of *Pan paniscus*. The small-bodied Gombe population demonstrates that the two species of *Pan* have distinct morphological patterns not affected by body size (Morbeck & Zihlman, 1989).

Further investigation of species and sex differences in *Pan* by quantitative anatomical dissections show that relative weight of body segments differs in the two species, with *Pan paniscus* having heavier lower limbs (Zihlman, 1984). *Pan paniscus* females and males differ significantly in body weight but conform to the pattern derived from an earlier skeletal sample in showing no significant difference in cranial capacity and long bone lengths (Zihlman, unpublished data).

Growth and development of extant hominoids and extinct hominids: life history connections

Adolph Schultz's pioneering work on fetal, infant and juvenile growth in body size, skeleton and dentition provided important data on the sequence of skeletal and dental development, its variation among hominoids and the similarity in growth patterns between humans and African apes (Schultz, 1956). Information from recent studies on growth and development link chimpanzees and hominids. New techniques and methods enable direct observation of the incremental development of bone and tooth microstructure. Growth lines are laid down at regular intervals, leaving daily incremental markings (Retzius lines and perikymata, the cross-striations of the prisms) in the enamel and dentine (Bromage & Dean, 1985; Dean, 1989; Boyde, 1990). Methods such as scanning electron micrograph (SEM) and tandem scanning reflected light microscopy (TSRLM) can directly measure the growth layers to establish the sequence and rate of crown and root formation. CAT scans allow direct observation of the stages of tooth calcification (i.e. the stage of mineralization) in both extinct and extant species (Conroy & Vannier, 1987).

These techniques directly measure growth layers, and like the molecular data, can be quantified. Direct measurements of the growth lines on teeth make it possible to infer accurately chronological age in fossil hominids. Furthermore, comparisons between extant hominoids and fossil hominids can be made using the same methods. This information can reveal landmarks in the life history of individual fossils. As Boyde (1990) notes, teeth are born fossils and each tooth is a sort of 'diary of events in an individual's life history' (p. 230).

These studies on dental development have shown that in terms of growth and life history, the early hominids are more similar to the great apes than to modern humans. In particular, the early hominids show many similarities to chimpanzees in facial growth, dental eruption and the overall pattern of development (Bromage, 1987; Smith, 1991). For example, *Australopithecus africanus* (Taung) erupted its first permanent molar between 3 and 3.5 years, an age comparable to that of chimpanzees (Bromage, 1987). Chronological age at death for an australopithecine who died at or near emergence of first permanent molar is now estimated as 3.25 years based on incremental lines in the teeth (Dean *et al.*, 1993). This differs substantially from expectations based on human growth schedules of 5.5–6 years and suggests parallels with life history features of chimpanzees rather than with those of *Homo sapiens*.

Aspects of dental development, such as age of emergence of the first permanent molar, correlate with other growth phenomena such as degree of brain growth or body size. In turn, these events can be associated with weaning time and thus provide landmarks in the life history of the individuals. The integration of life history features offers new dimensions for interpreting early hominids and understanding the evolution of human growth and ageing (Smith 1993).

Issues in hominid evolution: the chimpanzee framework

New and continuing studies on both species contribute important insights into the ape–human transition and early hominid behavior by expanding our knowledge of chimpanzee anatomy, locomotion, tool using, hunting, anti-predator defense and life history (see, for example, chapters in this volume and those in Wrangham *et al.*, 1994). This new information directs attention to new questions and new interpretations of hominid evolution.

Hominid locomotion

The data cited compel close examination of interpretations of early hominid locomotion, for example, whether *Australopithecus* was an efficient terrestrial biped or remained adept at climbing trees (e.g. Lovejoy, 1981 vs. Susman *et al.*, 1984). This debate perpetuates a traditional arboreal-terrestrial dichotomy, one that Washburn (1951) questioned when he pointed out the similarities in joint motions and muscle groups between ape climbing and hominid bipedality. However, when new data showed the genetic closeness of humans to African apes, and field research showed that chimpanzees and gorillas do climb in trees and do knuckle-walk on the ground, he modified his position (Washburn, 1967) and proposed that the ancestral hominid probably became bipedal from being a knuckle-walker on the ground. Based on studies of free-ranging African apes, Doran (Chapter 16) documents the high frequency of terrestrial knuckle-walking. These findings, combined with the newest molecular data on chimpanzee–human relatedness suggest again the possibility that the loco-motor behavior of hominid ancestors included knuckle-walking (Zihlman, 1990).

Species or sex?

Another debate in paleoanthropology centers on the issue of the species designation of *Australopithecus afar-ensis*, and whether one highly sexually dimorphic homi-nid – rather than multiple species – existed between 3 and 4 million years ago. Here too the chimpanzee data provide a basis for questioning current assumptions. In justifying the designation of a new species, Johanson & White (1979) argued that it consisted of one species and that the extreme range in size of the limb bones could be attributed to the existence of small females and large males of the same species. Johanson, White and colleagues (Clark *et al.*, 1984; White *et al.*, 1993; Kimbel *et al.*, 1994) continue to maintain this position even as larger and more robust cranium and limb bones are found that contrast with the smaller and more gracile AL 288 (or 'Lucy') and that show morphological similarities with the robust australopithecines.

The assumption that the prehominid ancestor was highly sexually dimorphic conflicts with the moderate body size dimorphism in both species of *Pan*. As a com-parative anatomist is likely to admit, it is difficult to sort chimpanzee postcranial skeletons into distinct sex or species categories. Limb bones in the two species are similar in size and shape; small statistical differences in limb proportions emerge only with complete skeletons of known sex and species and with a large enough sample size. Given the information available on the skeletal anat-omy of living species, the greater size difference attri-buted to sexual dimorphism in the newest fossils is unlikely, and the marked variation in the sample most likely represents at least two species.

Hominid social behavior

The interpretation of hominid social behavior has been a continuous source of discussion and debate (e.g. Fedi-gan, 1986; Kinzey, 1987). Which species of primate offers more insight into hominid origins and early homi-nid life – chimpanzees, gorillas, baboons, other apes or monkeys? Disagreements stem from the kinds of ques-tions posed, comparative methods used and evidence cited. For example, in their discussion of models and the evolution of hominid behavior, Tooby & DeVore (1987) dismiss the use of chimpanzees as models for early homi-nid behavior as proposed by Tanner & Zihlman (1976). The authors imply that the choice of chimpanzee behavior, as opposed to baboons, was made on ideologi-cal grounds. 'When male dominance became less popular as a research perspective, the putatively more peaceful chimpanzee became a popular referential species' (Tooby & DeVore, 1987, p. 188). Tooby and DeVore not only ignore the range of anatomical and fossil evi-dence that support a chimpanzee–type of model, but also ignore the conclusions of a number of primatologists who have supported a chimpanzee model (e.g. Goodall & Hamburg, 1975; McGrew, 1979, 1981; Kano, 1980; Kuroda, 1984); they ignore the molecular data as well. Reconstructions of hominid evolution cannot be reliably achieved if based upon one line of evidence alone, such as social behavior, while neglecting the influence of gen-etic potential, morphology and life history features. It is not that chimpanzees are primitive versions of ourselves, but they offer the possibility for understanding the kind of organization that may be characteristic of the ancestral forms, as noted early by Washburn (1951) and by Fedi-gan (1986) in her review of models.

New fossil discoveries

The fossil record is now even more in agreement with predictions made explicit in the pygmy chimpanzee model. Fossil discoveries in Ethiopia, along with geo-chronological dates, extend the hominid record from 3.2 million years ago (Hadar, Ethiopia) and 3.7 mya (Laetoli, Tanzania) to 3.9 mya (Awash, Ethiopia) (Clark *et al.*, 1984) and most recently to 4.4 mya (Aramis, Ethiopia). The newest fossils are said to represent a new species, *Australopithecus ramidus*, from 'ramid' meaning 'root' to represent 'the long-sought potential root species for the Hominidae' (White *et al.*, 1994, p. 306).

The new material from Aramis is fragmentary – parts of three arm bones, cranial pieces, associated and isolated teeth, and a piece of a child's mandible with a complete deciduous first molar. In the article reporting this find, the authors compare the Aramis deciduous molar with *Dryopithecus, Pan troglodytes* and *Pan paniscus, Australopithecus afarensis, Australopithecus africanus, Australopithecus robustus, Australopithecus boisei* and *Homo sapiens.* Of the array, the fossil *Australopithecus aramis* molar is in its morphology strikingly similar to *Pan paniscus*, and metrically it is 'centered in the chimpanzee ranges' (White *et al.*, 1994, p. 307). Whether or not *Australopithecus aramis* turns out as the ancestor to all later hominids, it does strongly suggest that the earliest hominids were likely derived from an ancestor resembling the genus *Pan.*

SUMMARY

Chimpanzees have long been used as possible models of the common ancestor of apes and hominids, but only within recent decades have various lines of evidence converged to validate the choice of chimpanzees (*Pan*) over other apes or monkeys in reconstructions. Comparisons of hominoid DNA have shown chimpanzees to be the closest living relatives of humans. Field studies have revealed behavior such as food sharing, tool use and sophisticated social interactions that once had been thought exclusively human. The oldest human fossils are markedly chimpanzee-like in overall size, brain size and proportions. The pattern of tooth eruption in infant australopithecines is more like that of chimpanzees than like that of humans.

Of the two chimpanzee species, *Pan troglodytes* and *Pan paniscus*, several anatomical features associated with potential bipedal behavior slightly favor a *Pan paniscus*-like model for the human–chimpanzee ancestor. In body proportions, the shorter upper limbs relative to lower limbs of *Pan paniscus* might facilitate the transition from knuckle-walking to habitual bipedalism. Another 'human' characteristic of smaller canine teeth and little sex difference found in *Pan paniscus* is important as well. The recently discovered fossils identified as *Australopithecus ramidus*, which at 4.4. million years are the oldest hominids yet found, have teeth that in size and shape resemble those of *Pan paniscus*.

In past decades, a variety of known primates have been proposed as models for the common ancestor of humans and apes. Now the choice of such a model is constrained by our growing knowledge of molecular genetic relationships, primate anatomy and behavior, and the fossil record. Obviously, no living primate is identical in all respects to the hominid ancestor that lived 5–6 million years ago, but *Pan paniscus* has been useful in visualizing the transition from quadrupedalism to bipedalism, and how a common progenitor could have given rise to three such distinct lineages as humans, chimpanzees and pygmy chimpanzees. In addition, the *Pan paniscus* model offers another way to view the social life of early hominids, given their sociability, lack of male dominance and the female-centric features of their society (Kano, 1992). With increasing information from both species of *Pan*, what it means to be a chimpanzee has been expanded. For the reconstruction of human evolution this means more possibilities within a *Pan* model for proposing hypotheses about hominid evolution.

ACKNOWLEDGMENTS

This paper was originally prepared for the Wenner-Gren conference 'The Great Apes Revisited,' held November 12–19 in Cabo San Lucas, Baja, Mexico. I thank the organizers for the opportunity to participate in the conference and for their editorial assistance. I appreciate the discussions with and comments from Jerry Lowenstein, Robin McFarland and Kim Nichols and the help of Josh Snodgrass in completing the manuscript. I thank the University of California, Santa Cruz, Division of Social Sciences and Faculty Research Committee, for financial support of my research.

REFERENCES

Badrian, A. & Badrian, N. 1977. Pygmy chimpanzees. *Oryx*, **13**, 463–8.

Bailey, W. J., Hayasaka, K., Skinner, C. G., Kehoe, S., Sieu, L. C., Slightom, J. L. & Goodman, M. 1992. Re-examination of the African hominoid trichotomy with additional sequences from the primate beta-globin gene cluster. *Molecular Phylogeny and Evolution*, **1**, 97–135.

Boyde, A. 1990. Developmental interpretations of dental microstructure. In: *Primate Life History and Evolution*, ed. J. DeRousseau, pp. 229–67. New York: Alan Liss.

Bromage, T. G. 1987. The biological and chronological maturation of early hominids. *Journal of Human Evolution*, 16, 257–72.

Bromage, T. G. & Dean, M. C. 1985. The re-evaluation of the age at death of Plio-Pleistocene fossil hominids. *Nature*, 317, 525–8.

Caccone, A. & Powell, J. R. 1989. DNA divergence among hominoids. *Evolution*, 43, 925–42.

Clark, J. D., Asfaw, B., Assefa, G., Harris, J. W. K., Kurashina, H., Walter, R. C., White, T. D. & Williams, M. A. J. 1984. Palaeoanthropological discoveries in the Middle Awash Valley, Ethiopia. *Nature*, 307, 423–8.

Conroy, G. C. & Vannier, M. W. 1987. Dental development of the Taung skull from computerized tomography. *Nature*, 329, 625–7.

Coolidge, H. J. 1933. *Pan paniscus*. Pygmy chimpanzee from south of the Congo River. *American Journal of Physical Anthropology*, 18, 1–57.

Cramer, D. L. 1977. Craniofacial morphology of *Pan paniscus*: a morphometric and evolutionary appraisal. In: *Contributions to Primatology, Vol. 10*, Basel: Karger.

Cramer, D. L. & Zihlman, A. L. 1978. Sexual dimorphism in *Pan paniscus*. In: *Recent Advances in Primatology, Vol. 3, Evolution*, ed. D. J. Chivers & K. A. Joysen, pp. 487–90. London: Academic Press.

Curtis, G. H. 1981. Establishing a relevant time scale in anthropological and archaeological research. *Philosophical Transactions of the Royal Society of London, Series B*, 292, 7–20.

Dean, M. C. 1989. The developing dentition and tooth structure in hominoids. *Folia Primatologica*, 53, 160–73.

Dean, M. C., Beynon, A. D., Thackeray, J. F. & Macho, G. A. 1993. Histological reconstruction of dental development and age at death of a juvenile *Paranthropus robustus* specimen, SK 53, from Swartkrans, South Africa. *American Journal of Physical Anthropology*, 91, 401–19.

Doolittle, R. & Mross, G. A. 1970. Identity of chimpanzee and human fibrinopeptides. *Nature*, 225, 643–4.

Fedigan, L. M. 1986. The changing role of women in models of hominid evolution. *Annual Review of Anthropology*, 15, 25–66.

Goodall, J. 1968. The behaviour of free-living chimpanzees in the Gombe Stream Reserve. *Animal Behaviour Monographs*, 1, 161–311.

Goodall, J. & Hamburg, D. A. 1975. Chimpanzee behavior as a model for the behavior of early man. New evidence on possible origins of human behavior. *American Handbook of Psychiatry*. New York: Basic Books.

Goodman, M. 1963. Man's place in the phylogeny of the primates as reflected in serum proteins. In: *Classification and Human Evolution*, ed. S. L. Washburn, pp. 204–34. Chicago: Aldine.

Goodman, M., Moore, W. G., Farris, W. & Poulik, E. 1970. The evidence from genetically informative macromolecules on the phylogenetic relationship of the chimpanzees. In: *The Chimpanzee, Vol. 1*, ed. G. H. Bourne, pp. 318–60. Basel: Karger.

Groves, C. P. 1986. Systematics of the great apes. In: *Comparative Primate Biology, Vol. 1, Systematics, Evolution and Anatomy*, pp. 187–217. New York: Alan R. Liss.

Hamburg, D. A. & McCown, E. R. (eds.) 1979. *The Great Apes*. Menlo Park, CA: Benjamin/ Cummings.

Hayasaka, K., Gojobori, T. & Horai, S. 1988. Molecular phylogeny and evolution of primate mitochondrial DNA. *Molecular Biology and Evolution*, 5, 626–44.

Horai, S., Satta, Y., Hayasaka, R., Kondo, R., Inoue, T., Ishida, T., Hayoshi, S. & Takahata, N. 1992. Man's place in Hominoidea revealed by mitochondrial DNA geneology. *Journal of Molecular Evolution*, 35, 32–43.

Johanson, D. C. 1974. Some metric aspects of the permanent and deciduous dentition of the pygmy chimpanzee (*Pan paniscus*). *American Journal of Physical Anthroplogy*, 41, 39–48.

Johanson, D. C. & White, T. D. 1979. A systematic assessment of early African hominids. *Science*, 202, 321–30.

Johnson, S. C. 1981. Bonobos: generalized hominid prototypes or specialized insular dwarfs? *Current Anthropology*, 22, 363–75.

Jungers, W. L. & Susman, R. L. 1984. Body size and skeletal allometry in African apes. In: *The Pygmy Chimpanzee*, ed. R. L. Susman, pp. 131–77. New York: Plenum Press.

Kano, T. 1979. A pilot study on the ecology of pygmy chimpanzees, *Pan paniscus*. In: *The Great Apes*, ed. D. A. Hamburg & E. R. McCown, pp. 123–35. Menlo Park, CA: Benjamin/Cummings.

Kano, T. 1980. Social behavior of wild pygmy chimpanzees (*Pan paniscus*) of Wamba: a preliminary report. *Journal of Human Evolution*, 9, 243–60.

Kano, T. 1992. *The Last Ape*. Stanford, CA: Stanford University Press.

Kimbel, W. H., Johanson, D. C. & Rak, Y. 1994. The first

skull and other new discoveries of *Australopithecus afarensis* at Hadar, Ethiopia. *Nature*, **368**, 449–51.

Kinzey, W. G. (ed.) 1987. *The Evolution of Human Behavior: Primate Models*. Albany, NY: State University of New York Press.

King, M. C. & Wilson, A. C. 1975. Evolution at two levels in humans and chimpanzees. *Science*, **188**, 107–16.

Kuroda, S. 1979. Grouping of the pygmy chimpanzees. *Primates*, **20**, 161–80.

Kuroda, S. 1984. Interactions over food among pygmy chimpanzees. In: *The Pygmy Chimpanzee*, ed. R. L. Susman, pp. 301–24. New York: Plenum Press.

Le Gros Clark, W. E. 1955. *Fossil Evidence for Human Evolution*. Chicago: University of Chicago Press.

Lewis, G. E. 1934. Preliminary notice of man-like apes from India. *American Journal of Science*, **27**, 161–79.

Lovejoy, C. O. 1981. The origin of man. *Science*, **211**, 341–50.

Marks, J., Schmid, C. W. & Sarich, V. M. 1988. DNA hybridization as a guide to phylogenetic relationships of the Hominoidea. *Journal of Human Evolution*, **17**, 769–86.

McGrew, W. C. 1979. Evolutionary implications of sex differences in chimpanzee predation and tool use. In: *The Great Apes*, ed. D. A. Hamburg & E. R. McCown, pp. 441–63. Menlo Park, CA: Benjamin/ Cummings.

McGrew, W. C. 1981. The female chimpanzee as a human evolutionary prototype. In: *Woman the Gatherer*, ed. F. Dahlberg, pp. 35–73. New Haven, CT: Yale University Press.

Miller, R. A. 1932. The musculature of *Pan paniscus*. *American Journal of Anatomy*, **91**, 183–232.

Morbeck, M. E. & Zihlman, A. L. 1989. Body size and proportions in chimpanzees, with special reference to *Pan troglodytes schweinfurthii* from Gombe National Park, Tanzania. *Primates*, **30**, 369–82.

Morin, P., Moore, J., Chakraborty, R., Jin, L., Goodall, J. & Woodruff, D. S. 1994. Kin selection, social structure, gene flow, and the evolution of chimpanzees. *Science*, **265**, 1193–201.

Nishida, T. 1968. The social structure of chimpanzees of the Mahale Mountains. In: *The Great Apes*, ed. D. A. Hamburg & E. R. McCown, pp. 73–121. Menlo Park, CA: Benjamin/Cummings.

Pilbeam, D. 1982. New hominoid skull material from the Miocene of Pakistan. *Nature*, **295**, 232–4.

Reynolds, V. 1966. Open groups in hominid evolution. *Man*, **1**, 441–52.

Ruvolo, M. 1994. Molecular evolutionary processes and conflicting gene trees: the hominoid case. *American Journal of Physical Anthropology*, **94**, 89–113.

Ruvolo, M., Disotell, T. R., Allard, M. W., Brown, W. M. & Honeycutt, R. F. 1991. Resolution of the African hominoid trichotomy using a mitochondrial gene sequence. *Proceedings of the National Academy of Sciences, U.S.A.*, **88**, 1570–4.

Ruvolo, M., Pan, D., Zehr, S., Goldberg, T., Disotell, T. & Dornum, M. 1994. Gene trees and hominoid phylogeny. *Proceedings of the National Academy of Sciences, U.S.A.*, **91**, 8900–4.

Sarich, V. M. & Cronin, J. E. 1976. Molecular systematics of the primates. In: *Molecular Evolution*, ed. M. Goodman & R. E. Tashian, pp. 141–70. New York: Plenum.

Sarich, V. M. & Wilson, A. C. 1967. Immunological time scale for hominid evolution. *Science*, **158**, 1200–3.

Savage-Rumbaugh, E. S. & Wilkerson, B. J. 1978. Socio-sexual behavor in *Pan paniscus* and *Pan troglodytes*: a comparative study. *Journal of Human Evolution*, **7**, 327–44.

Schultz, A. H. 1956. Postembryonic changes. *Primatologica Handbuch der Primatenkunde*, **1**, 887–964.

Schultz, A. H. 1969. The skeleton of the chimpanzee In: *The Chimpanzee: Anatomy, Behavior and Diseases of Chimpanzees*, pp. 50–103. Basel: Karger.

Sibley, C. G. & Ahlquist, J. E. 1984. The phylogeny of the hominoid primates, as indicated by DNA–DNA hybridization. *Journal of Molecular Evolution*, **20**, 2–15.

Simons, E. L. 1967. The earliest apes. *Scientific American*, **217**(6), 28–35.

Simons, E. L. 1976. The fossil record of primate phylogeny. In: *Molecular Anthropology*, ed. M. Goodman & R. E. Tashian, pp. 35–62. New York: Plenum.

Smith, B. H. 1991. Dental development and the evolution of life history in Hominidae. *American Journal of Physical Anthropology*, **86**, 157–74.

Smith, B. H. 1993. The physiological age of KMN-WT 15000. In: *The Nariokotome Homo erectus Skeleton*, ed. A. Walker & R. E. Leakey, pp. 195–220. Cambridge, MA: Harvard University Press.

Susman, R. L., Stern, J. T. Jr & Jungers, W. L. 1984. Arboreality and bipedality in the Hadar hominids. *Folia Primatologica*, **43**, 113–56.

Tanner, N. & Zihlman, A. 1976. Women in evolution. Part I. Innovation and selection in human origins. *Signs*, **1**, 585–608.

Tooby, J. & DeVore, I. 1987. The reconstruction of

hominid behavioral evolution through strategic modeling. In: *The Evolution of Human Behavior: Primate Models*, ed. W. G. Kinzey, pp. 183–237. Albany, NY: State University of New York Press.

Walker, A. 1976. Splitting times among hominoids deduced from the fossil record. In: *Molecular Anthropology*, ed. M. Goodman & R. Tashian, pp. 63–77. New York: Plenum.

Washburn, S. L. 1951. The analysis of primate evolution with particular reference to the origin of man. *Cold Spring Harbor Symposium on Quantitative Biology*, **15**, 67–78.

Washburn, S. L. 1967. Behaviour and the origin of man. Huxley Memorial Lecture. *Proceedings of the Royal Anthropology Institute of Great Britain*, pp. 21–7.

White, T. D., Suwa, G., Hart, W. K., Walter, R. C., WoldeGabriel, G., Heinzelin, J., Clark, J. D., Asfaw, B. & Vrba, E. 1993. New discoveries of *Australopithecus* at Maka in Ethiopia. *Nature*, **366**, 261–5.

White, T. D., Suwa, G. & Asfaw, B. 1994. *Australopithecus ramidus*, a new species of early hominid from Aramis, Ethiopia. *Nature*, **371**, 306–12.

Wilson, A. C., Carlson, S. S. & White, T. J. 1977. Biochemical evolution. *Annual Review of Biochemistry*, **46**, 573–639.

Wrangham, R. W., McGrew, W. C., de Waal, F. B. M. & Heltne, P. G. (eds.) 1994. *Chimpanzee Cultures*. Cambridge, MA: Harvard University Press.

Zihlman, A. L. 1981. Women as shapers of the human adaptation. In: *Woman the Gatherer*, ed. F. Dahlberg, pp. 75–120. New Haven, CT: Yale University Press.

Zihlman, A. L. 1984. Body build and tissue composition in *Pan paniscus* and *Pan troglodytes* with comparison to other hominoids. In: *The Pygmy Chimpanzee*, ed. R. L. Susman, pp. 179–200. New York: Plenum.

Zihlman, A. L. 1990. Knuckling under: controversy over hominid origins. In: *From Apes to Angels*, ed. G. Sperber, pp. 185–96. New York: Wiley-Liss.

Zihlman, A. L. & Cramer, D. L. 1978. Skeletal differences between pygmy (*Pan paniscus*) and common chimpanzees (*Pan troglodytes*). *Folia Primatologica*, **29**, 86–94.

Zihlman, A. L., Cronin, J. E., Cramer, D. L. & Sarich, V. M. 1978. Pygmy chimpanzee as a possible prototype for the common ancestor of humans, chimpanzees and gorillas. *Nature*, **275**, 744–6.

Zuckerkandl, E. & Pauling, L. 1962. Molecular disease, evolution and genetic heterogeneity. In: *Horizons in Biochemistry*, ed. M. Kasha & B. Pullman, pp. 189–225. New York: Academic Press.

NOTE ADDED IN PROOF

In a Corrigendum to *Nature* (1995, **376**, 88), T. D. White *et al.* proposed a new genus, *Ardipithecus*, for the previously named *Australopithecus ramidus*. Meave G. Leakey announced a new species of 4-million-year-old hominid, *Australopithecus anamensis*, from Kanapoi and Allia Bay, Kenya (*Nature*, 1995, **376**, 565–71).

Afterword: a new milestone in great ape research

JUNICHIRO ITANI

Each manuscript in this volume brings back to me the memory of a scene I witnessed with A. Suzuki in 1965. We were in western Tanzania, 50 km east of the escarpment in no man's land, when we came across an orderly procession of 43 chimpanzees, marching across the Miombo woodland with scarcely a leaf left on the boughs in the dry season (Itani & Suzuki, 1967).

It seems now that everything we must know about chimpanzees was condensed and suggested in the procession we witnessed, but we were unable to decipher the mystery at that moment. The manuscripts in this volume may present the possible answers for those who read this book; and this must have been the reason the manuscripts made me hark back to the scene of 30 years ago.

It was Thomas Huxley (1863), who first articulated the prologue to a plausible scenario recognizing the significance of great apes, especially the importance of inquiring into their relation to the human species, in the first chapter of *Evidence as to Man's Place in Nature*.

Later, well into the twentieth century, the first milestone was laid along the path towards the composition of a comprehensive scenario on the relationship between humanity and the apes. The milestone was *The Great Apes* by Robert & Ada Yerkes (1929). Just as Huxley extensively covered the insights on great apes accumulated by 1863, and analyzed them in great detail, the Yerkes also devoted their 650-page tome to the data amassed by 1929. However, most of the studies on apes at that time were analyses of behavior in the laboratory, including the memorable experiments on chimpanzee mentality by W. Köhler (1925) and N. Köhts (1928).

Field studies on non-human primates began immediately after the publication of *The Great Apes*. The Yerkes strongly supported such developments and acknowledged such field studies as being 'planned in supplementation of the experimental studies in the Yale Laboratories of Comparative Psychobiology.' The first results came from H. W. Nissen (1931), from his 1930 study of chimpanzees in French Guinea, now Guinea. Second, H. C. Bingham (1932) studied mountain gorillas in the Virunga Volcanoes of the Belgian Congo during 1929, in what is now Zaïre. The third such endeavor was C. R. Carpenter's (1934, 1935) naturalistic studies in 1931–33 on spider and howler monkeys. That the non-human primates to be studied first were the great apes should never be forgotten by science.

The Second World War temporarily hampered the burgeoning field studies of non-human primates. After Carpenter's monograph on his field study of the gibbon was published in 1940, more than 15 years passed without any field studies being carried out. Finally, towards the end of the 1950's, four scholars from four countries independently rekindled scientific interest in the great apes: L. S. B. Leakey in Kenya, R. Dart in South Africa, K. Imanishi of Japan and A. Kortlandt of the Netherlands. Three were anthropologists and one was an ethologist. Together, they triggered a new tidal wave of inquiry into human evolution, and one must marvel at the uncanny synchrony of their efforts. In 1956, Leakey sent R. Osborn (1963), and in 1957 Dart sent J. H. Donisthorpe (1958) to the Virunga Volcanoes to study mountain gorillas. Imanishi (1960) and I (Itani, 1991) reached the forests of Cameroon in 1958, after crossing equatorial Africa from east to west, having sighted a group of mountain gorillas in the Virungas and a single male western lowland gorilla in Cameroon. About the same time, Kortlandt (1962) pursued chimpanzees in the forests of the eastern Belgian Congo. All of us mentioned here represented a new scientific effort towards integrating inquiries into human evolution and studies on primates.

In 1959, a conference funded by the Wenner-Gren Foundation for Anthropological Research, was held on the 'Social Life of Early Man,' in an attempt to search for a more multifaceted theoretical and methodological basis for the new studies. In 1961, a volume was published under the same title as the conference, organized by S. L. Washburn. The conference and the subsequent publication initiated a torrent of primate studies in the field, which continues to this day, 35 years later.

In August 1968, the Wenner-Gren Foundation held another symposium in Burg Wartenstein in the Alps near Vienna, titled 'Social Organization and Subsistence in Primate Societies,' as an early compilation of the field studies. As the title indicated, special focus was placed on social organization and subsistence, and the topics presented reflected a wide range of scholarly interest, including studies on hunter–gatherer peoples. Nonetheless, no publication came forth as the result of the symposium.

Six years later Jane Goodall and David Hamburg organized a conference, this time specifically on great apes and again under the auspices of the Wenner-Gren Foundation. In the summer of 1974, researchers from various countries convened at Burg Wartenstein with their firsthand data from ongoing field studies. Five years later, the second milestone, *The Great Apes*, edited by D. Hamburg & E. McCown (1979), was published as the fifth volume of the series, 'Perspectives on Human Evolution.' The volume was dedicated to C. R. Carpenter, who had died in 1975.

It was 66 years after Huxley's publication that the first milestone on the great apes was laid by the husband and wife team of the Yerkes, and 50 years since that event came the second milestone by Hamburg and McCown. Fifteen years later, yet another memorable Wenner-Gren symposium was convened at Baja California Sur, Mexico, in November 1994, the results of which are published here. This marks the third milestone after Huxley's *Man's Place in Nature*.

I have had the opportunity to read over all of the manuscripts for this book, and what struck me most was the dedication with which the researchers conducted their field studies over the last 15 years. In Japanese,

there is an expression, 'crystallization of perspiration,' referring to achievements attained only through severe hardship. This book is filled with such crystallizations of the progression and achievements from the past 15 years of research, and lights the way to the future of our academic field.

Most of the topics chosen by researchers here are not so divergent from the ones we dealt with in the early days. Nesting, vocalization, party size and composition, diet, male coalitions seen in chimpanzee groups, laterality, sexual behavior, social ranks and adaptation to the savannah are all topics that early researchers grappled with. However, this volume offers various answers to various questions, each supported by ample data.

The third milestone reveals the enormous advances we have made in the study of great apes. For the first time, in this volume, the entire cast of characters among the great apes have gathered at center stage. In the second milestone, the bonobos were only fleetingly visible in the dense rainforests. In the third milestone, at last, the bonobos emerge distinctly from the mist as a research subject, along with the two subspecies of the giant, elusive lowland gorillas (see Fig. A.1). Moreover, the interactions between the sympatric lowland gorillas and chimpanzees have been observed and recorded. That the two species turned out to live in peaceful symbiosis is quite significant from the anthropological perspective. For example, one is reminded of the many debates that have taken place on the coexistence of *Australopithecus* spp. and *Homo habilis*, to which this volume offers an answer.

A second remarkable point is that true comparative analyses became possible with the success of long-term studies at many study sites. More than half the papers endeavor to gain new insights through comparative analyses and better clarify the characteristics of each of their research subjects. These analyses include comparisons between species, between the eastern and western regional populations of the same species, between groups with different habitats or environments, between individuals and between the sexes or age-groups. The numerous results attained through such comparisons as to behavior, ecology and society will undoubtedly converge towards an understanding of hominoid systematics, as well as the process of human evolution.

It goes without saying that parallel to field studies of

Fig. A.1. Adult male eastern lowland gorilla at Kahuzi-Biega. (Photo by J. Yamagiwa)

the great apes, behavioral studies in the laboratory setting have seen remarkable development. The two settings today play a complementary role, contrary to the comment Yerkes and Yerkes made on the field studies as simply the 'supplementation of the experimental study.' Furthermore, in the area pertaining to the origins of language, a topic particularly pertinent in the laboratory, one cannot but appreciate the fact that this volume contains two studies that either stepped out into the field from the laboratory or benefited both from experimental and field observations.

New hypotheses have been offered in this volume, on which I will not comment in detail here, but many, I believe, point to the new and future direction of this academic field of primate studies. I wish to point out just one issue: mainstream studies have long neglected various mechanisms of social interactions that construct the social structure through these interactions, exemplified in words describing the rich social capacities of primates, such as 'permit' and 'reject.' These more sophisticated concepts are now attracting new attention and leading to new research that fully appreciates the complexity of ape social systems.

It is sad that the great apes seem to be on the path to extinction even as research in the field has steadily progressed, a point Jane Goodall (1990) has continued to appeal to the world in all of her writings. I do not doubt that every author in *Great Ape Societies*, in his or her field, has striven greatly for the protection of the great apes, their subject of research. This must account for the depth and a kind of dignity in their papers, which inevitably strikes the reader.

I believe that we will not have to wait another 15 years for the fourth milestone, but that it will present itself much sooner. As we shall soon move into the twenty-first century whence the awaited milestone will materialize, the long endeavor of the study of the great apes will span three centuries, if not longer.

I offer my sincere admiration to the three editors of this book, William McGrew, Linda Marchant and Toshisada Nishida, who organized an international symposium with such a wide range of researchers and embarked on the publication of the 'new milestone.' I am grateful to them for giving me the opportunity to write the afterword to this book and to David Sprague and Kaoru Miyahara, who translated my manuscript as the deadline fast approached.

REFERENCES

Bingham, H. C. 1932. Gorillas in a native habitat. *Carnegie Institution of Washington Publications*, **426**, 1–66.

Carpenter, C. R. 1934. A field study of behavior and social relations of howling monkeys. *Comparative Psychology Monographs*, **10**, 1–168.

Carpenter, C. R. 1935. Behavior of red spider monkeys in Panama. *Journal of Mammalalogy*, **16**, 171–80.

Carpenter, C. R. 1940. A field study in Siam of the behavior and social relations of the gibbon, *Hylobates lar*. *Comparative Psychology Monographs*, **16**, 1–212.

Donisthorpe, J. H. 1958. A pilot study of the mountain gorilla (*Gorilla gorilla beringei*) in south-west Uganda, February to September 1957. *South African Journal of Science*, **54**, 195–217.

Goodall, J. 1990. *Through a Window. Thirty Years with the Chimpanzees of Gombe*. London: Weidenfeld and Nicolson.

Hamburg, D. A. & McCown, E. R. (eds.) 1979. *The Great Apes*. Menlo Park, CA: Benjamin/Cummings.

Huxley, T. H. 1863. *Man's Place in Nature*. Reprinted 1959. Ann Arbor Paperback Series. Ann Arbor: University of Michigan Press.

Imanishi, K. 1960. *Gorillas*. Tokyo: Bungeishunju. (in Japanese)

Itani, J. 1991. *Monkeys, Humans and Africa*. Tokyo: Nihon Keizai Shinbun. (in Japanese)

Itani, J. & Suzuki, A. 1967. The social unit of chimpanzees. *Primates*, **8**, 855–81.

Köhler, W. 1925. *The Mentality of Apes*. New York: Harcourt.

Köhts, N. 1928. Recherches sur l'intelligence du chimpanzé de 'choix d'après modèle'. *Journal de Psychologie Normal et Pathologie*, Paris, **25**, 255–75.

Kortlandt, A. 1962. Chimpanzees in the wild. *Scientific American*, **206**(5), 128–38.

Nissen, H. W. 1931. A field study of the chimpanzee. *Comparative Psychology Monographs*, **8**, 1–122.

Osborn, R. 1963. Observations on the behaviour of the mountain gorilla. *Symposia of the Zoological Society of London*, **10**, 29–37.

Washburn, S. L. (ed.) 1961. *Social Life of Early Man*. Chicago: Aldine.

Yerkes, R. M. & Yerkes, A. W. 1929. *The Great Apes*. New Haven, CT: Yale University Press.

Appendix: great ape study sites

The number of active research projects on (at least) wild apes continues to increase, as more are added than are dropped. This makes it harder to keep track of who is studying whom, what, when and where. Many of the authors in this book have worked on different species of ape at different places, often repeatedly. Furthermore, studies of apes increasingly involve direct comparison across sites, and such syntheses require specifically equivalent comparisons. For example, contrasts in day-to-day grouping patterns may reflect levels of predation pressure that range from non-existent to high. For these reasons, we have added an appendix to this book, in order to aid the reader who may feel overwhelmed at times in trying to keep things straight. Here, we attempt to record only the study sites focused on in this volume;

to do so comprehensively would require a monograph, which we hope that someone else will undertake!

Accordingly, each first author was sent a questionnaire with 20 items, as given in Table A.1. Items range from the simply descriptive to qualitatively evaluative. The 20 chosen topics are far from comprehensive, but we were mindful of limits on space in publication. The twentieth item provides references for readers seeking more details, and a contact person gives readers an opportunity to write for precise information.

One of the most difficult aspects of field primatological research to deal with is the extent of human influence exerted on a population of apes. None is wholly unaffected, even in the remotest wilderness, while, at the other extreme, a few are hanging on, close to local extinction.

Table A.1. *Information sought on site reports*

1. **Name:** Full name(s) of site
2. **Location:** Country, province or state (if applicable), geographical coordinates
3. **Status:** Unprotected, faunal reserve, national park, etc.
4. **Area:** In kilometers squared; within entire area of reserve, park, etc.
5. **Altitude:** Range from lowest to highest, in meters
6. **Rainfall:** Mean in millimeters, range, years covered
7. **Vegetation:** Gross types, e.g. primary forest, open woodland, etc.
8. **Human Influence:** Selective logging, no hunting, provisioning (1980–85), etc.
9. **Closest Village:** Distance in kilometers
10. **Species Studied:** Species (or subspecies) of ape
11. **Other Primates:** Species sympatric with apes
12. **Predators:** Known or suspected, scientific name, including *Homo sapiens*
13. **Study Period:** Years of study from start, with gaps, e.g. 1972–78, 1990–
14 **Habituation:** None, poor, fair, good, excellent
15. **Research Presence:** Occasional, seasonal, permanent, etc.
16. **Conservation:** Ecotourism, education in local schools, guards' training, etc.
17. **Current Research:** By topic, e.g. feeding ecology, tool-use, etc.
18. **Methods:** Focal-subject sampling, radio tracking, indirect evidence, nest-to-nest, etc.
19. **Contact Person:** Name and address of best person for further information about site
20. **Bibliography:** Three best published references to provide information on site and research

Table A.2. *Five scales of human influence (modified from Bishop* et al. *1981)*

Home range

1. Undisturbed habitat – no human habitation (other than research camp) or economic activities; no livestock or dogs within 5 km of group's range
2. Disturbed 1 – moderate exploitation of habitat (e.g. limited woodcutting, fodder cutting, some livestock present) immediately surrounding study area; edge effect
3. Disturbed 2 – similar exploitation to Disturbed 1 but extends throughout study area
4. Mosaic habitat – home range includes both human-made (i.e. fields, cleared areas) and Disturbed 1 or 2 habitat

Harassment

1. Undisturbed – contact between humans and apes rare (likely to see humans <1/month)
2. Contact – apes likely to see humans occasionally (e.g. *c.* 1/week)
3. Minimal harassment – apes are chased or harassed only when they enter fields to raid crops
4. Occasion-specific harassment – apes are harassed when they steal food and when they enter areas of human habitation

Hunted

1. Never hunted as best as can be ascertained
2. Rare or episodic hunting (includes revenge killings by local people unhappy about research or park policy; fewer than 5/decade)
3. Hunting part of local tradition in or near study area

Habituation

1. Unafraid – apes show curiosity, not fear, when first contacted; appear never to have seen humans before
2. Wild – apes flee when humans appear
3. Semi-habituated – apes move away if humans actually approach
4. Habituated – apes are accustomed to human presence, although they may not tolerate approach

Predators

1. Full complement of predators – large cats and other predators protected (or at near-natural populations)
2. Partial complement of predators – predators are represented by a few individuals of some species, but numbers are diminished due to hunting or habitat destruction
3. Impoverished complement – most major predators eliminated; village dogs may harass primates
4. No predators – neither feral predators nor dogs present a threat to apes

---

Such a key factor demands systematic consideration. Bishop *et al.* (1981) attempted such an analysis for two species of South Asian monkey: *Macaca mulatta*, rhesus macaque; and *Presbytis entellus*, Hanuman langur, and showed that such data could be informative and useful. Jim Moore has tailored Bishop *et al.*'s (1981) scales for application to great apes, and these are detailed in Table A.2. No wild ape populations are urban (unlike some monkeys), but some live cheek-to-jowl with agriculturalists. Consequently, no wild population is harassed daily. Hunting is treated separately as a special kind of human harassment and predation. Apes who are naturally unafraid of humans through lack of contact are here distinguished from those who are unafraid after learning to tolerate humans through benign experience.

In general, the lower the number on each of the five scales, the more pristine are the apes' circumstances. If the five scores from each site are summed, the range across sites is from 5 for Ndoki to 17 for Bossou, with a median

of 10. Few populations are undisturbed by human harassment or live in undisturbed habitats, but almost all have at least some non-human predators. None of the study populations is regularly hunted, but several suffer mortality and morbidity from being caught in snares intended for other prey. Even the most intelligent apes cannot solve all the problems encountered in the forest.

REFERENCE

Bishop, N., Hrdy, S. B., Teas, J. & Moore, J. 1981. Measures of human influence in habitats of South Asian monkeys. *International Journal of Primatology*, **2**, 153–67.

Site report

Hemelrijk

1. Name: Arnhem Burgers' Zoo

2. Location: The Netherlands, Arnhem, 52° N 6° E
3. Status: Zoological garden
4. Area: 0.007 km²
5. Altitude: 50 m
6. Rain fall: 800 mm
7. Vegetation: Meadow with protected trees
8. Human influence: Indoor cages, outdoor enclosure
9. Closest village: In town
10. Species studied: *Pan troglodytes*
11. Other primates: 11 spp.
12. Predators: None
13. Study period: 1976–present
14. Habituation: Excellent
15. Research presence: Seasonal
16. Conservation: Education, schools and university classes
17. Current research: Social relationships, male strategies, tension regulation and social homeostasis
18. Methods: Observational, experimental
19. Contact person: J. van Hooff, Ethology and Socio-ecology, Department of Comparative Physiology, Postbus 80.086, 3508 TB Utrecht, The Netherlands
20. Bibliography:

Adang, O. M. J., Wensing, J. A. B. & van Hooff, J. A. R. A. M. 1987. The Arnhem Zoo colony of chimpanzees, *Pan troglodytes*: development and management techniques. *International Zoo Yearbook*, **26**, 236–48.

Hemelrijk, C. K. & Ek, A. 1991. Reciprocity and interchange of grooming and 'support' in captive chimpanzees. *Animal Behaviour*, **41**, 923–35.

de Waal, F. B. M. 1982. *Chimpanzee Politics*. London: Jonathan Cape.

Site report

Matsuzawa

1. Name: Bossou
2. Location: Guinea, Lola Prefecture
3. Status: Protected
4. Area: 30 km²
5. Altitude: 500–700 m
6. Rain fall: 2000–3000 mm
7. Vegetation: Primary and secondary forest, cultivated field, savanna
8. Human influence: no hunting, some burning, cultivation, recent provisioning
   Disturbance ratings: 4, 4, 2, 3, 4

9. Closest village: Overlaps chimpanzees' range
10. Species studied: *Pan troglodytes verus*
11. Other primates:
12. Predators:
13. Study period: 1976–77, 1979–80, 1982–83, 1985–86, 1987-present
14. Habituation: Good
15. Research presence: Permanent
16. Conservation: Guards' training
17. Current research: Social organization, feeding ecology, tool use, growth and development, cognition
18. Methods: Focal-subject sampling, ad lib observation, field experiment
19. Contact person: Y. Sugiyama, Kyoto University Primate Research Insitute, Inuyama, Japan 484
20. Bibliography:

Matsuzawa, T. 1994. Field experiments on use of stone tools by chimpanzees in the wild. In: *Chimpanzee Cultures*, ed. R. W. Wrangham, W. C. McGrew, F. B. M. de Waal & P. G. Heltne, pp. 351–70. Cambridge, MA: Harvard University Press.

Sugiyama, Y. 1993. The flora of Bossou: its utilization by chimpanzees and humans. *African Study Monographs*, **13**, 127–69.

Sugiyama, Y., Fushimi, T., Sakura, O. & Matsuzawa, T. 1993. Hand preference and tool use in wild chimpanzees. *Primates*, **34**, 151–9.

Site report

Goodall

1. Name: Gombe Stream Wildlife Research Centre
2. Location: Tanzania, Kigoma Region, 4°40′ S 29°38′ E
3. Status: National park
4. Area: 32 km²
5. Altitude: 775–1500 m
6. Rain fall: 1600 mm
7. Vegetation: Evergreen riverine forest, deciduous dry forest, thicket, Miombo woodland, grassland, moorland
8. Human influence: Provisioning, some wood cutting
   Disturbance ratings: 2, 3, 1, 2, 3
9. Closest village: At boundary
10. Species studied: *Pan troglodytes schweinfurthii*
11. Other primates: *Papio anubis, Cercopithecus mitis, C. ascanius, C. aethiops, Colobus badius, Galago* sp.
12. Predators: *Panthera pardus*

13. Study period: 1960–present
14. Habituation: Excellent for Kasakela community, fair for Mitumba community
15. Research presence: Permanent
16. Conservation: Tourism, conservation education in local schools
17. Current research: Long-term monitoring of individuals; mother–infant development
18. Methods: Nest-to-nest follows, focal-subject sampling
19. Contact person: Dr. Anthony Collins, PO Box 185, Kigoma, Tanzania
20. Bibliography:

Bygott, D. 1992. *Gombe Stream National Park*. Arusha: Tanzania National Parks – African Wildlife Foundation, 72 pp.

Goodall, J. 1968. The behaviour of free-living chimpanzees in the Gombe Stream Reserve. *Animal Behaviour Monographs*, **1**, 161–311.

Goodall, J. 1986. *The Chimpanzees of Gombe: Patterns of Behavior*. Cambridge, MA: Harvard University Press.

Site report

Yamagiwa

1. Name: Kahuzi–Biega
2. Location: Zaïre, Province de Kivu, 2°S 28°E
3. Status: National park
4. Area: 6000 km²
5. Altitude: 600–3308 m
6. Rain fall: 1800 mm, range 1500–1900 mm
7. Vegetation: Primary forest, secondary forest, swamp forest, bamboo, subalpine
8. Human influence: hunting, cattle encroachment, slash and burn horticulture, no provisioning
   Disturbance ratings: 2, 3, 2, 2, 1
9. Closest village: 30 km
10. Species studied: *Gorilla gorilla graueri, Pan troglodytes schweinfurthii*
11. Other primates: *Perodicticus potto, Galago demidovii, G. crassicaudatus, Papio anubis, Colobus angolensis, C. badius, Cercopithecus ascanius, C. wolfi, C. l'hoesti, C. hamlyni, C. mitis, Cercocebus albigena*
12. Predators: *Homo sapiens, Panthera pardus*
13. Study period: 1978–79, 1987–present
14. Habituation: Lowland poor, highland fair or excellent for gorillas, fair for chimpanzees

15. Research presence: Permanent
16. Conservation: Ecotourism, education for local people
17. Current research: Feeding ecology (gorillas & chimpanzees), behavioral ecology and social organization (gorillas)
18. Methods: Group tracking, focal-subject sampling, indirect evidence
19. Contact person: Juichi Yamagiwa, Kyoto University Primate Research Institute, Inuyama, Japan 484
20. Bibliography:

Yamagiwa, J., Mwanza, N., Spangenberg, A., Maruhashi, T., Yumoto, T., Fischer, A. & Stenhauer, B. B. 1993. A census of the eastern lowland gorillas *Gorilla gorilla graueri* in Kahuzi-Biega National Park with reference to mountain gorillas *G. g. beringei* in the Virunga Region, Zaïre. *Biological Conservation*, **64**, 83–9.

Yamagiwa, J., Yumoto, T., Maruhashi, T. & Mwanza, N., 1993. Field methodology for analyzing diets of eastern lowland gorillas in Kahuzi-Biega National Park, Zaïre. *Tropics*, **2**, 209–18.

Yamagiwa, J., Mwanza, N., Yumoto, T. & Maruhashi, T. 1994. Seasonal change in the composition of the diet of eastern lowland gorillas. *Primates*, **35**, 1–14.

Site report

Watts

1. Name: Karisoke Research Centre, Parc National des Volcans
2. Location: Rwanda, 29° E 1°30′ S
3. Status: National park (and adjoining national parks in Zaïre, Uganda)
4. Area: 400 km²
5. Altitude: 2500–4500 m
6. Rain fall: 1800 mm
7. Vegetation: Montane rainforest and woodland
8. Human influence: Hunting, wood cutting, agricultural encroachment
   Disturbance ratings: 2, 2, 2, 2, 1
9. Closest village: 5 km
10. Species studied: *Gorilla gorilla beringei*
11. Other primates: *Perodicticus potto, Cercopithecus mitis kandtii, C. ascanius, Papio anubis*
12. Predators: *Homo sapiens*, previously *Panthera leo, P. pardus*
13. Study period: 1967–present
14. Habituation: Excellent

15. Research presence: Permanent 1967–94, now intermittent
16. Conservation: Ecotourism, education, training, park protection
17. Current research: None at present
18. Methods: Focal-subject sampling
19. Contact person: David Watts, Department of Anthropology, University of Michigan, Ann Arbor, MI 48109, U.S.A.
20. Bibliography:

Fossey, D. 1983. *Gorillas in the Mist*. Boston: Houghton Mifflin.

Schaller, G. B. 1963. *The Mountain Gorilla*. Chicago: University of Chicago Press.

Weber, A. W. 1987. Socioecologic factors in the conservation of afromontane forest reserves. In: *Primate Conservation in the Tropical Rain Forest*, ed. C. W. Marsh & R. A. Mittermeier, pp. 205–29. New York: Alan R. Liss.

Site Report

van Schaik

1. Name: Ketambe, Gunung Leuser
2. Location: Indonesia, Sumatra, 3°41′ N 97°39′ E
3. Status: National park
4. Area: *c* 4.5 km² in 9000 km²
5. Altitude: 350–1000 m
6. Rain fall: 3229 mm, 2860–3763 mm, 11 years
7. Vegetation: Primary forest
8. Human influence: Rattan extraction, some hunting, some illegal logging
   Disturbance ratings: 2, 2, 1, 2, 2
9. Closest village: 0.5 km
10. Species studied: *Pongo pygmaeus abelii*
11. Other primates: *Macaca fascicularis, M. nemestrina, Presbytis thomasi, Nycticebus coucang, Hylobates lar, H. syndactylus*
12. Predators: *Panthera tigris, Neofelis nebulosa, Homo sapiens* formerly
13. Study period: 1971–present
14. Habituation: Excellent
15. Research presence: Permanent
16. Conservation: Ecotourism nearby, some education, training of field staff
17. Current research: Socio-ecology
18. Methods: Focal-subject sampling, nest-to-nest
19. Contact person: Sri Suci Utami, Ketambe Research Station, PO Box 4, Kutacane 24601, Acek Tenggara, Indonesia
20. Bibliography:

te Boekhorst, I. J. A., Schürmann, V. L. & Sugardjito, J. 1990. Residential status and seasonal movements of wild orang-utans in the Gunung Leuser Reserve (Sumatra, Indonesia). *Animal Behaviour*, **39**, 1098–109.

Rijksen, H. D. 1978. *A Fieldstudy on Sumatran Orang Utans* (Pongo pygmaeus abellii, *Lesson, 1827*). Wageningen: H. Veenman and Zonen.

van Schaik, C. P. 1986. Phenological changes in a Sumatran rain forest. *Journal of Tropical Ecology*, **2**, 327–47.

Site report

Wrangham

1. Name: Kanyawara, Kibale
2. Location: Uganda, Kabarole District, 0°34′ N 30°22′ E
3. Status: National park
4. Area: 560 km²
5. Altitude: 1390–1625 m
6. Rain fall: 1671 mm, 1561–1859 mm, 1984–91
7. Vegetation: Medium altitude moist evergreen forest, colonizing forest-grassland, swamp communities
8. Human influence: Agriculture (nineteenth century?), selective logging (1960s), forestry plantations
   Disturbance ratings: 3, 3, 1, 2, 2
9. Closest village: Borders forest
10. Species studied: *Pan troglodytes schweinfurthii*
11. Other primates: *Colobus badius, C. guereza, Cercocebus albigena, Cercopithecus mitis, C. ascanius, C. l'hoesti, Papio anubis, Perodicticus potto* (?)
12. Predators: *Panthera pardus (?)*
13. Study period: 1978–79, 1981 (Ghiglieri); 1983–85 (Basuta); 1987–present
14. Habituation: Good
15. Research presence: Permanent
16. Conservation: Ecotourism, conservation education, buffer zone studies
17. Current research: Social organization, ecology, culture
18. Methods: Ten-minute focal-subject samples during multi-hour party observations
19. Contact person: Richard Wrangham, Peabody

Museum, Harvard University, Cambridge, MA 02138, USA
20. Bibliography:

Ghiglieri, M. P. 1984. *The Chimpanzees of the Kibale Forest: A Field Study of Ecology and Social Structure*. New York: Columbia University Press.

Isabirye-Basuta, G. 1989. Feeding ecology of chimpanzees in the Kibale Forest, Uganda. In: *Understanding Chimpanzees*, ed. P. G. Heltne & L. A. Marquardt, pp. 116–27. Cambridge, MA: Harvard University Press.

Chapman, C. A. & Wrangham, R. W. 1993. Range use of the forest chimpanzees of Kibale: implications for the understanding of chimpanzee social organization. *American Journal of Primatology*, **31**, 263–73.

Site report

Savage-Rumbaugh

1. Name: Language Research Center, Georgia State University
2. Location: USA, Decatur, Georgia, 33°47′ N 84°17′ W
3. Status: Research laboratory
4. Area: 0.22 km²
5. Altitude: 872 m
6. Rain fall: 1204 mm
7. Vegetation: Deciduous broadleaf forest
8. Human influence: Multiple feeding sites, cleared trails, no hunting, buildings, confinement
9. Closest village:
10. Species studied: *Pan paniscus, P. troglodytes, Pongo pygmaeus*
11. Other primates: *Macaca mulatta*
12. Predators: None
13. Study period: 1981–present
14. Habituation: Excellent
15. Research presence: Permanent
16. Conservation: Bonobo Conservation and Protection Society
17. Current research: Language, planning and communication skills, organization in manual activity and competence in construction
18. Methods: Observational, computer, subject case studies
19. Contact person: Duane Rumbaugh, Georgia State University, Language Research Center, 3401 Panthersville Road, Decatur, GA 30034, USA
20. Bibliography:

Rumbaugh, D. M. & Savage-Rumbaugh, E. S. 1994. Language in comparative perspective. In: *Animal Learning and Cognition*, ed. N. J. Mackintosh, pp. 307–33. London: Academic Press.

Rumbaugh, D. M., Savage-Rumbaugh, E. S. & Sevcik, R. A. 1994. Biobehavioral roots of language: a comparative perspective of chimpanzee, child and culture. In: *Chimpanzee Cultures*, ed. R. W. Wrangham, W. C. McGrew, F. B. M. de Waal & P. G. Heltne, pp. 319–34. Cambridge, MA: Harvard University Press.

Savage-Rumbaugh, E. S. & Lewin, R. 1994. *Kanzi: The Ape at the Brink of the Human Mind*. New York: Wiley & Sons.

Site report

Fruth

1. Name: Lomako
2. Location: Zaïre, Province de Equateur, 0°51′ N 21°5′ E
3. Status: Unprotected; timber concession granted
4. Area: 30 km² in 3500 km² of undisturbed forest
5. Altitude: 390 m
6. Rain fall: 1850 mm, 1994
7. Vegetation: Polyspecific primary forest, primary evergreen forest, swamp forest
8. Human influence: Hunting mostly with bow and arrow, snaring, gathering of fruit, honey, leaves, medicinal plants, fishing, no provisioning Disturbance ratings: 2, 1, 1, 3, 1
9. Closest village: 2 km
10. Species studied: *Pan paniscus* (Eyengo community)
11. Other primates: *Cercocebus aterrimus, Colobus angolensis, Cercopithecus mona, C. ascanius, Galago demidovii, Perodicticus potto*
12. Predators: *Homo sapiens* known; *Panthera pardus* probably
13. Study period: 1974–91 intermittent by State University of New York at Stony Brook; Project Pan from 1990–present
14. Habituation: Good
15. Research presence: Seasonal
16. Conservation: None
17. Current research: Communication, socio-ecology of nest building, social organization, food sharing, play
18. Methods: Focal-subject sampling, nest-to-nest, *ad libitum*

19. Contact persons: Gottfried Hohmann/Barbara Fruth: Max-Planck-Institut für Verhaltensphysiologie, D-82319 Seewiesen, Germany

20. Bibliography:

Badrian, A. & Badrian, N. R. 1984. Social organization of *Pan paniscus* in the Lomako Forest, Zaïre. In: *The Pygmy Chimpanzee*, ed. R. L. Susman, pp. 325–46. New York: Plenum Press.

Hohmann, G. & Fruth, B. 1994. Structure and use of distance calls in wild bonobos (*Pan paniscus*). *International Journal of Primatology*, **15**, 767–82.

White, F. J. 1992. Activity budgets, feeding behavior and habitat use of pygmy chimpanzee at Lomako, Zaïre. *American Journal of Primatology*, **26**, 215–23.

Site report

Tutin

1. Name: Station d'Etudes des Gorilles et Chimpanzés, Reserve de la Lopé
2. Location: Gabon, Ogooué–Ivindo, 0°10′ S 11°35′ E
3. Status: Faunal reserve
4. Area: 50 km² in 5000 km²
5. Altitude: 250–600 m
6. Rain fall: 1531 mm, range 1168–1851 mm, 1984–94
7. Vegetation: Marantaceae forest, closed canopy forest
8. Human influence: No hunting, no provisioning, selective logging in parts (1960–74)
   Disturbance ratings: 1, 1, 1, 2, 1
9. Closest village: 12 km
10. Species studied: *Gorilla gorilla gorilla, Pan troglodytes troglodytes*
11. Other primates: *Mandrillus sphinx, Colobus satanus, Cercocebus albigena, Cercopithecus nictitans, C. pogonias, C. cephus*
12. Predators: *Panthera pardus*
13. Study period: 1984–present
14. Habituation: Poor
15. Research presence: Permanent
16. Conservation: Collaboration with ECOFAC, training and ecotourism
17. Current research: Feeding ecology, population structure, social organization, ecological monitoring, plant–animal interactions
18. Methods: Tracking, observation as possible, DNA fingerprinting
19. Contact person: C. E. G. Tutin, SEGC, B.P. 7847, Libreville, Gabon

20. Bibliography:

Tutin, C. E. G. & Fernandez, M. 1993. Composition of the diet of chimpanzees and comparisons with that of sympatric lowland gorillas in the Lopé Reserve, Gabon. *American Journal of Primatology*, **30**, 195–211.

Tutin, C. E. G., White, L. J. T., Williamson, E. A., Fernandez, M. & McPherson, G. 1994. List of plant species identified in the northern part of the Lopé Reserve, Gabon. *Tropics*, 3, 249–76.

Tutin, C. E. G., Parnell, R. J., White, L. J. T. & Fernandez, M. 1995. Nest building by lowland gorillas in the Lopé Reserve, Gabon: Environmental influences and implications for censusing. *International Journal of Primatology*, **16**, 55–76.

Site report

Nishida

1. Name: Mahale Mountains Wildlife Research Centre
2. Location: Tanzania, Kigoma Region, 6°07′ S 29°44′ E
3. Status: National park
4. Area: 1613 km²
5. Altitude: 773–2515 m
6. Rain fall: 1836 mm, 1975–88
7. Vegetation: Tropical semi-evergreen forest, Miombo woodland
8. Human influence: Provisioning 1966–87
   Disturbance ratings: 1, 2, 1, 2, 1,
9. Closest village: 20 km
10. Species studied: *Pan troglodytes schweinfurthii*
11. Other primates: *Colobus badius, C. angolensis, Papio cynocephalus, Cercopithecus ascanius, C. mitis, C. aethiops, Galago crassicaudatus, G. senegalensis*
12. Predators: *Panthera leo, P. pardus*
13. Study period: 1965–present
14. Habituation: Excellent
15. Research presence: Permanent
16. Conservation: Ecotourism
17. Current research: Social relationships, hunting, feeding ecology, tool-using, zoopharmacognosy, vocal communication
18. Methods: Focal-subject sampling
19. Contact person: Toshisada Nishida, Department of Zoology, Faculty of Science, Kyoto University, Kyoto, Japan 606

20. Bibliography:

Huffman, M. A. & Wrangham, R. W. 1994. Diversity of medicinal plant use by chimpanzees in the wild. In: *Chimpanzee Cultures*, R. W. Wrangham, W. C. McGrew, F. B. M. de Waal & P. G. Heltne, pp. 129–48. Cambridge, MA: Harvard University Press.

Nishida, T. (ed.) 1990. *The Chimpanzees of the Mahale Mountains. Sexual and Life History Strategies*. Tokyo: University of Tokyo Press.

Nishida, T. 1994. Review of recent findings on Mahale chimpanzees: Implications and future research directions. In: *Chimpanzee Cultures*, ed. R. W. Wrangham, W. C. McGrew, F. B. M. de Waal & P. G. Heltne, pp. 373–96. Cambridge, MA: Harvard University Press.

Site report

Matsuzawa

1. Name: Nimba Mountains
2. Location: Ivory Coast, Yealé via Danané; Guinea, Nion via Lola; Liberia, Lamko
3. Status: Integrated reserve in Ivory Coast, reserve in Guinea
4. Area: 50 km²
5. Altitude: 400–1752 m
6. Rain fall:
7. Vegetation: Primary forest
8. Human influence: Selective logging at lower altitudes, hunting
    Disturbance ratings:
9. Closest village:
10. Species Studied: *Pan troglodytes verus*
11. Other primates:
12. Predators: *Homo sapiens*
13. Study period: 1986, 1993–present
14. Habituation: None
15. Research presence: Occasional, now seasonal
16. Conservation: Education for local people
17. Current research: Tool-use, feeding ecology, nest building
18. Methods: All-occurrences sampling, indirect evidence
19. Contact person: Tetsuro Matsuzawa, Kyoto University Primate Research Institute, Inuyama, Japan 484
20. Bibliography:

Kortlandt, A. & Holzhaus, E. 1987. New data on the use of stone tools by chimpanzees in Guinea and Liberia. *Primates*, **28**, 473–96.

Matsuzawa, T. & Yamakoshi, G. 1995. Comparison of chimpanzee material culture between Bossou and Nimba. In: *Reaching into Thought*, ed. A. Russon, K. Bard, & S. Parker. Cambridge: Cambridge University Press.

Site report

Kuroda

1. Name: Nouabalé-Ndoki
2. Location: Congo, Region de la Sangha, 2°10′ – 3°00′ N 16°11′ – 17°00′ E
3. Status: National park
4. Area: 4000 km²
5. Altitude: 300–400 m ?
6. Rain fall: 1430–1650 mm, 1991–92
7. Vegetation: Seasonal rain forest, semi-deciduous forest, evergreen forest, swamp forest, and swamp clearing
8. Human influence: None since 1940
    Disturbance ratings: 1, 1, 1, 1, 1, 1
9. Closest village: 20 km
10. Species studied: *Pan troglodytes troglodytes*, *Gorilla gorilla gorilla*
11. Other primates: *Cercopithecus pogonias*, *C. nictitans*, *C. neglectus*, *C. cephus*, *Allenopithecus nigroviridis*, *Cercocebus albigena*, *C. galeritus*, *Colobus guereza*, *C. badius*, *Perodicticus potto*, *Galago elegans*
12. Predators: *Panthera pardus* (suspected)
13. Study period: 1989–92, 1994–present
14. Habituation: Good for chimpanzees, none for gorillas
15. Research presence: Seasonal
16. Conservation: Congolaise students' training, distribution of conservation posters
17. Current research: Feeding ecology of gorillas and chimpanzees, tool-use of chimpanzees
18. Methods: Ad lib sampling, indirect evidence, fix point observation, etc.
19. Contact person: S. Kuroda, School of Human Cultures, University of Siga Prefecture, Hikone, Japan 522
20. Bibliography:

Kuroda, S. 1992. Ecological interspecies relationships between gorillas and chimpanzees in the Ndoki-Nouabalé reserve, northern Congo. In: *Topics in Primatology, Vol. 2, Behavior, Ecology, and*

Conservation, N. Itoigawa, Y. Sugiyama, G. P. Sackett, & R. K. R. Thompson, pp. 385–94. Tokyo: University of Tokyo Press.

Moutsamboté, J. M., Yumoto, T., Mitani. M., Nishihara, T., Suzuki, S. & Kuroda, S. 1994. Vegetation and list of plant species identified in the Nouabalé-Ndoki Forest, Congo. *Tropics*, **3**, 277–94.

Nishihara, T. 1995. Feeding ecology of western lowland gorillas in the Nouabalé-Ndoki National Park, northern Congo. *Primates*, **36**, 151–68.

Site report

Matsuzawa

1. Name: Primate Research Institute, Kyoto University
2. Location: Japan, Aichi Prefecture, Inuyama City
3. Status: Research laboratory
4. Area: 0.04 km^2
5. Altitude: 70–90 m
6. Rain fall:
7. Vegetation:
8. Human influence: Cages, enclosures
9. Closest village:
10. Species studied: *Pan troglodytes*
11. Other primates: 25 spp.
12. Predators: None
13. Study period: 1967–present; 1977–present, Chimpanzee Cognition Research
14. Habituation: Excellent
15. Research presence: Permanent
16. Conservation:
17. Current research: Perception/cognition, language, intelligence, locomotion, morphology, reproduction and birth control
18. Method: Direct observation, video-recording, experiments
19. Contact person: Tetsuro Matsuzawa, Kyoto University Primate Research Center, Inuyama, Japan 484
20. Bibliography:

Matsuzawa, T. 1985. Use of numbers by chimpanzees. *Nature*, **315**, 57–9.

Matsuzawa, T. 1985. Color naming and classification in a chimpanzee (*Pan troglodytes*). *Journal of Human Evolution*, **14**, 283–91.

Matsuzawa, T. 1990. Form perception and visual acuity in a chimpanzee. *Folia Primatologica*, **55**, 24–32.

Site report

van Schaik

1. Name: Suaq Balimbing, Gunung Leuser
2. Location: Indonesia, Sumatra, 3°04′ N 97°26′ E
3. Status: National park
4. Area: 2.5 km^2 in 9000 km^2
5. Altitude: 5–200 m
6. Rain fall: 3560 mm, 1993
7. Vegetation: Freshwater swamp, peat swamp, hill dipterocarp forest
8. Human influence: Some rattan and bark extraction; fishing
    Disturbance ratings: 1, 2, 1, 2, 2
9. Closest village: 6 km
10. Species studied: *Pongo pygmaeus abelii*
11. Other primates: *Hylobates lar, H. syndactylus, Macaca fascicularis, M. nemestrina, Presbytis thomasi, P. cristata, Nycticebus coucang*
12. Predators: *Panthera tigris, Neofelis nebulosa*
13. Study period: 1992–present
14. Habituation: Fair to excellent
15. Research presence: Permanent
16. Conservation: Training of field staff, advice to park management
17. Current research: Socio-ecology, conservation biology, tool use
18. Methods: Focal-subject sampling, nest-to-nest; nest surveys
19. Contact person: C. van Schaik, Department of Biological Anthropology and Anatomy, Duke University, Durham, NC 27705, USA
20. Bibliography:

van Schaik, C., Priatna, A., & Priatna, D. in press. Population estimates and habitat preferences of orangutans based on line transects of nests. In: *Orangutans*, ed. R. Nadler *et al.*, New York: Plenum.

Site report

Boesch

1. Name: Taï
2. Location: Ivory Coast, Sous Prefecture de Taï, 5°52′ N 7°20′ W
3. Status: National park
4. Area: 30 km^2 in 4260 km^2
5. Altitude: 160–245 m
6. Rain fall: 1829 mm, range 1505–2150 mm, 1988–1993

7. Vegetation: Evergreen moist rainforest
8. Human influence: Rare selective logging until 1975, some poaching, no provisioning
   Disturbance ratings: 1, 1, 2, 2, 1
9. Closest village: 9 km, some fields 5 km away
10. Species studied: *Pan troglodytes verus* (2 communities)
11. Other primates: *Galago demidovii, Perodicticus potto, Cercopithecus petaurista, C. diana diana, C. nictitans, Cercocebus atys, Colobus badius badius, C. polykomos polykomos, Procolobus verus.*
12. Predators: *Panthera pardus, Homo sapiens?*
13. Study period: 1976, 1979–present
14. Habituation: Excellent
15. Research presence: Permanent
16. Conservation: WWF and GTZ integrated conservation projects
17. Current research: Social organization, cooperation, food sharing, hunting, tool use, cultural variation, ontogeny, feeding ecology
18. Methods: Focal-subject sampling, nest-to-nest
19. Contact person: Christophe Boesch, Institute of Zoology, University of Basel, Rheinsprung 9, 4051 Basel, Switzerland
20. Bibliography:

Boesch, C. 1994. Cooperative hunting in wild chimpanzees. *Animal Behaviour*, **48**, 653–67.

Boesch, C. & Boesch, H. 1989. Hunting behavior of wild chimpanzees in the Taï National Park. *American Journal of Physical Anthropology*, **78**, 547–73.

Boesch, C. & Boesch, H. 1990. Tool use and tool making in wild chimpanzees. *Folia Primatologica*, **54**, 86–99.

Site report

Moore

1. Name: Ugalla, Tongwe East Forest Reserve
2. Location: Tanzania, Kigoma-Mpanda Districts; main camp at 5°19′ S 30°37′ E
3. Status: Forest reserve (part), part unprotected
4. Area: 2800 km²; boundaries undefined
5. Altitude: 1100–1600 m
6. Rain fall: 1012 mm, range 745–1280 mm, 1973–88 (at Uvinza)
7. Vegetation: Grassy woodland, *Brachystegia–Julbernardia* miombo woodland
8. Human influence: Selective logging, honey collecting, some hunting (local and professional), foot traffic

Disturbance ratings: 3, 2, 1, 2, 1
9. Closest village: 5–10 km
10. Species studied: *Pan troglodytes schweinfurthii*
11. Other primates: *Galago* spp. (at least 2), *Papio cynocephalus/anubis* (hybrid zone), *Cercopithecus mitis, C. aethiops, C. ascanius*
12. Predators: *Panthera leo, P. pardus, Crocuta crocuta, Crocodilus niloticus, Lycaon pictus, Homo sapiens* (none confirmed).
13. Study period: 1985, 1988, 1992, 1993
14. Habituation: None
15. Research presence: Occasional
16. Conservation: Informal and minimal to date, e.g. subscriptions and books to local schools
17. Current research: Feeding ecology, nest distribution, sustainable forest use, range use, community structure via DNA
18. Methods: Indirect evidence
19. Contact person: Jim Moore, Department of Anthropology, University of California San Diego, La Jolla, CA 92093, USA
20. Bibliography:

Itani, J. 1979. Distribution and adaptation of chimpanzees in an arid area. In: *The Great Apes*, ed. D. A. Hamburg & E. R. McCown, pp. 55–71. Menlo Park, CA: Benjamin/Cummings.

Moore, J. 1994. Plants of the Tongwe East Forest Reserve (Ugalla), Tanzania. *Tropics*, **3**, 333–40.

Nishida, T. 1989. A note on the chimpanzee ecology of the Ugalla area, Tanzania. *Primates*, **30**, 129–38.

Site report

Kano

1. Name: Station de Wamba, Centre de Recherche en Ecologie et Forestiere
2. Location: Zaïre, Provincè Collectivite de Luo, Zone de Djolu, Sous-region de Tshuapa, 00°11′ N 22°28′ E
3. Status: Scientific reserve
4. Area: 350 km²
5. Altitude: 370–430 m
6. Rain fall: 2000 mm
7. Vegetation: Rainforest-derived mosaic of dry, swamp, and secondary forest, cultivated fields
8. Human influence: Selective logging, hunting, provisioning
   Disturbance ratings: 4, 4, 2, 3, 3
9. Closest village: Adjacent

10. Species studied: *Pan paniscus*
11. Other primates: *Perodicticus potto, Galago demidovii, Cercopithecus ascanius, C. wolfi, C. neglectus, C. salongo, Colobus badius, C. angolensis, Cercocebus aterrimus, Allenopithecus nigroviridis*
12. Predators: *Homo sapiens, Panthera pardus*
13. Study period: 1974–81, 1983–91, 1994
14. Habituation: Good for 3 groups, fair for 2 groups, poor for 1 group
15. Research presence: Seasonal, occasional
16. Conservation: Guards' training
17. Current research: Sexual development, intergroup relationships, adoption
18. Method: Focal-subject sampling, *ad libitum* sampling, video recording
19. Contact person: Takayoshi Kano, Kyoto University Primate Research Institute, Inuyama, Japan 484
20. Bibliography:

Furuichi, T. 1989. Social interactions and the life history of female *Pan paniscus* in Wamba, Zaïre. *International Journal of Primatology*, **10**, 173–97.

Kano, T. 1992. *The Last Ape. Pygmy Chimpanzee Behavior and Ecology*. Stanford, CA: Stanford University Press.

Kuroda, S. 1980. Social behavior of the pygmy chimpanzees. *Primates*, **21**, 181–97.

Index